TRANSPORT PHENOMENA

TRANSPORT PHENOMENA

An Introduction to Advanced Topics

LARRY A. GLASGOW
Professor of Chemical Engineering
Kansas State University
Manhattan, Kansas

WILEY
A JOHN WILEY & SONS, INC., PUBLICATION

Published by John Wiley & Sons, Inc., Hoboken, New Jersey.
Published simultaneously in Canada.

Library of Congress Cataloging-in-Publication Data

Glasgow, Larry A., 1950-
Transport phenomena : an introduction to advanced topics / Larry A. Glasgow.
p. cm.
Includes index.
ISBN 978-0-470-38174-8 (cloth)
1. Transport theory–Mathematics. I. Title.
TP156.T7G55 2010
530.4'75–dc22

2009052127

CONTENTS

PREFACE

This book is intended for advanced undergraduates and first-year graduate students in chemical and mechanical engineering. Prior formal exposure to transport phenomena or to separate courses in fluid flow and heat transfer is assumed. Our objectives are twofold: to learn to apply the principles of transport phenomena to unfamiliar problems, and to improve our methods of attack upon such problems. This book is suitable for both formal coursework and self-study.

In recent years, much attention has been directed toward the perceived "paradigm shift" in chemical engineering education. Some believe we are leaving the era of engineering science that blossomed in the 1960s and are entering the age of molecular biology. Proponents of this viewpoint argue that dramatic changes in engineering education are needed. I suspect that the real defining issues of the next 25–50 years are not yet clear. It may turn out that the transformation from petroleum-based fuels and economy to perhaps a hydrogen-based economy will require application of engineering skills and talent at an unprecedented intensity. Alternatively, we may have to marshal our technically trained professionals to stave off disaster from global climate change, or to combat a viral pandemic. What may happen is murky, at best. However, I do expect the engineering sciences to be absolutely crucial to whatever technological crises emerge.

Problem solving in transport phenomena has consumed much of my professional life. The beauty of the field is that it matters little whether the focal point is tissue engineering, chemical vapor deposition, or merely the production of gasoline; the principles of transport phenomena apply equally to all. The subject is absolutely central to the formal study of chemical and mechanical engineering. Moreover, transport phenomena are ubiquitous—all aspects of life, commerce, and production are touched by this engineering science. I can only hope that you enjoy the study of this material as much as I have.

It is impossible to express what is owed to Linda, Andrew, and Hillary, each of whom enriched my life beyond measure. And many of the best features of the person I am are due to the formative influences of my mother Betty J. (McQuilkin) Glasgow, father Loren G. Glasgow, and sister Barbara J. (Glasgow) Barrett.

Larry A. Glasgow

Department of Chemical Engineering, Kansas State University, Manhattan, KS

1

INTRODUCTION AND SOME USEFUL REVIEW

1.1 A MESSAGE FOR THE STUDENT

This is an advanced-level book based on a course sequence taught by the author for more than 20 years. Prior exposure to transport phenomena is assumed and familiarity with the classic, *Transport Phenomena*, 2nd edition, by R. B. Bird, W. E. Stewart, and E. N. Lightfoot (BS&L), will prove particularly advantageous because the notation adopted here is mainly consistent with BS&L.

There are *many* well-written and useful texts and monographs that treat aspects of transport phenomena. A few of the books that I have found to be especially valuable for engineering problem solving are listed here:

Transport Phenomena, 2nd edition, Bird, Stewart, and Lightfoot.

An Introduction to Fluid Dynamics and *An Introduction to Mass and Heat Transfer*, Middleman.

Elements of Transport Phenomena, Sissom and Pitts.

Transport Analysis, Hershey.

Analysis of Transport Phenomena, Deen.

Transport Phenomena Fundamentals, Plawsky.

Advanced Transport Phenomena, Slattery.

Advanced Transport Phenomena: Fluid Mechanics and Convective Transport Processes, Leal.

The Phenomena of Fluid Motions, Brodkey.

Fundamentals of Heat and Mass Transfer, Incropera and De Witt.

Fluid Dynamics and Heat Transfer, Knudsen and Katz.

Fundamentals of Momentum, Heat, and Mass Transfer, 4th edition, Welty, Wicks, Wilson, and Rorrer.

Fluid Mechanics for Chemical Engineers, 2nd edition, Wilkes.

Vectors, Tensors, and the Basic Equations of Fluid Mechanics, Aris.

In addition, there are many other more specialized works that treat or touch upon some facet of transport phenomena. These books can be very useful in proper circumstances and they will be clearly indicated in portions of this book to follow. In view of this sea of information, what is the point of yet another book? Let me try to provide my rationale below.

I taught transport phenomena for the first time in 1977–1978. In the 30 years that have passed, I have taught our graduate course sequence, Advanced Transport Phenomena 1 and 2, more than 20 times. These experiences have convinced me that no suitable *single* text exists in this niche, hence, this book.

So, the course of study you are about to begin here *is the course sequence* I provide for our first-year graduate students. It is important to note that for many of our students, *formal exposure* to fluid mechanics and heat transfer *ends with this course sequence*. It is imperative that such students leave the experience with, at the very least, some cognizance of the breadth of transport phenomena. Of course, this reality has profoundly influenced this text.

In 1982, I purchased my first IBM PC (personal computer); by today's standards it was a *kludge* with a very low clock rate,

Transport Phenomena: An Introduction to Advanced Topics, By Larry A. Glasgow

just 64K memory, and 5.25″(160K) floppy drives. The high-level language available at that time was interpreted BASIC that had severe limits of its own with respect to execution speed and array size. Nevertheless, it was immediately apparent that the decentralization of computing power would spur a revolution in engineering problem solving. By necessity I became fairly adept at BASIC programming, first using the interpreter and later using various BASIC compilers. Since 1982, the increases in PC capability and the decreases in cost have been astonishing; it now appears that Moore's "law" (*the number of transistors on an integrated circuit yielding minimum component cost doubles every 24 months*) may continue to hold true through several more generations of chip development. In addition, PC hard-drive capacity has exhibited exponential growth over that time frame and the estimated cost per G-FLOP has decreased by a factor of about 3 every year for the past decade.

It is not an exaggeration to say that a cheap desktop PC in 2009 has much more computing power than a typical university mainframe computer of 1970. As a consequence, problems that were pedagogically impractical are now routine. This computational revolution has changed the way I approach instruction in transport phenomena and it has made it possible to assign more complex exercises, even embracing nonlinear problems, and still maintain expectations of timely turnaround of student work. It was my intent that this computational revolution be reflected in this text and in some of the problems that accompany it. However, I have avoided significant use of *commercial software* for problem solutions.

Many engineering educators have come to the realization that computers (and the microelectronics revolution in general) are changing the way students learn. The ease with which complicated information can be obtained and difficult problems can be solved has led to a physical disconnect; students have far fewer opportunities to develop somatic comprehension of problems and problem solving in this new environment. The reduced opportunity to experience has led to a reduced ability to perceive, and with dreadful consequence. Recently, Haim Baruh (2001) observed that the computer revolution has led young people to "think, learn and visualize differently.... Because information can be found so easily and quickly, students often skip over the basics. For the most part, abstract concepts that require deeper thought aren't part of the equation. I am concerned that unless we use computers wisely, the decline in student performance will continue."

Engineering educators must remember that computers are merely tools and skillful use of a commercial software package does not translate to the type of understanding needed for the formulation and analysis of engineering problems. In this regard, I normally ask students to be wary of reliance upon commercial software for solution of problems in transport phenomena. In certain cases, commercial codes can be used for comparison of alternative models; this is particularly useful *if* the software can be verified with *known* results for that particular scenario. *But*, blind acceptance of black-box computations for an untested situation is foolhardy.

One of my principal objectives in transport phenomena instruction is to help the student develop physical insight and problem-solving capability simultaneously. This balance is essential because either skill set alone is just about useless. In this connection, we would do well to remember G. K. Batchelor's (1967) admonition: "By one means or another, a teacher should show the relation between his analysis and the behavior of real fluids; fluid dynamics is much less interesting if it is treated largely as an exercise in mathematics." I also feel strongly that how and why this field of study developed is not merely peripheral; one can learn a great deal by obtaining a historical perspective and in many instances I have tried to provide this. I believe in the adage that you cannot know where you are going if you do not know where you have been. Many of the accompanying problems have been developed to provide a broader view of transport phenomena as well; they constitute a unique feature of this book, and many of them require the student to draw upon other resources.

I have tried to recall questions that arose in my mind when I was beginning my second course of study of transport phenomena. I certainly hope that some of these have been clearly treated here. For many of the examples used in this book, I have provided details that might often be omitted, but this has a price; the resulting work cannot be as broad as one might like. There are some important topics in transport phenomena that are not treated in a substantive way in this book. These omissions include non-Newtonian rheology and energy transport by radiation. Both topics deserve far more consideration than could be given here; fortunately, both are subjects of numerous specialized monographs. In addition, both boundary-layer theory and turbulence could easily be taught as separate one- or even two-semester courses. That is obviously not possible within our framework. I would like to conclude this message with five observations:

1. Transport phenomena are pervasive and they impact upon *every* aspect of life.
2. Rote learning is ineffective in this subject area because the successful application of transport phenomena is directly tied to physical understanding.
3. Mastery of this subject will enable you to critically evaluate many physical phenomena, processes, and systems across many disciplines.
4. Student effort is paramount in graduate education. There are many places in this text where outside reading and additional study are not merely recommended, but expected.
5. Time has not diminished my interest in transport phenomena, and my hope is that through this book I can share my enthusiasm with students.

1.2 DIFFERENTIAL EQUATIONS

Students come to this sequence of courses with diverse mathematical backgrounds. Some do not have the required levels of proficiency, and since these skills are crucial to success, a brief review of some important topics may be useful.

Transport phenomena are governed by, and modeled with, differential equations. These equations may arise through mass balances, momentum balances, and energy balances. The main equations of change are second-order partial differential equations that are (too) frequently nonlinear. One of our principal tasks in this course is to find solutions for such equations; we can expect this process to be challenging at times.

Let us begin this section with some simple examples of ordinary differential equations (ODEs); consider

$$\frac{dy}{dx} = c \quad (c \text{ is constant}) \tag{1.1}$$

and

$$\frac{dy}{dx} = y. \tag{1.2}$$

Both are *linear*, first-order ordinary differential equations. Remember that linearity is determined by the dependent variable y. The solutions for (1.1) and (1.2) are

$$y = cx + C_1 \text{ and } y = C_1 \exp(x), \text{ respectively.} \tag{1.3}$$

Note that if $y(x=0)$ is specified, then the behavior of y is set for all values of x. If the independent variable x were time t, then the future behavior of the system would be known. This is what we mean when we say that a system is deterministic. Now, what happens when we modify (1.2) such that

$$\frac{dy}{dx} = 2xy? \tag{1.4}$$

We find that $y = C_1 \exp(x^2)$. These first-order linear ODEs have all been separable, admitting simple solution. We will sketch the (three) behaviors for $y(x)$ on the interval 0–2, given that $y(0) = 1$ (Figure 1.1). *Match each of the three curves with the appropriate equation.*

Note what happens to $y(x)$ if we continue to add additional powers of x to the right-hand side of (1.4), allowing y to remain. If we add powers of y instead—and make the equation inhomogeneous—we can expect to work a little harder. Consider this first-order *nonlinear* ODE:

$$\frac{dy}{dx} = a + by^2. \tag{1.5}$$

This is a type of Riccati equation (Jacopo Francesco Count Riccati, 1676–1754) and the nature of the solution will

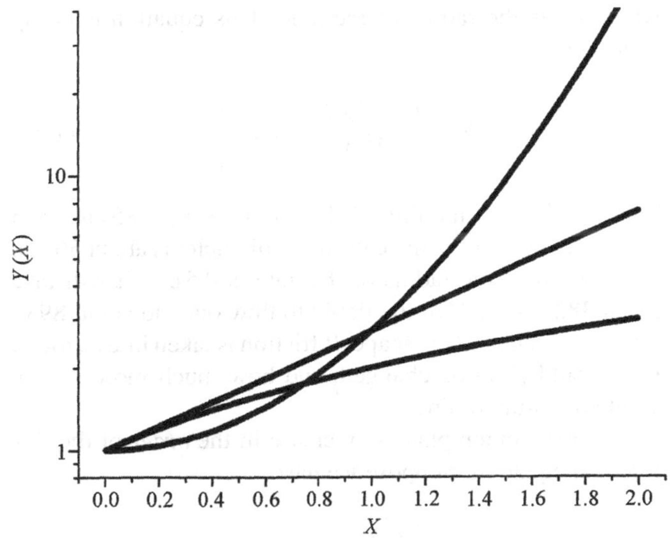

FIGURE 1.1. Solutions for $dy/dx = 1$, $dy/dx = y$, and $dy/dx = 2xy$.

depend on the product of a and b. If we let $a = b = 1$, then

$$y = \tan(x + C_1). \tag{1.6}$$

Before we press forward, we note that Riccati equations were studied by Euler, Liouville, and the Bernoulli's (Johann and Daniel), among others. How will the solution change if eq. (1.5) is rewritten as

$$\frac{dy}{dx} = 1 - y^2? \tag{1.7}$$

Of course, the equation is still separable, so we can write

$$\int \frac{dy}{1 - y^2} = x + C_1. \tag{1.8}$$

Show that the solution of (1.8), given that $y(0) = 0$, is $y = \tanh(x)$.

When a first-order differential equation arises in transport phenomena, it is usually by way of a macroscopic balance, for example, [Rate in] − [Rate out] = [Accumulation]. Consider a 55-gallon drum (vented) filled with water. At $t = 0$, a small hole is punched through the side near the bottom and the liquid begins to drain from the tank. If we let the velocity of the fluid through the orifice be represented by Torricelli's theorem (a frictionless result), a mass balance reveals

$$\frac{dh}{dt} = -\frac{R_0^2}{R_T^2}\sqrt{2gh}, \tag{1.9}$$

where R_0 is the radius of the hole. This equation is easily solved as

$$h=\left[-\sqrt{\frac{g}{2}\frac{R_0^2}{R_T^2}}t+C_1\right]^2. \tag{1.10}$$

The drum is initially full, so $h(t=0)=85\,\text{cm}$ and $C_1=9.21954\,\text{cm}^{1/2}$. Since the drum diameter is about 56 cm, $R_T=28\,\text{cm}$; if the radius of the hole is 0.5 cm, it will take about 382 s for half of the liquid to flow out and about 893 s for 90% of the fluid to escape. If friction is taken into account, how would (1.9) be changed, and how much more slowly would the drum drain?

We now contemplate an increase in the *order* of the differential equation. Suppose we have

$$\frac{d^2y}{dx^2}+a=0, \tag{1.11}$$

where a is a constant or an elementary function of x. This is a common equation type in transport phenomena for steady-state conditions with molecular transport occurring in one direction. We can immediately write

$$\frac{dy}{dx}=-\int a\,dx+C_1,\ \text{and if } a \text{ is a constant,}$$

$$y=-\frac{a}{2}x^2+C_1x+C_2.$$

Give an example of a specific type of problem that produces this solution. One of the striking features of (1.11) is the absence of a first derivative term. You might consider what conditions would be needed in, say, a force balance to produce *both* first and second derivatives.

The simplest second-order ODEs (that include first derivatives) are linear equations with constant coefficients. Consider

$$\frac{d^2y}{dx^2}+1\frac{dy}{dx}+y=f(x), \tag{1.12}$$

$$\frac{d^2y}{dx^2}+2\frac{dy}{dx}+y=f(x), \tag{1.13}$$

and

$$\frac{d^2y}{dx^2}+3\frac{dy}{dx}+y=f(x). \tag{1.14}$$

Using linear differential operator notation, we rewrite the left-hand side of each and factor the result:

$$(D^2+D+1)y \qquad (D+\frac{1}{2}+\frac{\sqrt{3}}{2}i)(D+\frac{1}{2}-\frac{\sqrt{3}}{2}i), \tag{1.15}$$

$$(D^2+2D+1)y \qquad (D+1)(D+1), \tag{1.16}$$

$$(D^2+3D+1)y \qquad (D+\frac{3+\sqrt{5}}{2})(D+\frac{3-\sqrt{5}}{2}). \tag{1.17}$$

Now suppose the forcing function $f(x)$ in (1.12)–(1.14) is a constant, say 1. What do (1.15)–(1.17) tell you about the nature of possible solutions? The complex conjugate roots in (1.15) will result in oscillatory behavior. Note that all three of these second-order differential equations have constant coefficients and a first derivative term. If eq. (1.14) had been developed by force balance (with x replaced by t), the dy/dx (velocity) term might be some kind of frictional resistance. We do not have to expend much effort to find second-order ODEs that pose greater challenges. What if you needed a solution for the nonlinear equation

$$\frac{d^2y}{dx^2}=a+by+cy^2+dy^3? \tag{1.18}$$

Actually, a number of closely related equations have figured prominently in physics. Einstein, in an investigation of planetary motion, was led to consider

$$\frac{d^2y}{dx^2}+y=a+by^2. \tag{1.19}$$

Duffing, in an investigation of forced vibrations, carried out a study of the equation

$$\frac{d^2y}{dx^2}+k\frac{dy}{dx}+ay+by^3=f(x). \tag{1.20}$$

A limited number of nonlinear, second-order differential equations can be solved with (Jacobian) elliptic functions. For example, Davis (1962) shows that the solution of the nonlinear equation

$$\frac{d^2y}{dx^2}=6y^2 \tag{1.21}$$

can be written as

$$y=A+\frac{B}{\text{sn}^2(C(x-x_1))}. \tag{1.22}$$

Tabulated values are available for the Jacobi elliptic sine, sn; see pages 175–176 in Davis (1962). The reader desiring an introduction to elliptic functions is encouraged to work problem 1.N in this text, read Chapter 5 in Vaughn (2007), and consult the extremely useful book by Milne-Thomson (1950).

The point of the immediately preceding discussion is as follows: The elementary functions that are familiar to us, such as sine, cosine, exp, ln, etc., are solutions to linear differential equations. Furthermore, when constants arise in the solution of linear differential equations, they do so linearly; for an example, see the solution of eq. (1.11) above. In nonlinear differential equations, arbitrary constants appear *nonlinearly.* Nonlinear problems abound in transport phenomena and we can expect to find analytic solutions only for a very limited number of them. Consequently, most nonlinear problems must be solved numerically and this raises a host of other issues, including existence, uniqueness, and stability.

So much of our early mathematical education is bound to linearity that it is difficult for most of us to perceive and appreciate the beauty (and beastliness) in nonlinear equations. We can illustrate some of these concerns by examining the elementary nonlinear difference (logistic) equation,

$$X_{n+1} = \alpha X_n(1 - X_n). \tag{1.23}$$

Let the parameter α assume an initial value of about 3.5 and let $X_1 = 0.5$. Calculate the new value of X and insert it on the right-hand side. As we repeat this procedure, the following sequence emerges: 0.5, 0.875, 0.38281, 0.82693, 0.5009, 0.875, 0.38282, 0.82694, ... Now allow α to assume a slightly larger value, say 3.575. Then, the sequence of calculated values is 0.5, 0.89375, 0.33949, 0.80164, 0.56847, 0.87699, 0.38567, 0.84702, 0.46324, 0.88892, 0.353, 0.8165, 0.53563, 0.88921, 0.35219, 0.81564, 0.53757, ... We can continue this process and report these results graphically; the result is a bifurcation diagram. How would you characterize Figure 1.2? Would you be tempted to use "chaotic" as a descriptor? The most striking feature of this logistic map is that a completely deterministic equation produces behavior that superficially appears to be random (*it is not*). Baker and Gollub (1990) described this map as having regions where the behavior is chaotic with windows of periodicity.

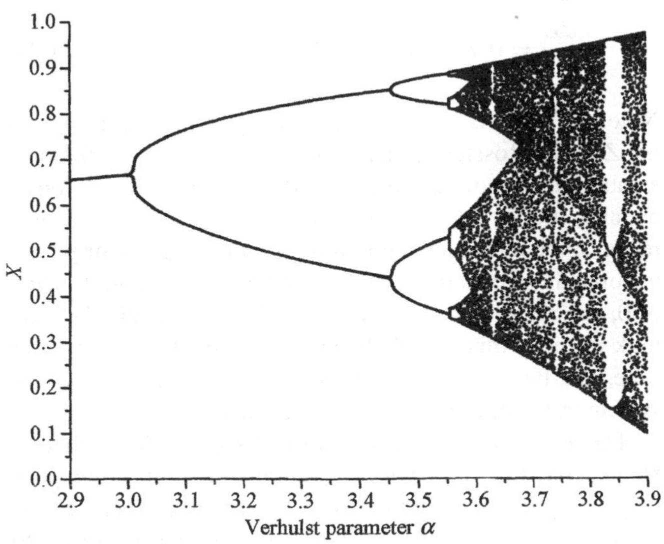

FIGURE 1.2. Bifurcation diagram for the logistic equation with the Verhulst parameter α ranging from 2.9 to 3.9.

Note that the chaotic behavior seen above is attained through a series of period doublings (or pitchfork bifurcations). Baker and Gollub note that many dynamical systems exhibit this path to chaos. In 1975, Mitchell Feigenbaum began to look at period doublings for a variety of rather simple functions. He quickly discovered that all of them had a common characteristic, a universality; that is, the ratio of the spacings between successive bifurcations was always the same:

$$4.6692016\ldots \qquad \text{(Feigenbaum number)}.$$

This leads us to hope that a relatively simple system or function might serve as a model (or at least a surrogate) for far more complex behavior.

We shall complete this part of our discussion by selecting two terms from the x-component of the Navier–Stokes equation,

$$\frac{\partial v_x}{\partial t} + v_x \frac{\partial v_x}{\partial x} + \cdots, \tag{1.24}$$

and writing them in finite difference form, letting i be the spatial index and j the temporal one. We can drop the subscript "x" for convenience. One of the possibilities (though not a very good one) is

$$\frac{v_{i,j+1} - v_{i,j}}{\Delta t} + v_{i,j} \frac{v_{i+1,j} - v_{i,j}}{\Delta x} + \cdots. \tag{1.25}$$

We might imagine this being rewritten as an explicit algorithm (where we calculate v at the new time, $j+1$, using velocities from the jth time step) in the following form:

$$v_{i,j+1} \approx v_{i,j} - \frac{\Delta t}{\Delta x} v_{i,j}(v_{i+1,j} - v_{i,j}) + \cdots. \tag{1.26}$$

Please make note of the dimensionless quantity $\Delta t v_{i,j}/\Delta x$; this is the Courant number, *Co*, and it will be extremely important to us later. As a computational scheme, eq. (1.26) is generally unworkable, but note the similarity to the logistic equation above. The nonlinear character of the equations that govern fluid motion guarantees that we will see unexpected beauty and maddening complexity, if only we knew where (and how) to look.

In this connection, a system that evolves in time can often be usefully studied using phase space analysis, which is an underutilized tool for the study of the dynamics of low-dimension systems. Consider a periodic function such as $f(t) = A \sin(\omega t)$. The derivative of this function is $\omega A \cos(\omega t)$. If we cross-plot $f(t)$ and df/dt, we will obtain a *limit cycle* in the shape of an ellipse. That is, the system trajectory in phase space takes the form of a closed path, which is expected

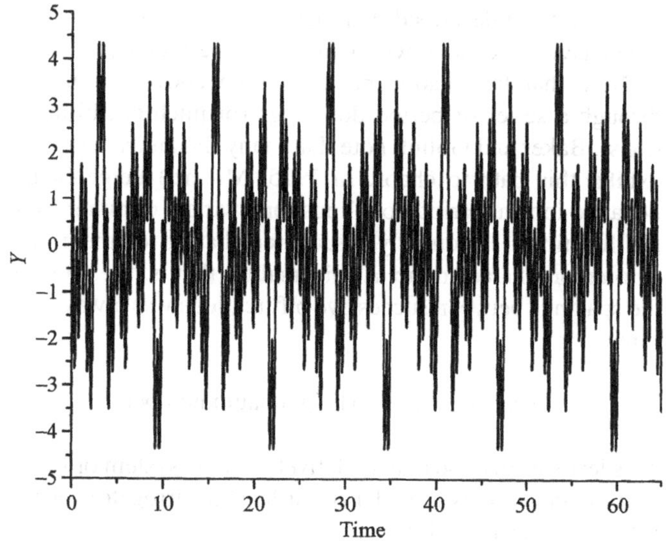

FIGURE 1.3. "Artificial" time-series data constructed from sinusoids.

behavior for a purely periodic function. If, on the other hand, we had an oscillatory system that was unstable, the amplitude of the oscillations would grow in time; the resulting phase-plane portrait would be an outward spiral. An attenuated (damped) oscillation would produce an inward spiral. This technique can be useful for more complicated functions or signals as well. Consider the oscillatory behavior illustrated in Figure 1.3.

If you look closely at this figure, you can see that the function $f(t)$ does exhibit periodic behavior—many features of the system output appear repeatedly. In phase space, this system yields the trajectory shown in Figure 1.4.

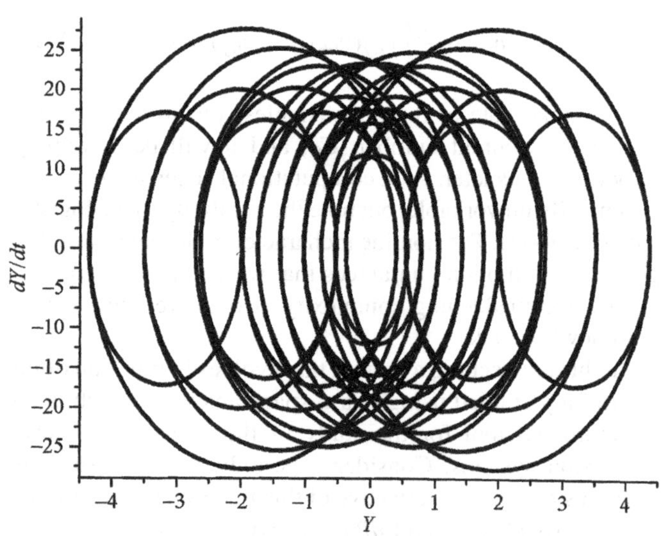

FIGURE 1.4. Phase space portrait of the system dynamics illustrated in Figure 1.3.

What we see here is the combination of a limited number of periodic functions interacting. Particular points in phase space are revisited fairly regularly. But, if the dynamic behavior of a system was truly chaotic, we might see a phase space in which *no point is ever revisited.* The implications for the behavior of a perturbed complex nonlinear system, such as the global climate, are sobering.

Another consequence of nonlinearity is sensitivity to initial conditions; to solve a general fluid flow problem, we would need to consider three components of the Navier–Stokes equation and the continuity relation simultaneously. Imagine an integration scheme forward marching in time. It would be necessary to specify initial values for v_x, v_y, v_z, and p. Suppose that v_x had the exact initial value, 5 cm/s, but your computer represented the number as 4.99999. . . cm/s. Would the integration scheme evolve along the "correct" pathway? Possibly not. Jules-Henri Poincaré(who was perhaps the last man to understand all of the mathematics of his era) noted in 1908 that "... small differences in the initial conditions produce very great ones in the final phenomena." In more recent years, this concept has become popularly known as the "butterfly effect" in deference to Edward Lorenz (1963) who observed that the disturbance caused by a butterfly's wing might change the weather pattern for an entire hemisphere. This is an idea that is unfamiliar to most of us; in much of the educational process we are conditioned to believe a model for a system (a differential equation), taken together with its present state, completely set the future behavior of the system.

Let us conclude this section with an appropriate example; we will explore the Rossler (1976) problem that consists of the following set of three (deceptively simple) ordinary differential equations:

$$\frac{dX}{dt} = -Y - Z, \quad \frac{dY}{dt} = X + 0.2Y, \quad \text{and}$$
$$\frac{dZ}{dt} = 0.2 + Z(X - 5.7). \tag{1.27}$$

Note that there is but one nonlinearity in the set, the product ZX. The Rossler model is synthetic in the sense that it is an abridgement of the Lorenz model of local climate; consequently, it does not have a direct physical basis. But it will reveal some unexpected and important behavior. Our plan is to solve these equations numerically using the initial values of 0, −6.78, and 0.02 for X, Y, and Z, respectively. We will look at the evolution of all three dependent variables with time, and then we will examine a segment or cut from the system trajectory by cross-plotting X and Y.

The main point to take from this example is that an elementary, low-dimensional system can exhibit unexpectedly complicated behavior. The system trajectory seen in Figure 1.5b is a portrait of what is now referred to in the literature as a "strange" attractor. The interested student is encouraged to read the papers by Rossler (1976) and Packard

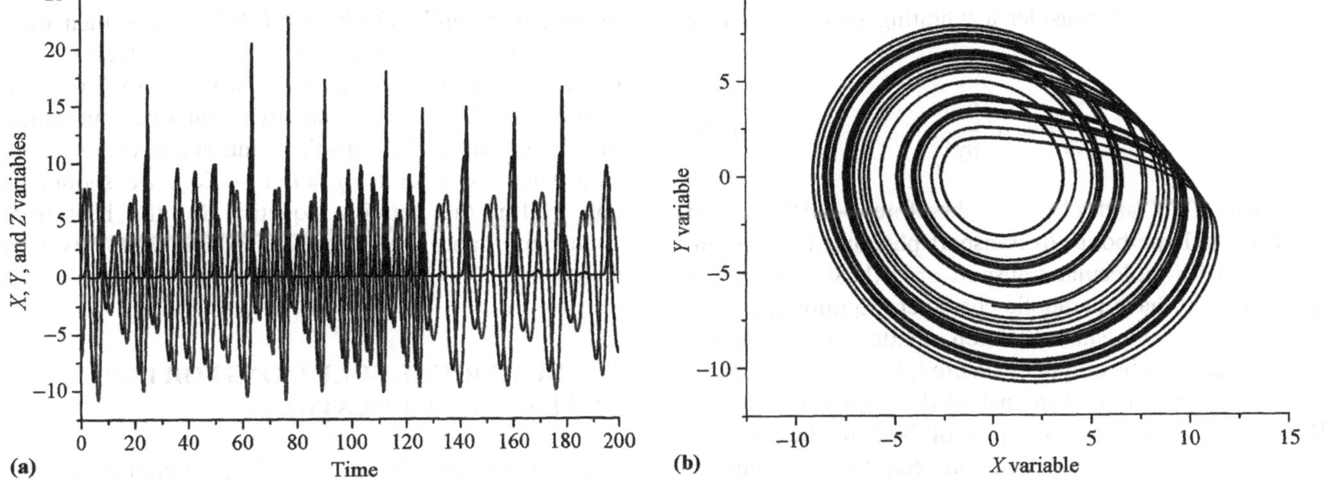

FIGURE 1.5. The Rossler model: $X(t)$, $Y(t)$, and $Z(t)$ for $0 < t < 200$ (a), and a cut from the system trajectory (Y plotted against X) (b).

et al. (1980). The formalized study of chaotic behavior is still in its infancy, but it has become clear that there are applications in hydrodynamics, mechanics, chemistry, etc.

There are additional tools that can be used to determine whether a particular system's behavior is periodic, aperiodic, or chaotic. For example, the rate of divergence of a chaotic trajectory about an attractor is characterized with Lyapunov exponents. Baker and Gollub (1990) describe how the exponents are computed in Chapter 5 of their book and they include a listing of a BASIC program for this task. The Fourier transform is also invaluable in efforts to identify important periodicities in the behavior of nonlinear systems. We will make extensive use of the Fourier transform in our consideration of turbulent flows.

The student with further interest in this broad subject area is also encouraged to read the recent article by Porter et al. (2009). This paper treats a historically significant project carried out at Los Alamos by Fermi, Pasta, and Ulam (Report LA-1940). Fermi, Pasta, and Ulam (FPU) investigated a one-dimensional mass-and-spring problem in which 16, 32, and 64 masses were interconnected with non-Hookean springs. They experimented (computationally) with cases in which the restoring force was proportional to displacement raised to the second or third power(s). FPU found that the nonlinear systems did not share energy (in the expected way) with the higher modes of vibration. Instead, energy was exchanged ultimately among just the first few modes, almost periodically. Since their original intent had been to explore the rate at which the initial energy was distributed among all of the higher modes (they referred to this process as "thermalization"), they quickly realized that the nonlinearities were producing quite unexpectedly localized behavior in phase space! The work of FPU represents one of the very first cases in which extensive computational experiments were performed for nonlinear systems.

1.3 CLASSIFICATION OF PARTIAL DIFFERENTIAL EQUATIONS AND BOUNDARY CONDITIONS

We have to be able to recognize and classify partial differential equations to attack them successfully; a book like Powers (1979) can be a valuable ally in this effort. Consider the generalized second-order partial differential equation, where ϕ is the dependent variable and x and y are arbitrary independent variables:

$$A\frac{\partial^2\phi}{\partial x^2}+B\frac{\partial^2\phi}{\partial x\partial y}+C\frac{\partial^2\phi}{\partial y^2}+D\frac{\partial\phi}{\partial x}+E\frac{\partial\phi}{\partial y}+F\phi+G=0. \tag{1.28}$$

A, B, C, D, E, F, and G can be functions of x and y, but *not* of ϕ. This *linear* partial differential equation can be classified as follows:

$$B^2-4AC<0 \quad \text{(elliptic)},$$
$$B^2-4AC=0 \quad \text{(parabolic)},$$
$$B^2-4AC>0 \quad \text{(hyperbolic)}.$$

For illustration, we look at the "heat" equation (transient conduction in one spatial dimension):

$$\frac{\partial T}{\partial t}=\alpha\frac{\partial^2 T}{\partial y^2}. \tag{1.29}$$

You can see that $A=\alpha$, $B=0$, and $C=0$; the equation is parabolic. Compare this with the governing (Laplace) equation for two-dimensional potential flow (ψ is the stream function):

$$\frac{\partial^2\psi}{\partial x^2}+\frac{\partial^2\psi}{\partial y^2}=0. \tag{1.30}$$

In this case, $A=1$ and $C=1$ while $B=0$; the equation is elliptic. Next, we consider a vibrating string (the wave equation):

$$\frac{\partial^2 u}{\partial t^2} = s^2 \frac{\partial^2 u}{\partial y^2}. \tag{1.31}$$

Note that $A=1$ and $C=-s^2$; therefore, $-4AC>0$ and eq. (1.31) is hyperbolic. In transport phenomena, transient problems with molecular transport only (heat or diffusion equations) will have parabolic character. Equilibrium problems such as steady-state diffusion, conduction, or viscous flow in a duct will be elliptic in nature (phenomena governed by Laplace- or Poisson-type partial differential equations). We will see numerous examples of both in the chapters to come. Hyperbolic equations are common in quantum mechanics and high-speed compressible flows, for example, inviscid supersonic flow about an airfoil. The Navier–Stokes equations that will be so important to us later are of mixed character.

The three most common types of boundary conditions used in transport phenomena are Dirichlet, Neumann, and Robin's. For Dirichlet boundary conditions, the field variable is specified at the boundary. *Two examples*: In a conduction problem, the temperature at a surface might be fixed (at $y=0$, $T=T_0$); alternatively, in a viscous fluid flow problem, the velocity at a stationary duct wall would be zero. For Neumann conditions, the flux is specified; for example, for a conduction problem with an insulated wall located at $y=0$, $(\partial T/\partial y)_{y=0}=0$. A Robin's type boundary condition results from equating the fluxes; for example, consider the solid–fluid interface in a heat transfer problem. On the solid side heat is transferred by conduction (Fourier's law), but on the fluid side of the interface we might have mixed heat transfer processes approximately described by Newton's "law" of cooling:

$$-k\left(\frac{\partial T}{\partial y}\right)_{y=0} = h(T_0 - T_f). \tag{1.32}$$

We hasten to add that the heat transfer coefficient h that appears in (1.32) is an empirical quantity. The numerical value of h is known only for a small number of cases, usually those in which molecular transport is dominant.

One might think that Newton's "law" of cooling could not possibly engender controversy. That would be a flawed presumption. Bohren (1991) notes that Newton's own description of the law as translated from Latin is "if equal times of cooling be taken, the degrees of heat will be in geometrical proportion, and therefore easily found by tables of logarithms." It is clear from these words that Newton meant that the cooling process would proceed exponentially. Thus, to simply write $q=h(T-T_\infty)$, without qualification, is "incorrect." On the other hand, if one uses a lumped-parameter model to described the cooling of an object, $mC_p(dT/dt)=-hA(T-T_\infty)$, then the oft-cited form does produce an exponential decrease in the object's temperature in accordance with Newton's own observation. So, do we have an argument over substance or merely semantics? Perhaps the solution is to exercise greater care when we refer to $q=h(T-T_\infty)$; we should probably call it the defining equation for the heat transfer coefficient h and meticulously avoid calling the expression a "law."

1.4 NUMERICAL SOLUTIONS FOR PARTIAL DIFFERENTIAL EQUATIONS

Many of the examples of numerical solution of partial differential equations used in this book are based on finite difference methods (FDMs). The reader may be aware that the finite element method (FEM) is widely used in commercial software packages for the same purpose. The FEM is particularly useful for problems with either curved or irregular boundaries and in cases where localized changes require a smaller scale grid for improved resolution. The actual numerical effort required for solution in the two cases is comparable. However, FEM approaches usually employ a separate code (or program) for mesh generation and refinement. I decided not to devote space here to this topic because my intent was to make the solution procedures as general as possible and nearly independent of the computing platform and software. By taking this approach, the student without access to specialized commercial software can still solve many of the problems in the course, in some instances using nothing more complicated than either a spreadsheet or an elementary understanding of any available high-level language.

1.5 VECTORS, TENSORS, AND THE EQUATION OF MOTION

For the discussion that follows, recall that temperature T is a scalar (zero-order, or rank, tensor), velocity V is a vector (first-order tensor), and stress τ is a second-order tensor. Tensor is from the Latin "tensus," meaning to stretch. We can offer the following, rough, definition of a tensor: It is a generalized quantity or mathematical object that *in three-dimensional space* has 3^n components (where n is the order, or rank, of the tensor). From an engineering perspective, tensors are defined over a continuum and transform according to certain rules. They figure prominently in mechanics (stress and strain) and relativity.

The del operator (∇) in rectangular coordinates is

$$\delta_x\frac{\partial}{\partial x} + \delta_y\frac{\partial}{\partial y} + \delta_z\frac{\partial}{\partial z}. \tag{1.33}$$

For a scalar such as T, ∇T is referred to as the gradient (of the scalar field). So, when we speak of the temperature gradient, we are talking about a vector quantity with both direction and magnitude.

A scalar product can be formed by applying ∇ to the velocity vector:

$$\nabla \cdot V = \frac{\partial v_x}{\partial x} + \frac{\partial v_y}{\partial y} + \frac{\partial v_z}{\partial z}, \qquad (1.34)$$

which is the divergence of the velocity, div(V). The physical meaning should be clear to you: For an incompressible fluid (ρ = constant), conservation of mass requires that $\nabla \cdot V = 0$; in 3-space, if v_x changes with x, the other velocity vector components must accommodate the change (to prevent a net outflow). You may recall that a mass balance for an element of *compressible* fluid reveals that the continuity equation is

$$\frac{\partial \rho}{\partial t} + \frac{\partial}{\partial x}(\rho v_x) + \frac{\partial}{\partial y}(\rho v_y) + \frac{\partial}{\partial z}(\rho v_z) = 0. \qquad (1.35a)$$

For a compressible fluid, a net outflow results in a change (decrease) in fluid density. Of course, conservation of mass can be applied in cylindrical and spherical coordinates as well:

$$\frac{\partial \rho}{\partial t} + \frac{1}{r}\frac{\partial}{\partial r}(\rho r v_r) + \frac{1}{r}\frac{\partial}{\partial \theta}(\rho v_\theta) + \frac{\partial}{\partial z}(\rho v_z) = 0 \qquad (1.35b)$$

and

$$\frac{\partial \rho}{\partial t} + \frac{1}{r^2}\frac{\partial}{\partial r}(\rho r^2 v_r) + \frac{1}{r \sin\theta}\frac{\partial}{\partial \theta}(\rho v_\theta \sin\theta) + \frac{1}{r \sin\theta}\frac{\partial}{\partial \phi}(\rho v_\phi) = 0. \qquad (1.35c)$$

In fluid flow, rotation of a suspended particle can be caused by a variation in velocity, even if every fluid element is traveling a path parallel to the confining boundaries. Similarly, the interaction of forces can create a moment that is obtained from the cross product or curl. This tendency toward rotation is particularly significant, so let us review the cross product $\nabla \times V$ in rectangular coordinates:

$$\nabla \times V = \begin{matrix} \dfrac{\partial v_z}{\partial y} - \dfrac{\partial v_y}{\partial z} & (1.36a) \\ \dfrac{\partial v_x}{\partial z} - \dfrac{\partial v_z}{\partial x} & (1.36b) \\ \dfrac{\partial v_y}{\partial x} - \dfrac{\partial v_x}{\partial y} & (1.36c) \end{matrix}$$

Note that the cross product of vectors is a vector; furthermore, you may recall that (1.36a)–(1.36c), the *vorticity* vector components ω_x, ω_y, and ω_z, are measures of the rate of fluid rotation about the x, y, and z axes, respectively. Vorticity is extremely useful to us in hydrodynamic calculations because in the interior of a homogeneous fluid vorticity is neither created nor destroyed; it is produced solely at the flow boundaries. Therefore, it often makes sense for us to employ the vorticity transport equation that is obtained by taking the curl of the equation of motion. We will return to this point and explore it more thoroughly later. In cylindrical coordinates, $\nabla \times V$ is

$$\nabla \times V = \begin{matrix} \dfrac{1}{r}\dfrac{\partial v_z}{\partial \theta} - \dfrac{\partial v_\theta}{\partial z} & (1.37a) \\ \dfrac{\partial v_r}{\partial z} - \dfrac{\partial v_z}{\partial r} & (1.37b) \\ \dfrac{1}{r}\dfrac{\partial}{\partial r}(r v_\theta) - \dfrac{1}{r}\dfrac{\partial v_r}{\partial \theta} & (1.37c) \end{matrix}$$

These equations, (1.37a)–(1.37c), correspond to the r, θ, and z components of the vorticity vector, respectively.

The stress tensor τ is a second-order tensor (nine components) that includes both tangential and normal stresses. For example, in rectangular coordinates, τ is

$$\begin{matrix} \tau_{xx} & \tau_{xy} & \tau_{xz} \\ \tau_{yx} & \tau_{yy} & \tau_{yz} \\ \tau_{zx} & \tau_{zy} & \tau_{zz} \end{matrix}$$

The normal stresses have the repeated subscripts and they appear on the diagonal. Please note that the sum of the diagonal components is the *trace* of the tensor (A) and is often written as tr(A). The trace of the stress tensor, $\Sigma\tau_{ii}$, is *assumed* to be related to the pressure by

$$p = -\frac{1}{3}(\tau_{xx} + \tau_{yy} + \tau_{zz}). \qquad (1.38)$$

Often the pressure in (1.38) is written using the Einstein summation convention as $p = -\tau_{ii}/3$, where the repeated indices imply summation. The shear stresses have differing subscripts and the corresponding off-diagonal terms are equal; that is, $\tau_{xy} = \tau_{yx}$. This requirement is necessary because without it a small element of fluid in a shear field could experience an infinite angular acceleration. Therefore, the stress tensor is symmetric and has just six independent quantities. We will temporarily represent the (shear) stress components by

$$\tau_{ji} = -\mu \frac{\partial v_i}{\partial x_j}. \qquad (1.39)$$

Note that this relationship (Newton's law of friction) between stress and strain is linear. There is little *a priori* evidence for its validity; however, known solutions (e.g., for Hagen–Poiseuille flow) are confirmed by physical experience.

It is appropriate for us to take a moment to think a little bit about how a material responds to an applied stress. Strain, denoted by e and referred to as displacement, is often written as $\Delta l/l$. It is a second-order tensor, which we will write as e_{ij}. We interpret e_{yx} as a shear strain, dy/dx or $\Delta y/\Delta x$. The normal strains, such as e_{xx}, are positive for an element of material that is stretched (extensional strain) and negative for one that is compressed. The summation of the diagonal components, which we will write as e_{ii}, is the volume strain (or dilatation). Thus, when we speak of the ratio of the volume of an element (undergoing deformation) to its initial volume, V/V_0, we are referring to dilatation. Naturally, dilatation for a real material must lie between zero and infinity. Now consider the response of specific material types; suppose we apply a fixed stress to a material that exhibits Hookean behavior (e.g., by applying an extensional force to a spring). The response is immediate, and when the stress is removed, the material (spring) recovers its initial size. Contrast this with the response of a Newtonian fluid; under a fixed shear stress, the resulting strain rate is constant, and when the stress is removed, the deformation remains. Of course, if a Newtonian fluid is incompressible, no applied stress can change the fluid element's volume; that is, the dilatation is zero. Among "real fluids," there are many that exhibit characteristics of *both* elastic solids and Newtonian fluids. For example, if a viscoelastic material is subjected to constant shear stress, we see some instantaneous deformation that is reversible, followed by flow that is not.

We now sketch the derivation of the equation of motion by making a momentum balance upon a cubic volume element of fluid with sides Δx, Δy, and Δz. We are formulating a vector equation, but it will suffice for us to develop just the x-component. The rate at which momentum accumulates within the volume should be equal to the rate at which momentum enters minus the rate at which momentum leaves (plus the sum of forces acting upon the volume element). Consequently, we write

accumulation

$$\Delta x \Delta y \Delta z \frac{\partial}{\partial t}(\rho v_x) = \tag{1.40a}$$

convective transport of x-momentum in the x-, y-, and z-directions

$$\begin{aligned} &+\Delta y \Delta z v_x \, \rho v_x|_x - \Delta y \Delta z v_x \, \rho v_x|_{x+\Delta x} \\ &+\Delta x \Delta z v_y \, \rho v_x|_y - \Delta x \Delta z v_y \, \rho v_x|_{y+\Delta y} \\ &+\Delta x \Delta y v_z \rho \, v_x|_z - \Delta x \Delta y v_z \, \rho v_x|_{z+\Delta z} \end{aligned} \tag{1.40b}$$

molecular transport of x-momentum in the x-, y-, and z-directions

$$\begin{aligned} &+\Delta y \Delta z \, \tau_{xx}|_x - \Delta y \Delta z \, \tau_{xx}|_{x+\Delta x} \\ &+\Delta x \Delta z \, \tau_{yx}|_y - \Delta x \Delta z \, \tau_{yx}|_{y+\Delta y} \\ &+\Delta x \Delta y \, \tau_{zx}|_z - \Delta x \Delta y \, \tau_{zx}|_{z+\Delta z} \end{aligned} \tag{1.40c}$$

pressure and gravitational forces

$$+\Delta y \Delta z (p|_x - p|_{x+\Delta x}) + \Delta x \Delta y \Delta z \rho g_x \tag{1.40d}$$

We now divide by $\Delta x \Delta y \Delta z$ and take the limits as all three are allowed to approach zero. The result, upon applying the definition of the first derivative, is

$$\begin{aligned} &\frac{\partial \rho v_x}{\partial t} + \frac{\partial}{\partial x}\rho v_x v_x + \frac{\partial}{\partial y}\rho v_y v_x + \frac{\partial}{\partial z}\rho v_z v_x \\ &= -\frac{\partial p}{\partial x} - \frac{\partial \tau_{xx}}{\partial x} - \frac{\partial \tau_{yx}}{\partial y} - \frac{\partial \tau_{zx}}{\partial z} + \rho g_x. \end{aligned} \tag{1.41}$$

This equation of motion can be written more generally in vector form:

$$\frac{\partial}{\partial t}(\rho v) + [\nabla \cdot \rho v v] = -\nabla p - [\nabla \cdot \tau] + \rho g. \tag{1.41a}$$

If Newton's law of friction (1.39) is introduced into (1.41) and if we take both the fluid density and viscosity to be constant, we obtain the x-component of the *Navier–Stokes* equation:

$$\begin{aligned} &\rho\left(\frac{\partial v_x}{\partial t} + v_x\frac{\partial v_x}{\partial x} + v_y\frac{\partial v_x}{\partial y} + v_z\frac{\partial v_x}{\partial z}\right) \\ &= -\frac{\partial p}{\partial x} + \mu\left[\frac{\partial^2 v_x}{\partial x^2} + \frac{\partial^2 v_x}{\partial y^2} + \frac{\partial^2 v_x}{\partial z^2}\right] + \rho g_x. \end{aligned} \tag{1.42}$$

It is useful to review the assumptions employed by Stokes in his derivation in 1845: (1) the fluid is continuous and the stress is no more than a linear function of strain, (2) the fluid is isotropic, and (3) when the fluid is at rest, it must develop a hydrostatic stress distribution that corresponds to the thermodynamic pressure. Consider the implications of (3): When the fluid is in motion, it is not in thermodynamic equilibrium, yet we still describe the pressure with an equation of state. Let us explore this further; we can write the stress tensor as Stokes did in 1845:

$$\tau_{ij} = -p\delta_{ij} + \mu\left(\frac{\partial v_i}{\partial x_j} + \frac{\partial v_j}{\partial x_i}\right) + \delta_{ij}\lambda \operatorname{div} V. \tag{1.43}$$

Now suppose we consider the three normal stresses; we will illustrate with just one, τ_{xx}:

$$\tau_{xx} = -p + 2\mu\left(\frac{\partial v_x}{\partial x}\right) + \lambda \operatorname{div} V. \tag{1.44}$$

We add all three together and then divide by (−)3, resulting in

$$-\frac{1}{3}(\tau_{xx} + \tau_{yy} + \tau_{zz}) = p - \left(\frac{2\mu + 3\lambda}{3}\right) \operatorname{div} V. \tag{1.45}$$

If we want the mechanical pressure to be *equal* to (negative one-third of) the trace of the stress tensor, then either div $V=0$, or alternatively, $2\mu+3\lambda=0$. If the fluid in question is incompressible, then the former is of course valid. But what about the more general case? If div $V\neq 0$, then it *would be extremely convenient* if $2\mu=-3\lambda$. This is Stokes' hypothesis; it has been the subject of much debate and it is almost certainly wrong except for monotonic gases. Nevertheless, it seems prudent to accept the simplification since as Schlichting (1968) notes, "... the working equations have been subjected to an unusually large number of experimental verifications, even under quite extreme conditions." Landau and Lifshitz (1959) observe that this second coefficient of viscosity (λ) is different in the sense that it is *not* merely a property of the fluid, as it appears to also depend on the frequency (or timescale) of periodic motions (in the fluid). Landau and Lifshitz also state that if a fluid undergoes expansion or contraction, then thermodynamic equilibrium must be restored. They note that if this relaxation occurs slowly, then it is possible that λ is large. There is some evidence that λ may actually be positive for liquids, and the student with deeper interest in Stokes' hypothesis may wish to consult Truesdell (1954).

We can use the substantial time derivative to rewrite eq. (1.42) more compactly:

$$\rho\frac{Dv}{Dt}=-\nabla p+\mu\nabla^2 v+\rho g. \tag{1.46}$$

We should review the meaning of the terms appearing above. On the left-hand side, we have the accumulation of momentum and the convective transport terms (these are the nonlinear inertial terms). On the right-hand side, we have pressure forces, the molecular transport of momentum (viscous friction), and external body forces such as gravity. Please note that the density and the viscosity are assumed to be constant. Consequently, we should identify (1.46) as the Navier–Stokes equation; it is inappropriate to refer to it as the generalized *equation of motion*. We should also observe that for the arbitrary three-dimensional flow of a *nonisothermal, compressible fluid*, it would be necessary to solve (1.41), along with the y- and z-components, the equation of continuity (1.35a), the equation of energy, and an equation of state simultaneously. In this type of problem, the six dependent variables are v_x, v_y, v_z, p, T, and ρ.

As noted previously, we can take the curl of the Navier–Stokes equation and obtain the vorticity transport equation, which is very useful for the solution of some hydrodynamic problems:

$$\frac{\partial\omega}{\partial t}=\nabla\times(v\times\omega)+\nu\nabla^2\omega, \tag{1.47}$$

or alternatively,

$$\frac{D\omega}{Dt}=\omega\cdot\nabla v+\nu\nabla^2\omega. \tag{1.48}$$

It is also possible to obtain an energy equation by multiplying the Navier–Stokes equation by the velocity vector v. We employ subscripts here, noting that i and j can assume the values 1, 2, and 3, corresponding to the x, y, and z directions:

$$\rho v_j\frac{\partial}{\partial x_j}\left(\frac{1}{2}v_i v_i\right)=\frac{\partial}{\partial x_j}(\tau_{ij}v_i)-\tau_{ij}\frac{\partial v_i}{\partial x_j}. \tag{1.49}$$

$\tau_{i,j}$ is the symmetric stress tensor, and we are employing Stokes' simplification:

$$\tau_{ij}=-p\delta_{ij}+2\mu S_{ij}. \tag{1.50}$$

δ is the Kronecker delta ($\delta_{ij}=1$ if $i=j$, and zero otherwise) and S_{ij} is the strain rate tensor,

$$S_{ij}=\frac{1}{2}\left[\frac{\partial v_i}{\partial x_j}+\frac{\partial v_j}{\partial x_i}\right]. \tag{1.51}$$

In the literature of fluid mechanics, the strain rate tensor is often written as it appears in eq. (1.51), but one may also find $S_{ij}=\left[\partial v_i/\partial x_j+\partial v_j/\partial x_i\right]$. Symmetric second-order tensors have three invariants (by invariant, we mean there is no change resulting from rotation of the coordinate system):

$$I_1(A)=\mathrm{tr}(A), \tag{1.52}$$

$$I_2(A)=\frac{1}{2}\left[(\mathrm{tr}(A))^2-\mathrm{tr}(A^2)\right] \tag{1.53}$$

(which for a symmetric A is $I_2=A_{11}A_{22}+A_{22}A_{33}+A_{11}A_{33}-A_{12}^2-A_{23}^2-A_{13}^2$), and

$$I_3(A)=\det(A). \tag{1.54}$$

The second invariant of the strain rate tensor is particularly useful to us; it is the double dot product of S_{ij}, which we write as $\sum_i\sum_j S_{ij}S_{ji}$. For rectangular coordinates, we obtain

$$I_2=2\left[\left(\frac{\partial v_x}{\partial x}\right)^2+\left(\frac{\partial v_y}{\partial y}\right)^2+\left(\frac{\partial v_z}{\partial z}\right)^2\right]+\left(\frac{\partial v_x}{\partial y}+\frac{\partial v_y}{\partial x}\right)^2+\left(\frac{\partial v_x}{\partial z}+\frac{\partial v_z}{\partial x}\right)^2+\left(\frac{\partial v_y}{\partial z}+\frac{\partial v_z}{\partial y}\right)^2. \tag{1.55}$$

You may recognize these terms; they are used to compute the production of thermal energy by viscous dissipation, and they can be very important in flow systems with large velocity gradients. We will see them again in Chapter 7.

We shall make extensive use of these relationships in this book. This is a good point to summarize the Navier–Stokes equations, so that we can refer to them as needed.

Rectangular coordinates

$$\rho\left(\frac{\partial v_x}{\partial t}+v_x\frac{\partial v_x}{\partial x}+v_y\frac{\partial v_x}{\partial y}+v_z\frac{\partial v_x}{\partial z}\right)=-\frac{\partial p}{\partial x}+\mu\left[\frac{\partial^2 v_x}{\partial x^2}+\frac{\partial^2 v_x}{\partial y^2}+\frac{\partial^2 v_x}{\partial z^2}\right]+\rho g_x, \tag{1.56a}$$

$$\rho\left(\frac{\partial v_y}{\partial t}+v_x\frac{\partial v_y}{\partial x}+v_y\frac{\partial v_y}{\partial y}+v_z\frac{\partial v_y}{\partial z}\right)=-\frac{\partial p}{\partial y}+\mu\left[\frac{\partial^2 v_y}{\partial x^2}+\frac{\partial^2 v_y}{\partial y^2}+\frac{\partial^2 v_y}{\partial z^2}\right]+\rho g_y, \tag{1.56b}$$

$$\rho\left(\frac{\partial v_z}{\partial t}+v_x\frac{\partial v_z}{\partial x}+v_y\frac{\partial v_z}{\partial y}+v_z\frac{\partial v_z}{\partial z}\right)=-\frac{\partial p}{\partial z}+\mu\left[\frac{\partial^2 v_z}{\partial x^2}+\frac{\partial^2 v_z}{\partial y^2}+\frac{\partial^2 v_z}{\partial z^2}\right]+\rho g_z. \tag{1.56c}$$

Cylindrical coordinates

$$\rho\left(\frac{\partial v_r}{\partial t}+v_r\frac{\partial v_r}{\partial r}+\frac{v_\theta}{r}\frac{\partial v_r}{\partial \theta}+v_z\frac{\partial v_r}{\partial z}-\frac{v_\theta^2}{r}\right)=-\frac{\partial p}{\partial r}+\mu\left[\frac{\partial}{\partial r}\left(\frac{1}{r}\frac{\partial}{\partial r}(rv_r)\right)+\frac{1}{r^2}\frac{\partial^2 v_r}{\partial \theta^2}+\frac{\partial^2 v_r}{\partial z^2}-\frac{2}{r^2}\frac{\partial v_\theta}{\partial \theta}\right]+\rho g_r, \tag{1.57a}$$

$$\rho\left(\frac{\partial v_\theta}{\partial t}+v_r\frac{\partial v_\theta}{\partial r}+\frac{v_\theta}{r}\frac{\partial v_\theta}{\partial \theta}+v_z\frac{\partial v_\theta}{\partial z}+\frac{v_r v_\theta}{r}\right)=-\frac{1}{r}\frac{\partial p}{\partial \theta}+\mu\left[\frac{\partial}{\partial r}\left(\frac{1}{r}\frac{\partial}{\partial r}rv_\theta\right)+\frac{1}{r^2}\frac{\partial^2 v_\theta}{\partial \theta^2}+\frac{\partial^2 v_\theta}{\partial z^2}+\frac{2}{r^2}\frac{\partial v_r}{\partial \theta}\right]+\rho g_\theta, \tag{1.57b}$$

$$\rho\left(\frac{\partial v_z}{\partial t}+v_r\frac{\partial v_z}{\partial r}+\frac{v_\theta}{r}\frac{\partial v_z}{\partial \theta}+v_z\frac{\partial v_z}{\partial z}\right)=-\frac{\partial p}{\partial z}+\mu\left[\frac{1}{r}\frac{\partial}{\partial r}\left(r\frac{\partial v_z}{\partial r}\right)+\frac{1}{r^2}\frac{\partial^2 v_z}{\partial \theta^2}+\frac{\partial^2 v_z}{\partial z^2}\right]+\rho g_z. \tag{1.57c}$$

Spherical coordinates

$$\rho\left(\frac{\partial v_r}{\partial t}+v_r\frac{\partial v_r}{\partial r}+\frac{v_\theta}{r}\frac{\partial v_r}{\partial \theta}+\frac{v_\phi}{r\sin\theta}\frac{\partial v_r}{\partial \phi}-\frac{v_\theta^2+v_\phi^2}{r}\right)=-\frac{\partial p}{\partial r}+\mu\left[\frac{1}{r^2}\frac{\partial^2}{\partial r^2}(r^2v_r)+\frac{1}{r^2\sin\theta}\frac{\partial}{\partial \theta}\left(\sin\theta\frac{\partial v_r}{\partial \theta}\right)+\frac{1}{r^2\sin^2\phi}\frac{\partial^2 v_r}{\partial \phi^2}\right]+\rho g_r, \tag{1.58a}$$

$$\rho\left(\frac{\partial v_\theta}{\partial t}+v_r\frac{\partial v_\theta}{\partial r}+\frac{v_\theta}{r}\frac{\partial v_\theta}{\partial \theta}+\frac{v_\phi}{r\sin\theta}\frac{\partial v_\theta}{\partial \phi}+\frac{v_r v_\theta-v_\phi^2\cot\theta}{r}\right)=-\frac{1}{r}\frac{\partial p}{\partial \theta}+\mu\left[\frac{1}{r^2}\frac{\partial}{\partial r}\left(r^2\frac{\partial v_\theta}{\partial r}\right)+\frac{1}{r^2}\frac{\partial}{\partial \theta}\left(\frac{1}{\sin\theta}\frac{\partial}{\partial \theta}(v_\theta\sin\theta)\right)+\frac{1}{r^2\sin^2\theta}\frac{\partial^2 v_\theta}{\partial \phi^2}+\frac{2}{r^2}\frac{\partial v_r}{\partial \theta}-\frac{2}{r^2}\frac{\cot\theta}{\sin\theta}\frac{\partial v_\phi}{\partial \phi}\right],+\rho g_\theta \tag{1.58b}$$

$$\rho\left(\frac{\partial v_\phi}{\partial t}+v_r\frac{\partial v_\phi}{\partial r}+\frac{v_\theta}{r}\frac{\partial v_\phi}{\partial \theta}+\frac{v_\phi}{r\sin\theta}\frac{\partial v_\phi}{\partial \phi}+\frac{v_\phi v_r+v_\theta v_\phi\cot\theta}{r}\right)=-\frac{1}{r\sin\theta}\frac{\partial p}{\partial \phi}+\mu\left[\frac{1}{r^2}\frac{\partial}{\partial r}\left(r^2\frac{\partial v_\phi}{\partial r}\right)+\frac{1}{r^2}\frac{\partial}{\partial \theta}\left(\frac{1}{\sin\theta}\frac{\partial}{\partial \theta}(v_\phi\sin\theta)\right)+\frac{1}{r^2\sin^2\theta}\frac{\partial^2 v_\phi}{\partial \phi^2}+\frac{2}{r^2\sin\theta}\frac{\partial v_r}{\partial \phi}+\frac{2\cot\theta}{r^2\sin\theta}\frac{\partial v_\theta}{\partial \phi}\right]+\rho g_\phi \tag{1.58c}$$

These equations have attracted the attention of many eminent mathematicians and physicists; despite more than 160 years of very intense work, only a handful of solutions are known for the Navier–Stokes equation(s). White (1991) puts the number at 80, which is pitifully small compared to the number of flows we might wish to consider. The Clay Mathematics Institute has observed that "... although these equations were written down in the 19th century, our understanding of them remains minimal. The challenge is to make substantial progress toward a mathematical theory which will unlock the secrets hidden in the Navier–Stokes equations."

1.6 THE MEN FOR WHOM THE NAVIER–STOKES EQUATIONS ARE NAMED

The equations of fluid motion given immediately above are named after Claude Louis Marie Henri Navier (1785–1836) and Sir George Gabriel Stokes (1819–1903). There was no professional overlap between the two men as Navier died in 1836 when Stokes (a 17-year-old) was in his second year at Bristol College. Navier had been taught by Fourier at the Ecole Polytechnique and that clearly had a great influence upon his subsequent interest in mathematical analysis. But in the nineteenth century, Navier was known primarily as a bridge designer/builder who made important contributions to structural mechanics. His work in fluid mechanics was not as well known. Anderson (1997) observed that Navier *did not understand shear stress* and although he did not intend to derive the equations governing fluid motion *with* molecular friction, he did arrive at the proper form for those equations. Stokes himself displayed talent for mathematics while at Bristol. He entered Pembroke College at Cambridge in 1837 and was coached in mathematics by William Hopkins; later, Hopkins recommended hydrodynamics to Stokes as an area ripe for investigation. Stokes set about to account for frictional effects occurring in flowing fluids and again the proper form of the equation(s) was discovered (but this time with intent). He became aware of Navier's work after completing his own derivation. In 1845, Stokes published "On the Theories of the Internal Friction of Fluids in Motion" recognizing that his development employed different assumptions from those of Navier. For a better glimpse into the personalities and lives of Navier and Stokes, see the biographical sketches written by O'Connor and Robertson2003 (*MacTutor History of Mathematics*). A much richer picture of Stokes the man can be obtained by reading his correspondence (especially between Stokes and Mary Susanna Robinson) in Larmor's memoir (1907).

1.7 SIR ISAAC NEWTON

Much of what we routinely use in the study of transport phenomena (and, indeed, in all of mathematics and mechanics) is due to Sir Isaac Newton. Newton, according to the contemporary calendar, was born on Christmas Day in 1642; by modern calendar, his date of birth was January 4, 1643. His father (also Isaac Newton) died prior to his son's birth and although the elder Newton was a wealthy landowner, he could neither read nor write. His mother, following the death of her second husband, intended for young Isaac to manage the family estate. However, this was a task for which Isaac had neither the temperament nor the interest. Fortunately, an uncle, William Ayscough, recognized that the lad's abilities were directed elsewhere and was instrumental in getting him entered at Trinity College Cambridge in 1661.

Many of Newton's most important contributions had their origins in the plague years of 1665–1667 when the University was closed. While home at Lincolnshire, he developed the foundation for what he called the "method of fluxions" (differential calculus) and he also perceived that integration was the inverse operation to differentiation. As an aside, we note that a fluxion, or differential coefficient, is the change in one variable brought about by the change in another, related variable. In 1669, Newton assumed the Lucasian chair at Cambridge (see the information compiled by Robert Bruen and also http://www.lucasianchair.org/) following Barrow's resignation. Newton lectured on optics in a course that began in January 1670 and in 1672 he published a paper on light and color in the *Philosophical Transactions of the Royal Society*. This work was criticized by Robert Hooke and that led to a scientific feud that did not come to an end until Hooke's death in 1703. Indeed, Newton's famous quote, "If I have seen further it is by standing on ye shoulders of giants," which has often been interpreted as a statement of humility appears to have actually been intended as an insult to Hooke (who was a short hunchback, becoming increasingly deformed with age).

Certainly Newton had a difficult personality with a dichotomous nature—he wanted recognition for his developments but was so averse to criticism that he was reticent about sharing his discoveries through publication. This characteristic contributed to the acrimony over who should be credited with the development of differential calculus, Newton or Leibniz. Indeed, this debate created a schism between British and continental mathematicians that lasted decades. But two points are absolutely clear: Newton's development of the "method of fluxions" predated Liebniz's work and each man used his own, unique, system of notation (suggesting that the efforts were completely independent). Since differential calculus ranks arguably as the most important intellectual accomplishment of the seventeenth century, one can at least comprehend the vitriol of this long-lasting debate. Newton used the Royal Society to "resolve" the question of priority; however, since he wrote the committee's report anonymously, there can be no claim to impartiality.

Newton also had a very contentious relationship with John Flamsteed, the first Astronomer Royal. Newton needed Flamsteed's lunar observations to correct the lunar theory he had presented in *Principia* (*Philosophiae Naturalis Principia Mathematica*). Flamsteed was clearly reluctant to provide these data to Newton and in fact demanded Newton's promise not to share or further disseminate the results, a restriction that Newton could not tolerate. Newton made repeated efforts to obtain Flamsteed's observations both directly and through the influence of Prince George, but without success. Flamsteed prevailed; his data were not published until 1725, 6 years after his death.

There is no area in optics, mathematics, or mechanics that was not at least touched by Newton's genius. No less a mathematician than Lagrange stated that Newton's *Principia* was the greatest production of the human mind and this evaluation was echoed by Laplace, Gauss, and Biot, among others. Two anecdotes, though probably unnecessary, can be used to underscore Newton's preeminence: In 1696, Johann Bernoulli put forward the *brachistochrone* problem (*to determine the path in the vertical plane by which a weight would descend most rapidly from higher point A to lower point B*). Leibniz worked the problem in 6 months; Newton solved it *overnight* according to the biographer, John Conduitt, finishing at about 4 the next morning. Other solutions were eventually obtained from Leibniz, l'Hopital, and both Jacob and Johann Bernoulli. In a completely unrelated problem, Newton was able to determine the path of a ray by (effectively) solving a differential equation in 1694; Euler could not solve the same problem in 1754. Laplace was able to solve it, *but in 1782*.

It is, I suppose, curiously comforting to ordinary mortals to know that truly rare geniuses like Newton always seem to be flawed. His assistant Whiston observed that "Newton was of the most fearful, cautious and suspicious temper that I ever knew."

Furthermore, in the brief glimpse offered here, we have avoided describing Newton's interests in alchemy, history, and prophecy, some of which might charitably be characterized as peculiar. It is also true that work he performed as warden of the Royal Mint does not fit the reclusive scholar stereotype; as an example, Newton was instrumental in having the counterfeiter William Chaloner hanged, drawn, and quartered in 1699. Nevertheless, Newton's legacy in mathematical physics is absolutely unique. There is no other case in history where a single man did so much to advance the science of his era so far beyond the level of his contemporaries.

We are fortunate to have so much information available regarding Newton's life and work through both his own writing and exchanges of correspondence with others. A select number of valuable references used in the preparation of this account are provided immediately below.

The Correspondence of Isaac Newton, edited by H. W. Turnbull, FRS, University Press, Cambridge (1961).

The Newton Handbook, Derek Gjertsen, Routledge & Kegan Paul, London (1986).

Memoirs of Sir Isaac Newton, Sir David Brewster, reprinted from the Edinburgh Edition of 1855, Johnson Reprint Corporation, New York (1965).

A Short Account of the History of Mathematics, 6th edition, W. W. Rouse Ball, Macmillan, London (1915).

See also http://www-groups.dcs.st-and.ac.uk and http://www.newton.cam.ac.uk.

REFERENCES

Anderson, J. D. *A History of Aerodynamics*, Cambridge University Press, New York (1997).

Baker, G. L. and J. P. Gollub. *Chaotic Dynamics*, Cambridge University Press, Cambridge (1990).

Baruh, H. Are Computers Hurting Education? *ASEE Prism*, p. 64 (October 2001).

Batchelor, G. K. *An Introduction to Fluid Dynamics*, Cambridge University Press, Cambridge (1967).

Bird, R. B., Stewart, W. E., and E. N. Lightfoot. *Transport Phenomena*, 2nd edition, Wiley, New York (2002).

Bohren, C. F. Comment on "Newton's Law of Cooling—A Critical Assessment," by C. T. O'Sullivan. *American Journal of Physics*, 59:1044 (1991).

Clay Mathematics Institute, www.claymath.org.

Davis, H. T. *Introduction to Nonlinear Differential and Integral Equations*, Dover Publications, New York (1962).

Fermi, E., Pasta, J., and S. Ulam. Studies of Nonlinear Problems, 1. *Report LA-1940* (1955).

Landau, L. D. and E. M. Lifshitz. *Fluid Mechanics*, Pergamon Press, London (1959).

Larmor, J., editor. *Memoir and Scientific Correspondence of the Late Sir George Gabriel Stokes*, Cambridge University Press, New York (1907).

Lorenz, E. N. Deterministic Nonperiodic Flow. *Journal of the Atmospheric Sciences*, 20:130 (1963).

Milne-Thomson, L. M. *Jacobian Elliptic Function Tables: A Guide to Practical Computation with Elliptic Functions and Integrals*, Dover, New York (1950).

O'Connor, J. J. and E. F. Robertson. MacTutor History of Mathematics, www.history.mcs.st-andrews.ac.uk (2003).

Packard, N. H., Crutchfield, J. P., Farmer, J. D., and R. S. Shaw. Geometry from a Time Series. *Physical Review Letters*, 45:712 (1980).

Porter, M. A., Zabusky, N. J., Hu, B., and D. K. Campbell. Fermi, Pasta, Ulam and the Birth of Experimental Mathematics. *American Scientist*, 97:214 (2009).

Powers, D, L. *Boundary Value Problems*, 2nd edition, Academic Press, New York (1979).

Rossler, O. E. An Equation for Continuous Chaos. *Physics Letters*, 57A:397 (1976).

Schlichting, H. *Boundary-Layer Theory*, 6th edition, McGraw-Hill, New York (1968).

Stokes, G. G. On the Theories of the Internal Friction of Fluids in Motion. *Transactions of the Cambridge Philosophical Society*, 8:287 (1845).

Truesdell, C. The Present Status of the Controversy Regarding the Bulk Viscosity of Liquids. *Proceedings of the Royal Society of London*, A226:1 (1954).

Vaughn, M. T. *Introduction to Mathematical Physics*, Wiley-VCH, Weinheim (2007).

White, F. M. *Viscous Fluid Flow*, 2nd edition, McGraw-Hill, New York (1991).

2

INVISCID FLOW: SIMPLIFIED FLUID MOTION

2.1 INTRODUCTION

In the early years of the twentieth century, Prandtl (1904) proposed that for flow over objects the effects of viscous friction would be confined to a thin region of fluid very close to the solid surface. Consequently, for incompressible flows in which the fluid is accelerating, *viscosity should be unimportant for much of the flow field.* This hypothesis might (in fact, did) allow workers in fluid mechanics to successfully treat some difficult problems in an approximate way. Consider the consequences of setting viscosity μ equal to zero in the x-component of the Navier–Stokes equation:

$$\rho\left(\frac{\partial v_x}{\partial t}+v_x\frac{\partial v_x}{\partial x}+v_y\frac{\partial v_x}{\partial y}+v_z\frac{\partial v_x}{\partial z}\right)=-\frac{\partial p}{\partial x}+\rho g_x. \tag{2.1}$$

The result is the x-component of the Euler equation and you can see that the order of the equation has been reduced from 2 to 1. Of course, this automatically means a loss of information; we can no longer enforce the no-slip condition. We will also require that the flow be irrotational so that $\nabla \times V=0$; consequently,

$$\frac{\partial v_x}{\partial z}=\frac{\partial v_z}{\partial x} \quad \text{and} \quad \frac{\partial v_x}{\partial y}=\frac{\partial v_y}{\partial x}. \tag{2.2}$$

Now we introduce the velocity potential ϕ. We can obtain the fluid velocity in a given direction by differentiation of ϕ in that direction; for example,

$$v_x=\frac{\partial \phi}{\partial x}. \tag{2.3}$$

These steps allow us to rewrite the Euler equation as follows:

$$\frac{\partial^2 \phi}{\partial t \partial x}+v_x\frac{\partial v_x}{\partial x}+v_y\frac{\partial v_y}{\partial x}+v_z\frac{\partial v_z}{\partial x}=-\frac{1}{\rho}\frac{\partial p}{\partial x}+\frac{\partial \Omega}{\partial x}, \tag{2.4}$$

where Ω is a potential energy function. Of course, this result can be integrated with respect to x:

$$\frac{\partial \phi}{\partial t}+\frac{v_x^2}{2}+\frac{v_y^2}{2}+\frac{v_z^2}{2}+\frac{p}{\rho}-\Omega=F_1. \tag{2.5}$$

Note that F_1 *cannot* be a function of x. The very same process sketched above can also be carried out for the y- and z-components of the Euler equation; when the three results are combined, we get the Bernoulli equation:

$$\frac{\partial \phi}{\partial t}+\frac{1}{2}|V|^2+\frac{p}{\rho}+gZ=F(t). \tag{2.6}$$

This is an inviscid energy balance; it can be very useful in the preliminary analysis of flow problems. For example, one could use the equation to *qualitatively* explain the operation of an airfoil or a Frisbee™ flying disk. For the latter, consider a flying disk with a diameter of 22.86 cm and mass of 80.6 g, given an initial velocity of 6.5 m/s. The airflow across the

Transport Phenomena: An Introduction to Advanced Topics, By Larry A. Glasgow

top of the disk (along a center path) must travel about 26 cm, corresponding to an approximate velocity of 740 cm/s. This increased velocity over the top gives rise to a pressure difference of about 75 dyn/cm^2, generating enough lift to partially offset the effect of gravity.

We emphasize that the Bernoulli equation does not account for dissipative processes, so we cannot expect quantitative results for systems with significant friction. We are, however, going to make direct use of potential flow theory a little later when we begin our consideration of *boundary-layer flows*.

2.2 TWO-DIMENSIONAL POTENTIAL FLOW

We now turn our attention to two-dimensional, inviscid, irrotational, incompressible (potential) flows. The descriptor "potential" comes from analogy with electrostatics. In fact, Streeter and Wylie (1975) note that the flow net for a set of fixed boundaries can be obtained with a voltmeter using a nonconducting surface and a properly bounded electrolyte solution. The student seeking additional background and detail for inviscid fluid motions should consult Lamb (1945) and Milne-Thomson (1958). The continuity equation for these two-dimensional flows is

$$\frac{\partial v_x}{\partial x} + \frac{\partial v_y}{\partial y} = 0. \tag{2.7}$$

Using the velocity potential ϕ to represent velocity vector components in eq. (2.7), we obtain the Laplace equation:

$$\frac{\partial^2 \phi}{\partial x^2} + \frac{\partial^2 \phi}{\partial y^2} = 0, \quad \text{or simply } \nabla^2 \phi = 0. \tag{2.8}$$

We define the stream function such that

$$v_x = -\frac{\partial \psi}{\partial y} \quad \text{and} \quad v_y = \frac{\partial \psi}{\partial x}. \tag{2.9}$$

This choice means that for a case in which ψ increases in the vertical (y) direction, flow with respect to the x-axis will be right-to-left. We can reverse the signs in (2.9) if we prefer the flow to be left-to-right. If we couple (2.9) with the irrotational requirement (2.2), we find

$$\frac{\partial^2 \psi}{\partial x^2} + \frac{\partial^2 \psi}{\partial y^2} = 0. \tag{2.10}$$

Note that the velocity potential and stream function must be related by the equations

$$\frac{\partial \phi}{\partial x} = -\frac{\partial \psi}{\partial y} \quad \text{and} \quad \frac{\partial \phi}{\partial y} = \frac{\partial \psi}{\partial x}. \tag{2.11}$$

These are the Cauchy–Riemann relations and they guarantee the existence of a complex potential, a mapping between the ϕ–ψ plane (or flow net) and the x–y plane. This simply means that *any* analytic function of z ($z = x + iy$) corresponds to the solution of *some* potential flow problem. This branch of mathematics is called *conformal mapping* and there are compilations of conformal representations that can be used to "solve" potential flow problems; see Kober (1952), for example. Alternatively, we can simply assume a form for the complex potential; suppose we let

$$W(z) = z + z^3 = (x + iy) + (x + iy)^3; \tag{2.12}$$

therefore,

$$\phi + i\psi = x + iy + x^3 + 3ix^2y - 3xy^2 - iy^3$$

and

$$\psi = y + 3x^2y - y^3. \tag{2.13}$$

What does this flow look like? It is illustrated in Figure 2.1.

Note that the general form of the complex potential for flow in a corner is $W(z) = Vh(z/h)^{\pi/\theta}$, where θ is the included angle. Therefore, for a 45° corner (taking the reference length to be 1), $\theta = \pi/4$ and $W(z) = Vz^4$.

Let us now consider the vortex, whose complex potential is given by

$$\phi + i\psi = \frac{\Gamma i}{2\pi} \ln(x + iy), \tag{2.14}$$

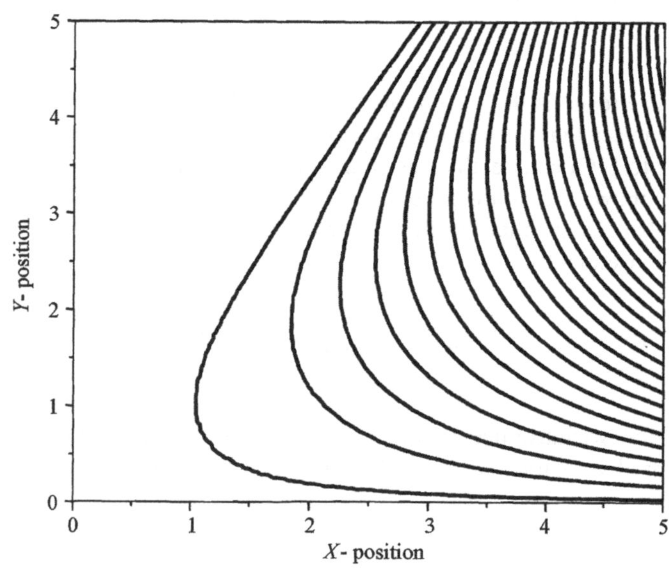

FIGURE 2.1. Variation of flow in a corner obtained from the complex potential $W(z) = z + z^3$.

where Γ is the circulation around a closed path. It is convenient in such cases to write the complex number in polar form, that is, $x+iy=re^{i\theta}$. The stream function and the velocity potential can then be written as

$$\psi = \frac{\Gamma}{2\pi}\ln r \quad \text{and} \quad \phi = -\frac{\Gamma\theta}{2\pi}. \tag{2.15}$$

Note that the stream function assumes very large negative values as the center of the vortex is approached. What does this tell you about velocity at the center of an ideal vortex?

Many interesting flows can be constructed by simple combination. For example, if we take uniform flow,

$$\phi + i\psi = V(x+iy), \tag{2.16}$$

and combine it with a source,

$$\phi + i\psi = \frac{Q}{2\pi}\ln(x+iy), \tag{2.17}$$

we can get the stream function for flow about a two-dimensional half-body:

$$\psi = Vr\sin\theta - \frac{Q}{2\pi}\theta. \tag{2.18}$$

This is illustrated in Figure 2.2. The radius of the body at the leading edge, or nose, is $Q/(2\pi V)$.

The complex potential for flow around a cylinder is

$$W(z) = -V\left(z + \frac{a^2}{z}\right), \tag{2.19}$$

and the stream function is

$$\psi = -V\left[y - \frac{a^2 y}{x^2+y^2}\right]. \tag{2.20}$$

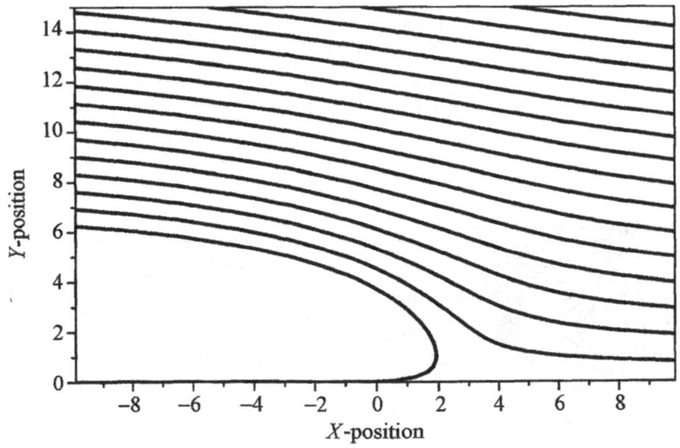

FIGURE 2.2. Two-dimensional potential flow around a half-body. The flow is symmetric about the *x*-axis, so only the upper half is shown.

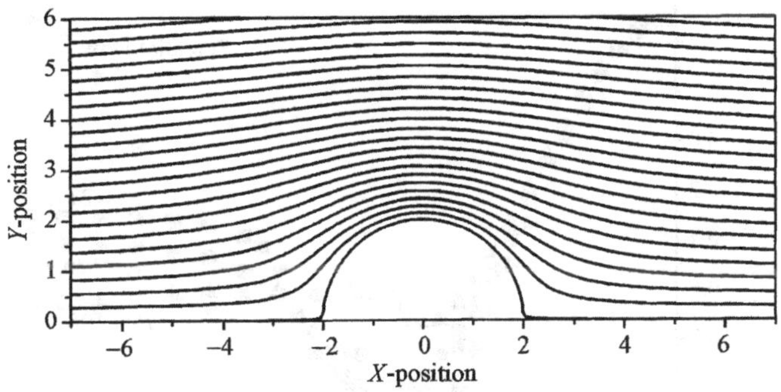

FIGURE 2.3. Potential flow past a circular cylinder. Note the fore-and-aft symmetry, which of course means that there is no form drag. This feature of potential flow is the source of d'Alembert's paradox and it was an enormous setback to fluid mechanics since many hydrodynamicists of the era concluded that the Euler equation(s) was incorrect.

This stream function is plotted in Figure 2.3. Note that there is *no difference* in the flow between the upstream and downstream sides. In fact, the pressure distribution at the cylinder's surface is perfectly symmetric:

$$p - p_\infty = \frac{1}{2}\rho V_\infty^2(1 - 4\sin^2\theta). \tag{2.21}$$

Make sure you understand how this result is obtained using eq. (2.6)! At $\theta = 0, p - p_\infty$ is the dynamic head, $\frac{1}{2}\rho V_\infty^2$. Note also that the pressure at 90° corresponds to $-3(\frac{1}{2}\rho V_\infty^2)$ and that the recovery is complete as one moves on to 180°. Experimental measurements of pressure on the surface of circular cylinders show that the minimum is usually attained at about 70° or 75° and the pressure recovery on the downstream side is far from complete. The potential flow solution gives a reasonable result only to about $\theta \cong 60°$ for large Reynolds numbers. This is evident from the pressure distributions shown in Figure 2.4.

If we combine a uniform flow with a doublet (a source and a sink combined with zero separation) and a vortex, we obtain flow around a cylinder with circulation (by *circulation* we mean the integral of the tangential component of velocity around a closed path):

$$\psi = V\sin\theta\left(r - \frac{R^2}{r}\right) + \frac{\Gamma}{2\pi}\ln r. \tag{2.22}$$

The pressure at the surface of the cylinder is

$$p = \frac{\rho V^2}{2}\left[1 - \left(2\sin\theta + \frac{\Gamma}{2\pi RV}\right)^2\right]. \tag{2.23}$$

Obviously, since this is inviscid flow there is no frictional drag, but might we have form drag? That is, is there a net force in the direction of the uniform flow? Consult Figure 2.5; note that the flow is symmetric fore and aft (upstream and

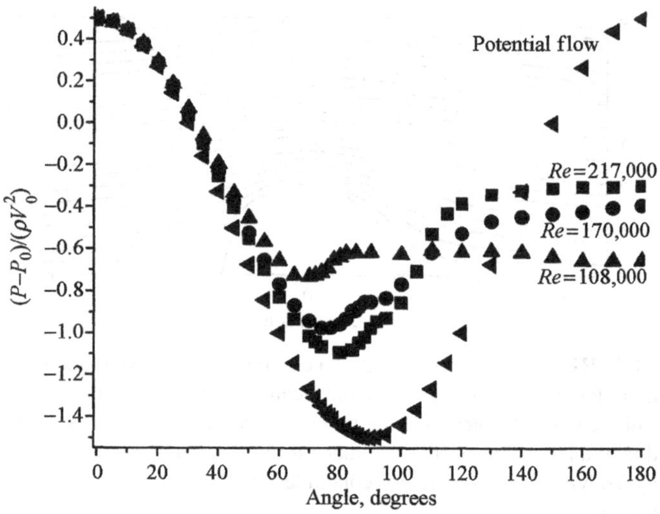

FIGURE 2.4. Dimensionless pressure $(p - p_\infty)/(\frac{1}{2}\rho V^2)$ distributions for flow over a cylinder; the potential flow case is clearly labeled and the experimental data points are from Fage and Falkner (1931) for $Re = 108{,}000$, 170,000, and 217,000.

downstream). Of course, this means that there is *no net force in the horizontal direction*, and hence, no drag. But suppose we look at the vertical component, that is, $-p \sin\theta$. When this quantity is integrated over the surface, the result is *not* zero; the rotating cylinder is *generating lift*. This phenomenon is known as the Magnus effect.

The lift being generated by the cylinder is $\rho V\Gamma$, which is equivalent to $2\pi\rho RVV_\theta$. For example, suppose air is approaching a circular cylinder (from the left) at 30 m/s. The cylinder is rotating in the clockwise direction at 1500 rpm (157 rad/s). If the cylinder diameter is 50 cm, then

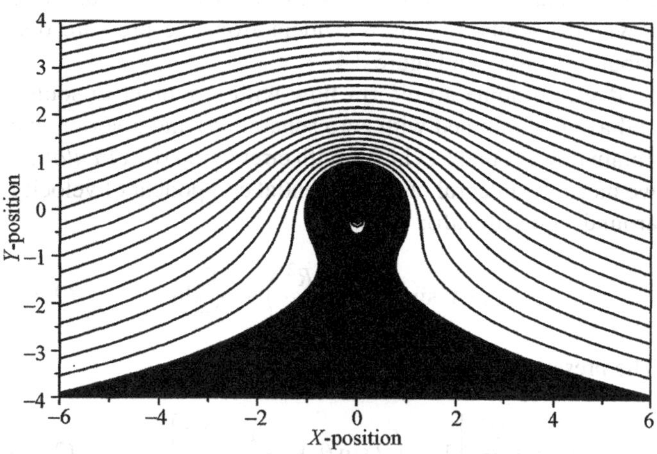

FIGURE 2.5. Two-dimensional potential flow about a cylinder with circulation. Note how the fluid is wrapped up and around the rotating cylinder. This generates lift since the pressure is larger across the bottom of the cylinder than across the top; the (Magnus) effect is significant for rotating bodies with large translational velocities.

V_θ is 3927 cm/s and the cylinder is generating a lift of 2.22×10^6 dyn per cm of length. This phenomenon is familiar to anyone who has played a sport in which sidespin and translation are simultaneously imparted to a ball; soccer, tennis, golf, and baseball come immediately to mind. Schlichting (1968) points out that an attempt was made to utilize the effect commercially with the Flettner "rotor" ship in the 1920s. More details regarding these efforts are provided by Ahlborn (1930). The first full-scale efforts to exploit the phenomenon were carried out with the steamship *Buckau*. This vessel made 7.85 knots in trials with 134 hp using its screw propeller; under favorable conditions in early 1925, it attained 8.2 knots using only 33.4 hp to turn the rotors (no propeller). Ahlborn noted that although wind tunnel tests indicated that the rotors might be considerably more efficient than canvas sails of comparable surface area, the Flettner rotor was a nautical and economic failure. In more recent years, spinning cylinders have been incorporated into experimental airfoils to promote lift and control the boundary layer; see Chapter 5 in Chang (1976). A modern computational study of steady, uniform flow past rotating cylinders has been carried out by Padrino and Joseph (2006).

Among other particularly interesting complex potentials are the infinite row of vortices and the von Karman vortex street. For the former,

$$W(z) = i\kappa \ln\left(\sin\frac{\pi z}{a}\right) \tag{2.24}$$

and

$$\psi = \frac{\kappa}{2}\ln\left(\frac{1}{2}\left(\cosh\frac{2\pi y}{a} - \cos\frac{2\pi x}{a}\right)\right). \tag{2.25}$$

The row of vortices is illustrated in Figure 2.6.

For the von Karman vortex street, the complex potential is

$$W(z) = i\kappa \ln\left(\sin\left(\frac{\pi}{a}\left(z - \frac{ib}{2}\right)\right)\right) - i\kappa \ln\left(\sin\left(\frac{\pi}{a}\left(z - \frac{a}{2} + \frac{ib}{2}\right)\right)\right), \tag{2.26}$$

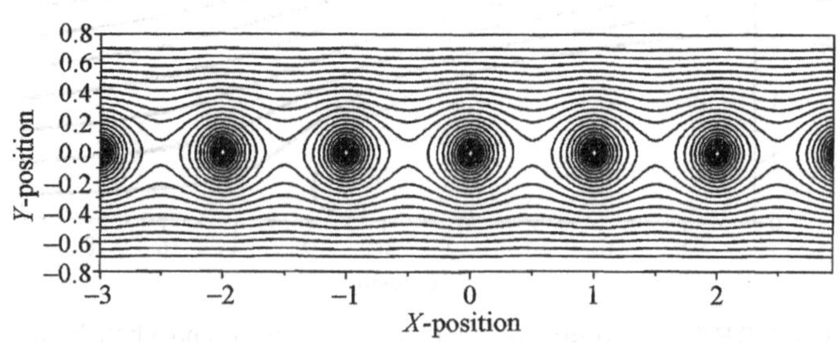

FIGURE 2.6. An infinite row of vortices each with the same strength and spaced a distance a apart.

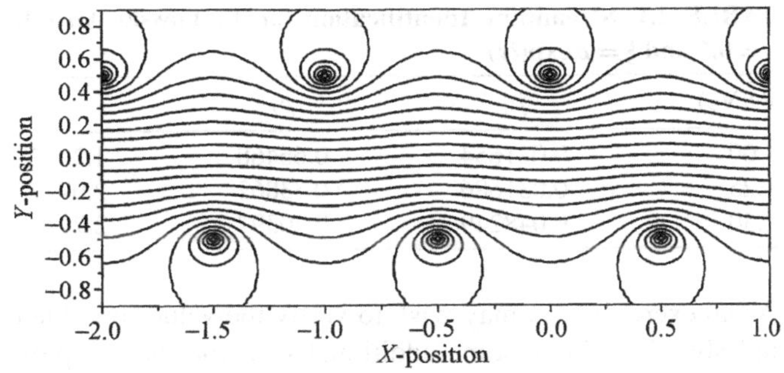

FIGURE 2.7. von Karman vortex street.

and the corresponding stream function is

$$\frac{\psi}{\kappa} = \frac{1}{2}\ln\left[\frac{\cosh(2\pi(y/a - k/2)) - \cos(2\pi x/a)}{\cosh(2\pi(y/a + k/2) - \cos(2\pi(x/a - 1/2))}\right], \tag{2.27}$$

where $k = b/a$. This flow field is illustrated in Figure 2.7. Many other interesting potential flows have been compiled by Kirchhoff (1985).

Complex potentials are also known for a variety of airfoils, including flat plate and Joukowski type (with and without camber), at different angles of attack; see Currie (1993) for additional examples. The complex potentials for these flows are linked to the z-plane through the Joukowski transformation; the Joukowski transformation between the z-plane and the ξ-plane is generally written as

$$z = \xi + \frac{L^2}{\xi}, \tag{2.28}$$

where L is a real constant. One of the features of this choice is that for very large ξ, $z \cong \xi$. Consequently, points that are far from the origin are unaffected by the mapping. Let us now illustrate how this works. Consider concentric circles located at the origin of the ξ-plane. Since the distance from the origin (O) to the point P_1 is a constant, then for the z-plane, $SP + HP = \text{constant}$. Accordingly, circles (with their centers at the origin) in the ξ-plane will map into confocal ellipses in the z-plane as demonstrated by Milne-Thomson (1958) and illustrated in Figure 2.8.

We should explore this process with an example. We take (2.28) and substitute $\xi = \alpha\, e^{i\lambda}$; therefore,

$$z = \alpha e^{i\lambda} + \frac{L^2}{\alpha e^{i\lambda}} = \left(\alpha + \frac{L^2}{\alpha}\right)\cos\lambda + i\left(\alpha - \frac{L^2}{\alpha}\right)\sin\lambda. \tag{2.29a}$$

This yields

$$x = \left(\alpha + \frac{L^2}{\alpha}\right)\cos\lambda \quad \text{and} \quad y = \left(\alpha - \frac{L^2}{\alpha}\right)\sin\lambda. \tag{2.29b}$$

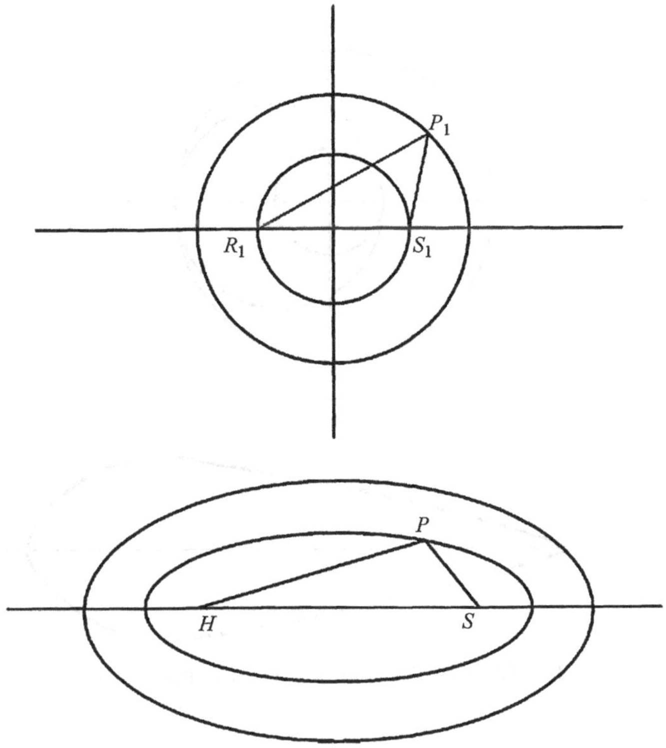

FIGURE 2.8. Concentric circles that map into confocal ellipses.

We can make use of the identity $\sin^2\lambda + \cos^2\lambda = 1$ to obtain

$$\left(\frac{x}{\alpha + L^2/\alpha}\right)^2 + \left(\frac{y}{\alpha - L^2/\alpha}\right)^2 = 1. \tag{2.29c}$$

If we let $\alpha = 3$ and $L = 2$, this equation produces an ellipse and the right half of this conic section is shown in Figure 2.9.

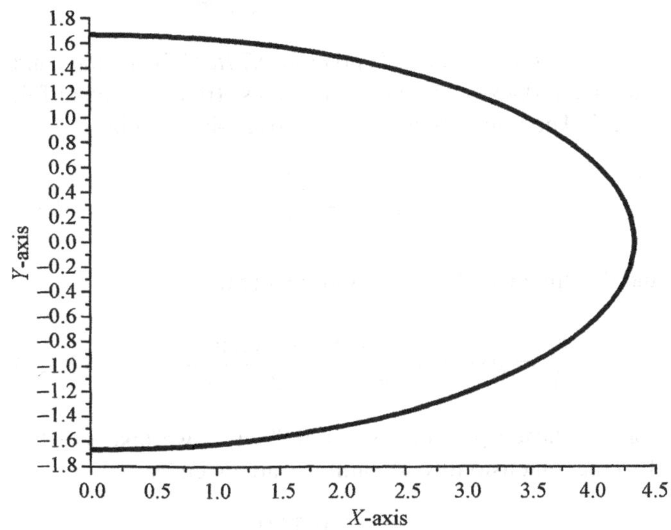

FIGURE 2.9. An ellipse constructed with eq. 2.29c.

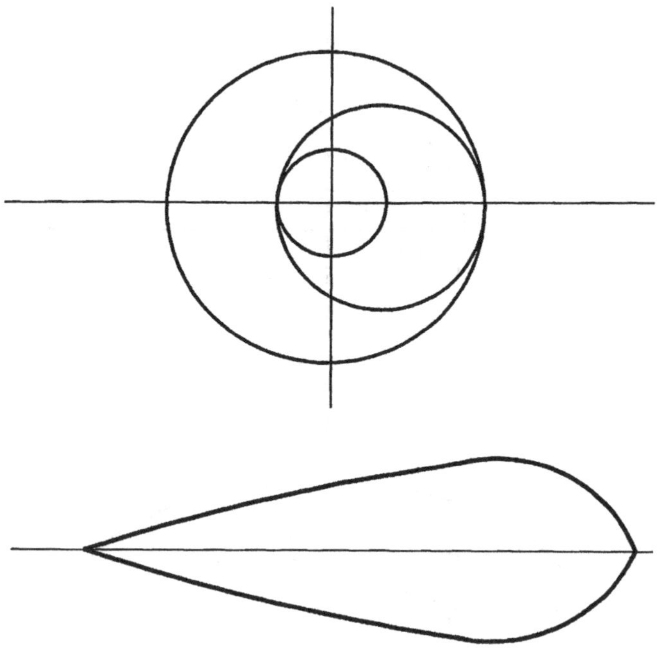

FIGURE 2.10. Mapping of an "off-center" circle.

In contrast, if we start with a circle whose center is on the real axis to the right of the origin as illustrated in Figure 2.10, we should get a map that lies between that of the concentric circles (with centers at the origin).

The "off-center" circle maps into the z-plane as a symmetric shape with a blunt nose on the right and a point (cusp) on the left. This technique can be used to generate potential flows about shapes that approximate a rudder or airfoil. For an airfoil with a chord of 4 and a thickness of 0.48, we can start with the complex potential

$$F(\xi) = V\left((\xi + m) + \frac{a^2}{\xi + m}\right), \tag{2.30}$$

where $a = l/4 + 0.77tc/l$ and $m = 0.77tc/l$. Note that l and t are 4 and 0.48, such that the thickness (ratio) of the airfoil is 12%. The transformation—as above—is given by

$$z = \xi + \frac{c^2}{\xi}, \tag{2.31}$$

and the dimensionless equation for streamlines is

$$\frac{\psi}{Vc} = \rho \sin \nu + \frac{\rho(1 + e)^2 \sin \nu}{\rho^2 + e^2 + 2\rho e \cos \nu}. \tag{2.32a}$$

For the chosen parameters, $e = 0.0924$; if we take $\nu = 0.5$, then the dimensionless streamlines are given by

$$\frac{\psi}{Vc} = 0.47943\rho + \frac{0.57212\rho}{\rho^2 + 0.16218\rho + 0.008538}. \tag{2.32b}$$

TABLE 2.1. Streamline Identification for Joukowski Airfoil; $\nu = 0.5$ and $\xi = \rho \exp(i\nu)$

$\psi/(Vc)$	ρ_1	ρ_2	ρ_3
1.00	0.021654	0.66468	1.23728
1.25	0.031686	0.33862	2.07478
1.50	0.048249	0.20397	2.71431

As an exercise, you may wish to verify the values provided in Table 2.1, obtain some additional sets, and then employ (2.31) to transform them to the physical plane.

2.3 NUMERICAL SOLUTION OF POTENTIAL FLOW PROBLEMS

Although hundreds of complex potentials (conformal mappings) have been developed over the years, we are not limited to flows that have been cataloged for us. Recall that both the velocity potential and the stream function satisfy the Laplace equation in ideal flows. We now employ a simple numerical procedure that will allow us to examine inviscid, irrotational, incompressible flows about nearly any object of our choice. We begin by writing the Laplace equation

$$\nabla^2 \psi = 0 \tag{2.33}$$

in finite difference form using second-order central differences:

$$\frac{\psi_{i+1,j} - 2\psi_{i,j} + \psi_{i-1,j}}{(\Delta x)^2} + \frac{\psi_{i,j+1} - 2\psi_{i,j} + \psi_{i,j-1}}{(\Delta y)^2} \cong 0. \tag{2.34}$$

The index i refers to the x-direction and j to the y-direction. Now we assume a square mesh such that $\Delta x = \Delta y$; we isolate the term with the largest coefficient, which is $\psi_{i,j}$. Consequently, we obtain a simple algorithm for computation of the central nodal point:

$$\psi_{i,j} = \frac{1}{4}(\psi_{i+1,j} + \psi_{i-1,j} + \psi_{i,j+1} + \psi_{i,j-1}). \tag{2.35}$$

The solution of such a problem is easy, in principle. We can apply (2.35) at every interior nodal point and solve the resulting system of equations iteratively, or we can solve the set of simultaneous algebraic equations directly using an elimination scheme (if the number of nodal points is not too large). We now illustrate the numerical procedure for flow over a reverse step; we will use the very simple Gauss–Seidel iterative method. The principal parts of the computation are as follows:

- initialize ψ throughout the flow field and on the boundary;

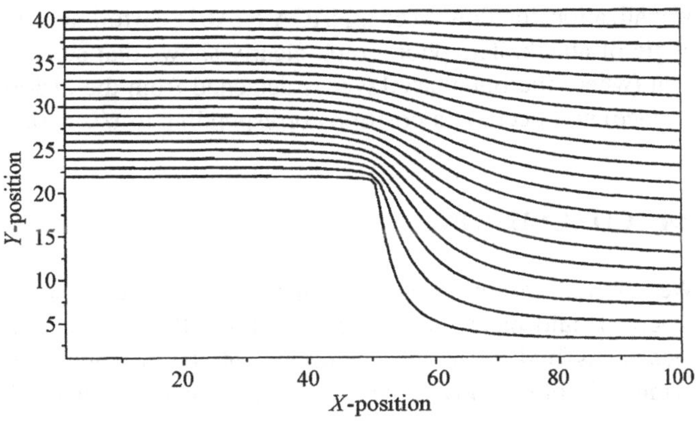

FIGURE 2.11. Potential flow over a reverse step where the flow area doubles.

- perform iterative computation row-by-row in the interior using the latest computed values as soon as they are available;
- test for convergence;
- output results to a suitable file.

The result of the computation is shown in Figure 2.11.

Note that the result in Figure 2.11 is *not what one would expect for a similar flow with a viscous fluid*; the decrease in velocity as the fluid comes off the step is accompanied by an increase in pressure. This situation usually results in the formation of a region of recirculation (a vortex) at the bottom of the step. There are several illustrations of this phenomenon in Van Dyke (1982); see pages 13–15.

A closely related problem is flow over an overhang and computed results are shown in Figure 2.12; again the resulting streamlines do not correspond to what one would expect from the flow of a viscous fluid.

For larger problems, the rate of convergence of the Gauss–Seidel method can be increased significantly through use of successive over-relaxation (SOR). SOR is also known as the extrapolated Liebmann method and it is described in detail

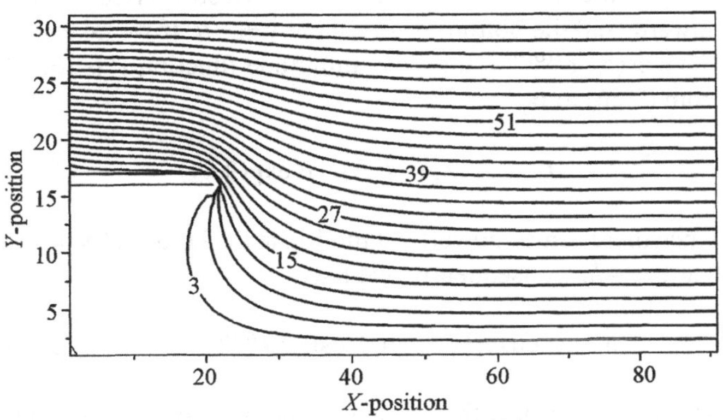

FIGURE 2.12. Numerical solution (Gauss–Seidel) for potential flow over an overhang.

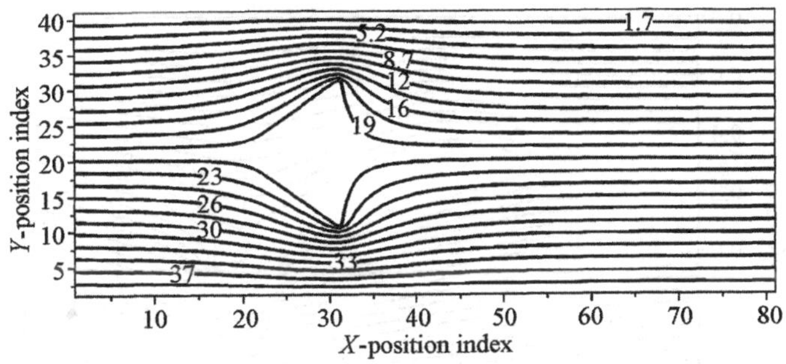

FIGURE 2.13. Confined potential flow about a triangular wedge placed at the centerline.

in the appendices; in essence, the size of the change made by one Gauss–Seidel iteration is increased by (typically) about 80%. In well-conditioned problems, the number of iterations can be reduced by a factor of roughly 10–100.

This method can also be used to compute the flow fields around arbitrary shapes; for example, consider a triangular wedge placed in the center of a confined flow. The stagnation streamline is incident upon the leading vertex and the flow is exactly split by the wedge. The iterative solution appears as shown in Figure 2.13. Note how the flow accelerates to the position of maximum thickness and then adheres to the wedge during deceleration at the trailing edge (a region of increasing pressure).

We conclude this chapter with an example in which flow about an airfoil is computed with the technique described immediately above. This case will illustrate two very important complications that one must take into account while solving such problems. An airfoil, with an angle of attack of 14°, is placed in a uniform potential flow. Because of the shape of the object, the nodal points of a square mesh will not necessarily coincide with the airfoil surface. We have a few options in computational fluid dynamics (CFD) for dealing with this problem: We might use an adaptive mesh generating program (if available), a transformed coordinate system that conforms to the surface of the body (if one could be found), or a node-by-node approximation to compute mesh points near (but not on) the surface. The latter was employed here. Now consider the computed result shown in Figure 2.14.

Pay particular attention to the stagnation streamline at the leading edge of the airfoil; now find the stagnation streamline that *leaves* the body. This will require that the fluid flowing underneath the airfoil turns sharply at the trailing edge and flows up the surface. This is untenable because the required fluid velocities at the trailing edge would be enormous; certainly, no viscous fluid can behave this way, although the phenomenon can be reproduced with a Hele-Shaw apparatus (see Van Dyke (1982), p. 10). It is necessary that the stagnation streamline leaving the upper surface in Figure 2.14 actually leaves the body smoothly at the trailing edge. A circulation about the airfoil is required to satisfy this criterion

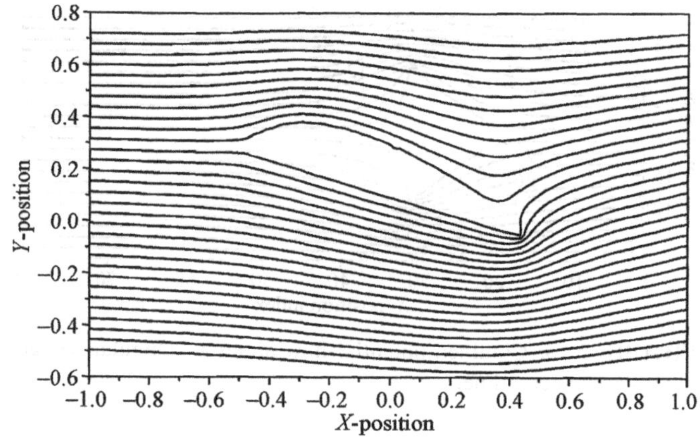

FIGURE 2.14. Computed inviscid flow about an airfoil with an angle of attack of 14° and *no circulation*. Note the nasty turn in the flow underneath the wing at the trailing edge.

(the Kutta–Joukowski condition). Therefore, the stagnation streamline value must be adjusted such that the computed flow appears as shown in Figure 2.15. You will note at once that the flow over the upper surface of the airfoil is now much faster; that is, through the addition of circulation, the flow about the airfoil is generating *lift*. This phenomenon has an interesting consequence: When circulation about the airfoil is established, a strong vortex with opposing circulation is generated by—and shed from—the wing. Such vortices can be persistent (due to conservation of angular momentum) and they can pose control problems for other aircraft that are unlucky enough to encounter them.

Once again it is important that we make the essential distinction between the ideal flow shown in Figure 2.15 and the movement of a real, viscous fluid past the same shape. For example, Van Dyke (1982) provides an example of flow

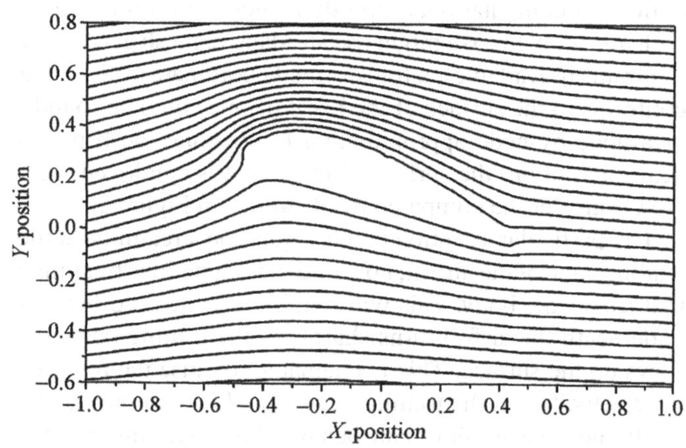

FIGURE 2.15. Computed flow about the same airfoil with circulation. The flow leaves the trailing edge of the wing smoothly and a significant difference in local velocities now exists between the top and bottom surfaces. The reduced pressure on top, relative to the pressure acting upon the bottom, produces lift.

visualization for flow over a NACA 64A015 airfoil at a 5° angle of attack. The photograph clearly shows that separation (where the boundary layer is detached from the airfoil surface) will occur at a position corresponding to $x/L \approx 0.5$.

2.4 CONCLUSION

We referred earlier to the schism that developed between practical fluid mechanics (hydraulics) and theoretical fluid mechanics (hydrodynamics). Since potential flow around any symmetric bluff body looks exactly the same fore and aft (see Figure 2.3), there are no pressure differences. And without pressure differences, there can be no form drag. This, of course, is contrary to common physical experience (i.e., d'Alembert's paradox). A student of fluid mechanics might therefore conclude (based on a cursory examination of the subject) that potential flow is a mere curiosity, a footnote to be appended to the history of fluid mechanics. That is an unwarranted characterization. There is a wonderful unattributed quote in de Nevers (1991) that clearly captures the situation: "Hydrodynamicists calculate that which cannot be observed; hydraulicians observe that which cannot be calculated." At the very least, potential flow theory allows us to think rationally about complicated flows *that cannot be easily calculated.*

In reality, there are many types of problems where viscous friction is quite unimportant, including flow through orifices and nozzles and flows into channel entrances. Another significant example is the behavior of waves on the surface of deep water. Indeed, this is a case where potential flow theory is reasonably accurate. Lamb (1945) devotes an entire chapter (IX) to this type of problem. For the case of "standing" waves in two dimensions, he notes that the velocity potential is governed by

$$\frac{\partial^2\phi}{\partial x^2} + \frac{\partial^2\phi}{\partial y^2} = 0. \tag{2.36}$$

The y-coordinate is measured from the (resting) free surface upward, and the bottom is located at $y = -h$. If we take $\phi = P(y)\cos(kx)e^{\mathrm{1}(\sigma t+\varepsilon)}$, then the amplitude function P is found from (2.36) to be

$$P = A\exp(-ky) + B\exp(+ky). \tag{2.37}$$

Since there can be no vertical motion at the bottom, $\partial\phi/\partial y = 0$ at $y = -h$. Consequently, we have

$$\phi = C\cosh[k(y+h)]\cos(kx)e^{i(\sigma t+\varepsilon)}. \tag{2.38}$$

At the free surface, the vertical velocity v_y must be related to the rate of change of the *position* of the surface: $\partial\phi/\partial y = \partial\eta/\partial t$, where η is the surface elevation (and a function of x

and t). If the pressure above the water surface is constant, then the Bernoulli equation can be used to close the set of equations at the free surface. Lamb shows that the stream function ψ for the standing waves is given by

$$\psi = \frac{g\alpha}{\sigma}\frac{\sinh[k(y+h)]}{\cosh(kh)}\sin(kx)\cos(\sigma t + \varepsilon), \tag{2.39}$$

where α is the vertical amplitude of the wave. The reader is invited to plot some streamlines for this example and then observe how $\partial\psi/\partial y$ behaves with increasing depth. You will note immediately that the motion is rapidly attenuated in the negative y-direction; this is one case where the model obtained from potential flow theory corresponds nicely with physical experience.

REFERENCES

Ahlborn, F. The Magnus Effect in Theory and in Reality, *NACA Technical Memorandum* 567 (1930).

Chang, P. K. *Control of Flow Separation*, Hemisphere Publishing, Washington, DC (1976).

Currie, I. G. *Fundamental Mechanics of Fluids*, 2nd edition, McGraw-Hill, New York (1993).

de Nevers, N. *Fluid Mechanics for Chemical Engineers*, 2nd edition, McGraw-Hill, New York (1991).

Fage, A. and V. M. Falkner. Further Experiments on the Flow Around a Circular Cylinder. *British Aeronautical Research Commission, R&M*, 1369 (1931).

Kirchhoff, R. H. *Potential Flows*, Marcel Dekker, Inc., New York (1985).

Kober, H. *Dictionary of Conformal Representations*, Dover Publications, New York (1952).

Lamb, H. *Hydrodynamics*, 6th edition, Dover Publications, New York (1945).

Milne-Thomson, L. M. *Theoretical Aerodynamics*, 4th edition, Dover Publications, New York (1958).

Padrino, J. C. and D. D. Joseph. Numerical Study of the Steady-State Uniform Flow Past a Rotating Cylinder. *Journal of Fluid Mechanics*, 557:191 (2006).

Prandtl, L. Uber Flussigkeitsbewgung bei sehr kleiner Reibung. *Proceedings of the 3rd International Mathematics Congress, Heidelberg* (1904).

Schlichting, H. *Boundary-Layer Theory*, 6th edition, McGraw-Hill, New York (1968).

Streeter, V. L. and E. B. Wylie. *Fluid Mechanics*, 6th edition, McGraw-Hill, New York (1975).

Van Dyke, M. *An Album of Fluid Motion*, Parabolic Press, Stanford, CA (1982).

3

LAMINAR FLOWS IN DUCTS AND ENCLOSURES

3.1 INTRODUCTION

Laminar fluid motion is atypical; it is a very highly ordered phenomenon in which viscous forces are dominant and momentum is transported by molecular friction. Disturbances that arise in, or are imposed upon, stable laminar flows are rapidly damped by viscosity. One can see some of the essential differences between laminar and turbulent flows with simple experiments; please examine Figure 3.1.

There are a couple of important inferences that can be drawn from these images:

1. Turbulent flows are three dimensional and the transverse velocity vector components will significantly increase momentum transfer normal to the direction of the mean flow.
2. In a duct of constant cross section, the highly ordered nature of laminar flow means that every fluid particle will travel a path parallel to the confining boundaries, so the transverse transport of momentum is a molecular (diffusional) process.

We begin our study of laminar flows in ducts with one of the most important flows of this class, pressure-driven flow in a cylindrical tube (the Hagen–Poiseuille flow).

3.2 HAGEN–POISEUILLE FLOW

Consider a cylindrical tube in which a viscous fluid moves in the z-direction in response to an imposed pressure difference. The appropriate Navier–Stokes equation for the steady flow case is

$$0 = -\frac{\partial p}{\partial z} + \mu\left[\frac{1}{r}\frac{\partial}{\partial r}\left(r\frac{\partial v_z}{\partial r}\right)\right]. \tag{3.1}$$

We should recognize that the entire left-hand side of the z-component (Navier–Stokes) equation has been reduced to 0. This means that there are *no inertial forces*. Consequently, the Reynolds number

$$Re = \frac{d\langle v_z\rangle\rho}{\mu}, \tag{3.2}$$

which is the ratio of inertial and viscous forces, *is not a natural parameter* for Hagen–Poiseuille flow. In a duct of constant cross section, the pressure must decrease linearly in the flow direction; therefore,

$$v_z = \frac{1}{4\mu}\frac{dp}{dz}r^2 + C_1 \ln r + C_2. \tag{3.3}$$

C_1 is 0 since the maximum velocity occurs at the centerline, and since $v_z = 0$ at $r = R$, we find that

$$v_z = \frac{1}{4\mu}\frac{dp}{dz}(r^2 - R^2) \quad \text{or} \quad \frac{(p_0 - p_L)R^2}{4\mu L}\left(1 - \frac{r^2}{R^2}\right), \tag{3.4}$$

which is the familiar parabolic velocity distribution. The shear stress for this problem is $\tau_{rz} = -\mu(dv_z/dr) = -(1/2)(dp/dz)r$. The volumetric flow rate Q is found by

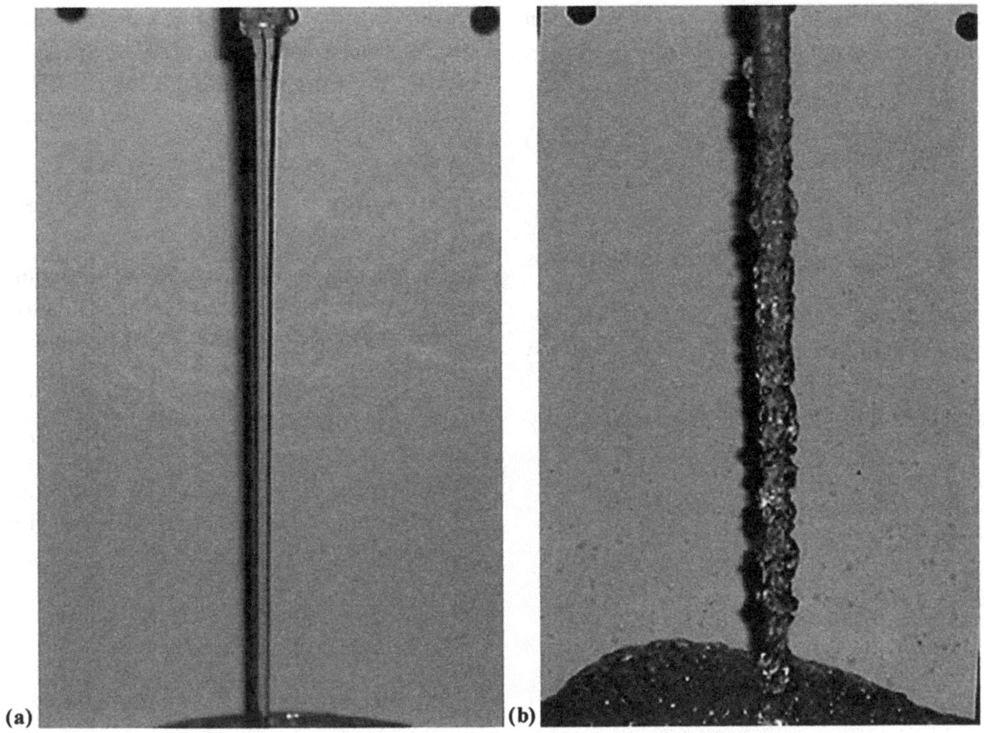

FIGURE 3.1. Digital images (using a short-duration flash) of water jets obtained at low (a) and high speed (b). Note the distorted surface of the high-speed jet.

integration across the cross section,

$$Q = \frac{1}{4\mu}\frac{dp}{dz}\int_0^R 2\pi r(r^2 - R^2)dr = -\frac{\pi}{8\mu}\frac{dp}{dz}R^4, \quad (3.5)$$

and the average velocity $\langle v_z \rangle$ is then simply

$$\langle v_z \rangle = \frac{(p_0 - p_L)}{8\mu L}R^2. \quad (3.6)$$

Thus, if water is to be pushed through a 1 cm diameter tube at 20 cm/s, we would need a pressure drop of about 6.4 dyn/cm^2 per cm. If the tube was 100 m long, then $p_0 - p_L \cong 64{,}000$ dyn/cm^2, which is equivalent to a head of about 65 cm of water (not a very large Δp for a tube of such length).

3.3 TRANSIENT HAGEN–POISEUILLE FLOW

The unsteady variant of the preceding example has some important practical implications. Consider a viscous fluid, initially at rest, in a cylindrical tube. At $t = 0$, a fixed pressure gradient (dp/dz) is imposed and the fluid begins to move in the z-direction. How long will the fluid take to attain, say, 50 or 90% of its ultimate centerline velocity? You can see immediately that such questions are crucial to process dynamics and control—especially in situations with intermittent flow (e.g., sample withdrawal or additive injection). The governing equation is

$$\rho\frac{\partial v_z}{\partial t} = -\frac{\partial p}{\partial z} + \mu\left[\frac{1}{r}\frac{\partial}{\partial r}\left(r\frac{\partial v_z}{\partial r}\right)\right]. \quad (3.7)$$

This problem has been solved by Szymanski (1932); it is a worthwhile exercise to reproduce the analysis. We begin by eliminating the inhomogeneity (dp/dz); let the fluid velocity be represented by the sum of transient and steady functions:

$$v_z = V_1 + v_{zSS}, \quad (3.8)$$

where v_{zSS} is the steady-state velocity distribution for the Hagen–Poiseuille flow (3.4). This ensures that $V_1 \to 0$ as $t \to \infty$. The result of this substitution is

$$\frac{\partial V_1}{\partial t} = \nu\left[\frac{\partial^2 V_1}{\partial r^2} + \frac{1}{r}\frac{\partial V_1}{\partial r}\right]. \quad (3.9)$$

The operator on the right-hand side is an indicator; we can expect to see some form of Bessel's differential equation here. Using the product method, with $V_1 = f(r)g(t)$, we confirm that

$$V_1 = A\exp(-\nu\lambda^2 t)J_0(\lambda r). \quad (3.10)$$

Since V_1 must disappear at the wall, $J_0(\lambda R)=0$. There are an infinite number of λ's that can satisfy this relation; therefore,

$$V_1 = \sum_{n=1}^{\infty} A_n \exp(-\nu\lambda_n^2 t) J_0(\lambda_n r). \tag{3.11}$$

Now one must impose the initial condition so that A_n's that cause the series to converge properly can be identified. Note that at $t=0$,

$$V_1 = -v_{z\mathrm{SS}}. \tag{3.12}$$

The interested reader should complete this analysis by demonstrating that

$$\frac{v_z}{V_{\max}} = \left(1 - \frac{r^2}{R^2}\right) - \sum_{n=1}^{\infty} \frac{4J_2(\lambda_n R)}{(\lambda_n R)^2 J_1^2(\lambda_n R)} \exp(-\nu\lambda_n^2 t) J_0(\lambda_n r). \tag{3.13}$$

The results are displayed in Figure 3.2. We should explore some examples to get a better sense of the duration of the start-up, or acceleration, period.

Consider water initially at rest in a 10 cm diameter tube. At $t=0$, a pressure gradient is imposed and the fluid begins to move. When will the water at the centerline achieve 65% of its ultimate value?

$$\frac{\nu t}{R^2} \cong 0.2, \quad \text{therefore} \quad t \cong \frac{(25)(0.2)}{(0.01)} = 500\,\mathrm{s}.$$

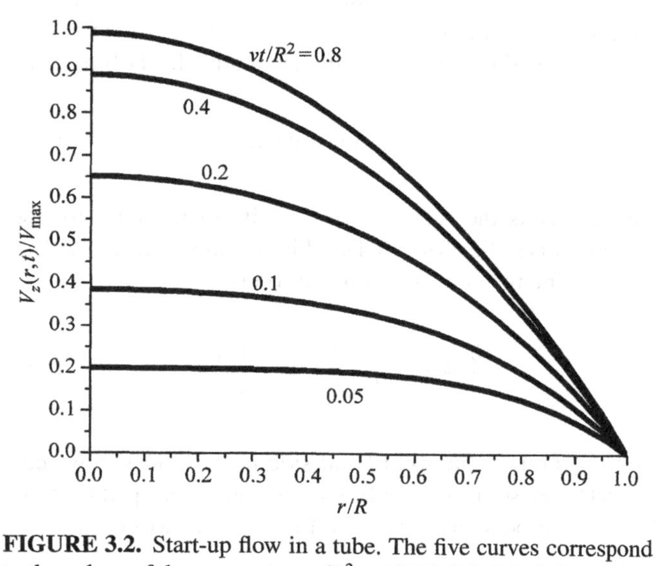

FIGURE 3.2. Start-up flow in a tube. The five curves correspond to the values of the parameter, $\nu t/R^2$, of 0.05, 0.1, 0.2, 0.4, and 0.8. These data were obtained by computation.

Contrast this result with the case in which glycol is at rest in a 1 cm diameter tube; again, a pressure drop is imposed at $t=0$. The time required for the centerline velocity to reach 65% of the ultimate value is only about 0.29 s.

3.4 POISEUILLE FLOW IN AN ANNULUS

The annulus is often employed in engineering applications and it warrants special attention. The governing equation for the pressure-driven flow in an annulus is

$$\rho\frac{\partial v_z}{\partial t} = -\frac{\partial p}{\partial z} + \mu\left[\frac{1}{r}\frac{\partial}{\partial r}\left(r\frac{\partial v_z}{\partial r}\right)\right]. \tag{3.14}$$

Let the cylindrical surfaces be located at $r=R_1$ (inner) and $r=R_2$ (outer). For the steady laminar flow, the velocity distribution is given by eq. (3.3):

$$v_z = \frac{1}{4\mu}\frac{dp}{dz}r^2 + C_1 \ln r + C_2, \tag{3.15}$$

but unlike the Hagen–Poiseuille case (where $C_1=0$),

$$C_1 = -\frac{(1/4\mu)(dp/dz)(R_2^2 - R_1^2)}{\ln(R_2/R_1)}. \tag{3.16}$$

The second constant of integration is found by applying the no-slip condition at either R_1 or R_2. Accordingly, we find

$$C_2 = -\frac{1}{4\mu}\frac{dp}{dz}R_2^2 - C_1 \ln R_2. \tag{3.17}$$

Note that the location of maximum velocity corresponds to

$$R_{\max} = \sqrt{\frac{(R_2^2 - R_1^2)}{2\ln(R_2/R_1)}}. \tag{3.18}$$

Therefore, if the inner and outer radii are 1 and 2, respectively, the position of maximum velocity is 1.47107—closer to the inner surface than the outer. As the radii become larger (with diminishing annular gap), the location of maximum velocity moves toward the center of the annulus. However, we must add some amplification to this remark; eq. (3.18) has been tested experimentally by Rothfus et al. (1955), who found that the radial position of maximum velocity deviates from eq. (3.18) for the Reynolds numbers (defined as) $Re = (2(R_2^2 - R_{\max}^2)\langle v_z\rangle)/\nu R_2$ between about 700 and 9000. This discrepancy is actually greatest at $Re \approx 2500$.

Suppose we consider an example (Figure 3.3) in which water is initially at rest in an annulus with R_1 and R_2 equal to 1 and 2 cm, respectively. At $t=0$, a pressure gradient of $-0.1\,\mathrm{dyn/cm^2}$ per cm is imposed and the fluid begins to move in the z-direction. This problem requires solution of eq. (3.14); the reader is encouraged to explore the alternatives.

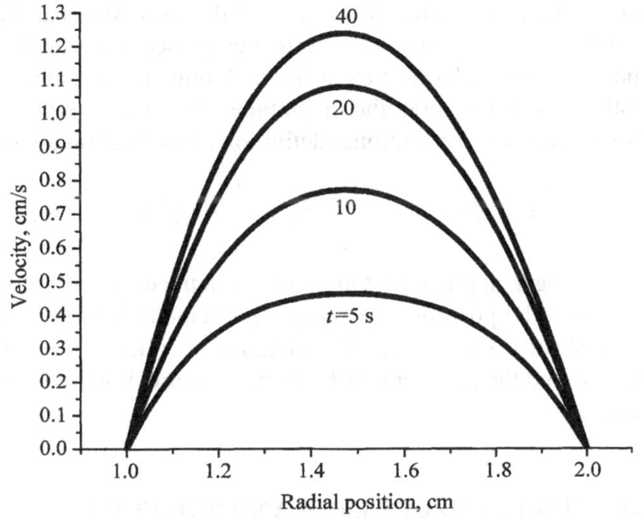

FIGURE 3.3. Velocity distributions for the example problem, start-up flow in an annulus, at $t = 5$, 10, 20, and 40 s. Note that the 50, 70, and 90% velocities will be attained in about 8, 14, and 28 s, respectively.

How long does it take for the velocity to approach V_{max}? In particular, when will the velocity at R_{max} attain 50, 70, and 90% of its ultimate value?

3.5 DUCTS WITH OTHER CROSS SECTIONS

We turn our attention to the steady pressure-driven flow in the z-direction in a generalized duct. The governing equation is

$$\frac{1}{\mu}\frac{dp}{dz} = \frac{\partial^2 v_z}{\partial x^2} + \frac{\partial^2 v_z}{\partial y^2}. \quad (3.19)$$

This is a Poisson (elliptic) partial differential equation; since the Newtonian no-slip condition is to be applied everywhere at the duct boundary, the problem posed is of the Dirichlet type. As one might expect, some analytic solutions are known; this group includes rectangular ducts, eccentric annuli, elliptical ducts, circular sectors, and equilateral triangles. White (1991) and Berker (1963) summarize solutions for these cross sections and others. We shall review the steps one might take to find an analytic solution for this type of problem in the case of a rectangular duct. Let

$$x^* = x/h, \quad y^* = y/h, \quad \text{and} \quad V = \frac{-\mu v_z}{h^2(dp/dz)}, \quad (3.20)$$

which when applied to (3.19) results in

$$\nabla^2 V = -1. \quad (3.21)$$

It is not surprising to find that the polynomial

$$a_0 + a_1 x + a_2 y + a_3 x^2 + a_4 y^2 + a_5 xy \quad (3.22)$$

can satisfy eq. (3.21). If we wish to apply the product method (separation of variables) to eq. (3.21), we must eliminate the inhomogeneity. Suppose we let $V^* = V + y^2/2$? The result is

$$\frac{\partial^2 V^*}{\partial x^{*2}} + \frac{\partial^2 V^*}{\partial y^{*2}} = 0. \quad (3.23)$$

In the usual fashion, we let $V^* = f(x)g(y)$, substitute it into (3.18), and then divide by fg. The result is two ordinary differential equations:

$$f'' - \lambda^2 f = 0 \quad \text{and} \quad g'' + \lambda^2 g = 0. \quad (3.24)$$

Since we choose to place the origin at the center of the duct, the solutions for (3.24) must be written in terms of even functions. Consequently,

$$V = -\frac{y^2}{2} + B\cos\lambda y \cosh\lambda x. \quad (3.25)$$

Of course, when $y = \pm h$, $V = 0$, so

$$V = \frac{h^2}{2} - \frac{y^2}{2} + \sum_{n=1,3,5,\ldots}^{\infty} B_n \cos\frac{n\pi y}{2h}\cosh\frac{n\pi x}{2h}. \quad (3.26)$$

V must also disappear for $x = \pm w$:

$$\frac{1}{2}(y^2 - h^2) = \sum_{n=1,3,5,\ldots}^{\infty} B_n \cos\frac{n\pi y}{2h}\cosh\frac{n\pi w}{2h}. \quad (3.27)$$

The leading coefficients can now be determined by Fourier theorem:

$$B_n = \frac{1}{h}\int_0^h \frac{(y^2 - h^2)}{\cosh(n\pi w/2h)}\cos\frac{n\pi y}{2h}\,dy. \quad (3.28)$$

You should verify that

$$B_n = -\frac{16h^2}{n^3\pi^3}\frac{\sin(n\pi/2)}{\cosh(n\pi w/2h)}. \quad (3.29)$$

An illustration of the computed velocity distribution is shown in Figure 3.4 for the case $h = 1$ and $w = 2h$.

The pressure-driven duct flows described by the elliptic partial differential equation (3.19) are also easily solved numerically either by iteration or by direct elimination. To illustrate this, we rewrite eq. (3.19) using the second-order central differences for the second derivatives; let the indices

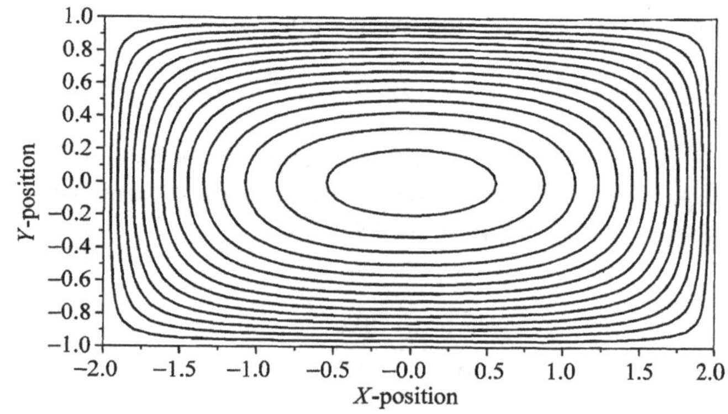

FIGURE 3.4. Velocity distribution for the steady flow in a rectangular duct obtained from the analytic solution (3.26), with $h=1$ and $w=2h$.

i and j correspond to the x- and y-directions, respectively. For the sake of legibility, we shall replace v_z with V:

$$\frac{V_{i+1,j}-2V_{i,j}+V_{i-1,j}}{(\Delta x)^2}+\frac{V_{i,j+1}-2V_{i,j}+V_{i,j-1}}{(\Delta y)^2}\cong\frac{1}{\mu}\frac{dp}{dz}. \tag{3.30}$$

We shall apply this technique to a duct with a cross section in the form of an isosceles triangle where the base is 15 cm and the height is 7.5 cm. This means that the flow area is 56.25 cm^2. The resulting velocity distribution is shown in Figure 3.5.

As one might expect, the vertices have a pronounced effect upon the velocity distribution in a duct of this shape. If the same Δp was applied to water in a cylindrical tube of equal flow area, the average velocity would be 3.55 cm/s and the Reynolds number 2530. That is, for the Hagen–Poiseuille flow in a tube with $R=4.23$ cm, the average velocity $\langle v_z\rangle$would be about 75% greater than in the triangular duct illustrated in Figure 3.5. There is another feature of both the rectangular and the triangular ducts illustrated above that is of interest. Observe that the shear stress at the wall τ_w is *not constant* on the perimeter. In fact, it is clear that the maximum value occurs at the midpoints of the sides in both cases. What about the magnitude of τ_w at the vertices? We see that our conventional definition of the friction factor

$$F=AKf \quad \text{or} \quad \frac{F}{A}=\tau_w=\frac{1}{2}\rho\langle v_z\rangle^2 f, \tag{3.31}$$

is no longer applicable. Obviously, f defined in this manner would be position dependent. One remedy is to use the mean shear stress in eq. (3.31), obtaining it either by integration around the perimeter or from the pressure drop by force balance.

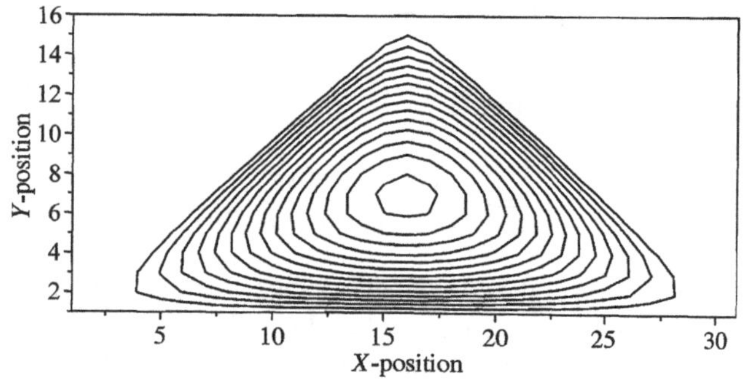

FIGURE 3.5. Computed velocity distribution for the steady laminar flow in a triangular duct; the fluid is water with *dp/dz* set equal to −0.0159 dyn/cm^2 per cm. The computed average velocity for this example is 2.03 cm/s.

3.6 COMBINED COUETTE AND POISEUILLE FLOWS

There are many physical situations in which fluid motion is driven simultaneously by both a moving surface and a pressure gradient. There are important lubrication problems of this type and we can also encounter such flows in coating and extrusion processes. We begin by examining a viscous fluid contained between parallel planar surfaces. The upper surface will move to the right ($+z$-direction) at constant velocity V and then dp/dz will be given a range of values (both negative and positive). Obviously, a negative dp/dz will support (augment) the Couette flow and a positive dp/dz will oppose it. The appropriate equation is

$$0=-\frac{\partial p}{\partial z}+\mu\frac{\partial^2 v_z}{\partial y^2}. \tag{3.32}$$

We choose to place the origin at the bottom plate and locate the top (moving) plate at $y=b$. Equation (3.32) can be integrated twice to yield

$$v_z=\frac{1}{2\mu}\frac{dp}{dz}y^2+C_1y+C_2. \tag{3.33}$$

Of course, $C_2=0$ by application of the no-slip condition at $y=0$. At $y=b$, $v_z=V$, so

$$v_z=\frac{1}{2\mu}\frac{dp}{dz}(y^2-by)+\frac{V}{b}y. \tag{3.34}$$

It is convenient to rewrite the equation as follows:

$$\frac{v_z}{V}=\frac{b^2}{2\mu V}\frac{dp}{dz}\left(\frac{y^2}{b^2}-\frac{y}{b}\right)+\frac{y}{b}. \tag{3.35}$$

What kinds of profiles can be represented by this velocity distribution? Depending upon the sign and magnitude of dp/dz, we can get a variety of forms, as illustrated in Figure 3.6;

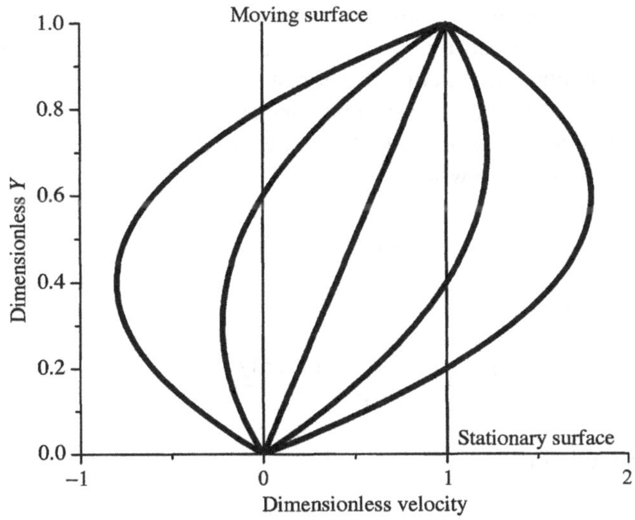

FIGURE 3.6. Velocity distributions for the combined Couette–Poiseuille flow occurring between parallel planes separated by a distance b. The upper surface moves to the right (positive z-direction) at constant velocity V.

in fact, we can adjust the pressure gradient to obtain zero net flow:

$$Q = \int_0^b \left(\frac{1}{2\mu} \frac{dp}{dz} (y^2 - by) + \frac{V}{b} y \right) W dy = 0. \quad (3.36)$$

Consequently, if dp/dz has the *positive* value of

$$\frac{dp}{dz} = \frac{6\mu V}{b^2}, \quad (3.37)$$

there will be no net flow in the duct.

The very same problem can arise in cylindrical coordinates when a rod or wire is coated by drawing it through a die (cylindrical cavity) containing a viscous fluid. We have

$$0 = -\frac{\partial p}{\partial z} + \mu \left[\frac{1}{r} \frac{\partial}{\partial r} \left(r \frac{\partial v_z}{\partial r} \right) \right]. \quad (3.38)$$

Accordingly,

$$v_z = \frac{1}{4\mu} \frac{dp}{dz} r^2 + C_1 \ln r + C_2. \quad (3.39)$$

The boundary conditions are $v_z = V$ at $r = R_1$ and $v_z = 0$ at $r = R_2$, therefore

$$C_1 = \left(V - \frac{1}{4\mu} \frac{dp}{dz} (R_1^2 - R_2^2) \right) / \ln(R_1/R_2) \quad (3.40)$$

and the shear stress τ_{rz} is

$$\tau_{rz} = -\mu \left[\frac{1}{2\mu} \frac{dp}{dz} r + \frac{C_1}{r} \right]. \quad (3.41)$$

Once again, dp/dz could be adjusted to produce zero net flow; the reader might wish to develop the criterion as an exercise.

3.7 COUETTE FLOWS IN ENCLOSURES

Shear flows driven solely by a moving surface are common in lubrication and viscometry. There is an important difference between this class of flows and the Poiseuille flows we examined previously. Consider a steady Couette flow between parallel planar surfaces—one plane is stationary and the other moves with constant velocity in the z-direction:

$$0 = \frac{d^2 v_z}{dy^2}, \text{ resulting in } v_z = C_1 y + C_2. \quad (3.42)$$

Note that the velocity distribution is *independent of viscosity*. A closely related problem, and one that is considerably more practical, is the Couette flow between concentric cylinders. The general arrangement is shown in Figure 3.7.

In this scenario, one (or both) cylinder(s) rotates and the flow occurs in the θ- (tangential) direction. Flows of this type were extensively studied by Rayleigh, Couette, Mallock, and others in the late nineteenth century; work continued throughout the twentieth century, and indeed there is still an active research interest in the case in which the flow is dominated by the rotation of the *inner* cylinder. This particular flow continues to attract attention because the transition process is *evolutionary*, that is, as the rate of rotation of the inner

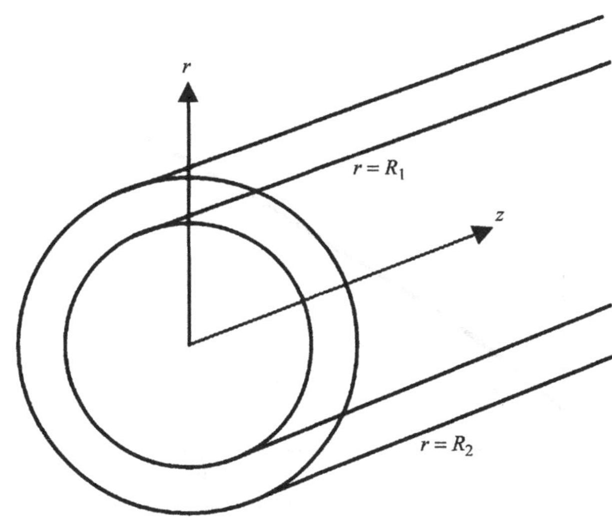

FIGURE 3.7. The standard Couette flow geometry for concentric cylinders.

cylinder is increased, a sequence of stable secondary flows develops in which the annular gap is filled with Taylor vortices rotating in opposite directions. We will examine this phenomenon in greater detail in Chapter 5. For present purposes, we will write down the governing equation for the Couette flow between concentric cylinders:

$$\rho \frac{\partial v_\theta}{\partial t} = \mu \left[\frac{\partial}{\partial r} \left(\frac{1}{r} \frac{\partial}{\partial r}(r v_\theta) \right) \right]. \tag{3.43}$$

For the steady flow case,

$$v_\theta = C_1 r + \frac{C_2}{r}. \tag{3.44}$$

If the outer cylinder is rotating at a constant angular velocity ω and the inner cylinder is at rest, then

$$v_\theta = \frac{\omega R_1^2 R_2^2}{R_1^2 - R_2^2} \left(\frac{1}{r} - \frac{r}{R_1^2} \right). \tag{3.45}$$

The shear stress for this flow is given by $\tau_{r\theta} = -\mu r(\partial/\partial r)(v_\theta/r) = (2\mu\omega R_1^2 R_2^2/R_1^2 - R_2^2)(1/r^2)$. Consider the case in which the radii R_1 and R_2 are 2 and 8 cm (a very wide annular gap), respectively, and the outer cylinder rotates at 30 rad/s. The resulting velocity distribution is illustrated in Figure 3.8.

Note the deviation from linearity apparent in Figure 3.8. If a Couette apparatus has large radii but a small gap, the velocity distribution can be accurately approximated with a straight line. In the case of the example above with the radii of 2 and 8 cm, $\tau_{r\theta}/\mu$ will range from about -34 to $-64\,\mathrm{s}^{-1}$ if $\omega = 30$ rad/s.

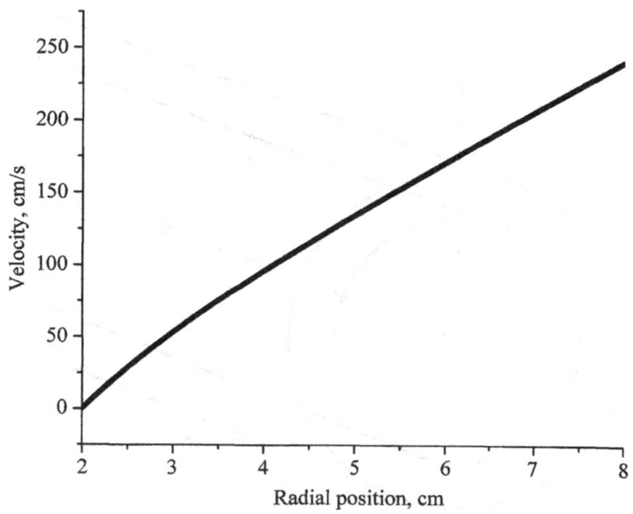

FIGURE 3.8. Velocity distribution in a concentric cylinder Couette device with a wide gap.

Now we turn our attention back to the more general problem as described by eq. (3.43); we assume that the fluid in the annular space is initially at rest. At $t = 0$, the outer cylinder begins to rotate with some constant angular velocity. The governing equation looks like a candidate for separation of variables, so we will try

$$v_\theta = f(r)g(t). \tag{3.46}$$

We find

$$\frac{g'}{\nu g} = \frac{f'' + (1/r)f' - (1/r^2)f}{f} = -\lambda^2, \tag{3.47}$$

resulting in

$$g = C\exp(-\nu\lambda^2 t) \quad \text{and} \quad f = AJ_1(\lambda r) + BY_1(\lambda r). \tag{3.48}$$

It clearly makes sense for us to combine the steady-state solution with this result:

$$v_\theta = C_1 r + \frac{C_2}{r} + C\exp(-\nu\lambda^2 t)[AJ_1(\lambda r) + BY_1(\lambda r)]. \tag{3.49}$$

Noting that our boundary conditions

$$r = R_1,\ v_\theta = 0 \quad \text{and} \quad r = R_2,\ v_\theta = \omega R_2 \tag{3.50}$$

must be satisfied by the steady-state solution, it is necessary that

$$0 = AJ_1(\lambda R_2) + BY_1(\lambda R_2) \quad \text{and}$$
$$0 = AJ_1(\lambda R_1) + BY_1(\lambda R_1). \tag{3.51}$$

Consequently,

$$0 = J_1(\lambda R_1)Y_1(\lambda R_2) - J_1(\lambda R_2)Y_1(\lambda R_1). \tag{3.52}$$

This transcendental equation has an infinite number of roots and it allows us to identify the λ_n's that are required for the series solution. However, we are still confronted with the constants A and B in eq. (3.49). There is a little trick that has been used by Bird and Curtiss (1959), among others, that allows us to proceed. We define a new function

$$Z_1 = J_1(\lambda_n r)Y_1(\lambda_n R_2) - J_1(\lambda_n R_2)Y_1(\lambda_n r) \tag{3.53}$$

that automatically satisfies the boundary conditions. We can now rewrite the solution for this problem as

$$v_\theta = C_1 r + \frac{C_2}{r} + \sum_{n=1}^{\infty} A_n \exp(-\nu\lambda_n^2 t) Z_1(\lambda_n r). \tag{3.54}$$

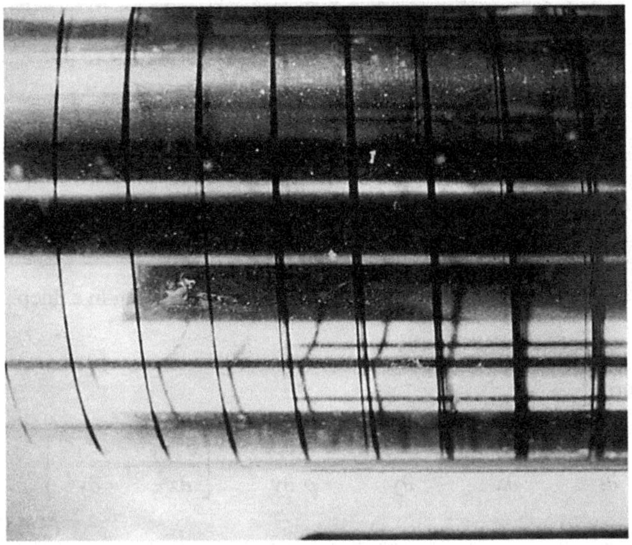

FIGURE 3.9. The helical Couette flow resulting from the rotation of the outer cylinder and the imposition of a small axial pressure gradient. For this case, $Ta = 245$ and $Re_z = 18$ (photo courtesy of the author).

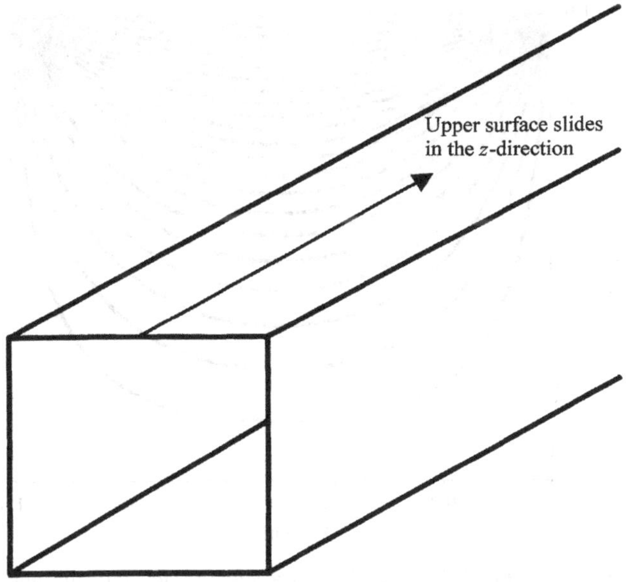

FIGURE 3.10. A square duct with upper surface sliding horizontally (in the z-direction) at a constant velocity.

The solution is completed by using the initial condition (with orthogonality) to find the A_n's:

$$A_n = \frac{\int_{R_1}^{R_2} (-C_1 r - (C_2/r)) Z_1(\lambda_n r) r dr}{\int_{R_1}^{R_2} Z_1^2(\lambda_n r) r dr}. \quad (3.55)$$

There is another important variation of Couette flow in the concentric cylinder apparatus; if an axial pressure gradient is added to the rotation, a helical flow results from the combination of the θ- and z-components. If the rotation of the outer cylinder is dominant relative to the axial flow, one can use dye injection to reveal the flow pattern shown in Figure 3.9.

The rotational motion is characterized with the Taylor number; for the case illustrated here (outer cylinder rotating), it is defined as

$$Ta = \frac{\omega R_2 (R_2 - R_1)}{\nu} \sqrt{\frac{R_2 - R_1}{R_2}}. \quad (3.56)$$

The axial component of the flow is driven by dp/dz and the resulting velocity distribution was given previously by (3.15). The rotational motion is described by (3.44).

The resultant point velocity is obtained from

$$V(r) = (v_\theta^2 + v_z^2)^{1/2}. \quad (3.57)$$

For the Poiseuille flow in plain annuli, Prengle and Rothfus (1955) found that the transition would occur at the axial Reynolds numbers between 700 and 2200. Glasgow and Luecke (1977) added rotation of the outer cylinder to the pressure-driven axial flow and discovered that the Reynolds number for the transition could be as low as about 350 for $Ta \approx 200$.

Of course, the Couette flows can also be generated in rectangular ducts. For example, suppose we have a square duct in which the top surface slides forward in the z-direction (Figure 3.10).

The governing Laplace equation for this flow is

$$0 = \frac{\partial^2 v_z}{\partial x^2} + \frac{\partial^2 v_z}{\partial y^2}. \quad (3.58)$$

We place the origin at the lower left corner and allow the square duct to have a width and height of 1. The no-slip condition applies at the sides and the bottom and the top surface has a constant velocity of 1 in the z-direction. This problem is readily solved with the separation of variables by letting $v_z = f(x)g(y)$; the resulting ordinary differential equations are

$$f'' + \lambda^2 f = 0 \quad \text{and} \quad g'' - \lambda^2 g = 0. \quad (3.59)$$

Due to our choice of location for the origin, the solution can only be constructed from odd functions. Therefore,

$$v_z = \sum_{n=1}^{\infty} A_n \sin n\pi x \sinh n\pi y. \quad (3.60)$$

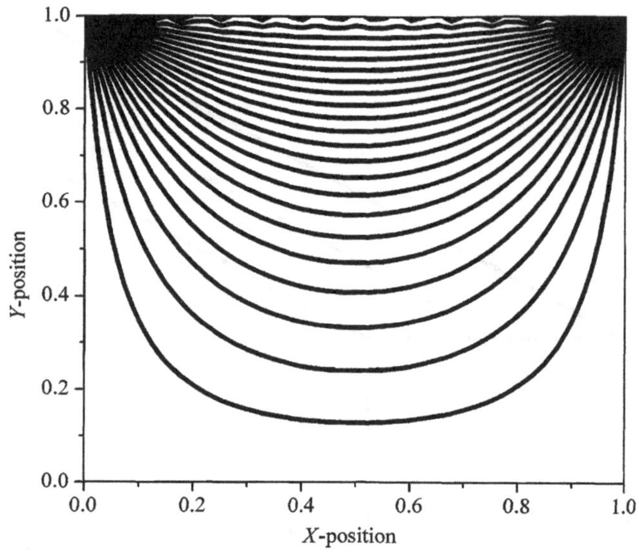

FIGURE 3.11. A laminar flow in a square duct with the upper surface sliding in the z-direction at a constant velocity of 1.

Of course, at $y=1$, $v_z=1$, so

$$1=\sum_{n=1}^{\infty} A_n \sin n\pi x \sinh n\pi. \tag{3.61}$$

This is a Fourier series, so the leading coefficients can be determined by integration:

$$A_n=\frac{2(1-\cos n\pi)}{n\pi \sinh n\pi}. \tag{3.62}$$

The solution is computed using eq. (3.60) and the result is shown in Figure 3.11.

3.8 GENERALIZED TWO-DIMENSIONAL FLUID MOTION IN DUCTS

We now turn our attention to a very common problem in which fluid motion occurs in two directions simultaneously. In a duct, this could result from a change in cross section, for example, flow over a step or obstacle. The conduit is assumed to be very wide in the z-direction such that the x- and y-components of the velocity vector are dominant. A typical problem type is illustrated in Figure 3.12.

For the most general case, the governing equations are

$$\frac{\partial v_x}{\partial t}+v_x\frac{\partial v_x}{\partial x}+v_y\frac{\partial v_x}{\partial y}=-\frac{1}{\rho}\frac{\partial p}{\partial x}+\nu\left[\frac{\partial^2 v_x}{\partial x^2}+\frac{\partial^2 v_x}{\partial y^2}\right] \tag{3.63}$$

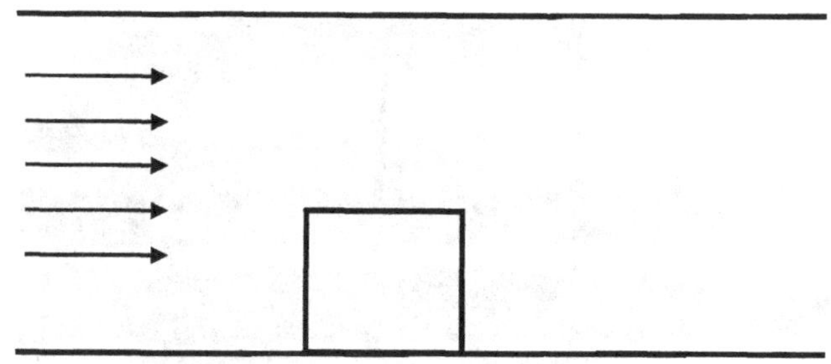

FIGURE 3.12. Flow over a rectangular obstruction in a duct.

and

$$\frac{\partial v_y}{\partial t}+v_x\frac{\partial v_y}{\partial x}+v_y\frac{\partial v_y}{\partial y}=-\frac{1}{\rho}\frac{\partial p}{\partial y}+\nu\left[\frac{\partial^2 v_y}{\partial x^2}+\frac{\partial^2 v_y}{\partial y^2}\right]. \tag{3.64}$$

You will note immediately that there are three dependent variables: v_x, v_y, and p. Of course we can add the continuity equation to close the system, but we now recognize a common dilemma in computational fluid dynamics (CFD). We cannot compute the correct velocity field without the correct pressure distribution $p(x,y,t)$. Let us examine an approach that will allow us to circumvent this difficulty. We cross-differentiate eqs. (3.63) and (3.64) and subtract one from the other, eliminating pressure from the problem. We also note that for this two-dimensional flow, the vorticity vector component is

$$\omega_z=\frac{\partial v_y}{\partial x}-\frac{\partial v_x}{\partial y}. \tag{3.65}$$

The stream function is defined such that continuity is automatically satisfied:

$$v_x=\frac{\partial \psi}{\partial y} \quad \text{and} \quad v_y=-\frac{\partial \psi}{\partial x}. \tag{3.66}$$

We can show that the result of this exercise is the *vorticity transport equation* (you may remember its introduction in Chapter 1):

$$\frac{\partial \omega}{\partial t}+v_x\frac{\partial \omega}{\partial x}+v_y\frac{\partial \omega}{\partial y}=\nu\left[\frac{\partial^2 \omega}{\partial x^2}+\frac{\partial^2 \omega}{\partial y^2}\right]. \tag{3.67}$$

In addition, the stream function and the vorticity are related through a Poisson-type equation:

$$-\omega=\frac{\partial^2 \psi}{\partial x^2}+\frac{\partial^2 \psi}{\partial y^2}. \tag{3.68}$$

We should recognize at this point that a powerful solution procedure for many two-dimensional problems is at hand. Given an initial distribution for vorticity, we can solve eq. (3.68) iteratively to obtain ψ. From the definition of ψ, we can then obtain v_x and v_y; eq. (3.67) can be solved explicitly to obtain the new distribution of ω at the new time $t+\Delta t$. This process can be repeated until the desired t is attained; this approach is appealing because the required numerical procedures are elementary. Before we proceed with an example, we should make an additional observation regarding the steady-state flows of this class. Such problems can be formulated entirely in terms of the stream function ψ. If we do not introduce vorticity, the governing equation can be written as

$$\frac{\partial\psi}{\partial y}\left[\frac{\partial^3\psi}{\partial x^3}+\frac{\partial^3\psi}{\partial y^2\partial x}\right]-\frac{\partial\psi}{\partial x}\left[\frac{\partial^3\psi}{\partial y^3}+\frac{\partial^3\psi}{\partial x^2\partial y}\right]$$
$$=\nu\left[\frac{\partial^4\psi}{\partial x^4}+2\frac{\partial^4\psi}{\partial x^2\partial y^2}+\frac{\partial^4\psi}{\partial y^4}\right]. \tag{3.69}$$

This is a fourth-order, nonlinear partial differential equation. Although it can be used to solve the steady two-dimensional flow problems by an iterative process, we should expect complications. Consider the fourth derivative of ψ with respect to x. After discretization, we write a finite difference approximation in the forward direction,

$$\frac{\partial^4\psi}{\partial x^4}\cong\frac{1}{h^4}\left[\psi_{i+4,j}-4\psi_{i+3,j}+6\psi_{i+2,j}-4\psi_{i+1,j}+\psi_{i,j}\right]. \tag{3.70}$$

You can see that the evaluation will require four nodal points (in addition to i,j) in the x-direction. For a Dirichlet problem in which the boundary values of the stream function are known, we would not be able to apply eq. (3.70) as we approach an obstacle or the right-hand boundary. In addition, since eq. (3.69) is nonlinear, familiar iterative methods may not necessarily converge to the desired solution. In some cases, underrelaxation might be required. And finally, there is another important point. Solution of eq. (3.69) would yield only the stream function ψ. In problems of this type, we are often interested in the velocity and pressure fields; we cannot determine the drag on an obstacle without them.

In view of these difficulties, we should turn our attention back to the solution of eq. (3.67). We isolate the time derivative on the left-hand side and for convenience, consider just two terms:

$$\frac{\partial\omega}{\partial t}=-v_x\frac{\partial\omega}{\partial x}+\cdots. \tag{3.71a}$$

In the finite difference form (letting $i\Rightarrow x$ and $j\Rightarrow t$),

$$\frac{\omega_{i,j+1}-\omega_{i,j}}{\Delta t}\cong -v_x\frac{\omega_{i,j}-\omega_{i-1,j}}{\Delta x}+\cdots. \tag{3.71b}$$

The derivative with respect to x appearing in (3.71a) is written in an *upwind* form. This is necessary to prevent disturbances in the flow field from being propagated upstream! Furthermore, note that when (3.71b) is rearranged for an explicit computation,

$$\omega_{i,j+1}\cong -\frac{\Delta t v_x}{\Delta x}(\omega_{i,j}-\omega_{i-1,j})+\cdots, \tag{3.72}$$

the dimensionless grouping $\Delta t\ v_x/\Delta x$ appears. It is the Courant number Co and the explicit algorithm will be stable *only* if $0<Co\le 1$. Finally, it is to be noted that the requirement that we use upwind differences on the convective transport terms means that we must keep track of the direction of flow (sign on the velocity vector components). Chow (1979) recommends the technique devised by Torrance (1968). This is *critically important* in flows with recirculation. To illustrate, consider the following convective transport term: $(\partial/\partial x)(v_x\phi)$, where v_x has the usual meaning and ϕ is the vector or scalar quantity being transported. Two average velocities are defined as follows (with V used in lieu of v_x):

$$V_f=\frac{1}{2}(V_{i+1,j}-V_{i,j})\quad\text{and}\quad V_b=\frac{1}{2}(V_{i,j}-V_{i-1,j}). \tag{3.73a}$$

The convective transport term at the point (i,j) is then written as follows:

$$\frac{\partial}{\partial x}(v_x\phi)\cong\frac{1}{2\Delta x}\left[\left(V_f-|V_f|\right)\phi_{i+1,j}+\left(V_f+|V_f|\right.\right.$$
$$\left.-V_b+|V_b|)\phi_{i,j}-(V_b+|V_b|)\phi_{i-1,j}\right]. \tag{3.73b}$$

We now apply the vorticity transport equation to laminar flow over an obstacle (a rectangular box). The fluid is initially at rest; at $t=0$, the upper surface begins to slide forward in the $+x$-direction. The evolution of the flow field is shown in the sequence in Figure 3.13 using a Courant number of 0.00525.

It is evident from Figure 3.13 that the vorticity transport equation gives us a powerful tool with which we can successfully analyze many two-dimensional flows. However, there is an additional point that requires our consideration. Look at the right-hand (outflow) boundary immediately above. In this problem, the flow areas for inflow and outflow are the same. Should the velocity fields (distributions) at those planes be identical? By specifying the flow on the outflow boundary, we may have placed an unwarranted constraint upon the entire flow field. Indeed, how can we avoid producing an undesirable artifact in the computation? In some types of flows, for example, in the entrance section of ducts, this is a critical consideration. Wang and Longwell (1964) transformed the x-variable in their study of entrance effects in the viscous flow between parallel plates by letting

$$\eta=1-\frac{1}{1+cx}. \tag{3.74}$$

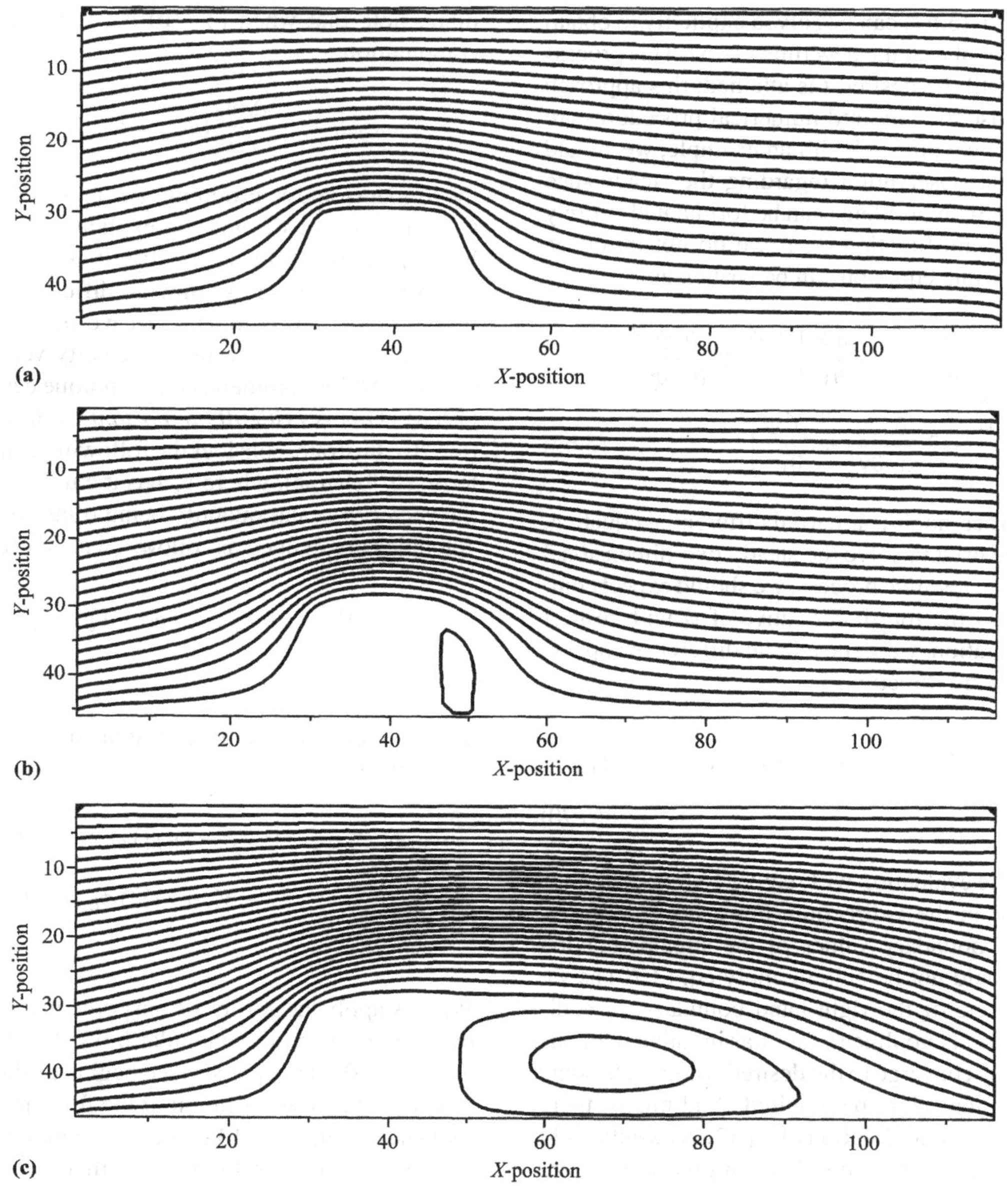

FIGURE 3.13. A transient, confined flow over a rectangular box at short, intermediate, and long times (top to bottom).

Consequently, as $x \to \infty$, $\eta \to 1$. They chose $c = 1.2$, such that when $x = 100$, $\eta = 0.99174$. Although some inconvenience is created by this process, for example,

$$\frac{\partial v_x}{\partial x} = \frac{\partial v_x}{\partial \eta}\frac{\partial \eta}{\partial x} = \frac{\partial v_x}{\partial \eta}\left(\frac{c}{(1+cx)^2}\right), \tag{3.75}$$

this transformation might allow us to circumvent problems stemming from the specification of velocity on the outflow boundary. We shall now consider another aspect of this same difficulty.

In duct flows for which an obstruction (or step) creates an area of recirculation, the length of the standing vortex or the "separation bubble" will increase with the flow rate. For a two-dimensional duct flow with a sudden increase in flow area (a reverse step), this phenomenon will produce results similar to those shown in Figure 3.14 (computations for the Reynolds numbers of 200, 300, and 400).

This illustration further emphasizes the problem created by a finite computational domain in CFD. As the Reynolds number is increased, the standing vortex increases in size, ultimately approaching the outflow boundary. At some point, the specified outflow condition will be violated and any solution obtained will be invalid. Of course, another possible "fix" is to simply increase the extent of the calculation in the downstream direction.

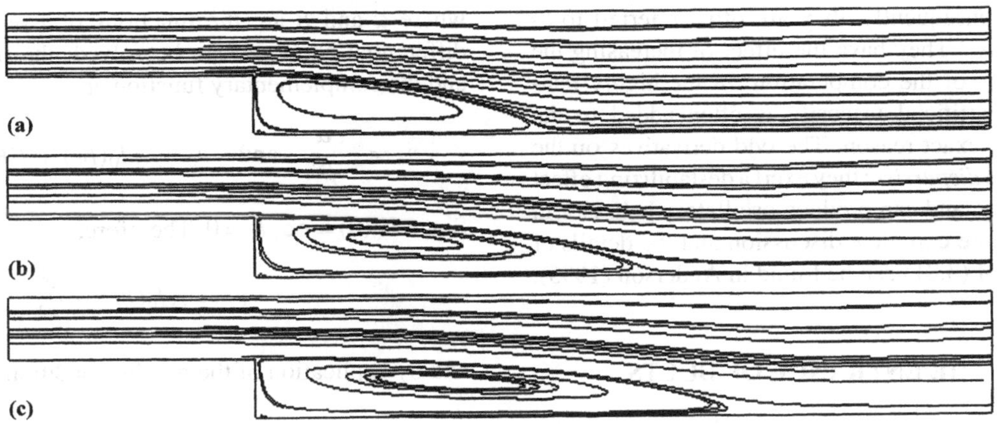

FIGURE 3.14. Increase in length of the recirculation area with the Reynolds number. These results were computed with COMSOL^{TM} at the Reynolds numbers of 200, 300, and 400 (top to bottom).

3.9 SOME CONCERNS IN COMPUTATIONAL FLUID MECHANICS

In the previous section, we indicated how many significant computational flow problems could be solved; we also recognized that the discretization process was an approximation. Consequently, the solutions obtained will have some "error." Actually we have two alternative viewpoints:

1. We are solving the original partial differential equation, but with some error resulting from the approximations.
2. We are solving a completely different partial differential equation that has been created by the discretization process.

We will illustrate the latter. Consider the following fragmentary partial differential equation:

$$\frac{\partial \phi}{\partial t} + V\frac{\partial \phi}{\partial x} = \cdots. \tag{3.76}$$

In this equation, ϕ is a generic-dependent variable (velocity, temperature, or concentration) and V is the velocity. We use finite difference approximations to rewrite this equation as

$$\frac{\phi_{i,j+1} - \phi_{i,j}}{\Delta t} + V\frac{\phi_{i,j} - \phi_{i-1,j}}{\Delta x} = \cdots. \tag{3.77}$$

We write the Taylor series expansions

$$\phi_{i,j+1} = \phi_{i,j} + \left(\frac{\partial \phi}{\partial t}\right)_{i,j}\Delta t + \left(\frac{\partial^2 \phi}{\partial t^2}\right)_{i,j}\frac{(\Delta t)^2}{2} + \left(\frac{\partial^3 \phi}{\partial t^3}\right)_{i,j}\frac{(\Delta t)^3}{6} + \cdots \tag{3.78}$$

and

$$\phi_{i-1,j} = \phi_{i,j} - \left(\frac{\partial \phi}{\partial x}\right)_{i,j}\Delta x + \left(\frac{\partial^2 \phi}{\partial x^2}\right)_{i,j}\frac{(\Delta x)^2}{2} - \left(\frac{\partial^3 \phi}{\partial x^3}\right)_{i,j}\frac{(\Delta x)^3}{6} + \cdots. \tag{3.79}$$

These expressions are introduced into eq. (3.77) with the result

$$\left(\frac{\partial \phi}{\partial x}\right)_{i,j} + V\left(\frac{\partial \phi}{\partial x}\right)_{i,j} = -\left(\frac{\partial^2 \phi}{\partial t^2}\right)_{i,j}\frac{\Delta t}{2} - \left(\frac{\partial^3 \phi}{\partial t^3}\right)_{i,j}\frac{(\Delta t)^2}{6} + \left(\frac{\partial^2 \phi}{\partial x^2}\right)_{i,j}\frac{V\Delta x}{2} - \left(\frac{\partial^3 \phi}{\partial x^3}\right)_{i,j}\frac{V(\Delta x)^2}{6} + \cdots. \tag{3.80}$$

If we differentiate this equation with respect to t, and separately differentiate it with respect to x, and subtract the latter (multiplied by V) from the former, we can eliminate the time derivatives on the right-hand side of the equation:

$$\frac{\partial \phi}{\partial t} + V\frac{\partial \phi}{\partial x} = \frac{V\Delta X}{2}(1 - Co)\frac{\partial^2 \phi}{\partial x^2} + \frac{V(\Delta x)^2}{6}(3Co - 2Co^2 - 1)\frac{\partial^3 \phi}{\partial x^3} + \cdots. \tag{3.81}$$

We recover eq. (3.76) on the left-hand side, but this exercise reveals that our finite difference approximation has actually produced a completely different partial differential equation. The even derivatives on the right-hand side are *dissipative*; consequently, they are often referred to as "artificial viscosity." They have the effect of increasing the numerical stability of the computation. In the highly nonlinear problems, artificial viscosity is often added to the algorithm for this exact reason. The odd derivatives on the right-hand side are *dispersive*; they exert a destabilizing effect upon the procedure and can produce oscillatory behavior in the solution. A more complete discussion and the details of the development of (3.81) can be found in Anderson (1995).

3.10 FLOW IN THE ENTRANCE OF DUCTS

As a fluid enters a duct, the retarding effect of the walls causes the velocity distribution to evolve; fluid motion near the walls is inhibited and the fluid on the centerline accelerates. Because the shear stress at the walls is abnormally large initially, the pressure drop in this region is excessive. For the laminar flow in cylindrical tubes, Prandtl and Tietjens (1931) found the entrance length to be a function of Reynolds number:

$$\frac{L_e}{d} \cong 0.05\,Re \tag{3.82}$$

Consequently, if $Re = 1000$, about 50 tube diameters would be required for the expected parabolic velocity distribution to develop. This is a critical phenomenon for cases in which a fluid enters a short pipe or tube; the Hagen–Poiseuille law will not give good results for such flows.

Consider a steady flow in the entrance of a tube, the z-component of the Navier–Stokes equation for this case is

$$\rho\left(v_r\frac{\partial v_z}{\partial r} + v_z\frac{\partial v_z}{\partial z}\right) = -\frac{\partial p}{\partial z} + \mu\left[\frac{1}{r}\frac{\partial}{\partial r}\left(r\frac{\partial v_z}{\partial r}\right) + \frac{\partial^2 v_z}{\partial z^2}\right]. \tag{3.83}$$

Although $v_z \gg v_r$, v_r is not negligible near the entrance. As a result, the nonlinear inertial terms must be retained on the left-hand side of the equation. This is a formidable problem and it was treated successfully in an approximate way by Langhaar (1942), who linearized this equation. A summary of his analysis follows.

We assume that eq. (3.83) can be written as

$$\frac{\partial^2 v_z}{\partial r^2} + \frac{1}{r}\frac{\partial v_z}{\partial r} - \beta v_z = \frac{1}{\mu}\frac{\partial p}{\partial z}. \tag{3.84}$$

Note that the viscous transport of momentum in the axial (z–) direction has been neglected and that the inertial terms are being approximated by βv_z. We therefore write

$$\frac{d^2 v_z}{dr^2} + \frac{1}{r}\frac{dv_z}{dr} - \beta^2 v_z = \alpha, \tag{3.85}$$

where α and β are the functions of z only. This ordinary differential equation can be solved; the particular integral and the complementary function are

$$v_z = -\frac{\alpha}{\beta^2} \quad\text{and}\quad v_z = AI_0(\beta r) + BK_0(\beta r). \tag{3.86}$$

Since $K_0(0) = \infty$, $B = 0$. Therefore,

$$v_z = AI_0(\beta r) - \frac{\alpha}{\beta^2} \tag{3.87}$$

and by application of the no-slip condition at $r = R$,

$$A = \frac{\alpha/\beta^2}{I_0(\beta r)}. \tag{3.88}$$

The function α (z) is eliminated in the following way:

$$\pi R^2\langle v_z\rangle = \int_0^R 2\pi r v_z(r)dr, \tag{3.89}$$

consequently,

$$\frac{1}{2}R^2\langle v_z\rangle = A\int_0^R rI_0(\beta r)dr - \frac{\alpha}{\beta^2}\int_0^R rdr. \tag{3.90}$$

This results in

$$\frac{v_z}{\langle v_z\rangle} = \frac{I_0[\phi] - I_0[\phi(r/R)]}{I_2[\phi]}, \tag{3.91}$$

where $\phi = \beta R$. For this result to be useful, of course, the function $\phi(z)$ must be determined. This is accomplished by developing an integral momentum equation from eq. (3.83)—a lengthy process! Langhaar's analysis produces the values shown in the table below. The modified Bessel functions I_0 and I_2 have been added for convenience.

$\phi(z)$	$\frac{z/d}{Re}$	$I_0(\phi)$	$I_2(\phi)$
20	0.000205	4.356×10^7	3.931×10^7
11	0.00083	7288	6025
8	0.001805	427.564	327.596
6	0.003575	67.234	46.787
5	0.00535	27.24	17.506
4	0.00838	11.302	6.422
3	0.01373	4.881	2.245
2.5	0.01788	3.29	1.276
2	0.02368	2.28	0.689
1.4	0.0341	1.553	0.288
1	0.04488	1.266	0.136
0.6	0.06198	1.092	0.046
0.4	0.076	1.04	0.02

This approximate treatment of the entrance length problem in cylindrical tubes results in velocity distributions shown in Figure 3.15.

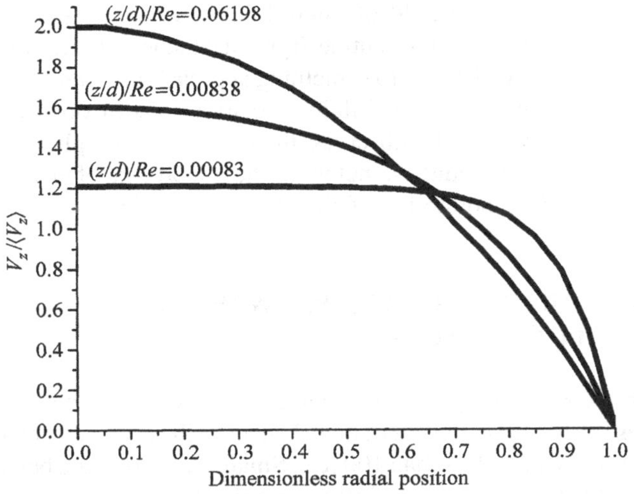

FIGURE 3.15. Velocity profiles in the entrance of a cylindrical tube for $(z/d)/Re = 0.00083$, 0.00838, and 0.06198. Note that the shear stress at the wall is about 3.5 times larger at $(z/d)/Re = 0.00083$ than it would be for the fully developed flow.

Langhaar's results suggest that

$$\frac{L_e}{d} \cong 0.0575 Re, \tag{3.92}$$

which is in accord with the previously cited result of Prandtl and Tietjens.

Much of the early work on laminar flows in entrance regions was based upon meshing a "boundary-layer" near the wall (where the fluid velocity is inhibited by viscous friction) with uniform (potential) flow in the central core. The interested reader should consult Sparrow (1955) for elaboration. However, the development of the digital computer made it possible to solve the entrance flow problems numerically; one of the simplest cases is the flow in the entrance between parallel planes, which was treated by Wang and Longwell (1964). The governing equations for this case are

$$\frac{\partial v_x}{\partial x} + \frac{\partial v_y}{\partial y} = 0, \tag{3.93}$$

$$v_x \frac{\partial v_x}{\partial x} + v_y \frac{\partial v_x}{\partial y} = -\frac{1}{\rho}\frac{\partial p}{\partial x} + \nu\left[\frac{\partial^2 v_x}{\partial x^2} + \frac{\partial^2 v_x}{\partial y^2}\right], \tag{3.94}$$

and

$$v_x \frac{\partial v_y}{\partial x} + v_y \frac{\partial v_y}{\partial y} = -\frac{1}{\rho}\frac{\partial p}{\partial y} + \nu\left[\frac{\partial^2 v_y}{\partial x^2} + \frac{\partial^2 v_y}{\partial y^2}\right]. \tag{3.95}$$

As we have seen previously, we can cross-differentiate eqs. (3.94) and (3.95) and subtract to eliminate pressure. Then by introducing the stream function ψ, continuity will automatically be satisfied and we obtain a fourth-order, nonlinear, partial differential equation for ψ. However, this does not offer us a practical route to solution of this problem since we must avoid specifying the stream function on the outflow boundary. Therefore, we choose to work with the vorticity transport equation and transform the x-coordinate as we discussed earlier:

$$\eta = 1 - \frac{1}{1+cx}. \tag{3.96}$$

Of course, this choice will yield $\eta = 1$ as $x \to \infty$. The equations employed by Wang and Longwell (in dimensionless form) are

$$\frac{d\eta}{dx}\left(\frac{\partial\psi}{\partial y}\frac{\partial\omega}{\partial\eta} - \frac{\partial\psi}{\partial\eta}\frac{\partial\omega}{\partial y}\right) = \frac{4}{Re}\left[\frac{d^2\eta}{dx^2}\frac{\partial\omega}{\partial\eta} + \left(\frac{d\eta}{dx}\right)^2\frac{\partial^2\omega}{\partial\eta^2} + \frac{\partial^2\omega}{\partial y^2}\right] \tag{3.97}$$

and

$$-\omega = \frac{d^2\eta}{dx^2}\frac{\partial\psi}{\partial\eta} + \left(\frac{d\eta}{dx}\right)^2\frac{\partial^2\psi}{\partial\eta^2} + \frac{\partial^2\psi}{\partial y^2}. \tag{3.98}$$

The origin ($y = 0$) is placed at the center of the duct such that at $y = 0$,

$$\frac{\partial v_x}{\partial y} = 0 \quad \text{and} \quad v_y = 0. \tag{3.99}$$

At the upper plane ($y = 1$), we have $v_x = v_y = 0$. Two different forms were used for the inlet boundary condition; they first took the velocity distribution at the inlet to be flat,

$$v_x = 1 \text{ for all } y \text{ at } x = 0. \tag{3.100}$$

Use of this condition led to an interesting result; the velocity distributions for small x show a central concavity. The earlier approximate solutions for this problem did not exhibit this behavior; however, recent work by Shimomukai and Kanda (2006) at $Re = 1000$ suggests that this (central concavity) is a real phenomenon and not a computational artifact. Figure 3.16 shows that the modern commercial CFD packages also lend credence to this result.

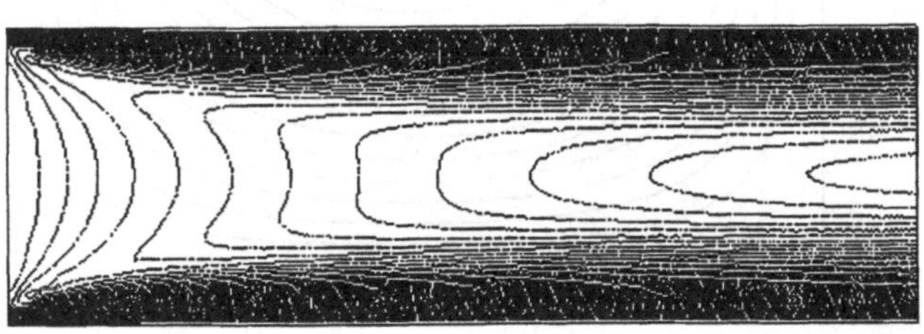

FIGURE 3.16. Contours of constant velocity for the two-dimensional entrance flow between parallel planes, as computed with COMSOL^{TM}. It is to be noted that the vertical axis has been greatly expanded.

3.11 CREEPING FLUID MOTIONS IN DUCTS AND CAVITIES

For flows with very small Reynolds numbers, the inertial forces can be neglected; this effects a considerable simplification since the governing partial differential equations are now linear. Consider a steady two-dimensional flow occurring at very small *Re*. Equation (3.69) is now

$$\frac{\partial^4 \psi}{\partial x^4} + 2\frac{\partial^4 \psi}{\partial x^2 \partial y^2} + \frac{\partial^4 \psi}{\partial y^4} = 0, \text{ or more simply, } \nabla^4 \psi = 0. \tag{3.101}$$

This is the biharmonic equation. It governs steady, slow, viscous flow in two dimensions. Similarly, we can also rewrite the vorticity transport equation for the transient problem with slow, viscous flow:

$$\frac{\partial \omega}{\partial t} = \nu \left[\frac{\partial^2 \omega}{\partial x^2} + \frac{\partial^2 \omega}{\partial y^2} \right]. \tag{3.102}$$

We should look at the following example (Figure 3.17). A viscous fluid, initially at rest, is contained in a square cavity. At $t = 0$, the upper surface begins to slide across the top at constant velocity V. Equation (3.102) is a parabolic partial differential equation and the vorticity will be transported throughout the cavity by molecular friction (diffusion).

As we noted above, creeping flow solutions are limited to the very low Reynolds numbers. While there are few circumstances in normal process engineering where $Re \ll 1$, there are many situations involving dispersed phases or particulate media where this condition is satisfied. The interested reader should consult Happel and Brenner (1965) as a starting point. Recently, problems of this type have also emerged in the developing field of microfluidics, where very small Reynolds numbers are routine. Typical channel sizes may be on the order of 100 nm to something approaching 1 mm; consequently, even a "large" fluid velocity results in small *Re*. But as Wilkes (2006) observes, there are some complicating factors in microfluidics, including the importance of electric fields and the possibility of slip at the boundaries.

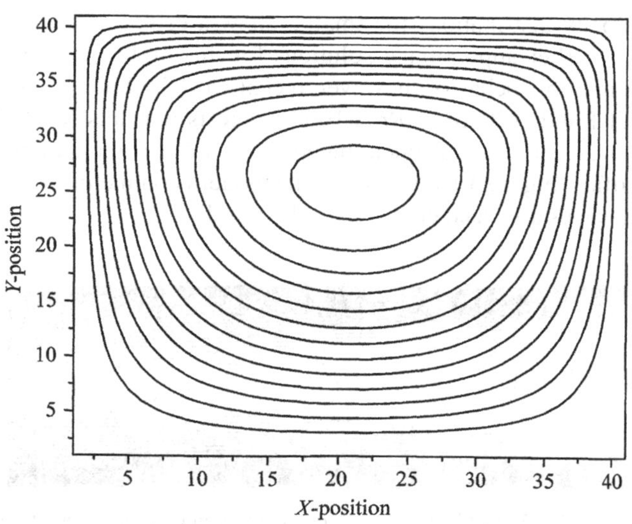

FIGURE 3.17. Slow viscous flow in a cavity. The flow is driven by the upper surface that slides across the top of the cavity at constant velocity V.

3.12 MICROFLUIDICS: FLOW IN VERY SMALL CHANNELS

In recent years, progress in biotechnology and biomedical testing has led to the use of flow devices with very small channel sizes, often less than 100 μm. Small-scale flows are being used for immunoassays, DNA analysis, flow cytometry, isoelectric focusing of proteins, analysis of serum electrolytes, and others. These analytic devices are being fabricated from glass, plastics, and silicon, and their operation presents a host of intriguing problems in transport phenomena. Although we cannot provide a comprehensive review of microfluidics, we can introduce the basics so that the reader has at least a starting point for further investigation.

First, let us recall the Hagen–Poiseuille law for laminar flow in a cylindrical tube:

$$\langle v_z \rangle = \frac{(P_0 - P_L)R^2}{8\mu L}. \tag{3.103}$$

Assume that the tube diameter is 30 μm and let $(P_0 - P_L)/L$ be 7500 dyn/cm^2 per cm. For an aqueous fluid, this means $\langle v_z \rangle \cong 0.21$ cm/s and $Re \cong 0.063$. What would the average velocity need to be in this tube to produce $Re = 2100$? Merely 70 m/s (230 ft/s), which is very unlikely! So for the most part, we can anticipate low Reynolds numbers in such devices.

In anticipation of other channel shapes, we shall define the Reynolds number as

$$Re = \frac{4R_h \langle v_z \rangle \rho}{\mu}, \tag{3.104}$$

where the hydraulic radius R_h is the quotient of the flow area and flow (wetted) perimeter: *A/P*. Now, consider a rectangular channel, 100 μm wide and 40 μm deep carrying an aqueous solution at an average velocity of 2 cm/s; R_h is 14.29 μm, so the Reynolds number for this flow is about 1.12. Since the flow is laminar, the only mixing taking place is by molecular diffusion. Of course, a solute molecule on the centerline will be transported through the channel much more rapidly than one located near the wall(s). This is illustrated clearly in Figure 3.18 that shows the velocity distribution for the pressure-driven flow described above.

Note the very significant variation in velocity with respect to transverse position; this produces axial dispersion, which

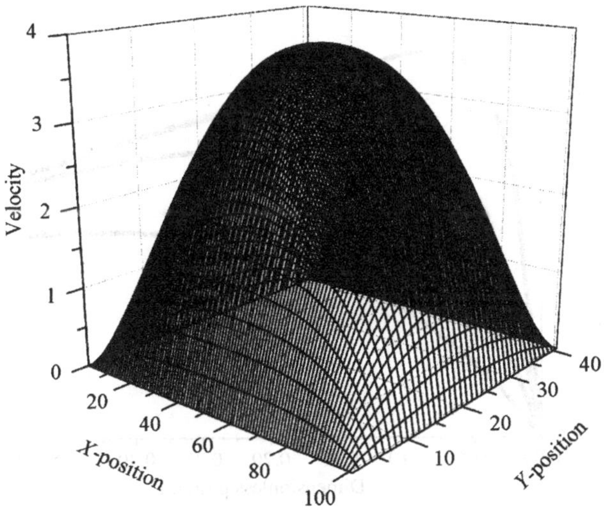

FIGURE 3.18. Variation of velocity in a rectangular channel, 100 μm × 40 μm, with an average velocity of 2 cm/s. The required *dp/dz* for this flow is about 20,100 dyn/cm^2 per cm.

is a potentially serious problem. Suppose a slug of reagent is introduced into the flow at $z=0$. This material will first appear at $z=L$ at time $t=L/V_{max}$. More important, it will continue to be found in the flow (in small amounts) for a very long time. Obviously, this dispersion phenomenon could be counter-productive; an additional discussion of dispersion is given in Chapter 9.

Some other concerns are raised as well: If the channel is very small, do we still have continuum mechanics? Are the no-slip boundary conditions still appropriate? For the first question, consider a cube, 1 μm on each side, filled with water. This very small container will hold about 3.3×10^{10} water molecules, a ridiculously large number that should ensure that fluctuations on a molecular level will be damped out. In the case of the second question, it has been suggested in the literature that nucleation might lead to a gas layer between the solid surface and the liquid being transported. This, or an atomically smooth surface, might produce slip at the boundary. Under such conditions, it may be necessary to replace the usual no-slip boundary condition with

$$V_0 = L_s \left.\frac{\partial v_z}{\partial y}\right|_{y=0}. \tag{3.105}$$

L_s is referred to as the slip, or extrapolation, length. The reader is cautioned that a physically sound basis for this relationship has not been established. In some types of systems, there is evidence that L_s is on the order of 1 μm. Application of this boundary condition yields $p(z)$ different from the one that would normally be expected for Poiseuille flow in a channel. We will examine a rectangular channel with a flow in the z-direction; the width W (x-direction) is much greater than the channel height h (y-direction). Therefore,

$$\frac{d^2 v_z}{dy^2} \cong \frac{1}{\mu}\frac{dp}{dz}. \tag{3.106}$$

Integrating twice (noting that the maximum velocity occurs at $y=h/2$, and applying the *slip condition* at the wall), we find

$$v_z = \frac{1}{2\mu}\frac{dp}{dz}(y^2 - hy - L_s h). \tag{3.107}$$

The volumetric flow rate is found by integration across the cross section yielding the following expression for pressure:

$$\frac{p_0 - p_L}{L} = \frac{2\mu Q}{Wh^2(h/6 + L_s)}. \tag{3.108}$$

The slip can have a profound impact upon flow rate under the right conditions. If $h=10$ μm and $L_s = 1$ μm, Q would be increased (at fixed Δp) by about 60%.

In the case of very small channels, it may be necessary to use very large Δp's to obtain reasonable flow rates. Bridgman (1949) suggested that for large pressures, $\mu = \mu(p)$:

$$\mu = \mu_0 \exp[\alpha(p - p_0)]. \tag{3.109}$$

Bridgman's data for diethylether and carbon disulfide reveal α's of about 3.63×10^{-4} and 2.48×10^{-4} cm^2/kg, respectively. He notes that in general, the more complicated the molecule, the greater the pressure effect upon μ. Water was found to behave a bit differently; at low temperatures (<10°C), μ initially *decreases* with increasing p (up to a pressure of about 1000 kg/cm^2). Suppose we have a pressure-driven flow in a very small cylindrical tube such that

$$\frac{1}{\mu}\frac{\partial p}{\partial z} = \left[\frac{1}{r}\frac{\partial}{\partial r}\left(r\frac{\partial v_z}{\partial r}\right)\right]. \tag{3.110}$$

Using the slip boundary condition at the wall,

$$v_z(r) = \frac{1}{4\mu}\left(r^2 - R^2\right)\frac{\partial p}{\partial z} - \frac{L_s R}{2\mu}\frac{\partial p}{\partial z}. \tag{3.111}$$

Therefore, the volumetric flow rate is related to the pressure by the equation

$$-\int_{z_1}^{z_2} \frac{8\mu_0 Q}{\pi R^4(1 + (4L_s/R))} dz = \int_{p_1}^{p_2} e^{\alpha(p-p_0)} dp. \tag{3.112}$$

3.12.1 Electrokinetic Phenomena

Consider water flowing through a 0.1 mm diameter glass capillary with a Δp of about 70 psi; Wilkes (2006) notes that these conditions will create a potential of about 1 V end-to-end. The situation can be reversed too; if we set $\Delta p = 0$ and apply a large voltage to the ends of the capillary, a flow of water will result. Both these effects result from the electrical double layer, as Wilkes observes.

When a charge-bearing surface is in contact with an electrolyte solution, the ions of opposite charge will be attracted and those of like charge will be repelled. The ionic "atmosphere" that occurs at interfaces is referred to throughout the literature as the double layer. Because of the thermal motion of the molecules, the distribution is *fuzzy*, that is, we should find more counterions near the charged surface, but some coions will be present as well. Naturally, at large distances from the surface, the numbers of positive and negative ions must be equal: $n^+ = n^-$. Consider a surface with a uniform charge distribution in contact with a ion-bearing solution. The distribution of ions in the solution is described by the Boltzmann equations:

$$n^+ = n_0 \exp\left(-\frac{ze\psi}{kT}\right) \quad \text{and} \quad n^- = n_0 \exp\left(\frac{ze\psi}{kT}\right). \tag{3.113}$$

The volumetric charge density ρ for a symmetric electrolyte is $\rho = ze(n^+ - n^-)$, and the electrostatic potential (ψ) in the double layer surrounding a charged, spherical entity is related to charge density by the Poisson equation:

$$\nabla^2\psi = \frac{4\pi\rho}{\varepsilon} = \frac{1}{r^2}\frac{d}{dr}\left(r^2\frac{d\psi}{dr}\right). \tag{3.114}$$

For a planar double layer, these equations can be combined to yield

$$\frac{d^2\psi}{dy^2} = \frac{2nze}{\varepsilon}\sinh\frac{ze\psi}{kT}. \tag{3.115}$$

We now transform the variables: $\psi^* = (ze\psi/kT)$ and $\eta = \kappa y$, where $\kappa = \sqrt{(2nz^2e^2/\varepsilon kT)}$.

The result is

$$\frac{d^2\psi^*}{d\eta^2} = \sinh\psi^*. \tag{3.116}$$

Note that $1/\kappa$ is the Debye length, an indicator of the extent of the ionic atmosphere; for an aqueous solution of a symmetric electrolyte (with $z=1$) and a 0.1 molar concentration, we find $1/\kappa \approx 1.07 \times 10^{-7}$ cm or 10.7 Å. Since

$$\sinh\psi^* = \psi^* + \frac{\psi^{*3}}{3!} + \frac{\psi^{*5}}{5!} + \frac{\psi^{*7}}{7!} + \cdots, \tag{3.117}$$

we can effect a considerable simplification in eq. (3.116) if ψ^* is small: $(d^2\psi^*/d\eta^2) \approx \psi^*$.

Consequently, $\psi^* \approx C_1\exp(\eta) + C_2\exp(-\eta)$. The potential must be bounded as $\eta \to \infty$ and have the surface value (ψ_0^*) at $\eta = 0$, so $\psi^* = \psi_0^*\exp(-\eta)$.

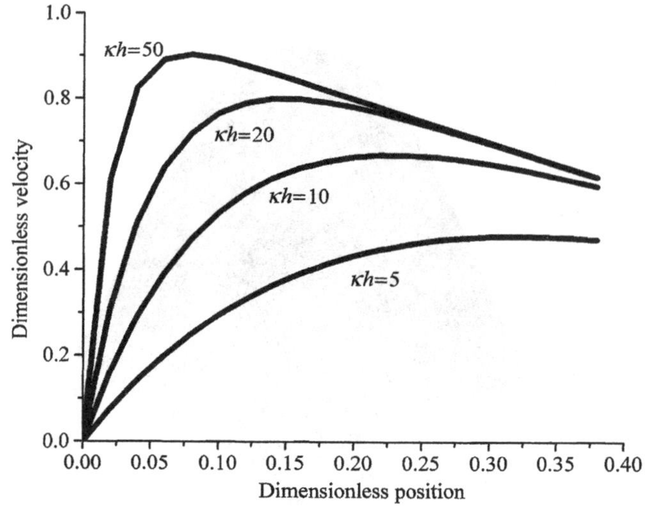

FIGURE 3.19. Velocity distribution in the vicinity of the wall with κh values ranging from 5 to 50.

In cases in which we have a flow of an electrolyte solution (in the z-direction) in the presence of an electric field, an additional force term must be included in the Navier–Stokes equation. For the steady flow in a rectangular channel at the low Reynolds numbers, we should expect

$$0 = -\frac{\partial p}{\partial z} + \mu\left[\frac{\partial^2 v_z}{\partial x^2} + \frac{\partial^2 v_z}{\partial y^2}\right] + F_E. \tag{3.118}$$

For a channel in which the depth is much less than the width ($l_y \ll l_x$), we have the approximation

$$\frac{1}{\mu}\frac{dp}{dz} = \frac{d^2 v_z}{dy^2} + \frac{F_E}{\mu}. \tag{3.119}$$

This provides us with an opportunity. If the channel depth (h) is much larger than the Debye length, we can use an electric field to square off the velocity distribution and flatten the profile over much of the channel. The implication, of course, is that the dispersion problem in Figure 3.17 could be ameliorated. Figure 3.19 shows how this electrokinetic phenomenon affects the velocity in the vicinity of the wall.

3.12.2 Gases in Microfluidics

Recall that we found that about 3.3×10^{10} water molecules occupy a cube 1 μm on each side. For an *ideal gas* at a pressure of 1 atm, this number is reduced to about 2.46×10^7 molecules—still a very large number. But, for the gas flows in very small channels at *lower* pressures, we may find that molecules are more likely to collide with the walls than with each other. The average distance traveled between molecule–

molecule collisions is the mean free path:

$$\lambda = \frac{1}{\sqrt{2}\pi N d^2}, \tag{3.120}$$

where N is the number of molecules per unit volume and d is the molecular diameter. Consider nitrogen at 0°C and a pressure of 1 atm: $\lambda = 600\text{Å}$ or $0.06\ \mu\text{m}$. If the temperature is raised to 300K and the pressure is reduced to 0.1 atm, $\lambda = 0.66\ \mu\text{m}$. What is the implication? We could possibly get to the point where continuum mechanics might not apply. This condition is assessed with the *Knudsen number Kn*:

$$Kn = \frac{\lambda}{h}, \tag{3.121}$$

where h is the characteristic size of the channel. If $Kn > 0.1$, the gas will not behave as a normal Newtonian fluid. Thus, if $h = 6\ \mu\text{m}$ and we use our example above of nitrogen at 300K and 0.1 atm pressure, we find

$$Kn = \frac{0.66}{6} = 0.11. \tag{3.122}$$

This suggests that a few microfluidic applications with gases may be on the threshold of Knudsen flow for which slip at the boundaries *must* be taken into account.

3.13 FLOWS IN OPEN CHANNELS

Liquids are often transported in open, two-, and three-sided channels; such flows are important to engineers concerned with pollution, drainage, irrigation, storm water runoff, and waste collection. Hydrologists use the Froude number Fr to characterize stream flows as *tranquil, critical, or rapid*, depending upon the value of Fr:

$$Fr = \frac{\langle v_z \rangle}{\sqrt{gh}}, \tag{3.123}$$

with

$Fr < 1 \Rightarrow$ tranquil
$Fr = 1 \Rightarrow$ critical
$Fr > 1 \Rightarrow$ rapid.

The characteristic depth of the channel is h. An open channel does not require much inclination or roughness for the flow to become disordered; even in a relatively smooth concrete channel, flow disturbances are nearly always apparent at the free surface.

Historically, uniform flows in open channels were represented with the Chezy equation (1769) for velocity:

$$V = C\sqrt{R_h s}, \tag{3.124}$$

where C is the Chezy discharge coefficient, R_h is the hydraulic radius of the channel, and s is the sine of the slope angle. If one assumes a parabolic velocity distribution in a wide channel, the value of C can be determined from $C = \left(\frac{(Re)(g)}{8}\right)^{1/2}$. Therefore, if $Re = 1000$, $C \approx 350\ \text{cm}^{1/2}/\text{s}$. About a century later, Manning tried to systematize existing data with the correlation:

$$V = \frac{1.5}{n} R_h{}^{2/3} s^{1/2}, \tag{3.125}$$

where n is the Manning roughness coefficient (n typically ranges from about $0.01\ \text{ft}^{1/6}$ for very smooth surfaces to about $0.035\ \text{ft}^{1/6}$ for winding natural streams with vegetative obstructions); see Chow (1964) for an extensive table of approximate roughness coefficients. We will be able to make an interesting comparison with these early results after we complete the following example.

Most open channel flows are at least intermittently turbulent. We will return to this point later, but for now we presume that such flows can be adequately described by the equation

$$\mu\left[\frac{\partial^2 v_z}{\partial x^2} + \frac{\partial^2 v_z}{\partial y^2}\right] = -\rho g \sin\theta. \tag{3.126}$$

This is an elliptic partial differential equation that can be solved rather easily for many different open channel flows. Consider a drainage channel (with reasonably smooth sides and bottom) with sloping sides. Water flows in this channel with a depth of 10 cm; the channel inclination is 0.001°. By computation, we find a maximum free surface velocity of about 66 cm/s and the velocity distribution shown in Figure 3.20.

For this illustration, the Manning correlation indicates a velocity of about 0.2 ft/s; this is about one-fourth of the computed average velocity (where we assumed the flow to be *very highly ordered*). We can also check the Froude number for this example:

$$Fr = \frac{(27.3)}{\sqrt{(980)(10)}} = 0.28, \tag{3.127}$$

which indicates tranquil flow in this small drainage channel. An average velocity of 99 cm/s would be required to attain the critical Fr ($Fr = 1$).

It is worthwhile to spend a little time considering boundary conditions for the previous problem. Naturally, we apply the no-slip condition at the bottom and sides. But at the free surface, we should be equating the momentum fluxes:

$$\tau_1 = -\mu_1 \left.\frac{\partial v_z}{\partial y}\right|_{y=y_0} = \tau_2 = -\mu_2 \left.\frac{\partial v_z}{\partial y}\right|_{y=y_0}. \tag{3.128}$$

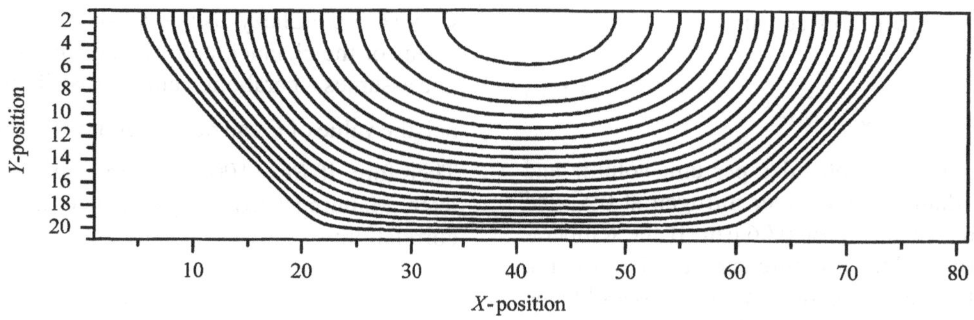

FIGURE 3.20. Velocity distribution in the drainage channel with a cross-sectional area of 350 cm^2 (0.377 ft^2) and an average velocity of about 27 cm/s (0.886 ft/s).

For water (1) and air (2) at normal ambient temperatures, we have

$\mu_1 \cong 1cp$ and $\mu_2 \cong 0.018cp$, respectively.

Accordingly, $\mu_1/\mu_2 \approx 56$, so little momentum is transported across the interface. In such cases it is reasonable to set $(\partial v_z/\partial y) = 0$ at the free surface. This brings another important situation to our attention: Suppose we have two immiscible liquids flowing in an open waste collection channel. Since the momentum fluxes are equated at the interface, we can use the first-order forward differences at the position of the interface (which we denote with the index *j*) to identify the velocity at the fluid–fluid boundary:

$$v_{i,j} = \frac{(\mu_2/\mu_1)v_{i,j+1} + v_{i,j-1}}{1 + (\mu_2/\mu_1)}. \tag{3.129}$$

In Figure 3.21, the interface between the light and heavy fluids is located at a *y*-position index of 26. The ratio of viscosities for this example is $\mu_1/\mu_2 = 4.5$, and the ratio of the fluid densities is $\rho_1/\rho_2 = 2.27$. The velocity profile at the interface is significantly distorted by the difference in viscosities.

3.14 PULSATILE FLOWS IN CYLINDRICAL DUCTS

Pulsatile flows created by the cardiac cycle are central to animal physiology and crucial to the understanding of hemodynamics. Since our initial discussion here is focused upon blood flow, we must note that blood is a Casson fluid; that is, the tendency for red blood cells to agglomerate leads to a definite yield stress. Consequently, we might anticipate the non-Newtonian behavior by writing the governing equation as

$$\rho\frac{\partial v_z}{\partial t} = -\frac{\partial p}{\partial z} - \frac{1}{r}\frac{\partial}{\partial r}(r\tau_{rz}). \tag{3.130}$$

However, the yield stress for blood is low (about 0.04 dyn/cm^2), so the flow is initiated by the small pressure drops. Furthermore, for strain rates above about 100 s^{-1}, blood exhibits a nearly linear relationship between stress and strain, so we can simplify by rewriting eq. (3.130) as

$$\frac{\partial v_z}{\partial t} = -\frac{1}{\rho}\frac{\partial p}{\partial z} + \nu\left[\frac{1}{r}\frac{\partial}{\partial r}\left(r\frac{\partial v_z}{\partial r}\right)\right]. \tag{3.131}$$

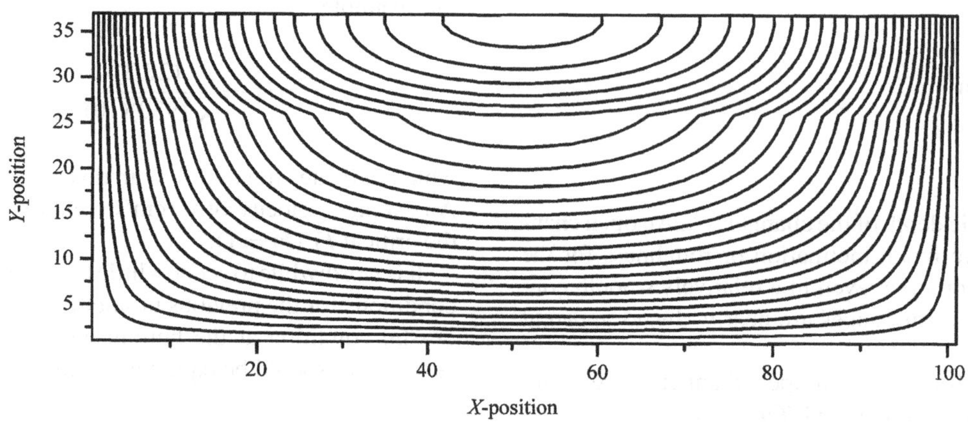

FIGURE 3.21. Flow of immiscible fluids in an open rectangular channel. Note the distortion of the velocity field in proximity to the interface (located at *y* position, or *j*-index, of 26).

Since pressure is periodic in blood flow, we write

$$-\frac{\partial p}{\partial z} = A\exp(2\pi i f t), \tag{3.132}$$

where f is the frequency in Hertz. We will also let the velocity be expressed as the product

$$v_z = \phi(r)\exp(2\pi i f t). \tag{3.133}$$

The consequence of these choices with respect to eq. (3.131) is

$$\frac{d^2\phi}{dr^2} + \frac{1}{r}\frac{d\phi}{dr} - \frac{2\pi i f}{\nu}\phi = -\frac{A}{\mu}. \tag{3.134}$$

Womersley (1955) found an analytic solution for this problem by making use of the fact that $i^2 = -1$; then

$$\frac{d^2\phi}{dr^2} + \frac{1}{r}\frac{d\phi}{dr} + \frac{2\pi i^3 f}{\nu}\phi = -\frac{A}{\mu} \tag{3.135}$$

and

$$\phi = \frac{A}{\rho}\frac{1}{2\pi i f}\left[1 - \frac{J_0\left(r\sqrt{(2\pi f/\nu)i^3}\right)}{J_0\left(R\sqrt{(2\pi f/\nu)i^3}\right)}\right]. \tag{3.136}$$

Womersley's work was crucial to the understanding of pulsatile flows and his contributions are remembered through a ratio of timescales (the characteristic time for molecular transport of momentum divided by the timescale of the periodicity) called the Womersley number *Wo*:

$$Wo = \sqrt{\frac{2\pi f R^2}{\nu}}. \tag{3.137}$$

We can use the pressure gradient data obtained by McDonald (1955) in the femoral artery of a dog to easily compute the dynamic flow behavior. For this example, $Wo \approx 3.3$; the duration of the cardiac cycle in the animal is about 0.360 s.

The curves provided in Figure 3.22 show the flow behavior for the late systolic phase and then for the diastolic where the reverse flow occurs. McDonald verified this phenomenon with high-speed cinematography of small oxygen bubbles injected into the dog's artery. We would not expect to see reverse flow throughout the circulatory system; Truskey et al. (2004) note that this phenomenon is observed only in certain arterial flows proximate to the heart. The flow in the venous system is nearly steady. The reader is urged to pay special attention to the shape of the velocity profiles at the larger times shown in Figure 3.22; the existence of points of inflection will be significant to us later as they call into question flow stability. It is well known that turbulence can arise easily in pulsatile flows despite the relatively low Reynolds

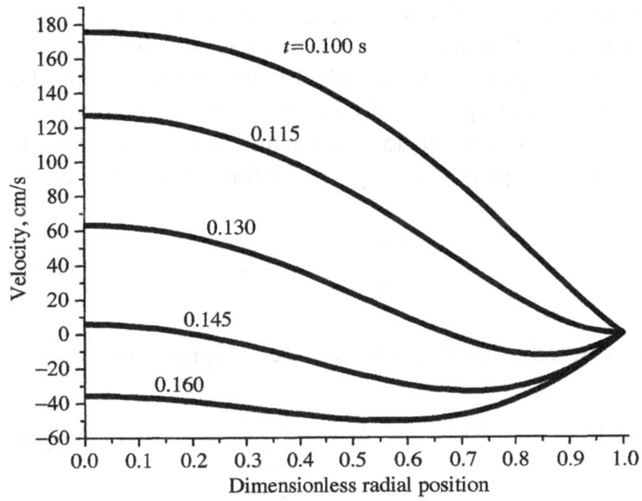

FIGURE 3.22. Computed velocity distributions for flow in the femoral artery of a dog at $t = 0.100$, 0.115, 0.130, 0.145, and 0.160 s using the pressure data obtained by McDonald (1955).

numbers. For the dog's artery example shown above, *Re* is generally less than 1000. Finally, we note that at present there is much interest in the exploitation of pulsatile flows for augmentation of heat and mass transfer; we will revisit this topic in Chapter 9.

3.15 SOME CONCLUDING REMARKS FOR INCOMPRESSIBLE VISCOUS FLOWS

We have only scratched the surface with respect to computational fluid dynamics and the interested reader should immediately turn to specialized monographs such as Anderson (1995) or Chung (2002). Also, we have not discussed compressible gas flows in ducts as the usual one-dimensional macroscopic treatments (assuming either isothermal or adiabatic pathways) are adequately treated in many elementary engineering texts. Our focus has been placed upon the flow of incompressible, viscous fluids in ducts and enclosures. The main difficulty with such flows is pressure: How do we find p accurately? Problems of this type have been attacked both through the primitive variables and with vortex methods. For the latter, you will recall that the development of the vorticity transport equation eliminated pressure. Chung (2002) notes that vortex methods are preferred, where applicable, because of their computational efficiency. They often provide a more accurate portrayal of the physical situation than primitive variable schemes. It is worthwhile for us to further consider this statement.

Consider a generalized two-dimensional flow. As we noted previously, we would *not normally know* $p(x,y,t)$. One possible approach is to estimate (guess) the pressure field, compute the resulting velocity field, and then check continuity to see if conservation of mass is upheld. Of course, our estimated pressure field would almost certainly need to be refined and one would presume that continuity might be used to produce a correction to $p(x,y,t)$. However, there is a pretty obvious problem that complicates this scheme. Suppose we write the continuity equation appropriate for this class of flows:

$$\frac{\partial v_x}{\partial x} + \frac{\partial v_y}{\partial y} = 0, \tag{3.138}$$

we discretize it with central difference approximations:

$$\frac{v_{x(i+1,j)} - v_{x(i-1,j)}}{2\Delta x} + \frac{v_{y(i,j+1)} - v_{y(i,j-1)}}{2\Delta y} \cong 0. \tag{3.139}$$

We can now imagine a saw-tooth or oscillating velocity field in which the nodal values of velocity appeared as follows:

4	20	4	20	4
5	2	5	2	5
4	20	4	20	4
2	5	2	5	2
4	20	4	20	4
5	2	5	2	5

The upper numbers (in this array) are v_x and the lower numbers (staggered below) are values for v_y. Applying the approximated continuity equation at the center point immediately above,

$$\frac{20 - 20}{2\Delta x} + \frac{5 - 5}{2\Delta y} = 0.$$

Though the velocity field makes no sense, continuity is satisfied. Following the procedure that we sketched above, it is clear that an oscillatory pressure field must result. *It is to be noted that the same problem could not arise in compressible flows because the velocity fluctuations would be absorbed by changes in density.* In 1972, Patankar and Spalding devised an algorithm known as SIMPLE (semi-implicit method for pressure-linked equations) to deal with this difficulty. In this method, a staggered grid is employed and a predictor–corrector approach is employed in which the estimated pressure field is adjusted as

$$P = \overline{P} + p', \tag{3.140}$$

where $p\prime$ is the pressure correction and $\overline{P}$ is the estimated pressure. Similarly, for a two-dimensional flow,

$$v_x = \overline{V_x} + v'_x \quad \text{and} \quad v_y = \overline{V_y} + v'_y. \tag{3.141}$$

The pressure and velocity corrections are related by the approximate equations:

$$\rho\frac{\partial v'_x}{\partial t} = -\frac{\partial p'}{\partial x} \quad \text{and} \quad \rho\frac{\partial v'_y}{\partial t} = -\frac{\partial p'}{\partial y}. \tag{3.142}$$

These relations can be used to rewrite eq. (3.141) to yield

$$v_x = \overline{V_x} - \frac{\Delta t}{\rho}\frac{\partial p'}{\partial x} \quad \text{and} \quad v_y = \overline{V_x} - \frac{\Delta t}{\rho}\frac{\partial p'}{\partial y}. \tag{3.143}$$

These two equations are introduced into the continuity equation resulting in a Poisson-type partial differential equation for $p\prime$:

$$\frac{\partial^2 p'}{\partial x^2} + \frac{\partial^2 p'}{\partial y^2} = \frac{\rho}{\Delta t}\left[\frac{\partial \overline{V_x}}{\partial x} + \frac{\partial \overline{V_x}}{\partial y}\right]. \tag{3.144}$$

The technique can now be summarized as follows:

1. Estimate P at each grid point.
2. Find V_x and V_y using the momentum equations.
3. Use the Poisson equation above to find $p\prime$.
4. Correct P, v_x, and v_y, and repeat.

The scheme has a tendency to overestimate p' and this can lead to slow convergence. It is sometimes effective to underrelax the pressure correction:

$$P = \overline{P} + \alpha p', \tag{3.145}$$

where $\alpha = 0.8$ has been used successfully. Additional details can be found in Patankar (1980). Variations of this technique have been incorporated into several commercial CFD programs.

REFERENCES

Anderson, J. D. *Computational Fluid Dynamics*, McGraw-Hill, New York (1995).

Berker, A. R. *Encyclopedia of Physics*, Vol. 8 (S. Flugge, editor), Springer, Berlin (1963).

Bird, R. B. and C. F. Curtiss. Tangential Newtonian Flow in Annuli-I, Unsteady State Velocity Profiles. *Chemical Engineering Science*, 11:108 (1959).

Bridgman, P. W. *The Physics of High Pressure*, G. Bell & Sons, London (1949).

Chow, V. T. *Handbook of Applied Hydrology*, McGraw-Hill, New York (1964).

Chow, C. Y. *An Introduction to Computational Fluid Mechanics*, Wiley, New York (1979).

Chung, T. J. *Computational Fluid Dynamics*, Cambridge University Press, Cambridge (2002).

Glasgow, L. A. and R. H. Luecke. Stability of Centrifugally Stratified Helical Couette Flow. *I & EC Fundamentals*, 13:366 (1977).

Happel, J. and H. Brenner. *Low Reynolds Number Hydrodynamics*, Prentice-Hall, Englewood Cliffs, NJ (1965).

Langhaar, H. L. Steady Flow in the Transition Length of a Straight Tube. *Transactions of the ASME*, 64:A-55 (1942).

McDonald, D. A. The Relation of Pulsatile Pressure to Flow in Arteries. *Journal of Physiology*, 127:533 (1955).

Patankar, S. V. *Numerical Heat Transfer and Fluid Flow*, Hemisphere Publishing, Washington (1980).

Prandtl, L. and O. Tietjens. *Hydro- und Aeromechanik*, Vol. 2, Springer-Verlag, Berlin (1931).

Prengle, R. S. and R. R. Rothfus. Transition Phenomena in Pipes and Annular Cross Sections. *Industrial & Engineering Chemistry*, 47:379 (1955).

Rothfus, R. R., Monrad, C. C., Sikchi, K. G., and W. J. Heideger. Isothermal Skin Friction in Flow Through Annular Sections. *Industrial & Engineering Chemistry*, 47:913 (1955).

Shimomukai, K. and H. Kanda. Numerical Study of Normal Pressure Distribution in Entrance Flow Between Parallel Plates: Finite Difference Calculations. *Electronic Transactions on Numerical Analysis*, 23:202 (2006).

Sparrow, E. M. Analysis of Laminar Forced-Convection Heat Transfer in Entrance Region of Flat Rectangular Ducts. *NACA Technical Note 3331* (1955).

Szymanski, P. Quelques Solutions exactes des equations de l'hydrodynamiquie due fluide visqueux dan les cas d'un tube cylindrique. *Journal de Mathematiques Pures et Appliquies*, 11:67 (1932).

Torrance, K. E. Comparison of Finite-Difference Computations of Natural Convection. *Journal of Research, NBS-B*, 72B:281 (1968).

Truskey, G. A., Yuan, F., and D. F. Katz. *Transport Phenomena in Biological Systems*, Pearson Prentice Hall, Upper Saddle River, NJ (2004).

Wang, Y. L. and P. A. Longwell. Laminar Flow in the Inlet Section of Parallel Plates. *AIChE Journal*, 10:323 (1964).

White, F. M. *Viscous Fluid Flow*, 2nd edition, McGraw-Hill, New York (1991).

Wilkes, J. O. *Fluid Mechanics for Chemical Engineers*, 2nd edition, Prentice Hall, Upper Saddle River, NJ (2006).

Womersley, J. R. Method for the Calculation of Velocity, Rate of Flow and Viscous Drag in Arteries when the Pressure Gradient is Known. *Journal of Physiology*, 127:553 (1955).

4

EXTERNAL LAMINAR FLOWS AND BOUNDARY-LAYER THEORY

4.1 INTRODUCTION

Imagine the difficulties facing Orville and Wilbur Wright as they prepared for the first powered flight of their heavier-than-air machine in 1903. How much power would be required to sustain lift, overcome drag, and keep the machine airborne? That they were able to obtain an answer empirically speaks directly of their ingenuity and persistence. However, progress in aviation was painfully slow until a more complete understanding of drag forces could be brought to bear upon the problem. Through the first quarter of the twentieth century—and long after they should have known better—airplane designers continued to exhibit astonishing lack of comprehension of drag. Some concluded that the route to larger, more useful payloads was through the addition of wings and engines (along with more struts, braces, etc.). The state of the art at the beginning of World War I is illustrated by the Royal Aircraft Factory BE2c bomber/reconnaissance aircraft (which is on display at London's Imperial War Museum) (Figure 4.1).

FIGURE 4.1. Royal Aircraft Factory BE2c bomber/reconnaissance aircraft built in 1915. It was powered by a 9 L V-8 engine and capable of about 72 mph. *Source*: (picture courtesy of the author).

By no means was the BE2c among the worst designs to come to life. A strong candidate for that honor would be W. G. Tarrant's *Tabor* bomber of 1919 (see Yenne (2001), and also http://avia.russian.ee/air/england/tarrant_tabor.html). A complicated three-wing structure was chosen for the *Tabor*; it (would have) created lift to be sure, but at the cost of enormous drag. Furthermore, two of the Napier engines were mounted well above the aircraft's center of gravity. Rotation is always a danger when the thrust line is above the center of gravity and, indeed, when Tarrant's aircraft was on its maiden takeoff roll, there was insufficient control authority to arrest the forward rotation and thus it nosed over at an airspeed of 100 mph destroying the machine and killing the pilots.

By 1930, enough aerodynamic progress had been made that mistakes like the Tarrant *Tabor* were less frequent. Indeed, by the outbreak of World War II, tremendous strides had been made in aerodynamics, structures, and reciprocating engines. These efforts culminated in many remarkable aircraft, including what was almost certainly the finest *long-range* fighter of the 1940s, the North American P-51 *Mustang* (Figure 4.2). This aircraft is of particular interest to us because it incorporated the NACA-developed "laminar flow" wing. The difference between this airfoil design and other contemporary wing profiles is quite apparent in the comparison graphic provided in *P51 Mustang* by Grinsell and Watanabe (1980); the maximum wing thickness was moved aft for the P-51, delaying the effects of the *adverse pressure gradient*. Let us emphasize that this is quite different

Transport Phenomena: An Introduction to Advanced Topics, By Larry A. Glasgow

FIGURE 4.2. An example of the North American P-51D on display in London's Imperial War Museum. The P-51 was equipped with a "laminar flow" wing. That appellation is technically incorrect; the airfoil was designed to *delay separation* of the boundary layer, resulting in increased lift and decreased form drag. *Source*: (picture courtesy of the author).

from actually attaining the laminar flow! Consider the local Reynolds number Re_x at a position 10 cm downstream from the wing's leading edge: If the airspeed was 400 mph, Re_x would be about 1.18×10^6, well above the usual laminar flow threshold. In any event, inadequate manufacturing tolerances and the consequences of wartime flying precluded any chance of maintaining extensive regions of laminar flow.

This chapter owes much to the incomparable monograph *Boundary-Layer Theory* by Hermann Schlichting (1968) that every student of fluid mechanics should own. Schlichting's work (initially a series of lectures given at the GARI in Braunschweig) was known to a few fluid dynamicists in the United States during World War II (see Hugh Dryden's comments in the foreword to the first English edition). It first appeared in the United States as NACA TM 1249 in 1949, although its distribution was controlled. I suppose this effort to minimize dissemination was made through postwar paranoia. Perhaps there was fear that a foreign aerodynamicist might use the knowledge to build a "super" plane. In fact, a shockingly advanced aircraft was constructed by Germany *during* the war, which owed more to Willy Messerschmitt, his design team, and serendipity than to Schlichting's exposition of boundary-layer theory. Interested students of aviation should see *Messerschmitt Me 262, Arrow to the Future* by W. J. Boyne (1980). Similarly, after World War II (1947–1948), the Soviet Union (specifically the Mikoyan–Gurevich OKB) produced the MiG-15; this aircraft completely stunned United Nations forces when it first appeared in the Korean conflict in November 1950. Neither the Me 262 nor the MiG-15 was affected in the least by efforts to limit the distribution of boundary-layer theory.

4.2 THE FLAT PLATE

Ludwig von Prandtl established the foundation for a major advance in fluid mechanics in 1904 when he observed that the effects of viscous friction are confined to a relatively thin fluid layer immediately adjacent to the immersed surface. Prandtl (1928) employed a simplified version of the Navier–Stokes equation in the boundary layer and the appropriate potential flow solution outside. Of course, the distinction between these two layers is quite fuzzy; it is a standard practice to *assume* that the boundary-layer thickness (δ) corresponds to the transverse position where $v_x/V_\infty = 0.99$.

Let us consider a steady two-dimensional flow in the vicinity of a fixed surface. The appropriate equations are

$$v_x\frac{\partial v_x}{\partial x} + v_y\frac{\partial v_x}{\partial y} = -\frac{1}{\rho}\frac{\partial p}{\partial x} + \nu\left[\frac{\partial^2 v_x}{\partial x^2} + \frac{\partial^2 v_x}{\partial y^2}\right], \quad (4.1)$$

$$v_x\frac{\partial v_y}{\partial x} + v_y\frac{\partial v_y}{\partial y} = -\frac{1}{\rho}\frac{\partial p}{\partial y} + \nu\left[\frac{\partial^2 v_y}{\partial x^2} + \frac{\partial^2 v_y}{\partial y^2}\right], \quad (4.2)$$

and

$$\frac{\partial v_x}{\partial x} + \frac{\partial v_y}{\partial y} = 0. \quad (4.3)$$

Now suppose the surface in question is a flat plate, and the origin is placed at the leading edge as shown in Figure 4.3. The characteristic thickness of the boundary layer (in the y-direction) is δ and the length of the plate is L.

We recognize that, in general, $L \gg \delta$ and $v_x \gg v_y$, except for the region very near the leading edge of the plate. These considerations led Prandtl to disregard the viscous transport of x-momentum in the x-direction (obviously, $\delta^2 \ll L^2$); in addition, every term in the y-component equation will be smaller than its x-component counterpart. Therefore, it seems likely that the flow very near the plate's surface can be simply represented with

$$v_x\frac{\partial v_x}{\partial x} + v_y\frac{\partial v_x}{\partial y} = \nu\frac{\partial^2 v_x}{\partial y^2} \quad (4.4)$$

and

$$\frac{\partial v_x}{\partial x} + \frac{\partial v_y}{\partial y} = 0. \quad (4.5)$$

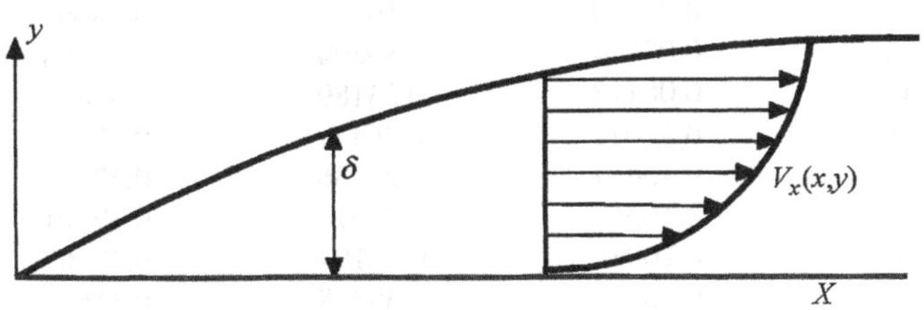

FIGURE 4.3. The boundary layer on a flat plate.

Observe that pressure has been removed from the problem. How can we justify this? We might also profit by considering the shape of the velocity profile(s) at various x-positions. We conclude that every distribution will bear similar features to the profile shown in Figure 4.3, that is, a scaling relationship may exist that would permit all the profiles to be represented by a single curve. If the appropriate *similarity* transformation can be found, we should be able to reduce the number of independent variables (from two to one). Blasius (1908) achieved this for the flat plate problem in 1908 by defining a new independent variable

$$\eta = y\sqrt{\frac{V_\infty}{\nu x}}. \tag{4.6}$$

Note that the scaling we were seeking is $y/\sqrt{x}$. The continuity equation can be satisfied automatically through the introduction of the stream function ψ that Blasius selected:

$$\psi = \sqrt{\nu x V_\infty}\, f(\eta). \tag{4.7}$$

In addition, if we choose to define the stream function such that $v_x = \partial\psi/\partial y$, then

$$v_x = \frac{\partial\psi}{\partial\eta}\frac{\partial\eta}{\partial y} = \sqrt{\nu x V_\infty}\, f'(\eta)\sqrt{\frac{V_\infty}{\nu x}} = V_\infty f'(\eta). \tag{4.8}$$

Clearly, we must have $f'(0)=0$ and $f'(\eta\to\infty)=1$. Since

$$v_y = -\frac{\partial\psi}{\partial x} = \frac{1}{2}\sqrt{\frac{\nu V_\infty}{x}}(\eta f' - f), \tag{4.9}$$

we find that $f(0)=0$ as well. The similarity transformation, with introduction of the stream function, results in the third-order nonlinear ordinary differential (Blasius) equation

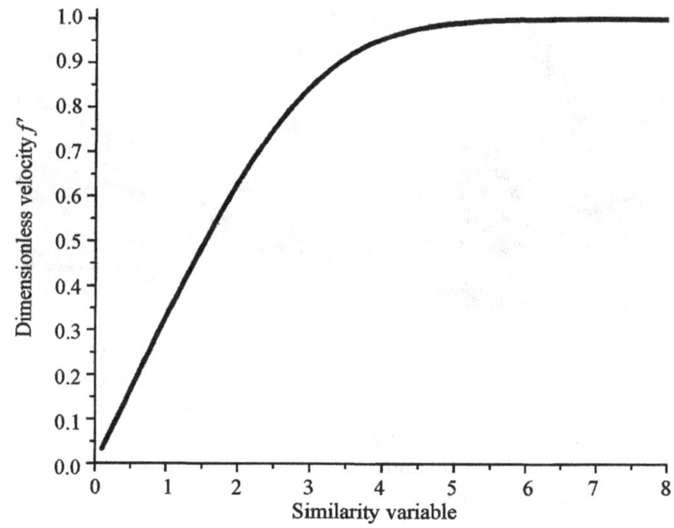

FIGURE 4.4. Velocity distribution for the laminar boundary layer on a flat plate, $f'(\eta)$.

$$f''' + \frac{1}{2}ff'' = 0. \tag{4.10}$$

Please note that the boundary conditions are split, two on one side (at $\eta=0$) and one on the other (as $\eta\to\infty$). This is characteristic of boundary-layer problems. No closed form solution has ever been found for the Blasius equation and the problem is usually solved numerically. The equation (4.10) presents no particular challenge, and a fourth-order Runge–Kutta algorithm with fixed step size will produce perfectly satisfactory results as shown in Figure 4.4. An extensive table of computed values for the Blasius problem for $0\le\eta\le 8$ is provided below.

Note that $v_x/V_\infty=0.99$ at $\eta\approx 5$; this is the position that corresponds to the boundary-layer thickness δ. Consequently, for air moving past a flat plate at 400 cm/s, 10 cm downstream

η	$f(\eta)$	$f'(\eta)$	$f'''(\eta)$	η	$f(\eta)$	$f'(\eta)$	$f'''(\eta)$
0.0	0.00000	0.00000	0.33206	4.1	2.40162	0.96159	0.05710
0.1	0.00166	0.03321	0.33205	4.2	2.49806	0.96696	0.05052
0.2	0.00664	0.06641	0.33199	4.3	2.59500	0.97171	0.04448
0.3	0.01494	0.09960	0.33181	4.4	2.69238	0.97588	0.03897
0.4	0.02656	0.13277	0.33147	4.5	2.79015	0.97952	0.03398
0.5	0.04149	0.16589	0.33091	4.6	2.88827	0.98269	0.02948
0.6	0.05974	0.19894	0.33008	4.7	2.98668	0.98543	0.02546
0.7	0.08128	0.23189	0.32892	4.8	3.08534	0.98779	0.02187
0.8	0.10611	0.26471	0.32739	4.9	3.18422	0.98982	0.01870
0.9	0.13421	0.29736	0.32544	5.0	3.28330	0.99155	0.01591
1.0	0.16557	0.32978	0.32301	5.1	3.38253	0.99301	0.01347
1.1	0.20016	0.36194	0.32007	5.2	3.48189	0.99425	0.01134
1.2	0.23795	0.39378	0.31659	5.3	3.58137	0.99529	0.00951
1.3	0.27891	0.42524	0.31253	5.4	3.68094	0.99616	0.00793

(continued)

η	$f(\eta)$	$f'(\eta)$	$f'''(\eta)$	η	$f(\eta)$	$f'(\eta)$	$f'''(\eta)$
1.4	0.32298	0.45627	0.30787	5.5	3.78060	0.99688	0.00658
1.5	0.37014	0.48679	0.30258	5.6	3.88032	0.99748	0.00543
1.6	0.42032	0.51676	0.29667	5.7	3.98009	0.99798	0.00446
1.7	0.47347	0.54611	0.29011	5.8	4.07991	0.99838	0.00365
1.8	0.52952	0.57476	0.28293	5.9	4.17976	0.99871	0.00297
1.9	0.58840	0.60267	0.27514	6.0	4.27965	0.99898	0.00240
2.0	0.65003	0.62977	0.26675	6.1	4.37956	0.99919	0.00193
2.1	0.71433	0.65600	0.25781	6.2	4.47949	0.99937	0.00155
2.2	0.78120	0.68132	0.24835	6.3	4.57943	0.99951	0.00124
2.3	0.85056	0.70566	0.23843	6.4	4.67939	0.99962	0.00098
2.4	0.92230	0.72899	0.22809	6.5	4.77935	0.99970	0.00077
2.5	0.99632	0.75127	0.21741	6.6	4.87933	0.99977	0.00061
2.6	1.07251	0.77246	0.20646	6.7	4.97931	0.99983	0.00048
2.7	1.15077	0.79255	0.19529	6.8	5.07929	0.99987	0.00037
2.8	1.23099	0.81152	0.18401	6.9	5.17928	0.99990	0.00029
2.9	1.31304	0.82935	0.17267	7.0	5.27927	0.99993	0.00022
3.0	1.39682	0.84605	0.16136	7.1	5.37927	0.99995	0.00017
3.1	1.48221	0.86162	0.15016	7.2	5.47926	0.99996	0.00013
3.2	1.56911	0.87609	0.13913	7.3	5.57926	0.99997	0.00010
3.3	1.65739	0.88946	0.12835	7.4	5.67926	0.99998	0.00007
3.4	1.74696	0.90177	0.11788	7.5	5.77925	0.99999	0.00006
3.5	1.83771	0.91305	0.10777	7.6	5.87925	0.99999	0.00004
3.6	1.92954	0.92334	0.09809	7.7	5.97925	1.00000	0.00003
3.7	2.02235	0.93268	0.08886	7.8	6.07925	1.00000	0.00002
3.8	2.11604	0.94112	0.08013	7.9	6.17925	1.00000	0.00002
3.9	2.21054	0.94872	0.07191	8.0	6.27925	1.00000	0.00001
4.0	2.30576	0.95552	0.06423				

from the leading edge,

$$\delta = \eta\left(\frac{\nu x}{V_\infty}\right)^{1/2} = (5)\left(\frac{(0.151)(10)}{(400)}\right)^{1/2} = 0.307\,\text{cm}. \tag{4.11}$$

We can also find the transverse (y-direction) velocity for this example by applying eq. (4.9):

$$v_y = \frac{1}{2}\left[\frac{(0.151)(400)}{(10)}\right]^{1/2}(\eta f' - f). \tag{4.12}$$

At $\eta = 2$, for example, $(\eta f' - f) = 0.6095$, as shown in Figure 4.5. Therefore, at this position, $v_y = 0.749$ cm/s. Contrast this with $v_x(\eta = 2)$, which is about 252 cm/s!

Now we will turn our attention back to the issue that was raised at the very beginning of this chapter; we need to find the drag force acting upon the plate. The shear stress at the wall is given by

$$\tau_{yx} = -\mu\left.\frac{\partial v_x}{\partial y}\right|_{y=0} = \tau_0 = \mu V_\infty\sqrt{\frac{V_\infty}{\nu x}}\,f''(0). \tag{4.13}$$

The minus sign has been dropped for convenience. We understand that momentum is being transferred in the negative y-direction. The value for $f''(0)$ must come from our numerical results; it is 0.33206. Of course, eq. (4.13) gives us just a local value. To find the total drag (F_D) on one side

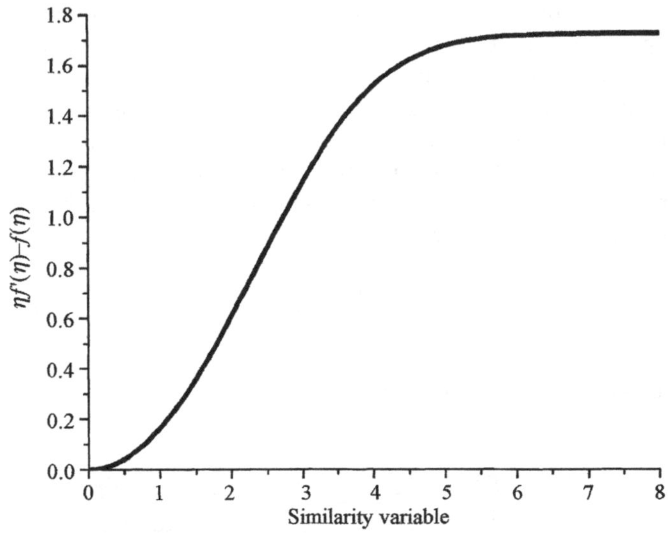

FIGURE 4.5. Transverse velocity component for the laminar boundary layer on a flat plate.

of a plate, we must integrate (4.13) over the surface area:

$$F_D = W \int_0^L \tau_0 dx, \tag{4.14}$$

where W and L are the width and length of the plate, respectively. The result of this integration is

$$F_D = 0.66412 W V_\infty \sqrt{\mu L \rho V_\infty}. \tag{4.15}$$

Consequently, for water flowing past one side of a plate (30.48 cm × 30.48 cm) at $V_\infty = 400$ cm/s, we have

$$F_D = (0.66412)(30.48)(400)[(0.01)(30.48)(1)(400)]^{1/2}$$
$$= 89,404 \text{ dyn } (0.894 \text{ N}).$$

Although the flat plate is of great practical importance, there are many other shapes of interest as well. Consider a curved body, for example, an airfoil. Continuity requires the fluid to accelerate between the leading edge and the location of maximum thickness normal to the chord. The Bernoulli equation indicates that the local pressure will decrease as the velocity increases. However, once the fluid flows past the position of maximum thickness and toward the trailing edge, it must decelerate, and in this region, the pressure is *increasing*. The character of the flow in the boundary layer is changed dramatically by this adverse, or unfavorable, pressure gradient. The changes are shown qualitatively in Figure 4.6.

Note that a point of inflection appears first; as the local pressure continues to increase, a region of reverse flow develops. In response to the unfavorable pressure gradient, the boundary layer actually detaches from the surface; this phenomenon is referred to as separation. We must recognize that Prandtl's equations will not be applicable near or beyond the point of separation because the velocity vector component normal to the surface (v_y) will no longer be small relative to v_x. At the point of separation,

$$\left(\frac{\partial v_x}{\partial y}\right)_{y=0} = 0, \tag{4.16}$$

as is apparent in Figure 4.6.

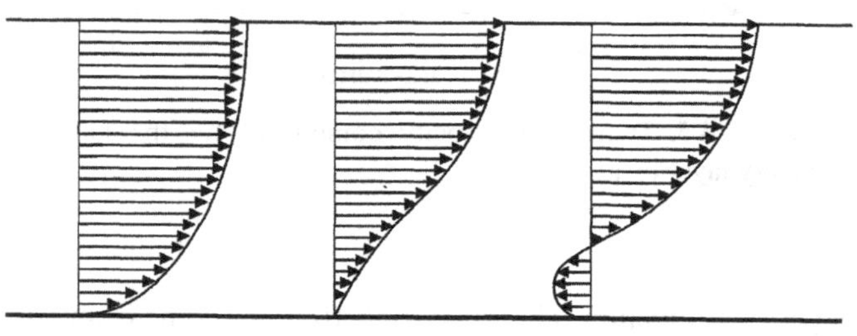

FIGURE 4.6. Progression of effects of an adverse pressure gradient upon the flow in the boundary layer.

4.3 FLOW SEPARATION PHENOMENA ABOUT BLUFF BODIES

Boundary-layer separation is usually undesirable because it results in a larger wake and increased form drag. In aviation, it diminishes the performance envelope of an aircraft; in critical flight regimes, the increased drag and decreased lift can work together catastrophically. In ground transportation, boundary-layer separation results in an increase in fuel consumption. In flow around structures such as bridges, power transmission lines, heat exchanger tubes, and skyscrapers, separation can lead to property damage and even loss of life. An example familiar to many engineering students is the failure of the Tacoma Narrows bridge in November 1940 (Ammann et al. 1941). A sustained 42 mph wind induced structural oscillations (both longitudinal and torsional) that ultimately put the center span at the bottom of the Narrows. The report of the disaster prepared for the Federal Works Agency in 1941 (published by the American Society of Civil Engineers in December, 1943) is fascinating reading, and it is now clear that this incident was a little more complex than a mere structural excitation caused by vortex shedding. For a more recent overview, see the article by Petroski (1991).

Readers interested in the control of boundary-layer separation may find the monograph *Control of Flow Separation* by Paul Chang (1976) quite useful. As one might imagine, a number of control techniques have been implemented on experimental aircraft, including suction (to remove the retarded fluid from the boundary layer) and incorporation of rotating cylinders at the wing surface to accelerate the retarded fluid. Both approaches have demonstrated effectiveness but at the cost of increased complexity and weight. Braslow (1999) gives a wonderful behind-the-scenes history of suction control.

As we observed in the previous section, the laminar boundary layer cannot withstand the significant adverse pressure gradients. Accordingly, a flow about any blunt object will produce separation phenomena; these may include the formation of fixed (standing) vortices at the trailing edge at the modest Reynolds numbers, or the formation of the von Karman vortex street (through periodic vortex shedding) as the Reynolds number is increased. We will continue this discussion by examining a flow about a circular cylinder since this case has been the focus of much attention.

Taneda (1959) conducted flow visualization experiments in which the model cylinders were towed through a tank of still water. Standing vortices were found to appear at $Re = 5$ and then increase in size with the increasing Reynolds number. At $Re = 10$, the fixed vortices have a streamwise size that is about 25% of the cylinder diameter (d); at $Re = 20$, they are about 90% of d. At $Re = 40$, the vortices extend in the downstream direction for about two cylinder diameters, and at about $Re \approx 45$, the flow becomes transient as the vortices are alternately shed from opposite sides of the cylinder. The

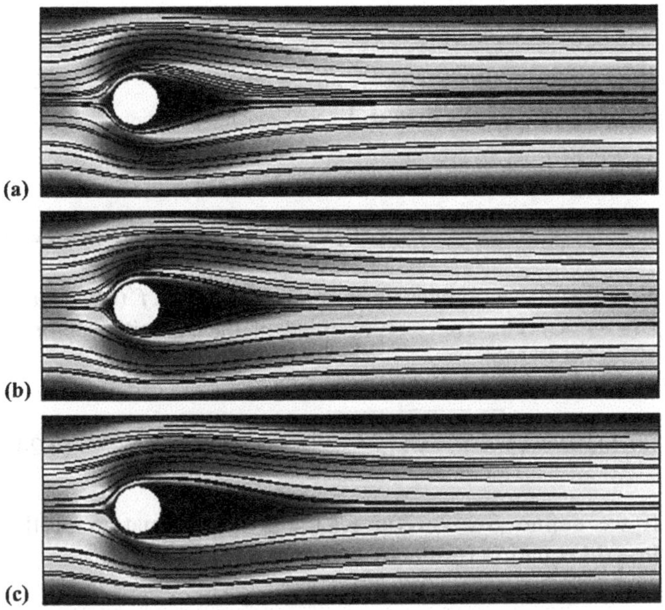

FIGURE 4.7. Fixed vortices behind a circular cylinder at the Reynolds numbers 15, 25, and 40. These results were obtained with COMSOL^{TM}.

growth of the fixed vortices is illustrated by the computational results shown in Figure 4.7.

Early calculations made using the potential flow pressure distribution showed that separation would occur at an angle (measured from the forward stagnation point) of about 109°. Experimental measurements of the pressure distribution indicated that separation occurred at about 80°.

As we observed previously, at larger Reynolds numbers, the vortices are shed alternately from the opposite halves of the cylinder. The resulting vortex street (at an instant in time) has the general appearance shown by the computational results in Figure 4.8. For experimentally recorded vortex streets, see Van Dyke (1982, pp. 56 and 57).

The dimensionless shedding frequency is characterized by the Strouhal number

$$St = \frac{df}{V}, \tag{4.17}$$

where d is the cylinder diameter, V is the velocity of approach, and f is the shedding frequency (from one side of the cylinder). The Strouhal number has been measured for many different shapes and Figure 4.9 compiles some of these results.

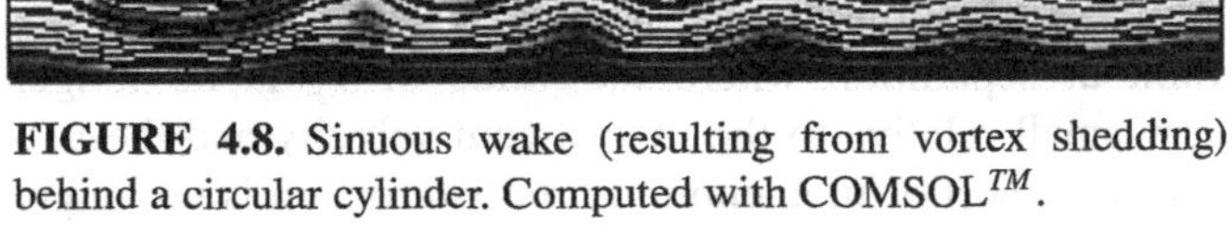

FIGURE 4.8. Sinuous wake (resulting from vortex shedding) behind a circular cylinder. Computed with COMSOL^{TM}.

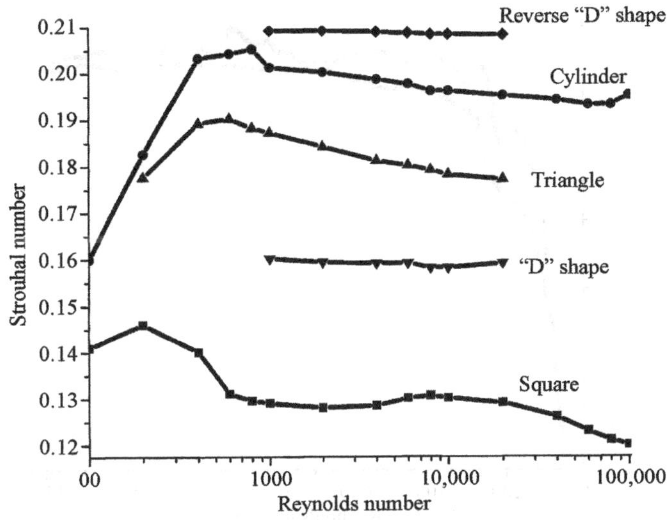

FIGURE 4.9. The Strouhal number for several different cross sections (flow from left to right) as (adapted from Blevins (1994) and Roshko (1954)).

To illustrate, consider air at a velocity of 700 cm/s flowing past a wire having a diameter of 3 mm. The Reynolds number is estimated as

$$\text{Re} = \frac{dV}{\nu} = \frac{(0.3)(700)}{(0.151)} = 1391. \tag{4.18}$$

Figure 4.9 indicates that $St \approx 0.2$, therefore, $f = 467$ Hz. Note that this is in the acoustic range; this phenomenon explains the humming telephone wire in the wind. A dangerous situation can arise when the frequency of vortex shedding matches the fundamental frequency of a structure or installation. The resulting oscillation can intensify the vortices resulting in an amplification of the motion; this phenomenon is known as "lock-in" and it has occurred in tubular air heaters, power transmission lines, highway signs, and so on. If left unchecked, vortex shedding with lock-in can lead to structural failure.

The reader is cautioned that the data shown in Figure 4.9 are approximate; they cannot be taken as crisp or precise. Extensive studies of transient vortex wake phenomena for cylinders have been conducted by Roshko (1954) and Tritton (1959) among others; an examination of Roshko's data, for example, at low Reynolds numbers shows regions of variability as seen in Figure 4.10.

Roshko reported a relationship between the Strouhal number (detected at fixed distance from the cylinder) and Reynolds number:

$$St = \frac{df}{V} = 0.212 - \frac{4.5}{\text{Re}} \quad (\text{for } 50 < \text{Re} < 150). \tag{4.19}$$

While exploring this relationship, Tritton discovered a discontinuity in the velocity–frequency curve, often occurring

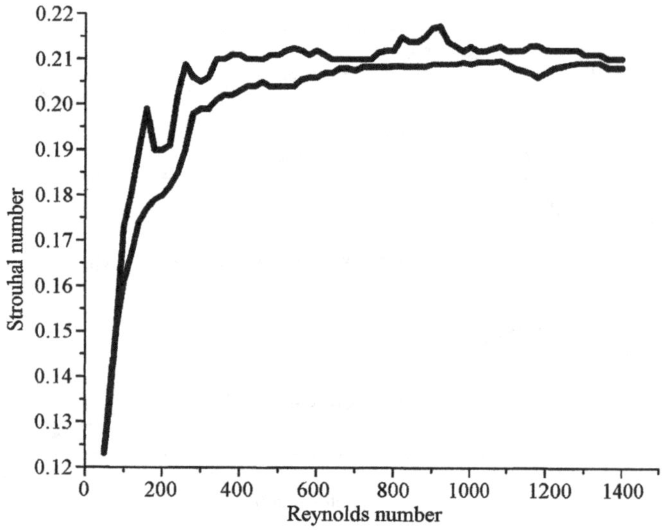

FIGURE 4.10. Upper and lower bounds for Roshko's data for the circular cylinder. Note the particularly broad range for *St*(*Re*) of $130 \leq Re \leq 300$.

in a Reynolds number range of 80 to 105. This discontinuity involved the transition between two clearly defined states; the amplitude of the fluctuations usually changed by 20–25% at the critical velocity. The exact location of the transition varied as we might expect from the phenomena governed by the nonlinear partial differential equations. Tritton observed, "...the exact behavior in the transition region that occurs on any particular occasion is governed by small unobserved deviations from the theoretical arrangement." Blevins provided additional emphasis by observing that vortex shedding from a fixed cylinder "...does not occur at a single distinct frequency, but rather it wanders over a narrow band of frequencies with a range of amplitudes and is not constant along the span."

4.4 BOUNDARY LAYER ON A WEDGE: THE FALKNER–SKAN PROBLEM

While the Blasius treatment of the flat plate was supremely important, one can imagine the circumstances in which the external flow must accelerate around some object. Consequently, it is not surprising that fluid dynamicists in the early years of the twentieth century sought solutions for such cases. Consider a potential flow in which the velocity is represented by

$$V_x = V_1 x^m. \tag{4.20}$$

If $m > 0$, then this is an accelerating flow with the pressure decreasing in the x-direction. If we assume that

$$\eta = y\sqrt{\frac{(m+1)}{2}\frac{V_1}{\nu}}\, x^{\frac{m-1}{2}} \tag{4.21}$$

and

$$\psi = \sqrt{\frac{2}{m+1}}\sqrt{\nu V_1}\, x^{\frac{m+1}{2}} f(\eta), \tag{4.22}$$

then

$$v_x = V_1 x^m f'(\eta). \tag{4.23}$$

Note that these choices once again ensure that $f'(0) = 0$ and $f'(\eta \to \infty) = 1$. The y-component of the velocity vector is given by

$$v_y = -\sqrt{\frac{m+1}{2}\nu V_1 x^{m-1}}\left[f + \frac{m-1}{m+1}\eta f'\right]. \tag{4.24}$$

If we define $\beta = 2m/(m+1)$, the transformation of the Prandtl equation results in

$$f''' + ff'' + \beta(1 - f'^2) = 0. \tag{4.25}$$

This nonlinear third-order differential equation is the Falkner–Skan (1931) equation for boundary-layer flow on a wedge. The included angle of the wedge is $\pi\beta$ radians; clearly, there are two limiting cases: $\beta = 0$, which is the Blasius problem, and $\beta = 1$, which is a two-dimensional stagnation flow. Although this ordinary differential equation received much attention following its discovery in 1930, there was resurgence in interest as a result of Stewartson's work in 1954 (Stewartson, 1954). Stewartson discovered that for some *increasing* pressures (negative included angles between −0.1988 and 0), additional solutions could be found that appeared to exhibit reverse flow. Three conventional solutions are illustrated in Figure 4.11.

Should the reader want to conduct his/her own exploration of the Falkner–Skan equation, a few values for $f''(0)$ are provided in the following table, which can help save time in dealing with the Falkner–Skan problem.

Included angle β	Correct value for $f''(0)$
1.0	1.2325876
0.2	0.68670
−0.16	0.19079
−0.0925	−0.138108
−0.0825	−0.1335869

Two pairs of solutions to the Falkner–Skan problem are shown in Figure 4.12, and reverse flow solutions are shown for β's of values −0.0825 and −0.12. We should be hesitant to assign too much meaning to these alternative solutions. Prandtl's equations for the laminar boundary layer are not valid at separation where the value of v_y is no longer very small relative to the mainstream velocity and the viscous transport of momentum in the x-direction is no longer

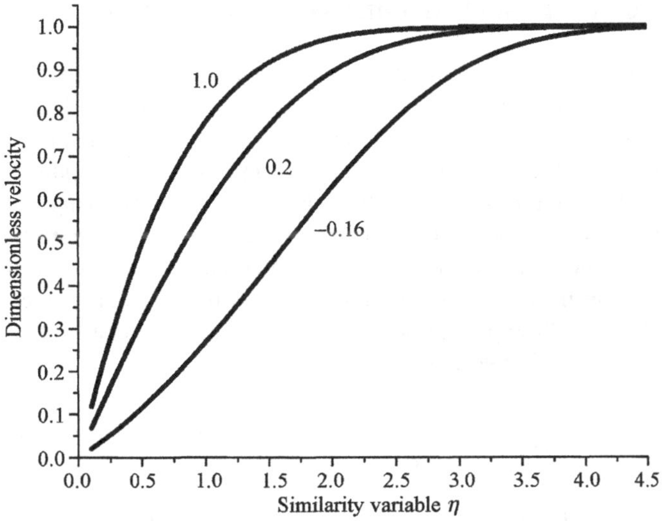

FIGURE 4.11. Some "conventional" solutions of the Falkner–Skan equation for β of values 1.0, 0.2, and −0.16. Note the point of inflection for the latter.

negligible. There also exists an additional class of solutions for values of $\beta < -0.19884$; these are called "overshoot" solutions because the dimensionless velocity f' exceeds 1 at some values of η. For example, at $\beta = -1.5, f'$ is greater than 3 at small η. This behavior has been compared with the effect of a jet issuing from the wall into the fluid (see White, 1991). Once again, however, these "overshoot" solutions are more of a mathematical curiosity than the representation of a physical phenomenon that could legitimately be expected from the Prandtl's boundary-layer equations.

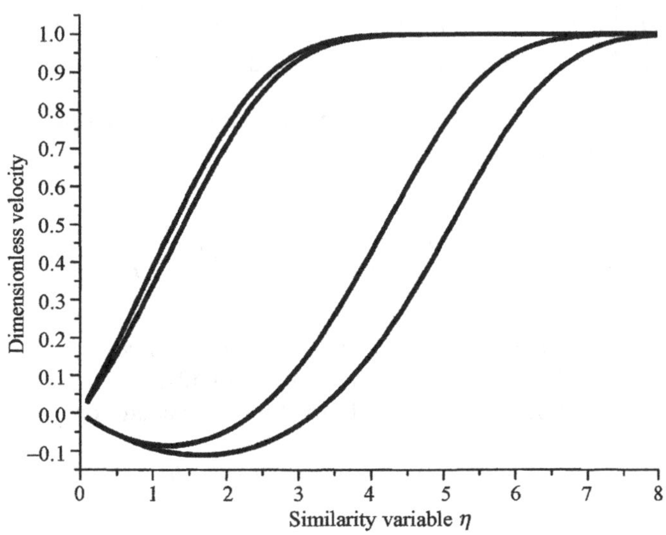

FIGURE 4.12. Pairs of solutions of the Falkner–Skan equation for β's of values −0.12 ($f''(0) = -0.142936$ and $+0.281765$) and −0.0825 ($f''(0) = -0.1335869$ and $+0.349384$).

4.5 THE FREE JET

The similarity transform approach employed above for the laminar boundary-layer flows can also be applied to the free jet even though there are no solid boundaries in play. We envision a jet emerging into an infinite fluid medium, through a small rectangular slit. By taking

$$\eta = \frac{y}{3\sqrt{\nu}x^{2/3}} \qquad \text{and} \qquad \psi = \nu^{1/2}x^{1/3}f(\eta), \tag{4.26}$$

the velocity vector components can be found:

$$v_x = \frac{1}{3x^{1/3}}f'(\eta) \tag{4.27}$$

and

$$v_y = -\frac{1}{3}\frac{\sqrt{\nu}}{x^{2/3}}(f - 2\eta f'). \tag{4.28}$$

The transformation is successful and a nonlinear ordinary differential equation results:

$$f''' + ff'' + f'^2 = 0. \tag{4.29}$$

Two boundary conditions for the jet centerline are $v_y = 0$ and, by symmetry, $(\partial v_x/\partial y)_{y=0} = 0$. Therefore, $f(0)$ and $f''(0)$ are both zero. At very large vertical distances (from the centerline), v_x must disappear, so $f'(\eta \to \infty) = 0$. Schlichting notes that eq. (4.29) can be integrated immediately to yield

$$f'' + ff' = 0. \tag{4.30}$$

The constant of integration is zero since both f and f'' are zero at $\eta = 0$. Schlichting points out that the transformations

$$\xi = \alpha\eta \qquad \text{and} \qquad f = 2\alpha F(\xi) \tag{4.31}$$

will introduce the necessary "2" into eq. (4.30), resulting in

$$F'' + 2FF' = 0. \tag{4.32}$$

We can now integrate again, getting

$$F' + F^2 = 1. \tag{4.33}$$

Since the unspecified constant α was introduced in (4.31), we can set the constant of integration here equal to 1. This is a form of the Riccati equation (which we saw in Chapter 1) named after Jacopo Francesco Count Riccati (1676–1754) who described it in 1724. Riccati equations were studied by notable mathematicians, including Euler, Liouville, and the Bernoullis. It is interesting to note that Johann Bernoulli examined a closely related equation ($dy/dx + y^2 + x^2 = 0$) in

1694 but was unable to find a solution. Our case is straightforward since

$$\int \frac{dF}{1-F^2} = \tanh^{-1} F \qquad \text{so} \qquad F = \tanh \xi. \qquad (4.34)$$

Working backward, we find that

$$v_x = \frac{2}{3}\alpha x^{-1/3}(1 - \tanh^2 \xi). \qquad (4.35)$$

A typical velocity distribution is shown in Figure 4.13.

The constant α is obtained from the total momentum of the jet:

$$M = \rho \int_{-\infty}^{-\infty} v_x^2 dy. \qquad (4.36)$$

Schlichting (1968) shows that

$$\alpha = 0.8255\left(\frac{M}{\rho \nu^{1/2}}\right)^{1/3}. \qquad (4.37)$$

For the example shown in Figure 4.13, $M/\rho = 1\ \text{cm}^3/\text{s}$.

It is essential that we recognize that laminar flow velocity profiles that contain a point of inflection are not very stable; we will clarify this observation later. Consequently, we should not expect the result presented above to be valid at large (or even modest) Reynolds numbers. Experimental work indicates that the stability limit for the laminar free jet is about $Re = 30$ where the characteristic length is taken as the size of the jet opening.

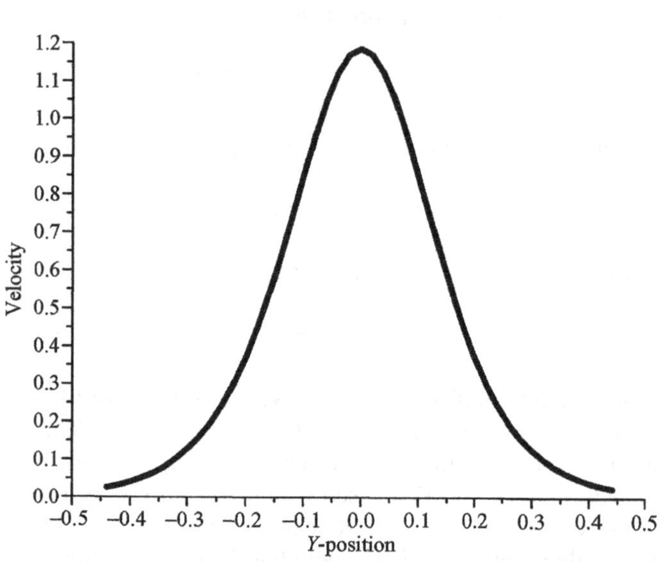

FIGURE 4.13. Laminar free jet example with $\alpha = 1.778$ and $\xi = 5.929(y/x^{2/3})$.

4.6 INTEGRAL MOMENTUM EQUATIONS

As we have seen, boundary-layer theory made drag calculations possible for a variety of surfaces moving through fluids. There is an enormous difference, however, between "possible" and "routine." Prior to the advent of digital computers, such calculations were anything but routine. Recognizing this problem, Theodor von Karman (1946) devised an approximate technique in the 1920s; he integrated the equation of motion in the normal direction, across the boundary layer. Consider flow past some surface that is (at least locally) flat. The governing equation is

$$v_x\frac{\partial v_x}{\partial x} + v_y\frac{\partial v_x}{\partial x} = -\frac{1}{\rho}\frac{\partial p}{\partial x} + \nu\frac{\partial^2 v_x}{\partial y^2}. \qquad (4.38)$$

We use the Bernoulli equation for the potential flow outside the boundary layer to write

$$-\frac{1}{\rho}\frac{dp}{dx} = V\frac{dV}{dx}. \qquad (4.39)$$

This is substituted into eq. (4.38) and the result is integrated (with respect to y) from the solid surface to a position across the boundary layer, say $y = h$:

$$\int_0^h \left[v_x\frac{\partial v_x}{\partial x} + v_y\frac{\partial v_x}{\partial y} - V\frac{dV}{dx}\right] dy = -\frac{\tau_0}{\rho}. \qquad (4.40)$$

Continuity for the two-dimensional flow requires that $v_y = -\int_0^y(\partial v_x/\partial x)dy$, so we can rewrite (4.40) as

$$\int_0^h \left[v_x\frac{\partial v_x}{\partial x} - \frac{\partial v_x}{\partial y}\int_0^y \frac{\partial v_x}{\partial x}dy - V\frac{dV}{dx}\right] dy = -\frac{\tau_0}{\rho}. \qquad (4.41)$$

By integrating the second term by parts, this equation is found to be equivalent to

$$\int_0^h \frac{\partial}{\partial x}[v_x(V - v_x)]\, dy + \frac{dV}{dx}\int_0^h (V - v_x)dy = \frac{\tau_0}{\rho}. \qquad (4.42)$$

How might we use this result? We could *assume* a rational form for $v_x(y)$ and introduce it into (4.42); naturally, the assumed function must satisfy the following conditions:

$$v_x(y = 0) = 0 \qquad \text{and} \qquad v_x(y = h) = V.$$

To illustrate, consider

$$\frac{v_x}{V} = \sin\frac{\pi y}{2h}. \qquad (4.43)$$

For a flat plate with a parallel potential flow, V is constant and (4.42) is rewritten as

$$\frac{\partial}{\partial x}\int_0^h v_x(V - v_x)dy = \frac{\tau_0}{\rho}. \tag{4.44}$$

By introducing (4.43) into (4.44) and noting that $\tau_0 = -\mu(\partial v_x/\partial y)_{y=0}$, we find

$$h = \delta = 4.795\sqrt{\frac{\nu x}{V}}. \tag{4.45}$$

This equation is in *fortuitous accord* with results from the Blasius solution.

4.7 HIEMENZ STAGNATION FLOW

What happens when a fluid stream impinges upon a flat surface that is perpendicular to the main flow direction? This is a scenario of practical importance; some CVD reactors used in semiconductor fabrication are operated in this manner. We might also consider mammalian cells grown on a support or the contaminant particles adhering to a surface that must be cleaned; perhaps we would need to examine the role of shear stress in the detachment of these entities from the surface.

Consider a flow approaching a plane surface as shown in Figure 4.14. The potential flow above the plate is described by

$$V_x = ax \quad \text{and} \quad V_y = -ay. \tag{4.46}$$

Close to the plate we *assume*

$$v_x = xf'(y) \quad \text{and} \quad v_y = -f(y). \tag{4.47}$$

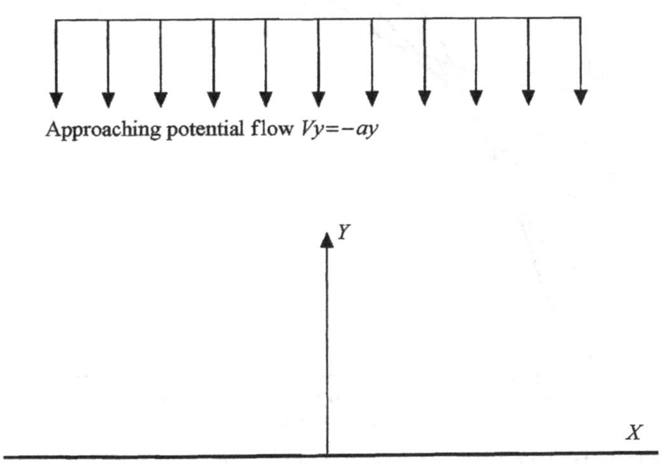

FIGURE 4.14. Two-dimensional stagnation flow at a plane surface.

Note that these choices guarantee that continuity will be satisfied for the two-dimensional flow. Obviously, when y is zero, both f and f' must be zero; at large distances above the surface, we must get the potential flow, so $f'(y \to \infty) = a$. The pressure distribution is

$$P_0 - P = \frac{1}{2}\rho a^2(x^2 + y^2), \tag{4.48}$$

which we rewrite as $P_0 - P = \frac{1}{2}\rho a^2[x^2 + F(y)]$.

We can introduce the assumed form for the velocity distribution into the Navier–Stokes equation(s) with the result

$$\nu f''' = f'^2 - ff'' - a^2. \tag{4.49}$$

The kinematic viscosity ν and the constant a can be eliminated from this equation by setting

$$\eta = \sqrt{\frac{a}{\nu}}y \quad \text{and} \quad f(y) = \sqrt{a\nu}\phi(\eta), \tag{4.50}$$

resulting in

$$\phi''' + \phi\phi'' - \phi'^2 + 1 = 0. \tag{4.51}$$

If we choose to solve (4.49), we can directly see the effects of a change in fluid viscosity upon the stagnation flow as shown in Figure 4.15.

Alternatively, we can solve (4.51), noting that

$$\phi(\eta = 0) = \phi'(\eta = 0) = 0 \quad \text{and} \quad \phi'(\eta \to \infty) = 1. \tag{4.52}$$

The solution for this equation is shown in Figure 4.16.

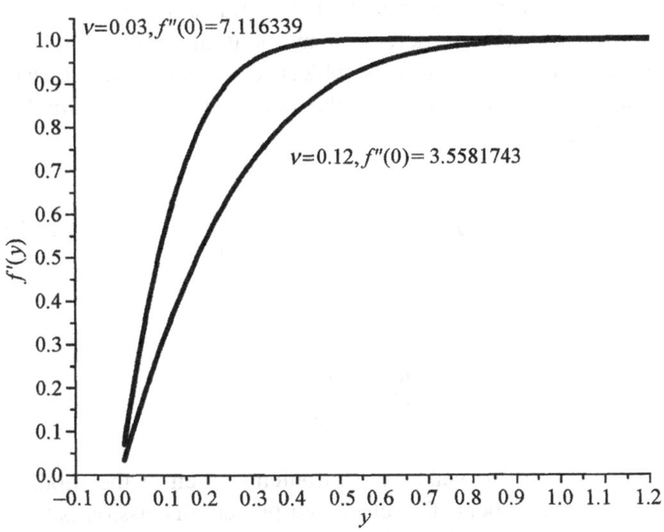

FIGURE 4.15. Computed Hiemenz profiles for $a = 1$ and the kinematic viscosity of values 0.03 and 0.12. Note that the increased kinematic viscosity has the effect of delaying the development of $f'(y)$.

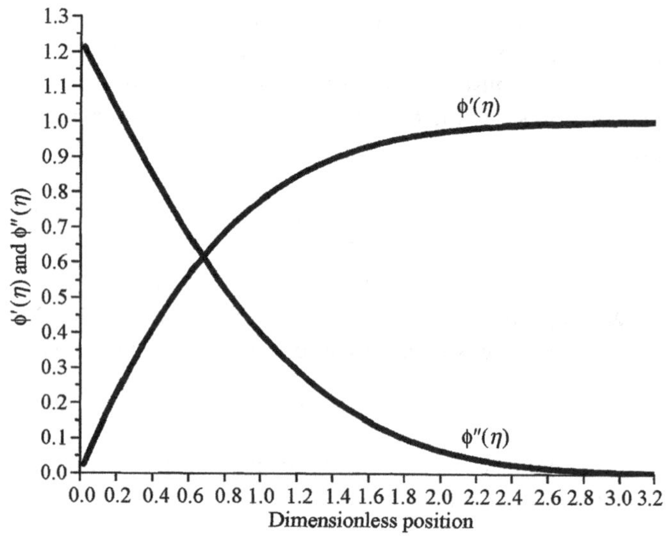

FIGURE 4.16. Solution of the dimensionless equation for Hiemenz stagnation flow, with $\phi''(0) = 1.2325877$.

The shear stress at the surface can be obtained for Hiemenz flow from the second derivative:

$$\left(\frac{\partial v_x}{\partial y}\right)_{y=0} = \frac{V_x}{a} f''(0). \tag{4.53}$$

4.8 FLOW IN THE WAKE OF A FLAT PLATE AT ZERO INCIDENCE

Flow around an object results in momentum transfer from the fluid to the surface, that is, drag. This transfer of momentum produces a velocity defect, or momentum deficit, immediately downstream from the object. Suppose we argue that Prandtl's equations apply in this near-wake region behind a flat plate such that

$$v_x \frac{\partial v_x}{\partial x} + v_y \frac{\partial v_x}{\partial y} = \nu \frac{\partial^2 v_x}{\partial y^2} \tag{4.54}$$

and

$$\frac{\partial v_x}{\partial x} + \frac{\partial v_y}{\partial y} = 0. \tag{4.55}$$

Of course, most wakes are turbulent—even at the modest Reynolds numbers. Therefore, our present discussion is limited to relatively slow viscous flows. We define a velocity difference in the wake as

$$V_1 = V_\infty - v_x(x, y) \tag{4.56}$$

and introduce this into eq. (4.54), resulting in

$$(V_\infty - V_1)\frac{\partial V_1}{\partial x} + v_y \frac{\partial V_1}{\partial y} = \nu \frac{\partial^2 V_1}{\partial y^2}, \tag{4.57}$$

with following boundary conditions: at $y=0$, $(\partial V_1/\partial y) = 0$ and $y \to \infty$, $V_1 = 0$. Schlichting (1968) argues that the quadratic terms in V_1 can be neglected; this leads to an analytic solution. Our approach will be a little different: Let us assume that v_y is much smaller than V_1, but the nonlinear term in V_1 is *not* negligible. We are left with

$$(V_\infty - V_1)\frac{\partial V_1}{\partial x} \approx \nu \frac{\partial^2 V_1}{\partial y^2}. \tag{4.58}$$

We can work through the following example: Suppose air flows past a flat plate (15 cm long) with a velocity of approach of 200 cm/s; the Reynolds number (Re_x) at the end of the plate will be about 20,000. We can solve eq. (4.58) numerically and compare our results with the Gaussian distribution curve obtained by Schlichting. Note that the boundary-layer thickness at the end of the plate will be about 0.53 cm.

An interesting exercise for the reader would be to take the data shown in Figure 4.17, determine the apparent momentum deficit, and then compare those results with the drag as computed from the Blasius solution. The drag can be obtained from (4.17) for one side of a plate (per unit width):

$$\frac{F_D}{W} = \rho \int_0^\infty v_x(V_\infty - v_x)dy. \tag{4.59}$$

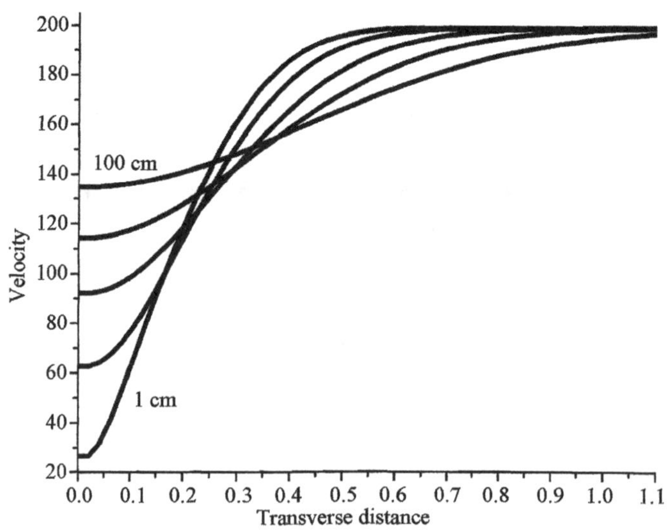

FIGURE 4.17. Velocity profiles in the wake of a flat plate at zero incidence for downstream positions of 1, 9, 25, 50, and 100 cm.

4.9 CONCLUSION

We do not want to leave the impression that the similarity transformation is the only tool available for external laminar flows. At the same time, it is to be recognized that it is a powerful technique through which some fairly difficult problems can be solved, or at least simplified. Often we can reduce our workload by noting that certain variables in a problem arise in combinations; examples include $y/\sqrt{x}$ for the Blasius problem and $y/\sqrt{4\alpha t}$ for some heat transfer problems. In such cases, the number of independent variables can be reduced through transformation. Systematic techniques exist to help identify the proper form of the transformation variable, and these include the free parameter, separation, group theory, and dimensional analysis methods. The interested reader should be aware that specialized monographs cover this area of fluid mechanics; an example is *Similarity Analyses of Boundary Value Problems in Engineering* by Arthur Hansen (1964).

But suppose we need to tackle a problem to which we do not want to apply a commercial CFD code and for which no similarity transformation exists. It is certainly possible that some of the methods described in the previous chapter might be applied, for example, we might be able to use vorticity transport. If we prefer to work strictly with the primitive variables, however, we will need something else. There is an explicit technique that is easy to employ and understand, however, the reader must remember that it *cannot be applied* to problems governed by the elliptic partial differential equations.

MacCormack (1969) devised a predictor–corrector approach in which new values of the primitive variables are obtained from an "average" time derivative, for example,

$$v_x(i, j, k+1) = v_x(i, j, k) + \left(\frac{\partial v_x}{\partial t}\right)_{\text{ave}} \Delta t, \tag{4.60}$$

where the indices i, j, and k refer to x, y, and t, respectively. In the predictor step, the time derivatives such as $(\partial v_x/\partial t)_{i,j,k}$ are computed using *forward* differences in the convective transport terms. These time derivatives are used to obtain "predicted" values for all the primitive variables. In the corrector step, these updated values are used to obtain the time derivatives at $t + \Delta t$ (or $k+1$) using *upwind* differences in the convective term, and the two values for the derivative are averaged:

$$\left(\frac{\partial v_x}{\partial t}\right)_{\text{ave}} = \frac{1}{2}\left[\left(\frac{\partial v_x}{\partial t}\right)_{i,j,k} + \left(\frac{\partial v_x}{\partial t}\right)_{i,j,k+1}\right]. \tag{4.61}$$

For the general case of a transient two-dimensional incompressible flow, the procedure can be summarized as follows:

The x- and y-components of the Navier–Stokes equation for a transient two-dimensional incompressible flow are written as

$$\frac{\partial v_x}{\partial t} = -v_x\frac{\partial v_x}{\partial x} - v_y\frac{\partial v_x}{\partial y} - \frac{1}{\rho}\frac{\partial p}{\partial x} + \nu\left[\frac{\partial^2 v_x}{\partial x^2} + \frac{\partial^2 v_x}{\partial y^2}\right] \tag{4.62a}$$

and

$$\frac{\partial v_y}{\partial t} = -v_x\frac{\partial v_y}{\partial x} - v_y\frac{\partial v_y}{\partial y} - \frac{1}{\rho}\frac{\partial p}{\partial y} + \nu\left[\frac{\partial^2 v_y}{\partial x^2} + \frac{\partial^2 v_y}{\partial y^2}\right]. \tag{4.62b}$$

On the predictor step, the time derivative is estimated using forward differences in the inertial terms and central differences for the viscous terms. As a general example,

$$\begin{aligned}\left(\frac{\partial v_x}{\partial t}\right)_{i,j,k} &= -v_x(i, j, k)\frac{v_x(i+1, j, k) - v_x(i, j, k)}{\Delta x} \\ &\quad - v_y(i, j, k)\frac{v_x(i, j+1, k) - v_x(i, j, k)}{\Delta y} \\ &\quad + \nu\left[\frac{v_x(i+1, j, k) - 2v_x(i, j, k) + v_x(i-1, j, k)}{(\Delta x)^2}\right. \\ &\quad \left. + \frac{v_x(i, j+1, k) - 2v_x(i, j, k) + v_x(i, j-1, k)}{(\Delta y)^2}\right].\end{aligned} \tag{4.63}$$

Now, the predicted values for the dependent variables are obtained with a truncated Taylor series using the time derivatives computed above:

$$v_x(i, j, k+1) = v_x(i, j, k) + \left(\frac{\partial v_x}{\partial t}\right)_{i,j,k} \Delta t. \tag{4.64}$$

Naturally, this is carried out for all the dependent variables. Next, we use these "new" predicted values to compute revised estimates for the time derivatives. *But*, we employ *backward* differences for the inertial terms:

$$\begin{aligned}&\left(\frac{\partial v_x}{\partial t}\right)_{i,j,k+1} \\ &= -v_x(i, j, k+1)\frac{v_x(i, j, k+1) - v_x(i-1, j, k+1)}{\Delta x} \\ &\quad - v_y(i, j, k+1)\frac{v_x(i, j, k+1) - v_x(i, j-1, k+1)}{\Delta y} \\ &\quad + \nu\left[\frac{v_x(i+1, j, k+1) - 2v_x(i, j, k+1) + v_x(i-1, j, k+1)}{(\Delta x)^2}\right] \\ &\quad + \nu\left[\frac{v_x(i, j+1, k+1) - 2v_x(i, j, k+1) + v_x(i, j-1, k+1)}{(\Delta y)^2}\right].\end{aligned} \tag{4.65}$$

Now we find the *average* of the two time derivatives for each dependent variable:

$$\left(\frac{\partial v_x}{\partial t}\right)_{\text{ave}} = \frac{1}{2}\left[\left(\frac{\partial v_x}{\partial t}\right)_{i,j,k} + \left(\frac{\partial v_x}{\partial t}\right)_{i,j,k+1}\right]. \quad (4.66)$$

This average derivative is used to calculate the corrected value for each dependent variable at time $t+\Delta t$:

$$v_x(i, j, k+1) = v_x(i, j, k) + \left(\frac{\partial v_x}{\partial t}\right)_{\text{ave}} \Delta t. \quad (4.67)$$

MacCormack's method is attractive because of its simplicity; the algorithm is easy to understand and to implement. Furthermore, it yields very acceptable results for some fairly complex flow problems; it has been used successfully for compressible (high-speed) flows as well. Indeed, MacCormack's approach was once one of the dominant strategies in CFD. However, it is to be kept in mind that MacCormack's technique *cannot be used* for the solution of elliptic partial differential equations. In cases where the procedure is to be applied to steady viscous flows, the unsteady equations are solved for large time t. Useful introductions to MacCormack's method can be found in Peyret and Taylor (1983), Anderson (1995), and Chung (2002).

REFERENCES

Ammann, O. H. , von Karman, T. , and G. B. Woodruff . The Failure of the Tacoma Narrows Bridge. *FWA Report* (1941).

Anderson, J. D. *Computational Fluid Dynamics*, McGraw-Hill, New York (1995).

Blasius, H. Grenzschicten in Flussigkeiten mit kleiner Reibung. *ZAMP*, 56:1 (1908).

Blevins, R. D. *Flow-Induced Vibration*, 2nd edition, Krieger Publishing, Malabar, FL (1994).

Boyne, W. J. *Messerschmitt Me 262, Arrow to the Future*, Smithsonian Institution Press, Washington (1980).

Braslow, A. L. A History of Suction-Type Laminar-Flow Control with Emphasis on Flight Research. *Monographs in Aerospace History*, No. 13, NASA History Division (1999).

Chang, P. K. *Control of Flow Separation*, Hemisphere Publishing, Washington (1976).

Chung, T. J. *Computational Fluid Dynamics*, Cambridge University Press, Cambridge (2002).

Falkner, V. M. and S. W. Skan . Some Approximate Solutions to the Boundary-Layer Equations. *Philosophical Magazine*, 12:856 (1931).

Grinsell, R. and R. Watanabe. *P51 Mustang*, Crown Publishers, New York (1980).

Hansen, A. G. *Similarity Analyses of Boundary Value Problems in Engineering*, Prentice-Hall, Englewood Cliffs, NJ (1964).

MacCormack, R. W. The Effect of Viscosity in Hypervelocity Impact Cratering. AIAA paper 69–354 (1969).

Petroski, H. Still Twisting. *American Scientist*, 79:398 (1991).

Peyret, R. and T. D. Taylor . *Computational Methods for Fluid Flow*, Springer-Verlag, New York (1983).

Prandtl, L. Motion of Fluids with Very Little Viscosity. *NACA TM 452* (1928).

Roshko, A. On the Development of Turbulent Wakes from Vortex Streets. *NACA Report 1191* (1954).

Schlichting, H. *Boundary-Layer Theory*, 6th edition, McGraw-Hill, New York (1968).

Stewartson, K. Further Solutions of the Falkner–Skan Equation. *Proceedings of the Cambridge Philosophical Society*, 50:454 (1954).

Taneda, S. Downstream Development of the Wakes Behind Cylinders. *Journal of the Physical Society of Japan*, 14:843 (1959).

Tritton, D. J. Experiments on the Flow Past a Circular Cylinder at Low Reynolds Numbers. *Journal of Fluid Mechanics*, 6:547 (1959).

Van Dyke, M. *An Album of Fluid Motion*, Parabolic Press, Stanford (1982).

von Karman, Th. On Laminar and Turbulent Friction. *NACA TM 1092*. (1946).

White, F. M. *Viscous Fluid Flow*, 2nd edition, McGraw-Hill, NewYork (1991).

Yenne, B. *The World's Worst Aircraft*, Barnes & Noble Books, New York (2001).

5

INSTABILITY, TRANSITION, AND TURBULENCE

5.1 INTRODUCTION

We have observed previously that laminar flow is atypical; turbulence is the *usual* state of fluid motion. The differences between the two are profound—consider flow through a cylindrical duct with constant diameter *d*. For the laminar fluid motion, the force exerted upon the tube wall is simply

$$\frac{F}{A} = \frac{8\mu\langle V\rangle}{d}. \tag{5.1}$$

But for the turbulent flow in rough tubes at the larger Reynolds numbers,

$$\frac{F}{A} = \frac{1}{2}\rho\langle V\rangle^2 f, \tag{5.2}$$

where the friction factor f is nearly constant. Thus, the rate at which momentum is transferred to the tube wall is proportional to the average velocity $\langle V\rangle$ in laminar flow, but to $\langle V\rangle^2$ for turbulent flow. There are other critical differences as well. We can compare timescales formulated for laminar and turbulent flows of water through a cylindrical tube:

$$\tau_L = \frac{R^2}{\nu} \quad \text{and} \quad \tau_K = \left(\frac{\nu}{\varepsilon}\right)^{1/2}. \tag{5.3}$$

The latter is the Kolmogorov timescale; it is a function of the kinematic viscosity ν and the dissipation rate per unit mass ε and it is the characteristic time for the small-scale (dissipative) structure of turbulence. If we assume that the fluid is water, that $R = 1$ cm, and that $\varepsilon = 100\,\text{cm}^2/\text{s}^3$, then

$$\tau_L \cong 100\,\text{s} \quad \text{and} \quad \tau_K \cong 0.01\,\text{s}.$$

Thus, it is clear that for the laminar flow, the characteristic time is large and in turbulence, the small-scale (viscous) eddies will have very small characteristic times and high (perhaps very high) frequencies. Obviously, the two flow regimes are very different. At this point we should be wondering: What is the pathway that leads from highly ordered to chaotic fluid motion?

Osborne Reynolds (1883) noted that there were two aspects of the question as to whether the motion of a fluid was *direct* (laminar) or *sinuous* (turbulent): There is a practical matter related to the nature of the resistance to flow, and the more "philosophical" question concerning the underlying principles of fluid motion. It is with regard to the latter where Reynolds' most important observations were made. First, he concluded that a critical velocity (at which eddies appear) existed and that

$$V_c \approx \frac{\mu}{d}. \tag{5.4}$$

This idea is recognized by every beginning student of fluid mechanics; for the flow in tubes, most will write reflexively:

$$Re_c = \frac{dV_c\rho}{\mu} = 2100. \tag{5.5}$$

Of course, the real situation is much less certain. For example, it is possible through special efforts to maintain laminar flow in tubes at the Reynolds numbers approaching 100,000. Reynolds also touched on this when he noted that "I had expected to see the eddies make their appearance as the velocity increased, at first in a slow or feeble manner, indicating

Transport Phenomena: An Introduction to Advanced Topics, By Larry A. Glasgow

that the water (the flow) was but slightly unstable. And it was a matter of surprise to me to see the sudden force with which the eddies sprang into existence, showing a highly unstable condition to have existed at the time the steady motion broke down." This observation was especially important, because Reynolds started his investigation from a viewpoint put forward by Stokes: That a steady (laminar) motion can become unstable such that an "…indefinitely small disturbance may lead to a change to sinuous motion." Reynolds further observed that efforts made to quell disturbances in the water prior to conduct of the experiment were critical; he found that transition could be triggered by the introduction of disturbances and he demonstrated this by placing an open coil of wire at the entrance of the test section.

In a very real, practical sense, the ability to delay the transition from laminar to turbulent flow would be enormously valuable. Consider, for example, the impact of maintaining laminarity (upon the friction factor) on flow through a hydraulically smooth tube at, say, $Re = 10{,}000$:

$$\text{Laminar flow}: \quad f(Re = 10,000) = 0.0016$$

$$\text{Turbulent flow}: \quad f(Re = 10,000) = 0.0079$$

Obviously, much more fluid could be delivered using fixed pressure drop under laminar flow conditions. And of course, this idea is not limited to flow through tubes. In transportation, any alteration that we could make to lessen the exchange of momentum between the fluid and the surface of a vehicle would be advantageous. To get a clearer picture of the scope of work being done in this area, the interested reader might begin with *Drag Reduction in Fluid Flows* (Sellin and Moses, 1989).

So far what we have seen is that laminar (or in Reynolds' description, direct) fluid motion will become unstable as the velocity increases. What is not clear is how this process evolves, in some cases however we can expect the nonlinear terms in the Navier–Stokes equation to play a critical role. We are going to turn our attention to a technique that was developed in the early twentieth century to analyze the laminar flow instability; it might be well imagined that these early efforts were focused upon finding a *linear* mode of attack. Consequently, we should not expect the approach to be universally successful unless the mechanism of instability (involving very small disturbances) is exactly the same for every flow. It is not, of course.

5.2 LINEARIZED HYDRODYNAMIC STABILITY THEORY

We begin by adopting Stokes' idea that under unstable conditions, a very small disturbance may grow, ultimately manifesting itself in sinuous (turbulent) fluid motion. The underlying principle is a simple one: We impose a small, periodic disturbance upon a laminar flow and then watch to see if the disturbance is either amplified or attenuated. It will be immediately recognized that this approach is at odds with Reynolds' experimental findings for flow in a cylindrical tube. For the Hagen–Poiseuille (HP) flow, turbulent eddies spring to life very dramatically: Either the crucial disturbances are not infinitesimally small, or the amplification rate is *very* large. Nevertheless, the approach we are about to describe has been used successfully for many other laminar flows. A beautiful introduction to the first 50 years of the "theory of small disturbances" has been provided by C. C. Lin (1955).

We start with a two-dimensional incompressible flow for which

$$\frac{\partial v_x}{\partial t} + v_x\frac{\partial v_x}{\partial x} + v_y\frac{\partial v_x}{\partial y} = -\frac{1}{\rho}\frac{\partial p}{\partial x} + \nu\left[\frac{\partial^2 v_x}{\partial x^2} + \frac{\partial^2 v_x}{\partial y^2}\right], \tag{5.6}$$

$$\frac{\partial v_y}{\partial t} + v_x\frac{\partial v_y}{\partial x} + v_y\frac{\partial v_y}{\partial y} = -\frac{1}{\rho}\frac{\partial p}{\partial y} + \nu\left[\frac{\partial^2 v_y}{\partial x^2} + \frac{\partial^2 v_y}{\partial y^2}\right], \tag{5.7}$$

and

$$\frac{\partial v_x}{\partial x} + \frac{\partial v_y}{\partial y} = 0. \tag{5.8}$$

The mean (base) flow is a *parallel* flow such that $V_x = V_x(y)$, and the total fluid motion is decomposed as the sum of mean flow and disturbance quantities:

$$v_x = V_x + v'_x, \quad v_y = v'_y, \quad \text{and} \quad p = P + p'. \tag{5.9}$$

Note that the disturbance is *assumed to be two-dimensional* (v'_x and v'_y). It is reasonable to question whether a two-dimensional disturbance has any real significance to laminar flow instability. This issue was addressed by Squire (1933), who demonstrated that a two-dimensional disturbance was actually more dangerous with respect to incompressible laminar flow stability than the one that was three dimensional. See Betchov and Criminale (1967) for elaboration on Squire's theorem.

We now introduce the decomposed quantities into eq. (5.6):

$$\begin{aligned}\frac{\partial v'_x}{\partial t} &+ V_x\frac{\partial V_x}{\partial x} + V_x\frac{\partial v'_x}{\partial x} + v'_x\frac{\partial V_x}{\partial x} + v'_x\frac{\partial v'_x}{\partial x} + v'_y\frac{\partial V_x}{\partial y}\\ &+ v'_y\frac{\partial v'_x}{\partial y} = -\frac{1}{\rho}\left(\frac{\partial P}{\partial x} + \frac{\partial p'}{\partial x}\right)\\ &+ \nu\left[\frac{\partial^2 V_x}{\partial x^2} + \frac{\partial^2 v'_x}{\partial x^2} + \frac{\partial^2 V_x}{\partial y^2} + \frac{\partial^2 v'_x}{\partial y^2}\right].\end{aligned} \tag{5.10}$$

Because this is a parallel flow, $V_x \neq f(x)$. Furthermore, it is assumed that the Navier–Stokes equation is satisfied identically for the mean flow such that

$$0 = -\frac{1}{\rho}\frac{\partial P}{\partial x} + \nu\left[\frac{\partial^2 V_x}{\partial y^2}\right], \qquad (5.11)$$

noting that both $(\partial V_x/\partial x)$ and $(\partial^2 V_x/\partial x^2)$ are zero. This equation is subtracted from (5.10), and we assume that the disturbance is small; consequently, the nonlinear terms in v'_x and v'_y are omitted. We are left with

$$\frac{\partial v'_x}{\partial t} + V_x\frac{\partial v'_x}{\partial x} + v'_y\frac{\partial V_x}{\partial y} = -\frac{1}{\rho}\frac{\partial p\prime}{\partial x} + \nu\left[\frac{\partial^2 v'_x}{\partial x^2} + \frac{\partial^2 v'_x}{\partial y^2}\right]. \qquad (5.12)$$

Similar steps for the y-component result in

$$\frac{\partial v'_y}{\partial t} + V_x\frac{\partial v'_y}{\partial x} = -\frac{1}{\rho}\frac{\partial p\prime}{\partial y} + \nu\left[\frac{\partial^2 v'_y}{\partial x^2} + \frac{\partial^2 v'_y}{\partial y^2}\right]. \qquad (5.13)$$

These equations are cross-differentiated; by subtraction, the pressure terms are eliminated. A form for the disturbance stream function is assumed:

$$\psi = \phi(y)e^{i(\alpha x - \beta t)}, \qquad (5.14)$$

which guarantees that continuity will be satisfied. $\phi(y)$ is the amplitude function, α is the wave number, and β is the frequency. β is, in general, complex (this is the temporal approach) and we define

$$\frac{\beta}{\alpha} = c = c_r + ic_i, \qquad (5.15)$$

where c_r is the velocity of propagation of the disturbance in the x-direction and c_i is the amplification (+) or damping (−) factor. Note that the exponential part of (5.14) can be rewritten as $e^{i\alpha[x-(c_r+ic_i)t]}$. A neutral disturbance, one for which the amplitude is not changing, corresponds to $c_i = 0$. Obviously, this condition is the demarcation between stability and instability. By defining

$$v'_x = \frac{\partial \psi}{\partial y}, \quad \text{we find} \quad v'_x = \phi\prime(y)e^{i(\alpha x - \beta t)}, \qquad (5.16)$$

and correspondingly,

$$v'_y = -\frac{\partial \psi}{\partial x} = -i\alpha\phi(y)e^{i(\alpha x - \beta t)}. \qquad (5.17)$$

These expressions for the fluctuations are introduced into disturbance equation (tedious), and the result is the Orr–Sommerfeld equation:

$$(V_x - c)(\alpha^2\phi - \phi'') + V''{}_x\phi = \frac{i\nu}{\alpha}\left[\phi'''' - 2\alpha^2\phi'' + \alpha^4\phi\right]. \qquad (5.18)$$

The reader is cautioned that the Orr–Sommerfeld equation pertains to *instability* and not to the transition to turbulence. What we can glean from this equation is a stability envelope, or possibly the amplification rate for a small disturbance; we *cannot* determine when or where the transition and turbulence will occur. The primes in (5.18), of course, refer to derivatives with respect to y; we have obtained a fourth-order, linear, ordinary differential equation. The disturbance velocities must disappear at the wall ($y = 0$), and they must also vanish far away from the wall (across the boundary layer, for example). Therefore, we have the following boundary conditions:

$$\text{for } y = 0,\ \phi = \phi' = 0 \text{ and as } y \to \infty,\ \phi = \phi' = 0. \qquad (5.19)$$

The characteristic value problem that we have described can be stated very succinctly:

$$F(\alpha, c, Re, \ldots) = 0. \qquad (5.20)$$

Given a particular parallel flow, the task is to find the eigenvalues that lead to solution of the Orr–Sommerfeld equation. This is not a trivial exercise; since instability can be expected to occur at large Reynolds numbers, the amplitude function will change rapidly with transverse position and a very small step size is required. Solutions of the Orr–Sommerfeld equation have been sought and found for boundary-layer flows, planar Poiseuille flows, free surface flows on inclined surfaces, free jets, wakes, and certain other flows as well. Linearized stability theory has failed in the case of Hagen–Poiseuille flow; numerous investigators have found that laminar pipe flow is stable to small axisymmetric and nonsymmetric disturbances. Stuart (1981) reviewed some of the attempts that have been made to identify the nature of the instability in the Hagen–Poiseuille flow, and, more recently, Walton (2005) examined the stability of the *nonlinear* neutral modes in the Hagen–Poiseuille flow. Walton found that by introducing unsteady effects into the critical layer, a threshold amplitude could be identified with amplification on one side and damping of the disturbance on the other.

We will examine a particular case (the Blasius profile on a flat plate) in greater detail (Figure 5.1). The pioneering work was performed by Tollmien (1929 also *NACA TM* 792, 1936) and Schlichting (1935, and summarized in *Boundary-Layer Theory*, 1968). Tollmien employed an analytic technique and demonstrated that viscosity was important not only near the wall (as expected) but also near the "critical layer" where the velocity of propagation of the disturbance was equal to the local velocity of the fluid. To honor their efforts, the two-dimensional traveling disturbances that arise in the boundary layer as precursors to transition are known as Tollmien–Schlichting waves.

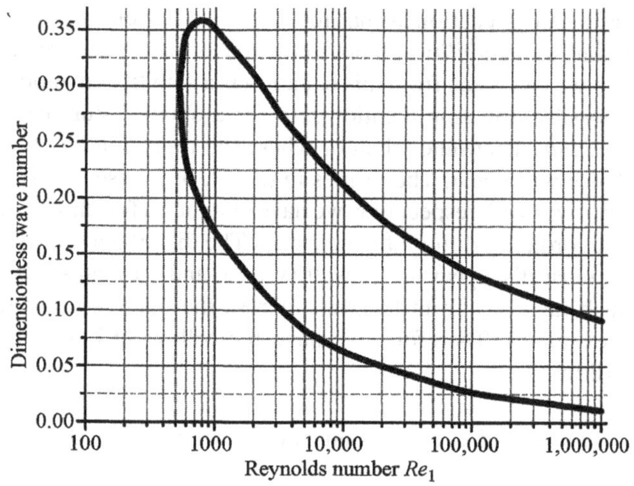

FIGURE 5.1. Curve of neutral stability for the Blasius profile on a flat plate. The Reynolds number is based upon the displacement thickness δ_1: $Re_1 = (\delta_1 V\rho/\mu)$. These results were adapted from Jordinson (1970). The characteristic shape explains why these stability envelopes are often referred to as "thumb" curves.

Modern calculations show that the critical Reynolds number (using the displacement thickness) for the Blasius profile is

$$Re_{1c} = \frac{\delta_1 V\rho}{\mu} = 520. \tag{5.21}$$

The displacement thickness is a measure of how far the external potential flow is moved away from the surface due to viscous friction:

$$\delta_1 = \int_0^\infty \left(1 - \frac{v_x}{V_\infty}\right) dy. \tag{5.22}$$

For the Blasius profile,

$$\delta_1 \cong 1.72\sqrt{\frac{\nu x}{V_\infty}}, \tag{5.23}$$

therefore, if we substitute this equation into (5.21), we find that the critical Reynolds number can be written in terms of Re_x:

$$Re_x(\text{critical}) = \frac{xV_\infty}{\nu} \cong 91,400. \tag{5.24}$$

Experimental studies, however, show that the laminarity can be maintained in the boundary layer on a flat plate up to a Reynolds number (Re_x) range of about

$$300,000 \leq Re_x \leq 3 \times 10^6. \tag{5.25}$$

The upper end of this range can only be approached in flows with *very low* levels of fluctuations (background turbulence).

The discrepancy between (5.24) and (5.25) is sizable. The explanation is that linearized hydrodynamic stability merely gives us the onset of instability; depending upon the amplification rate, some distance (in the x-direction) must pass before the instability is revealed as fully turbulent flow. Amplification rates for the initial disturbance have been computed by Shen (1954) among others. A good starting point for the interested reader is found in Chapter XVI of Schlichting (1968).

Although the Orr–Sommerfeld equation (the framework for linearized stability analyses) was known early in the twentieth century, no laboratory corroboration was available. In the case of the Blasius profile, the German workers had determined the stability envelope and some amplification rates, but their attempts to compare the theory with the experiment failed. However, with the approach of World War II improved wind tunnels were constructed and the background level of turbulence was finally low enough to permit fluid dynamicists to look for the signal of instability, the Tollmien–Schlichting waves. In August 1940, Schubauer and Skramstad conducted a series of measurements in the boundary layer on an aluminum plate using hot wire anemometry. Their work (Schubauer and Skramstad, 1948) validated the theory. In Figure 5.2, their hot wire data (as obtained from an oscilloscope) are shown at x-positions of 7, 8, 8.5, 9, 9.5, 10, 10.5, and 11 ft (measured from the leading edge). For these measurements, the free-stream velocity was 53 ft/s and the transverse (y) position was 0.023 in. above the surface. Note that the Tollmien–Schlichting waves begin to lose their organization at about $x = 9.5$–10 ft. By $x = 11$ ft, we see a hot wire signal characteristic of turbulent flow.

Consider the data shown in Figure 5.2 at $x = 9$ ft. At the measurement location ($y = 0.023$ in.) the local velocity was about 6.63 ft/s. The oscilloscope output shows a disturbance frequency of about 79 Hz; therefore, the wavelength of the disturbance was roughly 0.085 ft, which is three to four times the boundary-layer thickness at $x = 9$ ft, that is, the Tollmien–Schlichting waves are surprisingly long. In more recent years, photographs of the Tollmien–Schlichting waves have appeared in the literature; see Van Dyke (1982, pp. 62 and 63) and *Visualized Flow* (1988, p.19).

Schubauer and Skramstad also employed artificial excitation of the boundary layer using a phosphor bronze ribbon driven by an oscillator. In this manner, they were able to generate a periodic disturbance in the boundary layer of the desired frequency; the wavelength of the disturbance was determined from a Lissajous figure created by cross-plotting the signals from the oscillator and the output from the hot wire anemometer positioned downstream. It was also possible to compare oscillator amplitude with the mean square output from the hot wire and thus estimate the rates of damping or amplification of the disturbance. Their resulting locus of neutral points (where $c_i = 0$) confirmed Schlichting's calculations with remarkably good agreement.

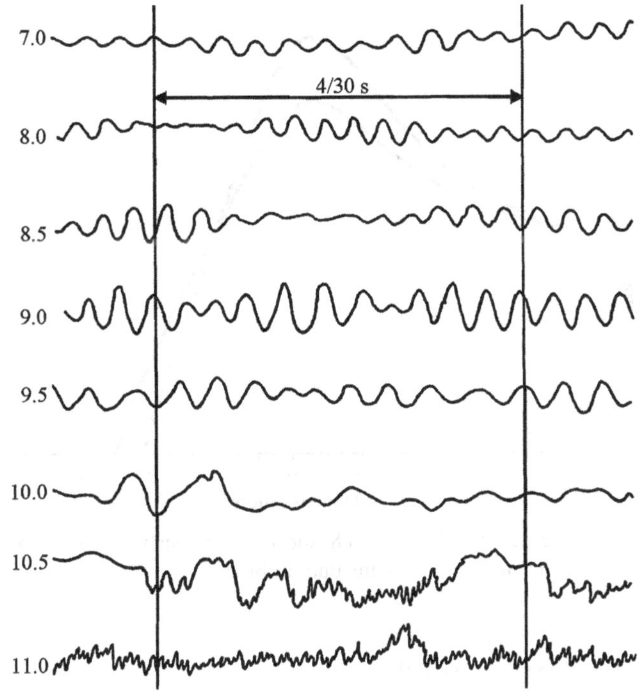

FIGURE 5.2. Hot wire measurements in the boundary layer on a flat plate, adapted from *NACA Report 909*. The Reynolds number Re_x at $x = 7$ ft was about $2.28 \times 10^6$and elapsed time between the light vertical lines was 4/30 s. Consequently, the very regular oscillations seen at 8–9 ft occur at about 80 Hz.

We noted previously that a number of other flows have been treated successfully with linearized hydrodynamic stability theory; many of the Falkner–Skan profiles have been examined by Schlichting and Ulrich (1942) and data are shown in Figure 5.3 for three cases (different included angles). These results indicate the profound influence that an adverse pressure gradient has upon the stability of flow in the boundary layer. Heeg et al. (1999) made stability calculations for the Falkner–Skan profiles with multiple inflection points and found, as expected, that the critical Reynolds number is dramatically reduced in such cases.

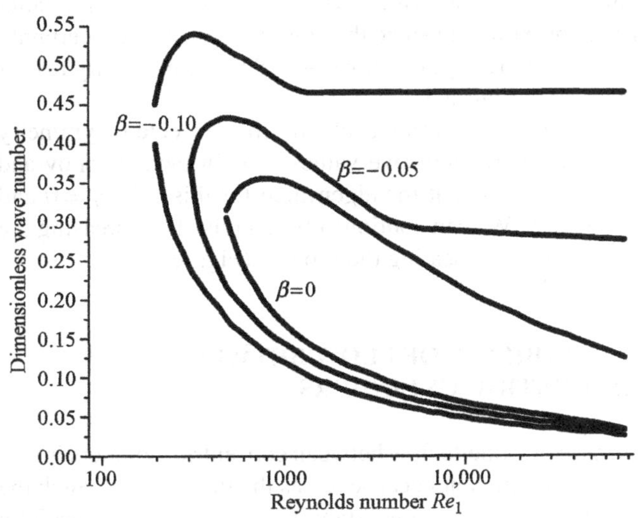

FIGURE 5.3. Curves of neutral stability for the Falkner–Skan velocity profiles with $\beta = -0.10, -0.05$, and 0. The Reynolds number is based upon the displacement thickness δ_1: $Re = (\delta_1 V\rho/\mu)$.

The effects of heating and cooling the wall upon the stability of boundary-layer profiles have also been investigated. Wazzan et al. (1968) studied the flow of water over heated and cooled plates; they modified the Orr–Sommerfeld equation to account for $\mu(T)$. Their results for water show that a heated wall stabilizes the flow. In fact, they found that for a free-stream water temperature of 60°F, a wall temperature of 130°F raises the critical Reynolds number to 15,700 (from 520 as shown in Figure 5.1).

There is a final point that must be made regarding the preceding discussion of the linearized theory of hydrodynamic stability: We have assumed that the base (or mean) flow is *parallel*. This is clearly incorrect for boundary-layer flows; for example, in the Blasius case, V_y is small but certainly not zero. Ling and Reynolds (1973) corrected the calculation of the "thumb" curve for the Blasius profile and they found that the neutral stability envelope was shifted very slightly toward the lower Reynolds numbers as a consequence of the nonparallel flow.

5.3 INVISCID STABILITY: THE RAYLEIGH EQUATION

If we set the kinematic viscosity ν equal to zero in the Orr–Sommerfeld equation and make a slight rearrangement,

$$\phi'' - \left(\frac{V_x''}{V_x - c} + \alpha^2\right)\phi = 0. \tag{5.26}$$

This is the stability equation for *inviscid* parallel flows and it bears Lord Rayleigh's name. Rayleigh (1899) found that if (5.26) was multiplied by the complex conjugate of ϕ, it was possible to show

$$c_i \int_0^\infty \frac{V_x''|\phi|^2}{|V_x - c|^2} dy = 0. \tag{5.27}$$

If we *do not have* a neutral disturbance (for which $c_i = 0$), then the integral in (5.27) must be zero. This will require that V_x'' change signs at least once; the velocity distribution must have a point of inflection. This led Rayleigh to conclude that it was *necessary* for instability that a velocity profile contain a point of inflection. This condition, known as the Rayleigh theorem, was strengthened to a sufficiency by Tollmien in 1929.

The Rayleigh equation can also be used to reveal the limiting behavior of the amplitude function ϕ. Suppose we consider a point just outside the boundary layer where $V''_x = 0$:

$$\phi'' - \alpha^2\phi = 0. \tag{5.28}$$

Clearly, we must have

$$\phi = C_1 e^{\alpha y} + C_2 e^{-\alpha y}. \tag{5.29}$$

The amplitude function cannot increase without bound in the y-direction, so $C_1 = 0$, and we find

$$\phi \approx e^{-\alpha y}. \tag{5.30}$$

Thus, the behavior of the amplitude function at large y (outside the boundary layer) is known.

We now turn our attention back to the Rayleigh equation (5.26). We note that there is a critical point if $V_x(y) = c$, that is, if the velocity of propagation equals the local velocity at position y_c (for a neutral disturbance), then we cannot obtain a regular solution unless $V''_x(y_c) = 0$. Lin (1955) notes that such difficulties do not arise for amplified or attenuated disturbances. Before proceeding, we also observe that eq. (5.26) will have particular value *if* the solution corresponds to the limiting case for the Orr–Sommerfeld equation when *Re* is very large (μ is very small). To give shape to this discussion, we examine the shear layer between two fluids moving in opposite directions; following Betchov and Criminale (1967), the velocity distribution is assumed to have the form

$$V_x = V_0 \tanh\left(\frac{y}{\delta}\right) \tag{5.31}$$

and it is shown in Figure 5.4.

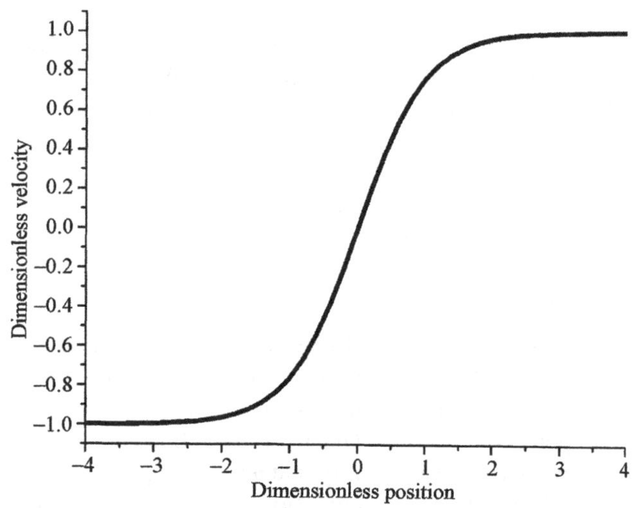

FIGURE 5.4. Shear layer at the interface between two fluids (dimensionless position zero) moving in opposite directions.

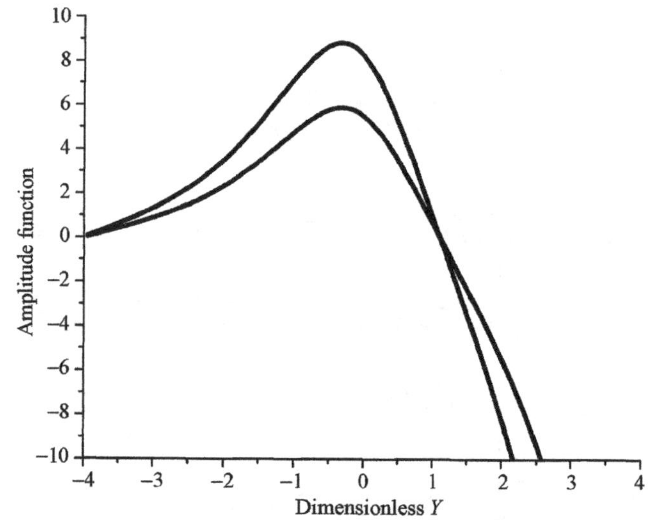

FIGURE 5.5. $\phi(y)$ for $\alpha = 0.8$ and $c = 0$. Clearly, we have not found a solution for this eigenvalue problem.

For this case, we have

$$\frac{dV_x}{dy} = \frac{V_0}{\delta}\frac{1}{\cosh^2(y/\delta)} \quad \text{and} \quad \frac{d^2V_x}{dy^2} = -\frac{8V_0}{\delta^2}\frac{e^X - e^{-X}}{(e^X + e^{-X})^3}, \tag{5.32}$$

where $X = y/\delta$. We can spend a little time profitably here by carrying out some numerical investigations of this problem. We arbitrarily set $\delta = 1, \alpha = 0.8$, and $V_0 = 1$; we start the integration at $y = -4$ and carry it out to $y = +4$. We know that the amplitude function must approach zero at large distances from the interface. If we can find a value of c that results in meeting these conditions, we will have identified an eigenvalue. We can start with $c = 0$ and let $\phi(-4) = 0$; the latter is an approximation since the amplitude function is certainly small but not really zero at $y = -4$. Some preliminary results are given in Figure 5.5.

Note that we cannot obtain the expected symmetry between negative/positive values of y. In fact, Betchov and Criminale show that the eigenvalue for this α is $c_r = 0$ and $c_i = 0.1345$. We can continue this exercise by increasing the value of α and repeating the process (Figure 5.6).

5.4 STABILITY OF FLOW BETWEEN CONCENTRIC CYLINDERS

The case of Couette flow between concentric cylinders is particularly significant because it was the first flow to which the linearized hydrodynamic stability theory was successfully applied. Moreover, a flow between the rotating concentric cylinders exhibits an array of behaviors that continues to intrigue investigators in the twenty-first century. Taylor

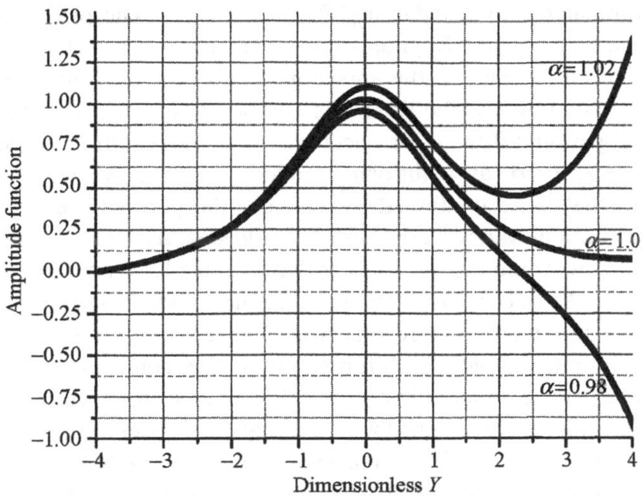

FIGURE 5.6. $\phi(y)$ for α 's of 0.98, 1.00, and 1.02. The interested reader might want to try $\alpha = 0.9986$.

(1923) determined the critical speed of rotation for the Couette flows dominated by the rotation of the *inner cylinder*. Before we provide a description of his analysis, we must note that there are two very different situations in the rotational Couette flows: motions that are driven primarily by the rotation of the inner cylinder, and those in which the outer cylinder provides the momentum. In the case of the former, the transition process has been described as *spectral evolution* by Coles (1965); the initial instability leads to a succession of stable secondary flows (the first is known as Taylor vortices). For the latter, the fluid is centrifugally stabilized, that is, the fluid with the greatest tendency to flee the center is already against the outermost surface. The transition process in this case has been described as *catastrophic*. Indeed, the theory of small disturbances has failed to find instability for this arrangement; this Couette flow is theoretically stable at any rate of rotation of the outer cylinder. Obviously, that cannot be correct; at some speed, bearing imperfections or eccentricities must create larger disturbances that are amplified through a nonlinear process.

We begin by noting that the velocity distribution for the steady cylindrical Couette flow is described by

$$V_\theta = Ar + \frac{B}{r} \tag{5.33}$$

and that the flow can be characterized with three dimensionless parameters:

$$\frac{R_2}{R_1}, \quad \frac{\omega_2}{\omega_1}, \quad \textit{and} \quad Re = \frac{\omega_1 R_1^2}{v}. \tag{5.34}$$

We are going to impose a three-dimensional disturbance upon the flow that is symmetric with respect to the θ-direction and periodic axially, such that

$$v_r{}' = \phi_1(r)e^{\sigma t}\cos\lambda z, \tag{5.35a}$$

$$v_\theta{}' = \phi_2(r)e^{\sigma t}\cos\lambda z, \tag{5.35b}$$

and

$$v_z{}' = \phi_3(r)e^{\sigma t}\sin\lambda z. \tag{5.35c}$$

The appropriate form for the continuity equation is $(\partial v_r/\partial r) + (v_r/r) + (\partial v_z/\partial z) = 0$, since $v_\theta \neq f(\theta)$. The imposed disturbance must satisfy continuity, so we find that

$$\phi'_1 + \frac{\phi_1}{r} + \lambda\phi_3 = 0. \tag{5.36}$$

The linearized disturbance equations become

$$(L - \lambda^2 - \sigma\, Re)(L - \lambda^2)\phi_1 = 2\lambda^2\, Re\frac{\omega}{\omega_1}\phi_2 \tag{5.37}$$

and

$$(L - \lambda^2 - \sigma\, Re)\phi_2 = 2\, Re\, A\phi_1, \tag{5.38}$$

where the operator L is $(d^2/dr^2) + (1/r)(d/dr) - (1/r^2)$ and

$$A = \frac{\left(\frac{R_2}{R_1}\right)^2\left(\frac{\omega_2}{\omega_1}\right) - 1}{\left(\frac{R_2}{R_1}\right)^2 - 1}. \tag{5.39}$$

These relations are to be solved with six boundary conditions obtained by requiring that the disturbances disappear at both cylindrical surfaces:

$$\phi_1 = \phi_2 = \phi_3 = 0 \quad \text{at both} \quad r = R_1 \text{ and } r = R_2. \tag{5.40}$$

The form of the operator L suggests Bessel functions, and Taylor developed a solution for this problem by using series expansions of the first-order Bessel functions, requiring the functions to disappear at the two cylindrical surfaces. Taylor devised an experimental test of this remarkable analysis and a comparison for the case in which the radii of the two cylinders were 3.8 and 4.035 cm is shown in Figure 5.7.

In honor of Taylor's achievements, flow in the Couette apparatus is often characterized with the Taylor number Ta:

$$Ta = \frac{R_1(R_2 - R_1)^3(\omega_1^2 - \omega_2^2)}{\nu^2}. \tag{5.41}$$

For devices with a small gap, Taylor's analysis revealed that $Ta_c = 1709$ for $\omega_2 = 0$.

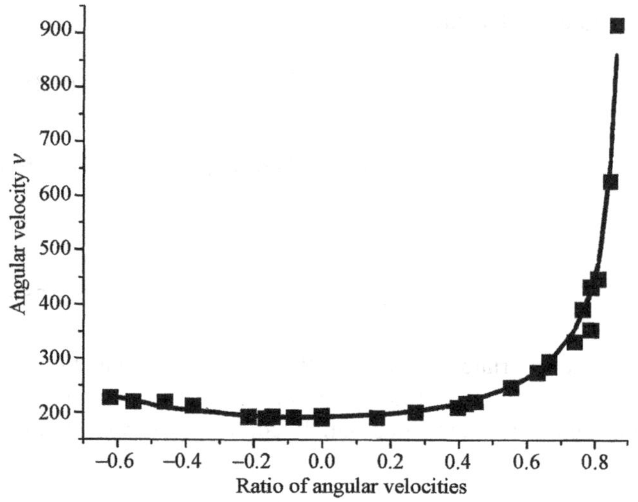

FIGURE 5.7. Comparison of Taylor's results for theory (curve) and experiment (filled squares). The abscissa is the ratio of angular velocities ω_2/ω_1, where "2" refers to the outer cylinder. The ordinate is the ratio ω_1/ν.

It is to be borne in mind that this threshold merely marks the initial instability, that is, the onset of Taylor vortices. Coles (1965) demonstrated that the behavior seen at higher speeds is highly complex with a succession of stable secondary states. He also noted the presence of hysteresis loops where the states attained during slow acceleration of the inner cylinder (outer cylinder at rest) do not correspond to those exhibited as the cylinder speed was decreased.

It is intriguing that after more than 100 years of investigation, Couette flow between concentric cylinders continues to elicit interest of fluid dynamicists around the world. Indeed, it seems that the more we learn about this flow, the more unexpected complexities emerge. To illustrate, let us consider the results of Burkhalter and Koschmieder (1974). They used impulsive starting of the rotation of the inner cylinder in which the supercritical Taylor numbers (Ta/Ta_c ranging from 1 to about 70) were achieved very rapidly (within about 0.5 s from rest). They found that the wavelength of the Taylor vortices varied in remarkable fashion (but always smaller than the critical wavelength), depending upon the value of Ta/Ta_c. Furthermore, these results were independent of fluid viscosity, end effects, and annular gap, and they were stable as long as the angular velocity achieved by the inner cylinder was maintained. This investigation showed very clearly that the stable secondary flow (Taylor vortices) is not unique, but quite dependent upon initial conditions.

5.5 TRANSITION

For many years a commonly accepted picture of transition was that put forward by L. D. Landau and conveniently summarized by Landau and Lifshitz (1959). The principal idea is that as the Reynolds number is increased, instability of the flow leads to the appearance of a new unsteady, but periodic flow. As the Reynolds number is further increased, this periodic flow in turn becomes unstable resulting in the emergence of an additional frequency, and so on. Landau felt that if *Re* continued to increase, the gap between the generations of new periodicities would steadily diminish and the flow would rapidly become "complicated and confused." As Yorke and Yorke (1981) noted, the Landau model suggests that turbulence results from a succession (they call it an infinite cascade) of bifurcations. If this conjecture were valid, then a suitable instrumental technique in which time-series data were obtained would reveal the Fourier transforms with an incrementally increasing number of discrete frequencies (i.e., a series of sharp spikes in the power spectra). Unfortunately, this very attractive concept of transition is incorrect for the Hagen–Poiseuille flow; no dominant frequencies appear in spectra intermediate to the development of fully turbulent flow. That said, there are some well-known cases where discrete frequencies do appear in power spectra en route to chaotic behavior. Examples include natural convection in enclosures and the Couette flow between cylinders.

5.5.1 Transition in Hagen–Poiseuille Flow

We observed previously that the classical linearized theory (of small perturbations) fails to find instability in the case of Hagen–Poiseuille flow. Since the time of Reynolds' work in the late nineteenth century, it has been apparent that *finite amplitude* disturbances drive the transition from laminar to turbulent flow in pipes. This is clear, because with special precautions, laminar flow can be maintained in tubes for the Reynolds numbers as high as about 10^5. Obviously, we are contemplating a very different situation than, say, the Rayleigh–Benard convection or the Taylor vortices in the Couette flow. Kerswell (2005) states the mathematical implication: No definitive bifurcation point can be identified for the Hagen–Poiseuille (HP) flow that might serve as a starting point in a search for additional solutions to the governing equations. The onset of turbulence in HP flow is sudden; there are no intermediate states or secondary flows, and the stability envelope is not crisply defined. Even though transition in HP flow remains as one of the most difficult problems in fluid mechanics, some exciting progress has been made in recent years.

In 1973, Wygnanski and Champagne identified "puffs" of turbulence in Hagen–Poisuille flow. These puffs appear for $1760 < Re < 2300$; they typically have a sharply defined trailing edge and a length of about 20*d*. Willis et al. (2009) noted that there is a lower (*Re*) bound for the existence of these "puffs." Should the Reynolds number fall below this value, the puffs can disappear very rapidly, even after traveling many hundreds of diameters downstream. When the Reynolds number exceeds about 2700, the turbulent "puffs" are replaced by "slugs." The front face of these slugs travels faster than the mean flow (about $1.5\langle V\rangle$), and the trailing edge slower (about $0.3\langle V\rangle$), so that the slug of turbulence expands as it moves downstream. At the University of Manchester, a unique test rig has been constructed with a length corresponding to $765d$. Mullin (2008) and coworkers have been using this apparatus to visualize and study puffs and slugs; they have the capability of introducing both jet puffs (through six azimuthal jets) and impulsive rotational disturbances at any point in the test section. They have identified the envelope (jet amplitude versus the Reynolds number) for which puffs and slugs either persist or decay. At the Delft University of Technology, J. Westerweel and coworkers have been developing a stereoscopic PIV (particle image velocimetry) technique to study turbulent slugs. They are refining their technique with the goal of obtaining data that can be used to support the latest theoretical studies. This method shows great promise, though it is necessary that errors near the pipe wall be minimized.

On the theoretical front, R. R. Kerswell's group at the University of Bristol and Eckhardt and Faisst at Marburg have been leaders in the discovery of alternative solutions (involving traveling waves) to the familiar Hagen–Poiseuille flow (Eckhardt and Faisst, 2008; Faisst and Eckhardt, 2003; Kerswell, 2005; Kerswell and Tutty, 2007). The waves appear for $Re > 773$, both with and without rotational symmetry. These transient traveling waves have been experimentally observed at Delft (UT), and Hof et al. (2004) show an intriguing comparison between the experimental and computed "streak" patterns (a streak is an anomaly created when a vortex moves fluid of higher velocity toward the wall and vice versa). Hof et al. (2004) have put the transition process for HP flow into the language of chaos theory: "...as the Reynolds number is increased further, this chaotic repellor is believed to evolve into a turbulent attractor, i.e., an attracting region in phase space, dynamically governed by the large number of unstable solutions, which sustains disordered turbulent flow indefinitely. The laminar state is still stable, but it is reduced from a global to a local attractor. As the Reynolds number increases, the basin of the turbulent attractor grows, whereas that of the laminar state diminishes."

The student interested in stability of the Hagen–Poiseuille flows should also be aware of some recent work reported by Trefethen et al. (1993). These authors noted that even in cases in which eigenvalues for a linearized system indicate stability, an input disturbance may be amplified at a large rate if the eigenfunctions are not orthogonal. It appears that the (stability) operator for the Hagen–Poiseuille flow may be in this category. Furthermore, Trefethen et al. observed that this "nonmodal amplification" applies to *three-dimensional* disturbances; therefore, the focus upon the two-dimensional disturbances for such cases appears to be inappropriate. They also offer a physical interpretation of the three-dimensional process: A streamwise vortex (a flow disturbance) moves fluid in a transverse direction to a region of higher or lower (streamwise) velocity. This movement of fluid results in a large, but local, discrepancy in streamwise velocity, referred to as a "streamwise streak." A good starting point for the reader interested in efforts to identify such disturbances is the contribution by Robinson (1991).

5.5.2 Transition for the Blasius Case

Even for the simplest of parallel flows, our understanding of the transition between laminar and turbulent flow regimes is incomplete. The reader is urged to consult Schlichting (1979), White (1991), and Bowles (2000) for general background and elaboration; some of D. Henningson's recent work at KTH in Stockholm is also useful in this context. It is to be noted that the immediately following observations apply *only* to the transition process occurring in the boundary layer on a flat plate; this case has probably seen the most comprehensive experimental investigations.

The apparent transition sequence is as follows: The laminar flow develops the unstable two-dimensional Tollmien–Schlichting waves. These disturbances become three-dimensional by a secondary instability and the "lambda" vortices (they have characteristic Λ-shape) appear. Bursts of turbulence (spikes in the disturbance velocity) appear in the regions of high vorticity. Turbulent (Emmons) spots show up in regions where the fluctuations are large. Finally, the turbulent spots coalesce into fully developed turbulent flow.

Formation of the Emmons spots is perhaps the most intriguing aspect of the transition process. These turbulent spots are roughly wedge shaped and were first observed on a water table by H. W. Emmons (1951); he noted that the spots tended to preserve their shape as they grew. Their migration downstream occurred in a straight line (aligned with the mean flow) and their lateral growth produced about the same angle as seen in a turbulent wake. Emmons also developed a functional representation for the fraction of time that flow at a particular point would be turbulent; obviously, this must involve rates of spot production, migration, and growth. In recent years, Emmons spots have been artificially triggered for study through flow visualization. There are some remarkable images of Emmons spots in Van Dyke (1982), in *Visualized Flow* (1988, p. 21), and on the KTH (Department of Mechanics) Web site.

5.6 TURBULENCE

Turbulence is one of the greatest unsolved mysteries of modern physics, and in the space available here, we can

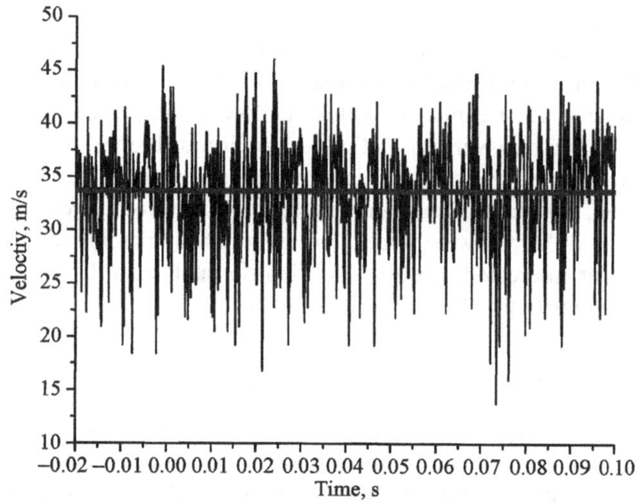

FIGURE 5.8. Point velocity measurement near the center of a deflected air jet.

do no more than provide an introduction to the subject. Fortunately, there are some wonderful books available for students beginning their exploration of turbulence. I particularly recommend Bradshaw (1975), Reynolds (1974), Tennekes and Lumley (1972), Hinze (1975), and Pope (2000). The latter provides a useful introduction to probability density function (PDF) methods, which are particularly valuable for turbulent reacting flows.

Suppose we measure the velocity at a single point in space in a turbulent flow; what are we likely to see? Consider Figure 5.8, which shows the signal obtained from a hot wire anemometer positioned near the center of a deflected jet of air.

You can see in Figure 5.8 that the mean velocity is about 33.6 m/s. You may also note that there are fluctuations occurring at frequencies at *least* as large as 1–2 kHz.

One might be tempted to describe the behavior in Figure 5.8 as *random*, but it is to be noted that care must be taken when using this word as a descriptor for turbulence. Statisticians would define a *random variable* as a real-valued function defined on a sample space (Hoel, 1971); this is appropriate for turbulence. But they might further relate the term *random variable* to a physical process with an uncertain outcome (which depends upon chance). When turbulence is viewed from the perspective of either an experimental or a computational ensemble, the outcome is neither uncertain nor the result of chance.

How might we represent such a process where fluctuations about the mean are occurring in both positive and negative directions? We use the Reynolds decomposition:

$$v_i = V_i + v_i', \qquad (5.42)$$

where V_i is the average (mean) velocity in the i-direction and v'_i is the fluctuation. Suppose we observe the fluctuating signal for a long period of time; it will be positive and negative equally if the flow is statistically stationary:

$$\text{limit as } (T \to \infty) \quad \left(\frac{1}{T}\right) \int_0^T v_i' dt = 0. \qquad (5.43)$$

Over the years, many investigators have defined a relative turbulence intensity (RTI) as the ratio of the root-mean square (rms) fluctuation to the mean velocity:

$$\text{RTI} = \frac{\sqrt{\overline{v_i'^2}}}{V_i}. \qquad (5.44)$$

For the fully developed turbulent flow in a pipe, the relative intensity will typically range from about 3 to 8% for the axial (z-direction) flow; it is usually larger near the wall with smaller values near the centerline. In free jets, the relative intensity can be much larger with typical values around 30% common on the centerline.

Naturally, the time average of a product of fluctuations, say $\overline{v_i'v_j'}$, will not be zero since the continuity equation will require that other velocity vector components react to a particular fluctuation. Consequently, the two fluctuations will be *correlated* if the observations are separated either by a small distance or by a short time (spatial or temporal separation). In the case of temporal separation, a correlation coefficient can be written as

$$\rho_{ij}(\tau) = \frac{\overline{v_i'(t)v_j'(t+\tau)}}{\left(\overline{v_i'^2 v_j'^2}\right)^{1/2}}. \qquad (5.45)$$

If $i = j$, $\rho(\tau)$ is referred to as the *autocorrelation coefficient*. Naturally, $\rho(\tau = 0) = 1$; with no time separation, the correlation is perfect. It is to be noted that the autocorrelation is an even function as this will be important to us later. We must also emphasize that some flows are turbulent only *intermittently*. For example, for a free jet or a wake, there is a mixing layer at the boundary between the bulk (undisturbed) free flow and the turbulent core. In this mixing region, the flow is turbulent for a fraction of the time and as we move away from the axis of the jet or the wake, that fraction approaches zero. Characterization of the turbulence in such areas would require conditional sampling, that is, data would be collected only when a turbulence criterion (usually a threshold value of vorticity) is satisfied. During quiescent periods, no data are recorded.

We now apply the Reynolds decomposition to the x-component of the Navier–Stokes equation for a "steady" turbulent flow:

$$\frac{\partial}{\partial x}\left[(V_x+v_x')(V_x+v_x')\right]+\frac{\partial}{\partial y}\left[(V_x+v_x')(V_y+v_y')\right]$$
$$+\frac{\partial}{\partial z}\left[(V_x+v_x')(V_z+v_z')\right]$$
$$=-\frac{1}{\rho}\frac{\partial}{\partial x}(P+p')+\nu\nabla^2\left[V_x+v_x'\right]. \tag{5.46}$$

We time average the result (indicated by an overbar) and note that any term that is linear in a fluctuation will be zero. We also make a slight rearrangement (convince yourself that this is appropriate) to get

$$\rho\left(V_x\frac{\partial V_x}{\partial x}+V_y\frac{\partial V_x}{\partial y}+V_z\frac{\partial V_x}{\partial z}\right)$$
$$=-\frac{\partial P}{\partial x}+\left[\frac{\partial}{\partial x}\left(\tau_{xx}-\rho\overline{v_x'v_x'}\right)+\frac{\partial}{\partial y}\left(\tau_{xy}-\rho\overline{v_x'v_y'}\right)\right.$$
$$\left.+\frac{\partial}{\partial z}\left(\tau_{xz}-\rho\overline{v_x'v_z'}\right)\right] \tag{5.47}$$

We see that three new terms have appeared on the right-hand side of the equation. The intent is clear, although the reasoning is flawed, that we are to interpret these quantities as some sort of stress. These Reynolds "stresses" are nine in number (three from each component of the Navier–Stokes equation), that is, we have discovered the second-order *turbulent inertia tensor*, which is symmetric. It is essential that we understand what these terms are really about: They represent the transport of turbulent momentum by the turbulence itself, and they are not stresses! Unfortunately, they are also unknowns (variables), so we now have 4 equations and 13 (10 by symmetry of the tensor) variables. This is the closure problem of turbulence and it is a characteristic of nonlinear stochastic systems. Much effort, and much of it wasted, has been devoted to "closing" systems of turbulent momentum and energy equations. Such work has usually entailed postulating new relationships, often with questionable underlying physics. We will return to this issue momentarily, but first we need to make the following observation regarding the time-averaging process. *Time averaging automatically results in a loss of information about the flow.* The averaging procedure must be long relative to the characteristic timescales of turbulence, but short relative to any transient or periodic phenomena of interest. In some applications, these requirements will be mutually incompatible.

We now turn our attention back to the closure problem. The simplest approach we could take would be to base our model on something familiar, for example, Newton's law of viscosity. This analogy is known as Boussinesq's eddy–viscosity model:

$$\tau_{ji}^T=-\rho\overline{v_j'v_i'}=-\rho\nu^T\frac{\partial V_i}{\partial x_j}, \tag{5.48}$$

note that ν^T is the "eddy viscosity." This is a gradient transport model; we imply that the turbulent transport of momentum is closely related to the gradient of the mean (time-averaged) velocity. There are two serious problems with this analogy: The eddy viscosity is a property of the *flow* and not of the fluid, and the coupling between the mean flow and the turbulence is generally weak. These deficiencies were recognized immediately, and Prandtl sought an improvement by introduction of mixing length theory, based loosely upon the kinetic theory of gases. We will see that the mixing length approach has had some important successes, but it is to be kept in mind that fluid flow is a continuum process and the interaction of gas molecules is not. The idea that a "particle" of fluid can be displaced a finite distance normal to the mean flow without immediately interacting with its neighbors is incorrect. Taylor addressed this point in 1935 when he wrote of "...the definite but quite erroneous assumption that lumps of air behave like molecules of a gas, preserving their identity till some definite point in their path, when they mix with their surroundings and attain the same velocity and other properties...." In spite of these clear objections, we note that the standard mixing length expression is

$$\tau_{ji}^T=-\rho l^2\left|\frac{dV_i}{dx_j}\right|\frac{dV_i}{dx_j}. \tag{5.49}$$

We will now apply this model to the turbulent flow in a tube; we rewrite the equation for convenience as

$$\tau_{rz}^T=\rho\kappa^2 s^2\left(\frac{dV_z}{ds}\right)^2, \tag{5.50}$$

where s is the distance measured from the wall into the fluid. Before we proceed with this development, we should familiarize ourselves with the experimental observations of time-averaged velocity in a tube. In Figure 5.9, data of Laufer (1954) for the flow of air through a tube at $Re=50{,}000$ and 500,000 are reproduced, along with a laminar (parabolic) profile for comparison.

Note how steep the gradients at the wall are relative to the laminar flow. It is evident that the rate at which momentum is transferred toward the wall has been dramatically increased. If the mixing length model were capable of describing this process, we should be able to determine $V_z(r)$.

The governing time-averaged equation of motion for the turbulent flow in a tube is simply

$$0=-\frac{dP}{dz}-\frac{1}{r}\frac{d}{dr}(r\bar{\tau}_{rz}). \tag{5.51}$$

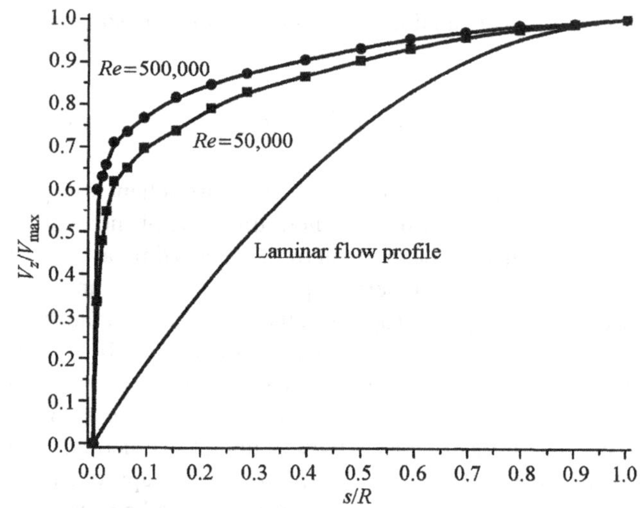

FIGURE 5.9. Typical velocity profiles for turbulent flow through a tube as adapted from Laufer's data. The laminar flow profile is shown to underscore important differences. The velocity profiles for turbulent flow are nicely represented by the *empirical* equation: $(V_z/V_{max}) = (s/R)^{1/n}$; for the data shown above, $n = 8.9$ at $Re = 500{,}000$, and $n = 6.54$ at $Re = 50{,}000$.

We integrate this equation and rewrite it as

$$\frac{P_0 - P_L}{L}\frac{r}{2} = \bar{\tau}_{rz}. \tag{5.52}$$

By force balance,

$$\frac{P_0 - P_L}{L} = \frac{2\tau_0}{R} \quad \text{so} \quad \tau_0\left(\frac{R-s}{R}\right) = \rho\kappa^2 s^2 \left(\frac{dV_z}{ds}\right)^2. \tag{5.53}$$

Note that $s = R - r$ and that the total time-averaged "stress" is being represented solely by Prandtl's mixing length expression. The latter, of course, means that molecular (viscous) friction is being discounted as small relative to turbulent momentum transport. We divide by the fluid density ρ and take the square root of both sides of the equation, noting that the shear (or friction) velocity is defined by $v^* = \sqrt{(\tau_0/\rho)}$. Thus,

$$(dV_z/ds) = v^* \frac{(1 - s/r)^{1/2}}{\kappa s}. \tag{5.54}$$

If we take s to be small relative to R, then we obtain the simple result:

$$V_z = \frac{v^*}{\kappa} \ln s + C_1. \tag{5.55}$$

This, of course, is the famous *logarithmic velocity profile* for the turbulent flow. It is also incorrect. The reader may wish to demonstrate that the "correct" result is

$$V_z = \frac{2v^*}{\kappa}\left[\left(1 - \frac{s}{R}\right)^{1/2} - \tanh^{-1}\left(1 - \frac{s}{R}\right)^{1/2}\right] + C_1. \tag{5.56}$$

Why do you suppose Prandtl would choose the result (5.55)? It is standard practice to define $v^+ = (V_z/v^*)$ and $s^+ = (sv^*\rho/\mu)$, and write the logarithmic equation as

$$v^+ = \frac{1}{\kappa} \ln s^+ + C_1. \tag{5.57}$$

In the turbulent core (away from the wall), it has been found that

$$v^+ \cong 2.5 \ln s^+ + 5.5. \tag{5.58}$$

Accordingly, Prandtl's "universal" constant has a value of approximately $\kappa \approx 0.4$. We will examine some experimental data for turbulent flow in a pipe in Figure 5.10 to see how well the logarithmic equation may work.

It is evident that a single logarithmic equation cannot describe the entire range of Laufer's data. Historically, the distribution of v^+ was broken into three pieces:

$$v^+ = s^+ \quad \text{for } 0 < s^+ < 5 \quad \text{(laminar sublayer)}, \tag{5.59a}$$

$$v^+ = 5 \ln s^+ - 3.047 \quad \text{for } 5 < s^+ < 30 \quad \text{(buffer region)}, \tag{5.59b}$$

$$v^+ = 2.5 \ln s^+ + 5.5 \quad \text{for } s^+ > 30 \quad \text{(turbulent core)}. \tag{5.59c}$$

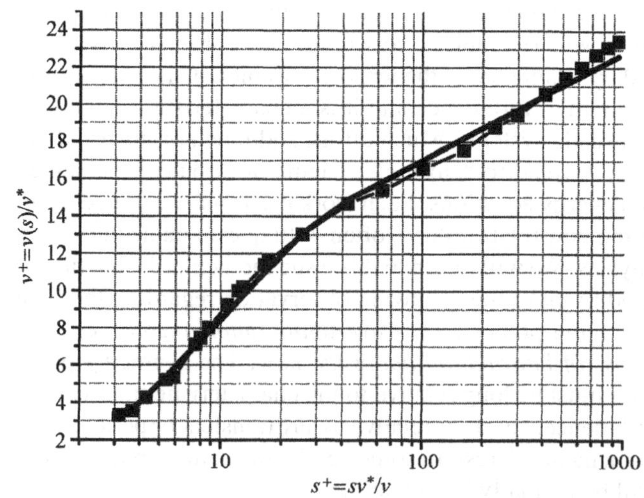

FIGURE 5.10. Laufer's data for the pipe flow at $Re = 50{,}000$, from *NACA Report 1174*. The comparison with the "model" is good enough for much of the range of s^+.

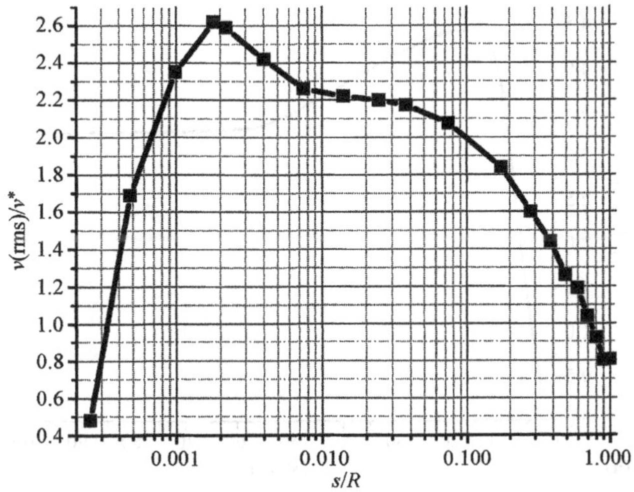

FIGURE 5.11. $\sqrt{\overline{v_z'^2}}$ (rms fluctuations) normalized with the shear or friction velocity v^* at a Reynolds number of 500,000 (Laufer based the Reynolds number upon the maximum or centerline velocity). Note that the largest fluctuations occur at *s/R* of about 0.002, quite close to the wall.

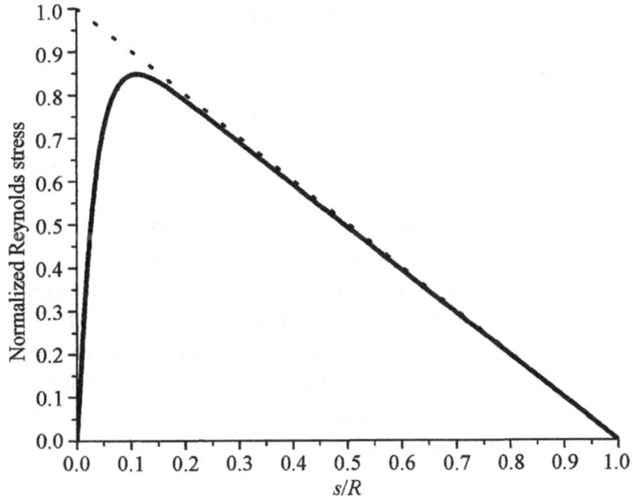

FIGURE 5.12. Variation of the normalized Reynolds stress $\overline{v_z'v_r'}/v^{*2}$ with dimensionless distance from the wall, according to Pai's (1953) relation.

The idea here is that viscous friction dominates momentum transfer very close to the wall, in the intermediate (buffer) region turbulent transport and molecular transport occur at comparable rates, and "far" from the wall the turbulent transport of momentum is dominant. Of course, this is completely synthetic; measurements have shown that turbulent eddies exist *very close* to walls. What we see here is a deeply flawed theory that happens to correlate well with (parts of) the empirical data as demonstrated in Figure 5.10.

We can also use Laufer's data to gain a greater appreciation for how velocity fluctuations behave as one moves from the wall into the interior of a turbulent pipe flow. Figure 5.11 portrays the axial (*z*-direction) rms fluctuations as a function of distance from the wall for a Reynolds number of 500,000.

Figure 5.11 shows that the largest rms value is about 2.6 times greater than v^*. Decreasing the Reynolds number for a turbulent pipe flow does not significantly change this ratio, but it does move the maximum value of *v(rms)/v* away from the wall* toward the interior of the flow. At $Re = 50{,}000$, Laufer found that the maximum is located at about $s/R = 0.015$.

The Reynolds stress for turbulent pipe flow is zero both at the wall and at the centerline; its behavior with *s/R* is very nicely described by the semiempirical relation given by Pai (1953): $\left(\frac{\overline{v_z'v_r'}}{v^{*2}} = 0.9835(1-\frac{s}{R})[1-(1-\frac{s}{R})^{30}]\right)$, which is in excellent accord with Laufer's data. The total stress appears in Figure 5.12 as a dashed line; note that by $s/R \approx 0.15$ or 0.2, the Reynolds stress accounts for nearly all the momentum transfer. The point where they are equal corresponds roughly to $s/R \approx 0.0232$.

5.7 HIGHER ORDER CLOSURE SCHEMES

When the Reynolds momentum equation is "solved" through the use of an eddy viscosity or mixing length model, we refer to the process as first-order modeling. This means that terms that were second order in the fluctuations (the Reynolds stresses) are determined through the first-order quantities like mean (time-averaged) velocity or gradients of mean velocity. Closure schemes have been classified by Mellor and Herring (1973) as either mean velocity field (MVF) or mean turbulent field (MTF). The former provides the time-averaged velocity and the Reynolds stresses, while the latter produces at least some of the characteristics of the turbulence. A well-known example of the latter (MTF) is the second-order modeling where the Navier–Stokes equation is multiplied by the instantaneous velocity; the result is time averaged, and the energy equation for the mean flow is subtracted, yielding the turbulent energy (k) equation:

$$V_j \frac{\partial}{\partial x_j}\left(\frac{1}{2}\overline{v_i v_i}\right) = -\frac{\partial}{\partial x_j}\left(\frac{1}{\rho}\overline{v_j p} + \frac{1}{2}\overline{v_i v_i v_j} - 2\nu\overline{v_i s_{ij}}\right) - \overline{v_i v_j}S_{ij} - 2\nu\overline{s_{ij}s_{ij}}. \qquad (5.60)$$

The turbulent kinetic energy is $k = (1/2)\sum\overline{v_i v_i}$ and the dissipation rate for homogeneous turbulence is defined as $\varepsilon = 2\nu\overline{s_{ij}s_{ij}}$, where the fluctuating strain rate is $s_{ij} = 1/2((\partial v_i/\partial x_j) + (\partial v_j/\partial x_i))$. The interaction between the *mean flow* strain rate and the turbulence produces turbulent energy (by vortex stretching); hence, $-\overline{v_i v_j}S_{ij} = P$.

Therefore, we may rewrite (5.60) as

$$V_j \frac{\partial k}{\partial x_j} = -\frac{\partial}{\partial x_j}\left(\frac{1}{\rho}\overline{v_j p} + \frac{1}{2}\overline{v_i v_i v_j} - 2\nu\overline{v_i s_{ij}}\right) + P - \varepsilon. \tag{5.61}$$

For a steady homogeneous flow in which all averaged quantities are independent of the position, we have the simple result: $P = \varepsilon$. For a more general flow situation, the terms appearing on the right-hand side of (5.61) must be "modeled" using some combination of theory and empiricism. Consider the application of (5.61) to the flow of an incompressible fluid in a turbulent (2D) boundary layer. We can achieve a little further economy by noting $\tau = -\rho\overline{v_i v_j}$, such that

$$V_x \frac{\partial k}{\partial x} + V_y \frac{\partial k}{\partial y} = \frac{\tau}{\rho}\frac{\partial V_x}{\partial y} - \frac{1}{\rho}\frac{\partial}{\partial y}\left(\overline{pv_y} + \rho\overline{kv_y}\right) - \varepsilon. \tag{5.62}$$

Turbulent KE models require some kind of postulated relationship between k and τ; two approaches appearing frequently in the literature have been attributed to Dryden (D) and Prandtl (P):

$$\text{(D)} \quad \tau = a_1 \rho k \quad \text{and} \quad \text{(P)}$$

$$\tau = \rho \nu^T \frac{\partial V_x}{\partial y} \quad \text{with} \quad \nu^T = C_\mu k^{1/2} l_k. \tag{5.63}$$

Of course, one must also have approximations for the sum $\left(\overline{pv_y} + \rho\overline{kv_y}\right)$ and the dissipation rate ε. Bradshaw and Ferriss (1972) used Dryden's relationship from (5.63), along with the empirical functions L and G:

$$L = \frac{(\tau/\rho)^{3/2}}{\varepsilon}, \qquad G = \frac{\overline{p'v'}/\rho + \overline{kv'}}{(\tau/\rho)(\tau_{max}/\rho)^{1/2}}. \tag{5.64}$$

They found that $a_1 = 0.15$, that the function L attained a maximum value at about $\delta/2$ and thereafter decreased to zero, and that G increased monotonically across the boundary layer (though at a reduced rate for $y > \delta$). One of the main concerns here is the pressure fluctuation term because the quantity $\overline{p'v'}$ is extremely difficult to measure. Harsha (1977) notes that it is thought to be small based upon available measurements of the other terms in (5.62). If one is to employ eq. (5.62) successfully, some knowledge of the behavior of the modeled quantities near the wall is necessary. In Figure 5.13, near-wall data compiled by Patel et al. (1985) for k^+, τ^+, and ε^+ are presented.

An inspection of these data shows that the Dryden relation $\tau = a_1 \rho k$, with $a_1 = 0.15$, is a very rough estimate indeed. It is also important that we note that the dissipation rate is difficult to measure accurately; you can gain greater appreciation for this problem by carefully reading the report by Laufer (1954).

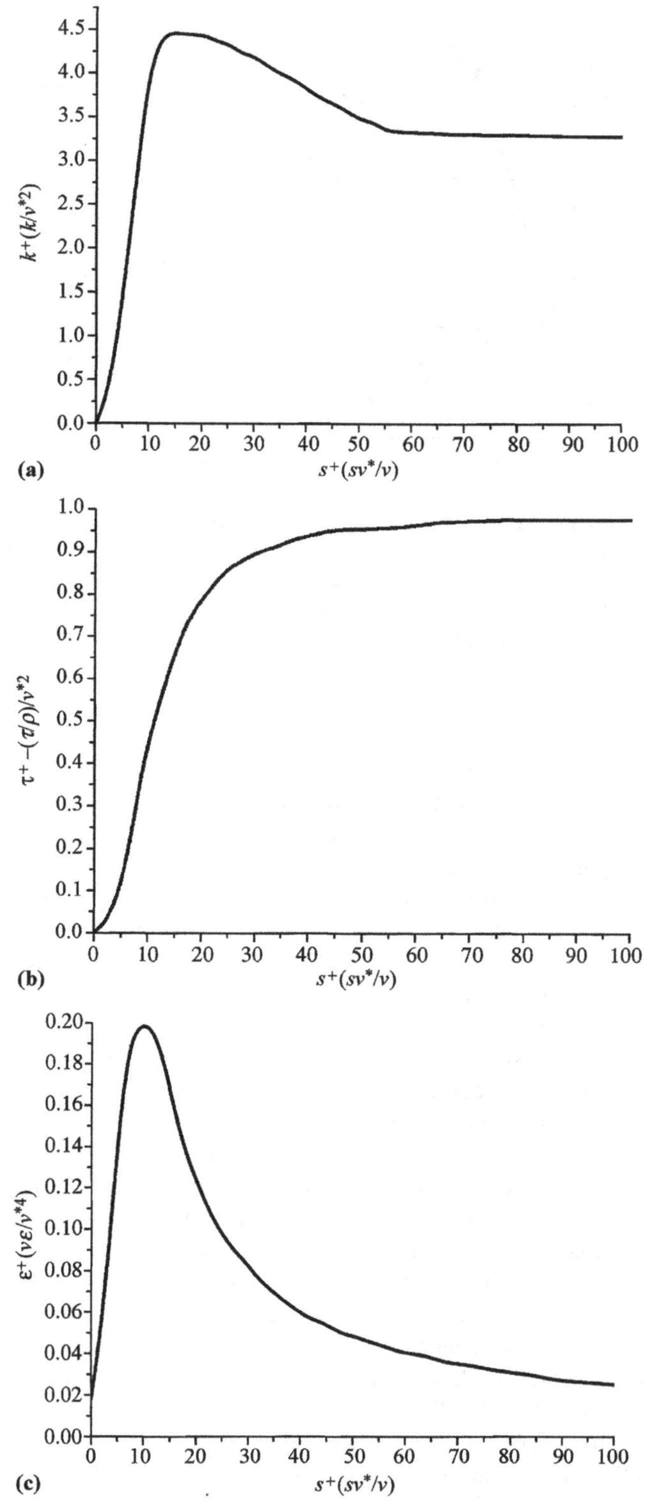

FIGURE 5.13. Near-wall values for the dimensionless turbulent kinetic energy k^+, the dimensionless Reynolds stress τ^+, and the dimensionless dissipation rate ε^+. These data were adapted from Patel et al. (1985) and have come from a variety of sources in the literature. The reader is cautioned that the scatter in the available data is large, often on the order of $\pm 25\%$ or more. The curves given here correspond *approximately* to the centroid of experimental data for each case.

Although the turbulent energy model (consisting of the momentum and continuity equations as well as eq. (5.62)) described above might seem to include a number of choices both empirical and arbitrary, its performance was evaluated critically at the Stanford Conference of 1968. Models were rated by a committee and the Bradshaw–Ferriss approach was scored "good" (the top category). The turbulent kinetic energy approach to the problem of closure has been intensively studied and used for simple shear layers; in the middle of 2007, a Google® search of "solutions of the turbulent energy equation" revealed about 10^6 hits. There is an important limitation however: In the complex turbulent flows, the length scale (l) distribution cannot be reliably specified. This is particularly problematic for turbulent flows in enclosures where large regions of recirculation may be set up. Flows with large coherent structures require a model that can reflect changes in l (which are dictated by initial size, dissipation, and vortex stretching). One possibility is to form a new dependent variable by *combining* k and l. Since $\varepsilon \approx Au^3/l$ (Taylor's inviscid relation) and $k \approx u^2$, the dissipation rate suits the requirements: $\varepsilon \approx k^{3/2}/l$. By the late 1970s, it was apparent that a greater computational adaptability could be achieved in terms of the broadest possible variety of turbulent flows, when the k-equation was coupled with a dissipation (ε) equation (hence the term, k–ε modeling). In the usual form seen in the literature, the two equations for the k–ε model are written as

$$\frac{\partial}{\partial x_j}(\rho V_j k) = -\rho\overline{v_i' v_j'}\frac{\partial V_i}{\partial x_j} - \rho\varepsilon + \frac{\partial}{\partial x_j}\left(\mu\frac{\partial k}{\partial x_j} - \frac{1}{2}\rho\overline{v_i' v_i' v_j'} - \overline{p' v_j'}\right) \tag{5.65}$$

and

$$\begin{aligned}\frac{\partial}{\partial x_j}(\rho V_j \varepsilon) = {}& \frac{\partial}{\partial x_j}\left[\mu\frac{\partial \varepsilon}{\partial x_j} - \rho\overline{v_j'\varepsilon'} - 2\nu\overline{\frac{\partial p'}{\partial x_i}\frac{\partial v_j'}{\partial x_i}}\right] \\ & -2\mu\left[\overline{\frac{\partial v_i'}{\partial x_i}\frac{\partial v_j'}{\partial x_i}} + \overline{\frac{\partial v_i'}{\partial x_i}\frac{\partial v_i'}{\partial x_j}}\right]\frac{\partial V_i}{\partial x_j} \\ & -2\mu\overline{\frac{\partial v_i'}{\partial x_i}\frac{\partial v_j'}{\partial x_i}\frac{\partial v_i'}{\partial x_j}} - 2\mu\overline{v_j'\frac{\partial v_i'}{\partial x_i}}\frac{\partial^2 V_i}{\partial x_j \partial x_i} \\ & -2\rho\overline{\left(\nu\frac{\partial^2 v_i'}{\partial x_j \partial x_i}\right)^2}.\end{aligned} \tag{5.66}$$

We can now better appreciate the circular nature of this enterprise; it is much like the small dog chasing his own tail. All the terms involving fluctuating pressure (p') and velocity (v') must be approximated with expressions containing k, ε, and mean field (time-averaged) values for velocity. For example, it is common practice to let

$$\mu\frac{\partial k}{\partial x_j} - \frac{1}{2}\rho\overline{v_i' v_i' v_j'} - \overline{p' v_j'} = \frac{\mu^T}{\sigma_k}\frac{\partial k}{\partial x_j}. \tag{5.67}$$

In its usual form, the k–ε model has five empirical constants. The eddy viscosity is usually approximated as

$$\nu^T = C_\mu\frac{k^2}{\varepsilon}, \tag{5.68}$$

where $C_\mu = 0.09$ for flows in which the production and dissipation of turbulent energy are in rough balance.

As we have come to expect, the convective transport terms in k–ε modeling pose a problem, especially in cases involving recirculating flows. Davidson and Fontaine (1989) have shown that the computed results for turbulence in a ventilated room are significantly affected by the type of difference scheme implemented. They examined HD (hybrid upwind/central difference), SUD (skewed-upwind difference), and QUICK (quadratic upstream interpolation for convective kinematics). Although the QUICK scheme is generally regarded to be more accurate, Davidson and Fontaine found that it did not work well with a coarse grid. The reader concerned with this aspect of k–ε modeling should definitely consult Leonard (1979) and Raithby (1976).

Jones and Launder (1973) extended the k–ε approach to turbulent pipe flows of (relatively) low Reynolds numbers. The equations they employed were

$$\rho\left(V_x\frac{\partial k}{\partial x} + V_y\frac{\partial k}{\partial y}\right) = \frac{\partial}{\partial y}\left[\left(\mu + \frac{\mu^T}{\sigma_k}\right)\frac{\partial k}{\partial y}\right] + \mu^T\left(\frac{\partial V_x}{\partial y}\right)^2 - \rho\varepsilon - 2\mu\left(\frac{\partial k^{1/2}}{\partial y}\right)^2 \tag{5.69}$$

for turbulent energy and

$$\begin{aligned}\rho\left(V_x\frac{\partial \varepsilon}{\partial x} + V_y\frac{\partial \varepsilon}{\partial y}\right) = {}& \frac{\partial}{\partial y}\left[\left(\mu + \frac{\mu^T}{\sigma_\varepsilon}\right)\frac{\partial \varepsilon}{\partial y}\right] \\ & + C_1\frac{\varepsilon}{k}\mu^T\left(\frac{\partial V_x}{\partial y}\right)^2 - \frac{C_2\rho\varepsilon^2}{k} \\ & + 2\frac{\mu\mu^T}{\rho}\left(\frac{\partial^2 V_x}{\partial y^2}\right)^2\end{aligned} \tag{5.70}$$

for dissipation. Note that the last term on the right-hand side of the energy equation has been added for computational reasons. Jones and Launder observed that it is convenient to let $\varepsilon = 0$ at the pipe wall; however, it is clear that the normal (y-direction) derivative of the tangential velocity fluctuations, when squared and time averaged, would not be zero. Therefore, the term in question was added to account for dissipation close to the wall. In pipe flow, of course, the normal gradients of both k and ε are set to zero at the centerline. As we noted previously, we have a model with five "constants:"

$$C_\mu \quad C_1 \quad C_2 \quad \sigma_k \quad \sigma_\varepsilon.$$

Jones and Launder found that at the low turbulence Reynolds numbers, defined as $Re^T = (\rho k^2/\mu\varepsilon)$, both C_μ and C_2 vary with Re^T. In fact, it appears that nearly every worker in this area of fluid mechanics has his/her own opinion about how low Re and near-wall turbulence problems should be handled. The work of Patel et al. (1985) is illuminating in this regard; their comparisons of computed results and data for relatively simple cases show that k–ε modeling is too often only semiquantitative.

We conclude this section on k–ε modeling by considering recently reported work in which turbulence inside a rectangular tank (100 cm long and 25 cm wide) was modeled. Schwarze et al. (2008) studied the practically important case in which water was fed into the tank at one end through a round jet. Water exited the enclosure through a round tube at the other end (and through the opposite wall). This is a formidable problem because the jet issuing into the tank impinges upon the opposite wall and generates large regions of recirculation. It is also a type of problem that is of immense significance to the process industries (consider the number of vessels, tanks, reactors, and basins that have continuous feed streams). The coherent structures formed in such situations can further complicate modeling efforts by exhibiting oscillatory (or periodic) behavior. These investigators used laser Doppler velocimetry to obtain experimental data for comparison and they used the SIMPLE algorithm with Fluent 6™ for their computations. They were able to compare both mean velocity and rms fluctuations along the planar cuts extracted from the tank. Although the computed mean velocities were in general agreement with the experimental data, the k–ε model did not produce results for the turbulence variables that were quantitatively reliable. They obtained somewhat better results by replacing the k-equation with transport equations for the Reynolds stress. The clear lesson here is that strongly anisotropic flows with coherent structures remain particularly challenging for k–ε modeling efforts.

5.7.1 Variations

There are other two-equation models for turbulent flows that have been used successfully. One of the more frequently cited is the k–ω model originally proposed by Kolmogorov, where $\omega = \varepsilon/k$ is the specific dissipation rate. Wilcox (1998) has been a developer and an advocate for this model and he notes that it offers greater promise for complex flows that include both free- and wall-bounded regions. The specific dissipation rate equation in the Wilcox model is

$$\frac{\partial \omega}{\partial t} + V_j \frac{\partial \omega}{\partial x_j} = \alpha \frac{\omega}{k} \tau_{ij} \frac{\partial V_i}{\partial x_j} - \beta \omega^2 + \frac{\partial}{\partial x_j}\left[(\upsilon + \sigma \nu^T)\frac{\partial \omega}{\partial x_j}\right]. \tag{5.71}$$

Wilcox's book (1998) contains both values for the empirical constants and the necessary closure relationships. More important, the book includes software that will allow a novice to compare the performance of different two-equation models of turbulence for pipe and channel flows, as well as for free shear flows.

5.8 INTRODUCTION TO THE STATISTICAL THEORY OF TURBULENCE

Our intent in this section is to provide a brief introduction to the statistical theory of turbulence; for a comprehensive treatment, readers will have to turn to Hinze (1975) and Monin and Yaglom (1975). Be forewarned: These books are extensive in coverage and difficult reading for newcomers to the subject. Nevertheless, they provide the *definitive* accounts of the development and status of statistical fluid mechanics.

When we ponder the observed fluctuations in turbulence, it is natural to think about statistical measures associated with random variables, like the mean, moments about the mean, and correlation coefficients. However, as we noted previously, turbulence is not *precisely* a random process; it is a nonlinear stochastic system. Bradshaw (1975) observes that most naturally occurring random processes are Gaussian (i.e., follow a normal distribution) but turbulence is not (and does not). In fact, he points out that the deviations from Gaussian behavior are often what make turbulence so interesting (and infuriating sometimes, too).

It is to be noted that at this point we will change our notation for velocity. Though we would prefer to let components of the velocity vector continue to be represented by V_i, this practice is both inconvenient and out of step with nearly all the literature of the statistical theory of turbulence. Consequently, for the balance of this chapter we will use U and u for mean and fluctuating (turbulent) velocities, respectively. This is common practice, and it precludes the possibility of confusion with the kinematic viscosity ν.

We should begin by thinking about what determines how eddy scales are distributed. We envision a process in which turbulent energy is transferred from large eddies to smaller eddies, to yet smaller eddies, and so on, by vortex stretching. At very small scales, this kinetic energy is dissipated by the action of molecular viscosity (the process is cutely summarized by L. F. Richardson's adaptation of Swift's poem: *Big whorls have little whorls, which feed on their velocity; And little whorls have lesser whorls, and so on to viscosity*). In 1941,

A. N. Kolmogorov used dimensional reasoning to estimate the characteristic eddy size for this dissipative structure using the kinematic viscosity (ν) and the dissipation rate per unit mass (ε), we now call this length the Kolmogorov microscale:

$$\eta = \left(\frac{\nu^3}{\varepsilon}\right)^{1/4}. \tag{5.72}$$

Clearly, characteristic time and the velocity scales can also be formed for these small-scale motions:

$$\tau_K = \left(\frac{\nu}{\varepsilon}\right)^{1/2} \quad \text{and} \quad u_K = (\nu\varepsilon)^{1/4}. \tag{5.73}$$

Therefore, for the turbulent flow of water with a dissipation rate of $1000\,\text{cm}^2/\text{s}^3$, we can estimate the three scales: 0.0056 cm, 0.0032 s, and 1.78 cm/s. Note also that if we use the Kolmogorov scales for length and velocity to form a Reynolds number, we get $Re = 1$; these small-scale motions are quite viscous in character. What about eddies at the other end of the scale? For a confined turbulent flow (in a duct, for example), this is pretty easy. For the pipe flow, the largest eddies have a size corresponding *roughly* to the radius R and a velocity comparable to U. The characteristic time is easily formulated: R/U; for the flow of water through a 6 in. pipe at $Re = 200{,}000$, $\langle U\rangle \approx 4.3$ ft/s and this quotient is about 0.058 s.

Recall that we previously defined the autocorrelation coefficient; we restate it here as

$$\rho(\tau) = \frac{\overline{u(t)u(t+\tau)}}{\overline{u^2}}. \tag{5.74}$$

Of course, for $\tau = 0$, $\rho(\tau) = 1$, and as $\tau \to \infty$, $\rho(\tau) \to 0$. We can define an integral timescale (which is a measure of the length of time we see connectedness in the signal behavior):

$$T_I = \int_0^\infty \rho(\tau)d\tau. \tag{5.75}$$

We can offer a crude interpretation for T_I: Suppose that a very large eddy (with a characteristic size of 10 cm) is being carried past the measurement point at the velocity of the mean flow, say 30 cm/s. The duration of the signal dynamic created by this large eddy will be about 1/3 s. Compare this with the Kolmogorov timescale computed in the example above—T_I is about 100 times larger than τ_K. We can also determine a time microscale (quite distinct from the Kolmogorov microscale τ_K) by fitting a parabola of *osculation* to $\rho(\tau)$. Assuming

$$\rho(\tau) = 1 - a\tau^2, \tag{5.76}$$

we note that this curve crosses the τ-axis at some $\tau = \lambda_T$, that is, $\rho(\tau = \lambda_T) = 0$. Consequently, $a = 1/\lambda_T^2$. Now we match the curvature of the autocorrelation coefficient at the origin:

$$\frac{d^2\rho(\tau)}{d\tau^2} = -2/\lambda_T^2. \quad \text{for} \quad \tau = 0. \tag{5.77}$$

The significance of this new timescale will be apparent soon.

As we saw earlier, the dissipation rate is defined as $\varepsilon = 2\nu\overline{s_{ij}s_{ij}}$. In 1935, Taylor noted that for *isotropic* turbulence, the product of the strain rates could be approximated by

$$\varepsilon = 2\nu\overline{s_{ij}s_{ij}} = 15\nu\overline{\left(\frac{\partial u_1}{\partial x_1}\right)^2} = 15\nu\frac{\overline{u^2}}{\lambda^2}, \tag{5.78}$$

where the length scale λ is now referred to as the Taylor microscale. Taylor also suggested that the dissipation rate could be estimated using the large-scale (inviscid) dynamics (the energy dissipated at the bottom of the cascade must come from vortex stretching at large scales); let u^2 be the kinetic energy of the large-scale motions and u/l represent the mean flow strain rate, then

$$\varepsilon \approx A\frac{u^3}{l}. \tag{5.79}$$

The integral length scale l appearing here is the size of the largest eddies and sometimes it can be estimated from the controlling dimension of the flow, a duct width, for example. Taylor referred to l as "some linear dimension defining the scale of the system." Studies of grid-generated turbulence in wind tunnels have shown that the constant A is on the order of 1. The two descriptions for dissipation rate can be equated:

$$A\frac{u^3}{l} = 15\nu\frac{u^2}{\lambda^2}. \tag{5.80}$$

Consequently, $(\lambda^2/l^2) = (15/A)(v/ul)$ and $(\lambda/l) = \sqrt{(15/A)}Re_l^{-1/2}$. Suppose we now assume that $Re_l = 10^5$ and $l = 20$ cm; then $\lambda/l \approx 0.0122$ and the Taylor microscale λ would be on the order of 0.25 cm. We can carry this one step further; since the Reynolds number and the integral length scale have been specified, we need only the kinematic viscosity to find the characteristic velocity u. Taking $\nu = 0.01\,\text{cm}^2/\text{s}$ and $u = 50$ cm/s, the dissipation rate and the Kolmogorov microscale can now both be estimated: $6250\,\text{cm}^2/\text{s}^3$ and 0.0036 cm, respectively. We are now in a position to examine the ratios of the length scales that will help us understand where the Taylor microscale fits into the range of eddy sizes:

$$\frac{l}{\lambda} \approx 80 \quad \text{and} \quad \frac{\lambda}{\eta} \approx 69. \tag{5.81}$$

It is now clear that while the Taylor microscale may be small, it is far larger than the Kolmogorov microscale.

For the discussion that follows, we will generally take the turbulence to be homogeneous and isotropic (the latter means that $\overline{u_1^2} = \overline{u_2^2} = \overline{u_3^2}$). Obviously, turbulent pipe flow is neither. But we can produce a decent approximation to these conditions with grid-generated turbulence in a wind tunnel. Indeed, in the 1930s, much progress was made in turbulence as a result of the development of improved wind tunnels and hot wire anemometry.

We need to recall our definition of the autocorrelation coefficient, which has the form shown in eq. (5.74). We now introduce the Fourier transform pair, consisting of the autocorrelation coefficient $\rho(\tau)$ and the power spectrum $S(f)$:

$$\rho(\tau) = \int_{-\infty}^{+\infty} \exp(i\tau f) S(f) df \quad \text{and}$$

$$S(f) = \frac{1}{2\pi} \int_{-\infty}^{+\infty} \exp(-i\tau f) \rho(\tau) d\tau. \tag{5.82}$$

Since negative frequencies hold no physical meaning for us and the autocorrelation coefficient is an *even function*, we usually rewrite (5.82) as the "one-sided" spectrum:

$$S_1(f) = \frac{1}{\pi} \int_{0}^{\infty} \cos(\tau f) \rho(\tau) d\tau. \tag{5.83}$$

The spectrum, or spectral density, tells us how the signal energy is distributed with respect to frequency. We obtain the spectrum from time-series data, for example, from measurements of velocity at a point in space with an instrument like a hot wire anemometer. We saw an example of this in Figure 5.8. The spectrum accompanying those data was obtained by the Fourier transformation (actually FFT) and it is shown in Figure 5.14.

We can gain a clearer picture of the relationship between the autocorrelation and the power spectrum by looking at some of the common Fourier transform pairs. In particular, we might propose some very simple functional forms for $\rho(\tau)$; what will the corresponding $S(f)$ look like?

$\rho(\tau)$	$S(f)$
1 (for $0 < \tau < a$)	$\frac{\sqrt{2}}{\pi} \frac{\sin(af)}{f}$
$\cos(f_0 \tau)$	$\frac{1}{2}[\delta(f - f_0) + \delta(f + f_0)]$
$\exp(-a\tau)$	$\frac{a}{f^2 + a^2}$
$\text{sech}(a\tau)$	$\frac{\pi}{2a} \text{sech} \frac{\pi f}{2a}$

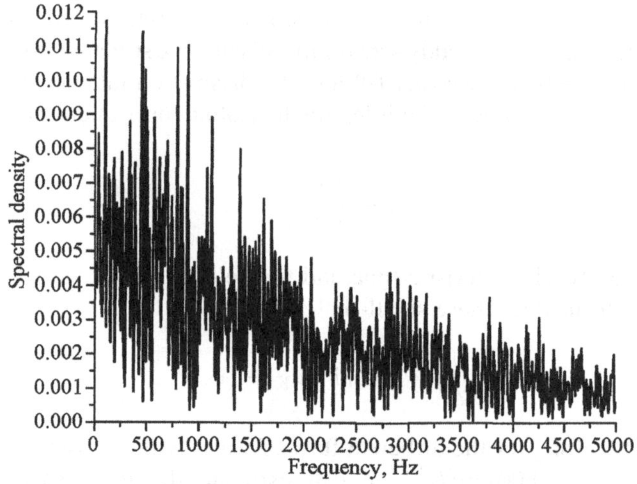

FIGURE 5.14. Power spectrum for time-series data (jet velocity) shown in Figure 5.8. Note that most of the signal energy is located at frequencies less than about 1500 Hz. The energy is broadly distributed up to about 900 Hz, and there are important contributions at about 1200–1700 Hz.

Note from this table that a *uniform* correlation coefficient produces an oscillating spectrum. Conversely, an oscillating (or *ringing*) correlation coefficient will produce a very sharp spike (a delta function) in the spectrum. Clearly, if turbulent energy was distributed among a few sharply discrete frequencies, the autocorrelation would oscillate with a limited of number of periodicities. This is not what we expect to see (generally) when we make measurements in turbulent flows. Usually the signal energy is distributed broadly over a wide range of frequencies; of course, there are exceptions. If we were to make measurements in the wake of a bluffbody, or in the impeller stream of a stirred tank, or in the discharge of an electric fan, we might obtain spectra with a small number of very dominant frequencies. Consider the Eulerian measurements made in the impeller stream of a stirred tank reactor: Every time a blade passes the measurement point, a spike in velocity ensues (investigators studying this problem have termed this *pseudo-turbulence*). This can completely obscure characteristics of the turbulence that are of interest, so it may be necessary to subtract the blade passage periodicity from the signal prior to further processing.

Let us now illustrate the outcome for an oscillatory autocorrelation. Suppose we let $\rho(\tau) = \cos((100 + 10n)\tau)/(1 + \tau^2)$, where n is a uniform random number between 0 and 1. We note that $\rho(\tau = 0)$ will be 1; furthermore, as τ becomes very large, $\rho \to 0$. For this example, we are working with radian frequency, so it is clear that f will be distributed between 100 and 110 rad/s. We can construct a very simple algorithm to determine the spectrum by integration. The main spectral feature will be a broad spike concentrated around 105 rad/s and this result is illustrated in Figure 5.15.

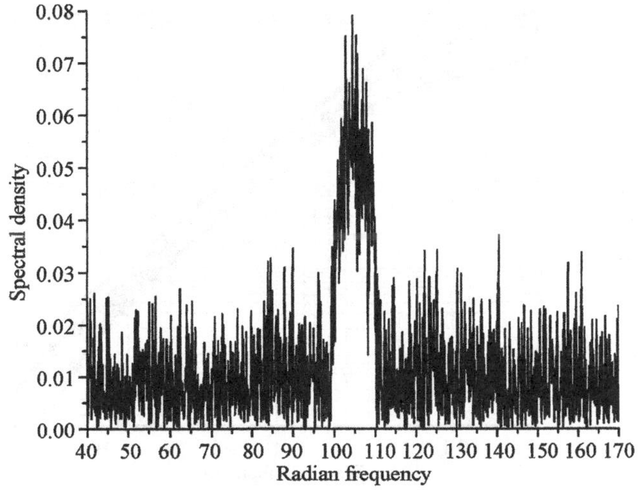

FIGURE 5.15. Frequency spectrum computed from a "fuzzy" autocorrelation coefficient with diminishing amplitude oscillations centered around 105 rad/s.

Although frequency spectra obtained from the time-series data are useful and pretty easy to obtain, wave number spectra computed from measurements with *spatial separation* can contribute more significantly to our understanding of energy transfer and the interactions of eddies of different scales. So, it is appropriate for us to consider the relationship between conventional time-series data and measurements made with spatial separation.

We noted previously that the grid-generated turbulence in wind tunnels has been very intensively studied. The eddies are created by fluid passage over a square array of rods with an on-center mesh spacing of M. The turbulence generated in this fashion is a decaying field of low intensity ($\sqrt{\overline{u^2}}/U$ is typically a few percent) and it is very nearly isotropic. There is an extensive accumulation of experimental data for such flows, with both spatial and temporal measurements available. These data allow us to critically evaluate Taylor's (frozen turbulence) hypothesis: Taylor (1938) suggested that $U\tau$ and x were directly equivalent for homogeneous isotropic turbulent flows with a constant mean velocity in the x-direction, that is, $(\partial/\partial t) \approx U(\partial/\partial x)$. This is important, because it implies that equivalent information could be obtained from either temporal or spatial measurements. The hypothesis has been tested many times; it is *approximately valid* for the low-intensity, grid-generated turbulence as demonstrated by Favre et al. (1955).

Let us begin our consideration of measurements with spatial separation by introducing the definition of the second-order correlation tensor R using notation similar to Tennekes and Lumley (1972):

$$R_{ij}(r) = \overline{u_i(x)u_j(x+r)}. \qquad (5.84)$$

In this equation, x represents a generic spatial position and r is the separation between measurement points. It will be convenient to let the principal directions x, y, and z be represented by 1, 2, and 3, respectively. Therefore, if we wanted to study the behavior of the x-component motions with spatial separation in the y-direction, we would write:

$$R_{11}(x, y+r, z) = \overline{u_1(x, y, z)u_1(x, y+r, z)}, \qquad (5.85)$$

which we will represent in the following discussion as $R_{11}(0,r,0)$. We can Fourier-transform the components of the correlation tensor just as we did in the case of the autocorrelation obtained from the time-series data. Recall that for the latter, we went from measurements with time separation τ to frequency f. Now for the correlation tensor, the transform will take us from the spatial separation r to the wave number κ; for example,

$$F_{22}(\kappa_1) = \frac{1}{2\pi}\int_{-\infty}^{+\infty} \overline{u_2(x)u_2(x+r)}\exp(-i\kappa_1 r_1)dr_1. \qquad (5.86)$$

Clearly, this density function, the one-dimensional transverse spectrum, is related to the turbulent kinetic energy in the "2" (or y) direction. We must keep in mind that although the one-dimensional measurements are relatively easy to make, they are subject to aliasing; larger eddies not aligned with the axis of the measurement will contribute to the measured signal. Consequently, most one-dimensional spectra will exhibit nonzero values as $\kappa \to 0$. This is energy that has been aliased from larger eddies (with lower wave numbers) from directions oblique to the measurement axis. See Tennekes and Lumley (1972) for elaboration.

It is advantageous to remove the directional information from one-dimensional spectra. A three-dimensional wave number spectrum ϕ_{ij} can be determined by analogy with (5.86); this requires integration with respect to r_1, r_2, and r_3. Then, the three-dimensional wave number magnitude spectrum $E(\kappa)$ is found from the integral of the diagonal components of ϕ_{ii} (1,1), (2,2), and (3,3) over a spherical surface:

$$E(\kappa) = \frac{1}{2}\oiint \phi_{ii}(\kappa)dA. \qquad (5.87)$$

$E(\kappa)$, the three-dimensional wave number spectrum, is the density function for turbulent energy (without directional information). Consequently, for isotropic turbulence,

$$\int_0^{\infty} E(\kappa)d\kappa = \frac{1}{2}\overline{u_i u_i} = \frac{1}{2}\overline{u_1 u_1} + \frac{1}{2}\overline{u_2 u_2} + \frac{1}{2}\overline{u_3 u_3} = \frac{3}{2}u^2. \qquad (5.88)$$

Bradshaw (1971) pointed out that it is impractical to try to determine $E(\kappa)$ directly, since that would require an array of measurement locations and devices operating simultaneously. Of course, in recent years, particle image velocimetry (PIV) has been used to obtain two- and three-dimensional data and one can expect as PIV resolution improves that more results from such measurements will become available.

Much work has been carried out over the past 70 years to deduce, infer, or derive the functional form of $E(\kappa)$. Naturally, due to the inverse relationship between κ and eddy size, small wave numbers correspond to large eddies and large wave numbers correspond to the small-scale (or dissipative) structure. There are several wave numbers of particular significance. The location of the maximum in the distribution is κ_e, which is roughly centered among the large energy-containing eddies. We define the threshold marking the beginning (actually upper end) of the dissipative structure by the reciprocal of the Kolmogorov microscale:

$$\kappa_d = \frac{1}{\eta}. \tag{5.89}$$

A qualitative portrait of the entire spectrum follows in Figure 5.16; please make note of the scaling that has been used in this illustration. Normally, we would not see spectra presented like this because both values on both axes, $E(\kappa)$ and κ, can vary over several orders of magnitude.

For *isotropic turbulence*, the relationship between $E(\kappa)$ and the easily measured one-dimensional longitudinal spectrum is simple:

$$E(\kappa) = \kappa^3 \frac{d}{d\kappa}\left(\frac{1}{\kappa}\frac{dF_{11}}{d\kappa}\right). \tag{5.90}$$

This is particularly significant because it means that if $E(\kappa) \approx \kappa^{-5/3}$, then

$$F_{11} \propto \frac{9}{55}\kappa^{-5/3}. \tag{5.91}$$

Consequently, we can use the experimentally measured spectra to confirm the existence of the inertial subrange. In Figure 5.17, two spectra measured by Laufer at $Re = 500{,}000$ are given. Note that there is a region of wave numbers for which the slope (on the log–log plot) is about -1.66.

As we observed previously, energy is transferred from large eddies to smaller ones by vortex stretching. The dynamic spectrum equation is

$$\frac{\partial}{\partial t}E(\kappa, t) = F(\kappa, t) - 2\nu\kappa^2 E(\kappa, t), \tag{5.92}$$

where $F(\kappa,t)$ is the spectral energy transfer function; refer to Chapter 3 in Hinze (1975) for the development of (5.92). *If* the functional form of F were known, then E could be obtained directly. As you might imagine, this approach has piqued the interest of *many* researchers; in the beginning, dimensional reasoning (which has proven so powerful in turbulence) was employed and Kovasznay (1948) was among the first to try this. Obviously, the transfer function must have the same

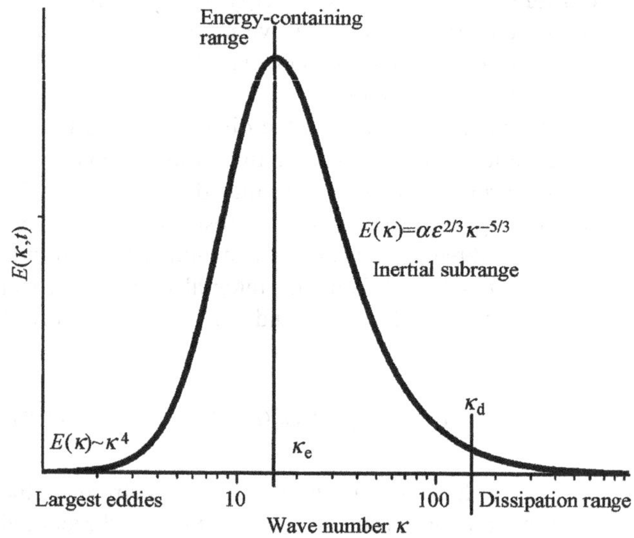

FIGURE 5.16. Three-dimensional wave number spectrum of turbulent energy $E(\kappa)$. Kolmogorov found that for the inertial subrange, $E(\kappa) = \alpha\varepsilon^{2/3}\kappa^{-5/3}$. It is to be noted that under transient circumstances (decaying turbulence, for example), the wave number spectrum is a function of time $E(\kappa,t)$. Indeed, under decaying conditions, the location of κ_e remains about the same, but the peak height decreases and the dissipative range (right-hand tail) moves to the left, toward lower wave numbers.

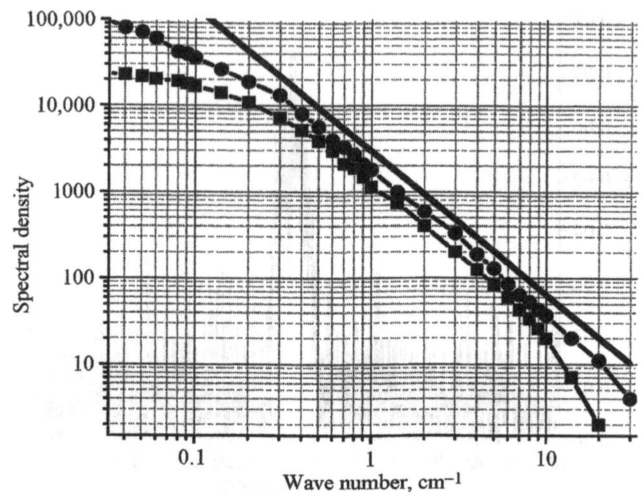

FIGURE 5.17. One-dimensional on-axis spectra measured in pipe flow at a Reynolds number of 500,000 as adapted from Laufer (1954). The squares are from measurements on the centerline and the filled circles correspond to $(1 - r/R) = 0.28$. An additional line with a slope of $-5/3$ has been added for comparison. You can see that an inertial subrange is present in the spectra and it is about 1 to $1\frac{1}{2}$ decades wide.

dimensions as the dissipation term,

$$2\nu\kappa^2 E(\kappa, t) \Rightarrow \left(\frac{\text{cm}^2}{\text{s}}\right)\left(\frac{1}{\text{cm}}\right)^2\left(\frac{\text{cm}^3}{\text{s}^2}\right) \Rightarrow \left(\frac{\text{cm}^3}{\text{s}^3}\right). \tag{5.93}$$

Therefore, if we suggest that $\int F d\kappa$ depends only upon E and κ, then

$$F(\kappa, t) \propto [E(\kappa)]^{3/2}\kappa^{5/2}. \tag{5.94}$$

It is to be noted that this result was obtained solely through dimensional reasoning—there is no physical basis.

Several of the world's luminaries in physics, including Heisenberg, proposed theories of spectral energy transfer. These ideas have run the gamut from a diffusion-like process modeled on neutron transport to the Boussinesq idea that turbulent transport can be represented with an eddy viscosity and the mean velocity field. One of the reasons this particular aspect of turbulence theory has attracted so much attention is that a functional form for F leads directly to E through the dynamic spectrum equation, as we noted previously. A hypothesis can be tested easily since one must obtain the Kolmogorov equation ($E \approx \kappa^{-5/3}$) in the inertial subrange. It has become apparent that spectral energy transfer is a much more difficult problem than many of these early efforts suggested, hence the relative lack of success in the development of a comprehensive model. Any student intrigued by this subset of fluid mechanics may want to begin by consulting the work by Kraichnan (1966) on the Lagrangian history of velocity correlations.

An important question in the context of spectral energy transfer concerns where energy passing a given wave number originates. Can large eddies interact directly with small ones? An appealing argument can be made (see Tennekes and Lumley, 1972, p. 260) that most of the energy passing κ comes from eddies that are just one or two "sizes" larger. Semiquantitative form can be given to this point with the following reasoning: We imagine that in wave number space, an eddy contribution is centered at κ, but extends from 0.62κ to 1.62κ. Characteristic velocity and size depend upon wave number such that

$$u(\kappa) \cong [\kappa E(\kappa)]^{1/2} \quad \text{and} \quad l(\kappa) \cong 2\pi/\kappa. \tag{5.95}$$

For an eddy at wave number located in the inertial subrange, the strain rate is estimated with u/l:

$$s(\kappa) \approx \frac{\sqrt{\alpha}}{2\pi}\varepsilon^{1/3}\kappa^{2/3} = B\kappa^{2/3}. \tag{5.96}$$

Now, suppose we look at the next three slightly larger eddies, with contributions centered at 0.38κ, 0.15κ, and 0.057κ; the strain rates for κ, 0.38κ, 0.15κ, and 0.057κ are then proportional to 1, 0.53, 0.28, and 0.148, respectively. This suggests that the influence of larger eddies in the energy cascade is not felt too "far away." That is, we are implying that the large and small eddies do not directly interact. Of course, the fact that the dissipative motions are at least nearly isotropic supports our conclusion that strains imposed by the large-scale motions do not affect eddies at large wave numbers. That said, there is some unsettling evidence to the contrary. Nelkin (1992), for example, observes that there are at least three reasons to question the idealized picture of spectral energy transfer described above:

1. In the isotropic turbulence, the spectrum obtained from the cross-correlation R_{12} (r) should be zero.
2. Anisotropy may not relax as rapidly as $\kappa^{-2/3}$.
3. Some direct numerical simulations have shown that anisotropy remains at the smallest scales even for very large *Re*.

We need to re-emphasize that the reader interested in this discussion must be aware of the contributions to this field by Robert H. Kraichnan, one of the greatest physicists of the twentieth century (Kraichnan passed away in 2008). Kraichnan championed the idea that direct interactions between the large and small eddies might not be negligible (direct interaction theory). In 1961, he published a paper, *Dynamics of Nonlinear Stochastic Systems* in which he addressed the many-body problem in both quantum mechanics and turbulence. The original theory (applied to turbulence) was flawed in that it failed to produce a Kolmogorov relation (−5/3 power law) in the inertial subrange of isotropic turbulence. Subsequently (in the mid-1960s), Kraichnan produced the Lagrangian history, direct interaction theory that resolved this defect. He also discovered that the energy cascade in certain two-dimensional flows could reverse, that is, turbulent energy could be transferred from smaller eddies to larger ones. This inverse cascade has been observed in the laboratory and it is thought to exist in some geophysical flows as well. Kraichnan's papers make for very dense reading but a novice can begin by consulting Hinze (1975) and Monin and Yaglom (1975, Vol. 2). The latter particularly gives nice historical context to the many Russian contributions to this area of fluid mechanics.

5.9 CONCLUSION

About 30 years ago, H. W. Liepmann gave an address at Georgia Tech as the Ferst Award honoree; his remarks were converted into a paper published in *American Scientist* entitled "The Rise and Fall of Ideas in Turbulence" (Liepmann, 1979). Liepmann noted that the questions in turbulence research always seem to outnumber the answers—a closure problem on a grand scale. Even the familiar accepted results can serve up perplexing questions. For example, why should the logarithmic velocity distribution work at all? The physical basis is extremely weak to say the least. And perhaps more important, the difficulties created by the Reynolds decomposition and time-averaging processes are alarming; for one thing, the process results in more variables than equations. One can apply the technique successively, but the resulting hierarchy of equations still cannot be closed. We are "chasing our own tail" but must wonder if we catch it, what have we caught? Liepmann also noted that some averaged quantities (an $x - y$ correlation coefficient, for example) exhibit "burst" behavior, that is, fluctuate chaotically between 0 and 1 but with an "average" value of, say, 0.4. Is averaging meaningful for such a quantity?

Of the higher order closure schemes, $k - \varepsilon$ modeling has matured into an industry all by itself. One can purchase commercial codes developed for turbulence modeling, and even "solve" some problems of practical importance. We must remember, however, that this approach to turbulence will not lead to breakthroughs in the understanding of underlying phenomena. Liepmann observed that much of this enormous computational effort "...will be of passing interest only." He further noted that this kind of modeling is rarely evaluated quantitatively. $k - \varepsilon$ modeling has become a "publication engine" for many fluid dynamicists, and while it may be driven by industrial needs, it is very unlikely that it will ever reveal much about the physics of turbulence.

It certainly *appears* that turbulence is contained within the framework of the Navier–Stokes equations, and this makes direct numerical simulation (DNS) fundamentally attractive. However, enthusiasm for this approach must be tempered for two reasons: (1) Many fluid dynamicists, including O. E. Lanford, have observed that no general existence theorem has been found for the initial value problems of the Navier–Stokes equation (it is possible that the theory is incomplete), and (2) we have a dreadful practical problem regarding eddy scale. Consider, for example, a turbulent flow occurring in a process vessel with a diameter of 5 m. If the dissipation rate per unit mass is $10^3\ \mathrm{cm^2/s^3}$ and the fluid has properties similar to water, then the smallest (dissipative) scales will be on the order of

$$\eta = \left(\frac{\nu^3}{\varepsilon}\right)^{1/4} \approx 0.0056\,\mathrm{cm}. \qquad (5.97)$$

Thus, there are about five decades of eddy sizes and a single planar cut from a discretization (that *could fully resolve* the flow) will involve about 2.5×10^9 nodal points.

We can look at this in a more general way as well. The minimum number of nodal points required for the simulation of a three-dimensional flow should scale as

$$\frac{l^3}{\eta^3} \Rightarrow \frac{l^3}{(\nu^3/\varepsilon)^{3/4}}. \qquad (5.98)$$

Since the dissipation rate can be estimated with the Taylor's relation $\varepsilon \approx (Au^3/l)$, we find (taking $A \approx 1$)

$$\frac{u^{9/4} l^{9/4}}{\nu^{9/4}} = Re_l^{9/4}. \qquad (5.99)$$

If the integral-scale Reynolds number is large, the required number of points for the discretization will be extremely large; for example, if $Re_l = 100{,}000$, then $Re_l{}^{9/4} \approx 3.16 \times 10^{13}$. It is evident that the storage requirements for a usefully complete computation will be prohibitive. Nevertheless, it is the opinion of this writer that direct numerical attack on the Navier–Stokes equations offers one of the better prospects for fundamental progress in turbulence. The interested reader is directed to Chapter 9 in Pope (2000).

Although DNS is both appealing and promising, we must be careful about being too optimistic regarding the results obtained solely from the increased computational power. The following quote from the physicist Peter Carruthers (regarding the work of Mitchell Feigenbaum and cited by Gleick) is probably all too accurate:

> "If you had set up a committee in the laboratory or in Washington and said, 'Turbulence is really in our way, we've got to understand it, the lack of understanding really destroys our chance of making progress in lots of fields,' then of course, you would hire a team. You'd get a giant computer. You'd start running big programs. And you would never get anywhere. Instead, we have this smart guy, sitting quietly—talking to people to be sure, but mostly working all by himself."

It is certainly possible that a breakthrough in turbulence may come from an unexpected direction. The emergence of nonlinear or chaotic physics over the last couple of decades is a cause for hope. Indeed, there are many investigators who share the opinion voiced by O. E. Lanford (1981):

> "The mathematical object which accounts for turbulence is an attractor or a few attractors, of reasonably small dimension, imbedded in the very-large-dimensional state space of the fluid system. Motion on the attractor depends sensitively on initial conditions, and this sensitive dependence accounts for the apparently stochastic time dependence of the fluid."

You can learn something about the interface between chaos theory and fluid mechanics by reading the very accessible popular book *Chaos* by Gleick (1987). For a more mathematical treatment of this subject area, see Berge et al. (1984).

REFERENCES

Berge,P., Pomeau, Y., and C. Vidal. *Order Within Chaos: Towards a Deterministic Approach to Turbulence*, John Wiley & Sons, New York (1984).

Betchov, R. and W. O. Criminale, Jr. *Stability of Parallel Flows*, Academic Press, New York (1967).

Bowles, R. I. Transition to Turbulent Flow in Aerodynamics. *Philosophical Transactions of the Royal Society of London A*, xxx:245 (2000).

Bradshaw, P. *An Introduction to Turbulence and Its Measurement*, Pergamon Press, Oxford (1975).

Bradshaw, P. and D. H. Ferriss. Applications of a General Method of Calculating Turbulent Shear Layers. *Journal of Basic Engineering, Transactions ASME*, 94:345 (1972).

Burkhalter, J. E. and E. L. Koschmieder. Steady Supercritical Taylor Vortices After Sudden Starts. *Physics of Fluids*, 17:1929 (1974).

Coles, D. Transition in Circular Couette Flow. *Journal of Fluid Mechanics*, 21:385 (1965).

Davidson, L. and J. R. Fontaine. Calculation of the Flow in a Ventilated Room Using Different Finite-Difference Schemes and Different Treatments of the Walls. *Proceedings of CLIMA 2000*, 3:219 (1989).

Eckhardt, B. and H. Faisst. Travelling Waves and Transition to Turbulence in Pipe Flow. http://www.physik.uni-marburg.ed/ksoy (2008).

Emmons, H. W. The Laminar–Turbulent Transition in a Boundary layer. Part I. *Journal of the Aeronautical Sciences*, 18:490 (1951).

Faisst, H. and B. Eckhardt. Traveling Waves in Pipe Flow. *Physical Review Letters*, 91:224502 (2003).

Favre, A., Gaviglio, J., and R. Dumas. Some Measurements of Time and Space Correlation in a Wind Tunnel. *NACA Technical Memorandum 1370* (1955).

Gleick, J. *Chaos*, Viking, New York (1987).

Harsha, P. T. Kinetic Energy Methods. *Handbook of Turbulence*, Vol. 1, Chapter 8, Plenum Press, New York (1977).

Heeg, R. S., Dijkstra, D., and P. J. Zandbergen. The Stability of Falkner–Skan Flows with Several Inflection Points. *ZAMP*, 50:82 (1999).

Hinze, J. O. *Turbulence*, 2nd edition, McGraw-Hill, New York (1975).

Hoel, P. G. *Introduction to Mathematical Statistics*, 4th edition, John Wiley & Sons, New York (1971).

Hof, B., van Doorne, C., Westerweel, J., Nieuwstadt, F., Faisst, H., Eckhardt, B., Wedlin, H., Kerswell, R., and F. Waleffe. Experimental Observations of Nonlinear Travelling Waves in Turbulent Pipe Flow. *Science*, 305:1594 (2004).

Japan Society of Mechanical Engineers. *Visualized Flow*, Pergamon Press, Oxford (1988).

Jones, W. P. and B. E. Launder. The Calculation of Low-Reynolds Number Phenomena with a Two-Equation Model of Turbulence. *International Journal of Heat and Mass Transfer*, 16:1119 (1973).

Jordinson, R. The Flat Plate Boundary Layer, Part 1. *Journal of Fluid Mechanics*, 43:810 (1970).

Kerswell, R. R. Recent Progress in Understanding the Transition to Turbulence in a Pipe. *Nonlinearity* 18:R17 (2005).

Kerswell, R. and O. R. Tutty. Recurrence of Travelling Waves in Transitional Pipe Flow. *Journal of Fluid Mechanics*, 584:69 (2007).

Kovasznay, L. S. G. Spectrum of Locally Isotropic Turbulence. *Journal of the Aeronautical Sciences*, 15:745 (1948).

Kraichnan, R. H. Dynamics of Nonlinear Stochastic Systems. *Journal of Mathematical Physics*, 2:124 (1961).

Kraichnan, R. H. Isotropic Turbulence and Inertial-Range Structure. *Physics of Fluids*, 9:1728 (1966).

Landau, L. D. and E. M. Lifshitz. *Fluid Mechanics*, Pergamon Press, London (1959).

Lanford, O. E. Strange Attractors and Turbulence. In: *Hydrodynamic Instabilities and the Transition to Turbulence* (H. L. Swinney and J. P. Gollub, editors), Springer-Verlag, Berlin (1981).

Laufer, J. The Structure of Turbulence in Fully-Developed Pipe Flow. *NACA Report 1174* (1954).

Leonard, B. P. A Stable and Accurate Convective Modeling Procedure Based on Quadratic Upstream Interpolation. *Computer Methods in Applied Mechanics and Engineering*, 19:59 (1979).

Liepmann, H. W. The Rise and Fall of Ideas in Turbulence. *American Scientist*, 67:221 (1979).

Lin, C. C. *Theory of Hydrodynamic Stability*, Cambridge University Press, London (1955).

Ling, C. H. and W. C. Reynolds. Non-Parallel Flow Corrections for the Stability of Shear Flows. *Journal of Fluid Mechanics*, 59:571 (1973).

Mellor, G. R. and H. J. Herring. A Survey of the Mean Turbulent Field Closure Models. *AIAA Journal*, 11:590 (1973).

Monin, A. S. and A. M. Yaglom. *Statistical Fluid Mechanics*, MIT Press, Cambridge, MA (1975).

Mullin, T. Experimental Study of the Transition to Turbulence in a Constant Mass-Flux Pipe. http://www.maths.manchester.ac.uk (2008).

Nelkin, M. In What Sense Is Turbulence an Unsolved Problem? *Science*, 255:566 (1992).

Pai, S. I. On Turbulent Flow Between Parallel Plates. *Journal of Applied Mechanics*, 20:109 (1953).

Patel, V. C., Rodi, W., and G. Scheuerer. Turbulence Models for Near-Wall and Low Reynolds Number Flows: A Review. *AIAA Journal*, 23:1308 (1985).

Pope, S. B. *Turbulent Flows*, Cambridge University Press, Cambridge (2000).

Raithby, G. D. Skew Upstream Differencing Schemes for Problems Involving Fluid Flow. *Computer Methods in Applied Mechanics and Engineering*, 9:153 (1976).

Rayleigh, L. On the Stability or Instability of Certain Fluid Motions. *Scientific Papers*, 1:474, Cambridge University Press, London (1899–1920).

Reynolds, O. An Experimental Investigation of the Circumstances Which Determine Whether the Motion of Water Shall Be Direct or Sinuous, and the Law of Resistance in Parallel Channels. *Philosophical Transactions of the Royal Society*, 174:935 (1883).

Reynolds, A. J. *Turbulent Flows in Engineering*, Wiley-Interscience, London (1974).

Robinson, S. K. The Kinematics of Turbulent Boundary Layer Structure. *NASA Technical Memorandum 103859* (1991).

Schlichting, H. *Boundary-Layer Theory*, 6th edition, McGraw-Hill, New York (1968).

Schlichting, H. *Boundary-Layer Theory*, 7th edition, McGraw-Hill, New York (1979).

Schlichting, H. and A. Ulrich. Zur Berechnung des Umschalages laminar-turbulent. *Jb. Dt. Luftfahrtforschung*, 1:8 (1942).

Schubauer, G. B. and H. K. Skramstad. Laminar Boundary-Layer Oscillations and Transition on a Flat Plate. *NACA Report 909* (1948).

Schwarze, R., Klostermann, J. and C. Brucker. Experimental and Numerical Investigations of a Turbulent Round Jet into a Cavity. *International Journal of Heat and Fluid Flow*, 29:1688 (2008).

Sellin, R. H. J. and R. T. Moses, editors. *Drag Reduction in Fluid Flows*, IAHR, Halsted Press, New York (1989).

Shen, S. F. Calculated Amplified Oscillations in Plane Poiseuille and Blasius Flows. *Journal of the Aeronautical Sciences*, 21:62 (1954).

Squire, H. B. On the Stability of Three-Dimensional Disturbances of Viscous Flow Between Parallel Walls. *Proceedings of the Royal Society of London A*, 142:621 (1933).

Stuart, J. T. Instability and Transition in Pipes and Channels. In *Transition and Turbulence* (R. E. Meyer, editor), Academic Press, New York (1981).

Taylor, G. I. Stability of a Viscous Liquid Contained Between Two Rotating Cylinders. *Philosophical Transactions of the Royal Society A*, 223:289 (1923).

Taylor, G. I. Statistical Theory of Turbulence. *Proceedings of the Royal Society of London A*, 151:421 (1935).

Taylor, G. I. The Spectrum of Turbulence. *Proceedings of the Royal Society of London A*, 164:476 (1938).

Tennekes, H. and J. L. Lumley. *A First Course in Turbulence*, MIT Press, Cambridge, MA (1972).

Tollmien, W. General Instability Criterion of Laminar Velocity Distributions. *NACA TM 792* (1936).

Trefethen, L. N., Trefethen, A. E., Reddy, S. C., and T. A. Driscoll. Hydrodynamic Stability Without Eigenvalues. *Science*, 261:578 (1993).

Van Dyke, M. *An Album of Fluid Motion*, Parabolic Press, Stanford (1982).

Walton, A. G. The Stability of Nonlinear Neutral Modes in Hagen–Poiseuille Flow. *Proceedings of the Royal Society of London A*, 461:813 (2005).

Wazzan, A. R., Okamura, T., and A. M. O. Smith. The Stability of Water Flow over Heated and Cooled Flat Plates. *Journal of Heat Transfer*, 90:109 (1968).

White, F. M. *Viscous Fluid Flow*, 2nd edition, McGraw-Hill, Boston (1991).

Wilcox, D. C. *Turbulence Modeling for CFD*, 2nd edition, DCW Industries, La Canada, CA (1998).

Willis, A., Peixinho, J., Kerswell, R., and T. Mullin. Experimental and Theoretical Progress in Pipe Flow Transition. *Proceedings of the Royal Society of London A*, 366:2671–2684 (2009).

Wygnanski, I. and F. Chanpagne. On Transition in a Pipe, Part 1. The Origin of Puffs and Slugs and the Flow in a Turbulent Slug. *Journal of Fluid Mechanics*, 59:281 (1973).

Yorke, J. A. and E. D. Yorke. Chaotic Behavior and Fluid Dynamics. In *Hydrodynamic Instabilities and the Transition to Turbulence* (H. L. Swinney and J. P. Gollub, editors), Springer-Verlag, Berlin (1981).

6

HEAT TRANSFER BY CONDUCTION

6.1 INTRODUCTION

We begin this chapter with a very brief sketch of the life of Jean Baptiste Joseph Fourier, who contributed much to the development of molecular heat transfer theory. Fourier was born on March 21, 1768 in Auxerre, Bourgogne, France, the ninth of 12 children of Joseph Fourier and Edmie Germaine LeBegue. At the age of 8, Fourier lost his father; fortunately, his formal education was initiated when the bishop of Auxerre succeeded in getting him admitted to the local military school. Later in 1794, Fourier was nominated to study at the Ecole Normale in Paris. At the age of 30, he was selected to accompany Napoleon to Egypt (in 1798) as a member of the scientific and literary commission. He fulfilled a variety of administrative tasks and began a study of Egyptian antiquities. He also acquired the habit of wrapping himself like a mummy, a practice that might have played a role in his death in Paris in 1830. The results of the French occupation (and exploration) of Egypt were mixed: The campaign was a military failure but it resulted in the publication of *Description of Egypt*, a product of the Institute founded by Bonaparte. And although Fourier gained valuable administrative experience that served him nicely later, the Rosetta stone was taken from the French (from J. F. de Menou), escorted to Britain, translated, and ensconced in the British Museum.

Upon Fourier's return to France, Napoleon appointed him Prefect of Isere where he accomplished what many had thought to be impossible: he persuaded the 40 surrounding communities of the benefits of draining the swamps of Bourgoin. The project cost about 1.2 million francs but it immeasurably improved the value of the land and the health of the inhabitants.

Herivel (1975) describes how Fourier survived Bonaparte's abdication—Fourier was transformed into a servant of the crown and was able to continue as prefect. Then came Napoleon's return from Elba, Fourier's embarrassing flight from Grenoble, and his surprising appointment as Prefect of the Rhone (a position he held from March until May). That Fourier was able to weather the "Hundred Days" debacle is a testament to his skills at negotiating and his popularity with both Napoleon and select royalists.

His contributions to both mathematics and physics were profound and "Fourier" is included in the list of 72 names inscribed on the Eiffel Tower (18 on each side). As an aside, students of transport phenomena should find the list of names intriguing; it includes Carnot, Cauchy, Coriolis, Fourier, Fresnel, Lagrange, Laplace, Navier, Poisson, and Sturm. By 1807, Fourier (Fourier, 1807) completed "On the Propagation of Heat in Solid Bodies," which was contested by Biot because Fourier did not cite Biot's earlier work. Fourier's development of the equations governing heat transfer became part of a submission in 1811 to a rigged contest held by the Paris Institute; the judges were Laplace, Lagrange, Malus, Hauy, and Legendre. Fourier was selected as the winner, but Herivel (1975) notes that there were mixed reactions to portions of the "Prize Essay." Fourier was stung and the experience heightened his animosity toward Biot and Poisson. Some perspective on the criticisms can be found in the Introduction to M. Gaston Darboux's *Oeuvres de Fourier* (available in English translation). Nevertheless, Fourier's contributions to mathematical physics are irrefutable, among his legacies are

Transport Phenomena: An Introduction to Advanced Topics, By Larry A. Glasgow

his book *Theorie analytique de la Chaleur* (1822), the Fourier transform, the theory of orthogonal functions, Fourier's law, and Fourier series. The latter has been described with a very nice mathematical/historical perspective by Carslaw (1950).

Let us now review the law of conduction (y-component) that carries Fourier's name:

$$q_y = -k\frac{\partial T}{\partial y}, \tag{6.1}$$

where q_y is the flux of thermal energy in the y-direction and k is the thermal conductivity of the medium. Note the linearity of the expression, that is, the flux is directly proportional to the temperature (gradient). This is obviously an advantageous form because it means that a thermal energy balance, in the absence of fluid motion, will lead generally to the second-order, linear, partial differential equations (PDE) of either parabolic (transient) or elliptic (equilibrium) character. So, for a pure conduction problem in a stationary medium with constant properties and no thermal energy production, we should expect to see

$$\rho C_p\frac{\partial T}{\partial t} = k\left[\frac{\partial^2 T}{\partial x^2} + \frac{\partial^2 T}{\partial y^2} + \frac{\partial^2 T}{\partial z^2}\right], \tag{6.2}$$

$$\rho C_p\frac{\partial T}{\partial t} = k\left[\frac{1}{r}\frac{\partial}{\partial r}\left(r\frac{\partial T}{\partial r}\right) + \frac{1}{r^2}\frac{\partial^2 T}{\partial \theta^2} + \frac{\partial^2 T}{\partial z^2}\right], \tag{6.3}$$

$$\rho C_p\frac{\partial T}{\partial t} = k\left[\frac{1}{r^2}\frac{\partial}{\partial r}\left(r^2\frac{\partial T}{\partial r}\right) + \frac{1}{r^2\sin\theta}\frac{\partial}{\partial \theta}\left(\sin\theta\frac{\partial T}{\partial \theta}\right) + \frac{1}{r^2\sin^2\theta}\frac{\partial^2 T}{\partial \phi^2}\right], \tag{6.4}$$

for the rectangular, cylindrical, and spherical coordinates, respectively. You should also note the parallel between Fourier's law, (6.1), and Newton's law of viscosity. It is apparent that instantaneously raising the temperature of one face of a semi-infinite slab of material is equivalent to Stokes' first problem (viscous flow near a wall suddenly set in motion).

Before we congratulate ourselves on the simplicity of the generalized conduction problem, we ought to examine the thermal conductivity k to see if a thermal energy balance will actually lead to eqs. (6.2)–(6.4). For example, the thermal conductivity of water increases by about 14.7% over the temperature range 280–340K. For type 347 stainless steel, k increases from 8.5 Btu/(h ft °F) at 100°F to 12.1 Btu/(h ft °F) at 800°F. Figure 6.1 shows the thermal conductivity of steel with 1% chrome for temperatures ranging from 0 to 800°C; the data were adapted from Holman (1997).

Between 0 and 600°C, the data in the figure are roughly represented by

$$k \approx 61.5 - 0.0425T\ \mathrm{W/(m\ ^\circ C)}. \tag{6.5}$$

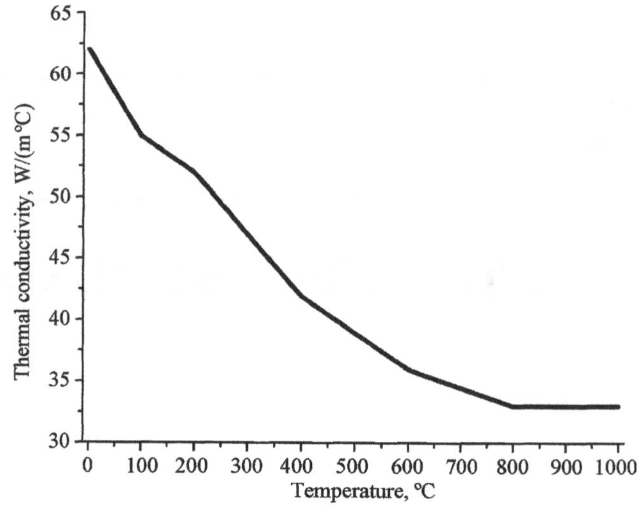

FIGURE 6.1. Thermal conductivity of chrome steel (1%) for temperatures ranging from 0 to 800°C. *Source*: These data were adapted from Holman (1997).

Now suppose we have transient conduction in one spatial dimension (y) in a chrome steel slab. If the product ρC_p is nearly constant and if we take $k = a + bT$, then the governing equation has the form

$$\rho C_p\frac{\partial T}{\partial t} = \frac{\partial}{\partial y}\left[k\frac{\partial T}{\partial y}\right] = b\left[\frac{\partial T}{\partial y}\right]^2 + (a + bT)\frac{\partial^2 T}{\partial y^2}. \tag{6.6}$$

Equation (6.6) presents an entirely different set of challenges, as it is a partial differential equation with two nonlinearities. This type of problem arises with some regularity and we will look at strategies for dealing with it a little later. But before we move on, there is another complication that is common enough to warrant some concern: There are many materials with thermal conductivities that vary with principal direction. Examples include common woods like pine and oak, composite materials, graphite, quartz, and so on. In the case of pine wood parallel to the grain, $k = 0.000834$ cal/(cm s °C), and perpendicular to the grain, $k = 0.000361$ cal/(cm s °C). In such cases, it may be necessary to write the conduction equation (6.2) as

$$\rho C_p\frac{\partial T}{\partial t} = \frac{\partial}{\partial x}\left(k_x\frac{\partial T}{\partial x}\right) + \frac{\partial}{\partial y}\left(k_y\frac{\partial T}{\partial y}\right) + \frac{\partial}{\partial z}\left(k_z\frac{\partial T}{\partial z}\right). \tag{6.7}$$

6.2 STEADY-STATE CONDUCTION PROBLEMS IN RECTANGULAR COORDINATES

We first consider equilibrium problems in one and two spatial dimensions. For a slab extending in the y-direction, from

$y=0$ to y, we have

$$\frac{d}{dy}\left(\frac{dT}{dy}\right)=0. \tag{6.8}$$

Note that the resulting temperature distribution is linear ($T=C_1y+C_2$) and independent of thermal conductivity. In this regard, it is completely analogous to the steady Couette (shear-driven) flow between planar surfaces, which is, of course, independent of viscosity. The generalized problem is governed by the Laplace equation:

$$\nabla^2 T=0. \tag{6.9}$$

Suppose that we have a two-dimensional slab with one edge maintained at an elevated temperature, say 200°C, and the other three edges maintained at 0°C. Let the slab have unit length in both the x- and y-directions, as shown in Figure 6.2.

We want to find the temperature distribution and perhaps the rate at which thermal energy must be withdrawn at the opposing (bottom) face. Dirichlet problems of this type lend themselves to finite difference and finite element solutions, but they can also be readily solved by separation of variables. We let $T=f(x)g(y)$ and apply this product to (6.9); this results in two ordinary differential equations:

$$f''+\lambda^2 f=0 \quad \text{and} \quad g''-\lambda^2 g=0. \tag{6.10}$$

Both these second-order equations are familiar to us, so we immediately write

$$T=(c_1\cos\lambda x+c_2\sin\lambda x)(a_1\cosh\lambda y+a_2\sinh\lambda y). \tag{6.11}$$

Since we have placed the origin at the lower left-hand corner of the slab, can even functions be part of the solution? It is clear that the boundary conditions can only be satisfied if

$$T=\sum_{n=1}^{\infty} C_n \sin n\pi x \sinh n\pi y. \tag{6.12}$$

Furthermore, we must have $T(x,1)=200$°C, so

$$200=\sum_{n=1}^{\infty} C_n \sinh n\pi \sin n\pi x,$$

$$C_n=2\int_0^1 \frac{200}{\sinh n\pi}\sin n\pi x dx. \tag{6.13}$$

You might want to verify that

$$C_n=\frac{400}{\sinh n\pi}\left[\frac{1-\cos n\pi}{n\pi}\right], \tag{6.14}$$

such that $C_1=22.0498$, $C_3=0.0137$, $C_5=1.535\times10^{-5}$, and $C_7=2.0476\times10^{-8}$. The even C's, of course, are all zero. We can now use eq. (6.12) to find the temperature at any point; if we choose the center of the slab, $T(x=0.5, y=0.5)=49.9997$°C. The series converges quickly at this position, which gives the analytic solution some practical value.

The problem described above can be solved other ways as well. For example, suppose we use the second-order central differences to discretize the elliptic equation (6.9). Let the i-index represent the x-direction and j represent y. We obtain

$$\frac{T_{i+1,j}-2T_{i,j}+T_{i-1,j}}{(\Delta x)^2}+\frac{T_{i,j+1}-2T_{i,j}+T_{i,j-1}}{(\Delta y)^2}\approx 0. \tag{6.15}$$

If we employ a square mesh, then $\Delta x=\Delta y$, and we have the computational algorithm:

$$T_{i,j}=\frac{1}{4}(T_{i+1,j}+T_{i-1,j}+T_{i,j+1}+T_{i,j-1}). \tag{6.16}$$

Thus, we have a set of simultaneous linear algebraic equations that are well suited to the Gauss–Seidel iterative solution. If we use 50 nodes in each direction with 1000 iterations, the computed solution will take the form shown in Figure 6.3. The rate of heat transfer at the bottom face ($y=0$) is obtained directly from numerical values of the derivative $\partial T/\partial y$.

Compare the temperature field shown in Figure 6.3 with the point values calculated with eq. (6.12). We should also note that the iterative solution procedure used to generate Figure 6.3 can be applied to three-dimensional problems just as easily.

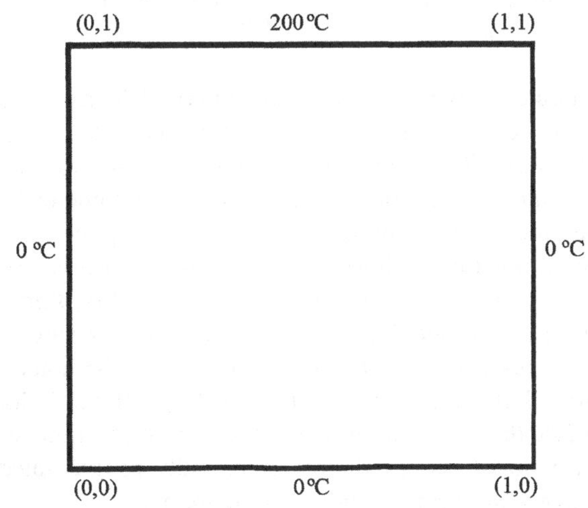

FIGURE 6.2. Two-dimensional slab extending from $(x,y)=(0,0)$ to (1,1). Three edges are maintained at 0°C and one at 200°C.

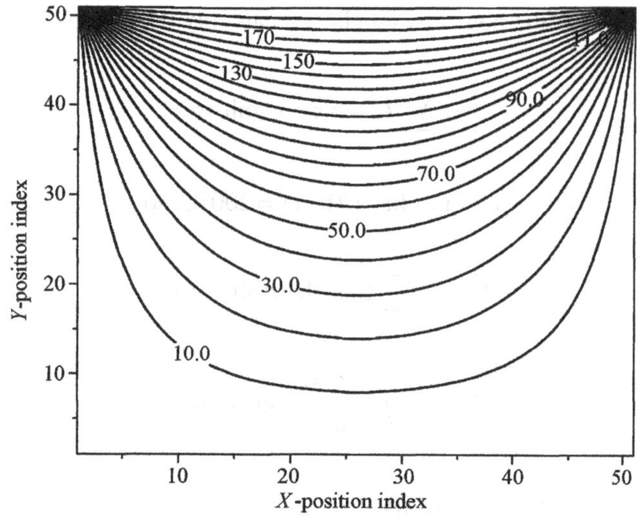

FIGURE 6.3. Temperature distribution in a slab with the top maintained at 200°C and the other three edges at 0°C.

6.3 TRANSIENT CONDUCTION PROBLEMS IN RECTANGULAR COORDINATES

We begin with a semi-infinite slab of material extending to very large distances in the y-direction. The slab is initially at some uniform temperature T_i. At $t=0$, a large thermal mass at elevated temperature is brought into contact with the front face (at $y=0$). This surface instantaneously attains T_0, and thermal energy begins to flow into the slab. If the thermal diffusivity α is constant, then the governing form of eq. (6.2) is

$$\frac{\partial T}{\partial t} = \alpha \frac{\partial^2 T}{\partial y^2}. \tag{6.17}$$

Now we define a dimensionless temperature θ and a new independent variable η:

$$\theta = \frac{T - T_i}{T_0 - T_i} \quad \text{and} \quad \eta = \frac{y}{\sqrt{4\alpha t}}.$$

We introduce these choices into eq. (6.17). The result is

$$\frac{d^2\theta}{d\eta^2} + 2\eta\frac{d\theta}{d\eta} = 0. \tag{6.18}$$

This ordinary differential equation is readily integrated if we reduce the order by letting $\phi = d\theta/d\eta$. A second integration leads to

$$\theta = C_1 \int_0^\eta \exp(-\eta^2)d\eta + C_2. \tag{6.19}$$

At $\eta=0$, $\theta=1$, and as $\eta \to \infty$, $\theta=0$. Consequently,

$$\theta = 1 - \frac{\int_0^\eta \exp(-\eta^2)d\eta}{\int_0^\infty \exp(-\eta^2)d\eta}, \tag{6.20}$$

or alternatively, $\theta = \text{erfc}(\eta)$. As you can see, this is completely equivalent to Stokes' first problem, viscous flow near a wall suddenly set in motion. The variable transformation allowed us to change the parabolic PDE, eq. (6.17), into a second-order ordinary differential equation that was easy to solve.

Many problems involving the conduction equation, eq. (6.17), are candidates for separation of variables. Consider the case of a solid tin bar with $\alpha = 0.38\ \text{cm}^2/\text{s}$ extending from $y=0$ to $y=3$ cm; the bar has an initial temperature of 25°C, but for all positive t's, the ends are maintained at $T=0$°C. Applying separation of variables to eq. (6.17), we obtain

$$T = C_1 \exp(-\alpha\lambda^2 t)[A \sin \lambda y + B \cos \lambda y]. \tag{6.21}$$

The boundary conditions lead us to conclude that $B=0$ and $\sin(3\lambda)=0$. The latter will occur for $\lambda = n\pi/3$, where $n = 1, 2, 3, \ldots$. Consequently, the solution can be written as

$$T = \sum_{n=1}^{\infty} A_n \exp\left(-\frac{\alpha n^2 \pi^2 t}{9}\right) \sin\frac{n\pi y}{3}. \tag{6.22}$$

Applying the initial condition,

$$25 = \sum_{n=1}^{\infty} A_n \sin\frac{n\pi y}{3}. \tag{6.23}$$

This is a half-range Fourier sine series, and by Fourier theorem,

$$A_n = \frac{2}{3}\int_0^3 25 \sin\frac{n\pi y}{3} dy = \frac{50}{n\pi}(1 - \cos n\pi). \tag{6.24}$$

You should recognize a familiar pattern: When we apply separation of variables (the product method) to parabolic equations like (6.17), we use the boundary conditions to get a constant of integration and the separation parameter λ. We then use the initial condition to eliminate the exponential part and determine the leading coefficients (the A_n's) either by the Fourier theorem or by application of orthogonality. Temperature profiles computed using eqs. (6.22) and (6.24) are given in Figure 6.4 for t's of 0.2, 1, and 4 s. It is to be noted that the flux $-k(\partial T/\partial y)$ is easily determined by differentiation of eq. (6.22); the exponential part of the solution guarantees, in this case, that the flux will decrease rapidly, as illustrated by the temperature profiles shown in Figure 6.4.

Equation (6.17) can also be applied to a slab of material (extending from $y=-b$ to $y=+b$ with the center positioned at $y=0$) for the case where the surface temperatures are

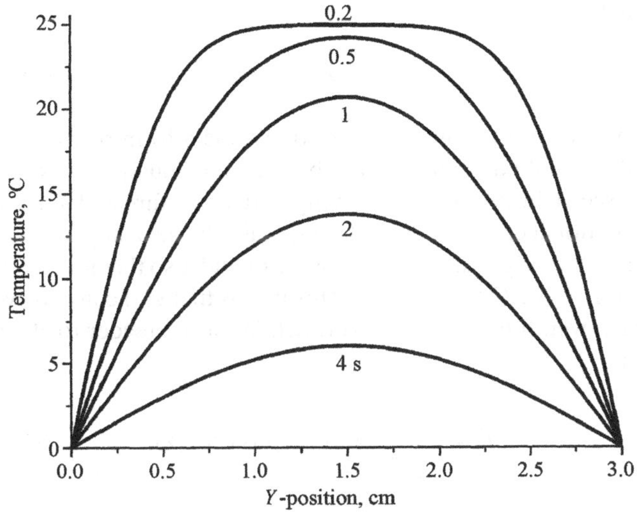

FIGURE 6.4. Temperature distributions in a 3 cm tin bar suddenly cooled at both ends for t's of 0.2, 0.5, 1, 2, and 4 s.

instantaneously elevated to some new value at $t = 0$. The solution (the reader should work it out) can be conveniently presented graphically as shown in Figure 6.5.

Figure 6.5 can be used to determine the temperature at any point in the material; to illustrate, consider an acrylic plastic slab 10 cm thick (so $b = 5$ cm), with an initial temperature of 5°C. At $t = 0$, the surfaces of the slab are instantaneously heated to 90°C. When will the temperature at $y = 2.5$ cm reach 50°C? We have

$$\frac{y}{b} = 0.5 \text{ and } \theta = \frac{T - T_i}{T_b - T_i} = \frac{50 - 5}{90 - 5} = 0.529.$$

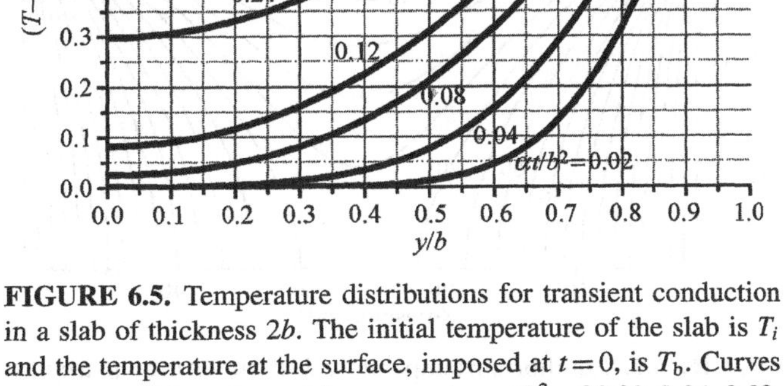

FIGURE 6.5. Temperature distributions for transient conduction in a slab of thickness $2b$. The initial temperature of the slab is T_i and the temperature at the surface, imposed at $t = 0$, is T_b. Curves are presented for values of the parameter, $\alpha t/b^2$, of 0.02, 0.04, 0.08, 0.12, 0.24, 0.36, 0.48, 0.60, 0.80, and 1.00. The left-hand side of the figure is the center of the slab. The temperature distributions appearing in this figure were computed.

Consulting Figure 6.5, we find

$$\frac{\alpha t}{b^2} \cong 0.27, \text{ therefore, } t = \frac{(25)(0.27)}{(0.0012)} = 5625 \text{ s}.$$

Earlier we introduced the possibility that $k = k(T)$; let us examine a transient problem with a variable thermal conductivity (as described in the introduction) to better understand the effects of the resulting nonlinear terms. Suppose we have a slab of chrome steel (1%) at an initial temperature of 30°C. Let the slab have a depth in the y-direction of 20 cm, and assume that the back edge is insulated. At $t = 0$, the front face is instantaneously heated to 550°C. We can get the constant k solution from eq. (6.20) for an infinite slab; we will find the nonlinear solution numerically for comparison. Let i be the position index and j represent the time; we use a first-order forward difference for time derivative and central differences elsewhere. An elementary explicit algorithm can be developed easily:

$$\frac{T_{i,j+1} - T_{i,j}}{\Delta t} \approx \frac{b}{\rho C_p}\left[\frac{T_{i+1,j} - T_{i-1,j}}{2\Delta y}\right]^2 + \frac{a + bT_{i,j}}{\rho C_p}\left[\frac{T_{i+1,j} - 2T_{i,j} + T_{i-1,j}}{(\Delta y)^2}\right]. \tag{6.25}$$

Note that only one temperature on the new $(j + 1)$ time-step row appears in eq. (6.25). If we isolate it on the left-hand side, we can compute the temperature distribution in the slab by merely forward marching in time. It will be necessary to make Δt small enough to provide numerical stability, however, for an explanation of this constraint, see Appendix D. The results of this computation are shown in Figure 6.6.

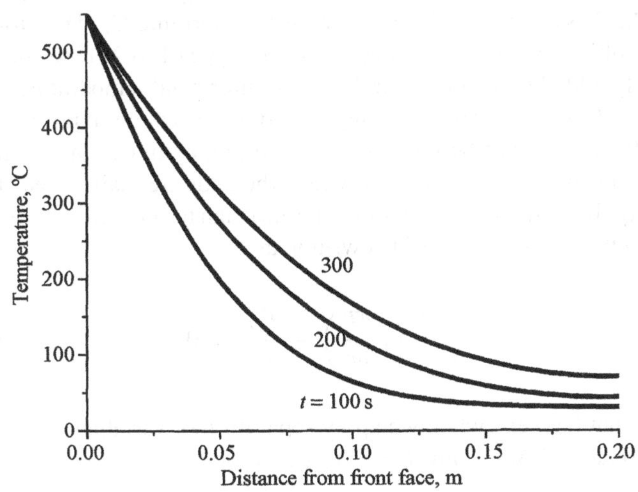

FIGURE 6.6. Temperature distributions computed for the nonlinear case using eq. (6.25). The three curves correspond to $t = 100$, 200, and 300 s.

The computed results shown in Figure 6.6 give us an opportunity to gauge the importance of the nonlinearities in eq. (6.25). We can compare these results with those obtained from eq. (6.20) for an infinite slab at modest t's. For example, using a fixed α of $1.566\times10^{-5}\,\mathrm{m^2/s}$ and setting $y=10$ cm with $t=200$ s, the error function solution shows that $T\cong139$°C. For $y=5$ cm with $t=300$ s, the error function solution produces $T\cong345$°C. If we get the corresponding results from Figure 6.6, we find $T(0.10,200)=121.8$°C and $T(0.05,300)=314.1$°C. Naturally, as the thermal energy penetrates more of the slab, the actual thermal conductivity will decrease and the discrepancy between models will become significantly greater.

6.4 STEADY-STATE CONDUCTION PROBLEMS IN CYLINDRICAL COORDINATES

The most commonly encountered problem of this type involves a radially directed flux with angular symmetry where the axial transport is negligibly small. Examples include insulated pipes and tanks, chemical reactors, current-carrying wires, nuclear fuel rods, and so on. With no production, we write eq. (6.3) as

$$\frac{d}{dr}\left(r\frac{dT}{dr}\right)=0. \tag{6.26}$$

If we integrate eq. (6.26) with specified temperatures T_1 and T_2 at radial positions R_1 and R_2, then

$$T_2-T_1=C_1\ln\left(\frac{R_2}{R_1}\right). \tag{6.27}$$

At any r-position, the product of the flux q_r and surface area $2\pi rL$ is a constant. This allows us to determine C_1. Then for multilayer cylinders, equations of the type of (6.27) are simply added together to eliminate the interfacial temperatures.

However, there are many situations in which axial conduction cannot be ignored, for example, cylinders in which L/d is not large or cases for which the ends are maintained at significantly different temperature(s) than the curved surface. In these cases, eq. (6.3) is written as

$$\frac{1}{r}\frac{\partial}{\partial r}\left(r\frac{\partial T}{\partial r}\right)+\frac{\partial^2T}{\partial z^2}=0. \tag{6.28}$$

What happens when we apply separation of variables to this equation? Assuming $T=f(r)g(z)$, we find

$$f''+\frac{1}{r}f'-\lambda^2f=0 \tag{6.29a}$$

and

$$g''+\lambda^2g=0. \tag{6.29b}$$

You should recognize that eq. (6.29a) is a form of Bessel's differential equation; as we observed previously, we expect to see it in problems involving a radially directed flux in cylindrical coordinates. Before we take the next step, we will place the origin at the center of the cylinder so that it extends from $z=-L/2$ to $z=+L/2$. This means that $g(z)$ can involve *only* even functions. It is worthwhile for the reader to show that

$$T=\sum_{n=1,2\ldots}^{\infty}A_nI_0(\lambda_nr)\cos(\lambda_nz), \tag{6.30}$$

with

$$\lambda_n=\frac{(2n-1)\pi}{L}. \tag{6.31}$$

To complete the solution, the A_n's must be determined using the Fourier theorem, which results in

$$A_n=\frac{200}{I_0(\lambda_nR)}\frac{\sin(\lambda_n(L/2))}{\lambda_n(L/2)}. \tag{6.32}$$

It is convenient in a case like this to have access to the numerical solution; it can provide a sense of confidence about the analysis. Equation (6.28) is suitable for iterative solution by, for example, the Gauss–Seidel method. The computed temperature distribution is shown in Figure 6.7.

Note that at the very center of the cylinder, where the z-position index is 26, the numerical solution yields $T=72.26$°C. Alternatively, we take eq. (6.30) and let both r and z be zero. The result obtained from the *first three* terms is $74.068-2.0125+0.3413=72.397$°C.

We conclude this section with an example in which we have production of thermal energy in a long cylinder. We

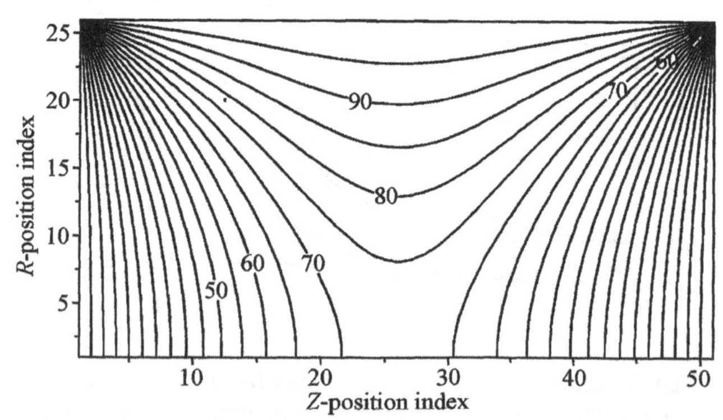

FIGURE 6.7. Equilibrium temperature distribution in a squat cylinder for which the ends are maintained at 0°C and the curved surface at 100°C. The bottom of the figure corresponds to the z-axis where we have $\partial T/\partial r=0$.

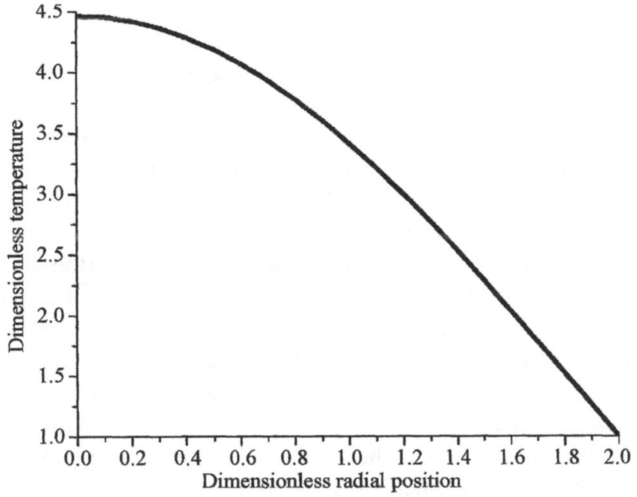

FIGURE 6.8. Temperature in a long cylinder with thermal energy production and the outer surface maintained at T_s.

take the production rate per unit volume, S, to be directly proportional to temperature: $S=\beta T$. Therefore, for steady-state conditions we have

$$r^2\frac{d^2T}{dr^2}+r\frac{dT}{dr}+\frac{\beta}{k}r^2T=0. \tag{6.33}$$

Note the similarity to eq. (6.29a); the solution for eq. (6.33) can be written in terms of Bessel functions of the first and second kind of order zero:

$$T=C_1J_0\left(\sqrt{\frac{\beta}{k}}r\right)+C_2Y_0\left(\sqrt{\frac{\beta}{k}}r\right). \tag{6.34}$$

The solution must be finite at the center and since $Y_0(0)=-\infty$, $C_2=0$. If the temperature of the outer surface is maintained at T_s, then the solution for this problem must be

$$\frac{T}{T_s}=\frac{J_0\left(\sqrt{(\beta/k)}r\right)}{J_0\left(\sqrt{(\beta/k)}R\right)}. \tag{6.35}$$

How does this solution behave? Suppose $\sqrt{(\beta/k)}R=2$; at the centerline ($r=0$), we should find that $T/T_s=4.466$. Naturally, when $r=R$, we obtain $T/T_s=1$. Figure 6.8 shows the dimensionless temperature T/T_s for this problem as a function of dimensionless radial position $\sqrt{(\beta/k)}r$.

6.5 TRANSIENT CONDUCTION PROBLEMS IN CYLINDRICAL COORDINATES

We begin this section with a heat transfer situation that presents some interesting challenges. Suppose we take a solid cylindrical billet at some uniform initial temperature and plunge it into a heated bath at $t=0$. If we record the emf produced by a copper-constantan thermocouple on the cylinder centerline, we can obtain a record of the approach of the sample's temperature to that of the heated bath. In the interior of the solid sample, heat transfer occurs solely by conduction; therefore, the appropriate form of eq. (6.3) is

$$\rho C_p\frac{\partial T}{\partial t}=k\left[\frac{1}{r}\frac{\partial}{\partial r}\left(r\frac{\partial T}{\partial r}\right)+\frac{\partial^2T}{\partial z^2}\right]. \tag{6.36}$$

If the cylinder is infinitely long, or practically speaking, if *L/D* is sufficiently large, then the axial conduction term can be neglected. Under what circumstances is this is a reasonable assumption, and how might we test its validity? We will find it useful to employ a dimensionless temperature, defined by

$$\theta=\frac{T-T_b}{T_i-T_b}, \tag{6.37}$$

where T_b is the temperature of the heated bath and T_i is the initial temperature of the specimen. Note that this definition means that $\theta=1$ initially, and that $\theta\to 0$ as $t\to\infty$. This proves to be quite convenient as we shall see shortly. We now introduce θ into eq. (6.36) and divide by ρC_p. The result is

$$\frac{\partial\theta}{\partial t}=\alpha\left[\frac{1}{r}\frac{\partial}{\partial r}\left(r\frac{\partial\theta}{\partial r}\right)\right]. \tag{6.38}$$

Of course, eq. (6.38) is also a candidate for application of the product method (separation of variables). We propose a solution of the form

$$\theta=f(r)g(t), \tag{6.39}$$

where f is a function solely of r and g is a function solely of t. Consider the consequences of introducing eq. (6.39) into (6.38):

$$fg'=\alpha\left(gf''+\frac{1}{r}gf'\right). \tag{6.40}$$

We now divide eq. (6.40) by the product fg. The result is

$$\frac{g'}{\alpha g}=\frac{f''+(1/r)f'}{f}. \tag{6.41}$$

Note that the left-hand side is a function *only* of time and the right-hand side is a function *only* of radial position. Obviously, both sides of eq. (6.41) must be equal to a constant; we will write this constant of separation as $-\lambda^2$. The rationale for this choice will become apparent momentarily. It should be evident to you that we now have two *ordinary* differential equations:

$$\frac{dg}{g}=-\alpha\lambda^2dt \quad\text{and}\quad \frac{d^2f}{dr^2}+\frac{1}{r}\frac{df}{dr}+\lambda^2f=0. \tag{6.42a,b}$$

The solution to eq. (6.42a) is $g = C_1 \exp(-\alpha\lambda^2 t)$. Equation (6.42b) is a form of Bessel's differential equation, and the solution for this case is

$$f = AJ_0(\lambda r) + BY_0(\lambda r), \qquad (6.43)$$

where J_0 and Y_0 are the zero-order Bessel functions of the first and second kind, respectively. According to our hypothesis put forward in eq. (6.39),

$$\theta = C_1 \exp(-\alpha\lambda^2 t)[AJ_0(\lambda r) + BY_0(\lambda r)]. \qquad (6.44)$$

It is easy enough to verify that eq. (6.44) is in fact a solution for eq. (6.38). We have two boundary conditions that must be satisfied, the first being that at $r=0$, θ must be finite. Since $Y_0(0) = -\infty$, it is necessary for us to set $B=0$. Now consider the boundary condition to be applied at $r=R$; if the cylinder surface attains the bath temperature *very* rapidly, then at $r=R$, $\theta=0$, and this will require that $J_0(\lambda R)=0$. However, J_0 has infinitely many zeros, and we have no reason to believe that at fixed time and radial position, any single *one* of the possible values of λ would result in solution. Therefore, we use superposition to rewrite eq. (6.44) as

$$\theta = \sum_{n=1}^{\infty} A_n \exp(-\alpha\lambda_n^2 t) J_0(\lambda_n r). \qquad (6.45)$$

Whether this instantaneous change of surface temperature is an appropriate boundary condition depends upon the relative rates of heat transfer on the two sides of the fluid–solid interface. If the cylinder has a (relatively) large thermal conductivity, then heat flow to the interior of the solid will occur at such a rate as to preclude use of this boundary condition. In fact, this will be the general situation with metals immersed in liquids or gases. For these cases, a Robin's-type boundary condition must be employed in which the thermal energy fluxes are equated on either side of the interface. We accomplish this by using Fourier's law and Newton's "law" of cooling:

$$-k\left.\frac{\partial T}{\partial r}\right|_{r=R} = h(T_{r=R} - T_b). \qquad (6.46)$$

After introducing our dimensionless temperature and performing the indicated differentiation (term-by-term), this boundary condition can be rewritten as

$$\lambda_n R J_1(\lambda_n R) = \frac{hR}{k} J_0(\lambda_n R). \qquad (6.47)$$

This transcendental equation occurs frequently in mathematical physics and the roots are widely available. Pay particular attention to the quotient hR/k. *This is not the Nusselt number, it is the Biot modulus.* It is essential that the reader make note of the difference. In the Nusselt number, *both h and k are on the fluid side of the interface.* Now, suppose that $hR/k = 1.5$; in this case the first six roots for $\lambda_n R$ are

1.4569, 4.1902, 7.2233, 10.3188, 13.4353, and 16.5612.

You should be aware that the use of eq. (6.46) as a boundary condition (with the introduction of the heat transfer coefficient h) has caused us an additional problem; we have no *a priori* means of determining h. The Robin's-type boundary condition has introduced an unknown parameter into the solution. Before we attempt to resolve this difficulty, we need to finish our analytic solution. This means choosing values for the leading coefficients (the A_n's) that cause our series to converge to the desired solution. Note that we have applied two boundary conditions; we now employ the initial condition: For all time up to $t=0$, the sample temperature is a uniform T_i such that $\theta = 1$. Therefore, we rewrite eq. (6.45) as

$$1 = \sum A_n J_0(\lambda_n r). \qquad (6.48)$$

We now take advantage of the orthogonality of Bessel functions by making use of the following relationship:

$$0 = \int_0^R r J_0(\lambda_n r) J_0(\lambda_m r) dr, \quad \text{for} \quad n \neq m. \qquad (6.49)$$

Thus, in principle, we multiply both sides of eq. (6.48) by $rJ_0(\lambda_n r)dr$ and integrate from 0 to R to determine the unknown coefficients. *It is to be noted that we will get a different result for each of the surface ($r=R$) boundary conditions discussed above. If the surface temperature attains the bath value rapidly then,*

$$A_n = \frac{2}{\lambda_n R J_1(\lambda_n R)}. \qquad (6.50)$$

This is correct *only for the case in which the λ_n's are the roots of $J_0(\lambda_n R) = 0$*, that is, for cylindrical solids with *low* thermal conductivities. Our situation with the metallic billets is more complicated since the separation constants have come from the Robin's-type boundary condition (6.46). It is a bit more difficult to show that for this case,

$$A_n = \frac{2\lambda_n R J_1(\lambda_n R)}{((h^2R^2/k^2) + \lambda_n^2 R^2) J_0^2(\lambda_n R)}. \qquad (6.51)$$

We see now that another important question has arisen: How fast does the series appearing as eq. (6.45) converge? If more than three or four terms are required, the analytic solution may be worthless. Note that if α and/or t are large, the exponential factor will certainly be dominant. It is useful to explore series convergence for a specific case; suppose we have a phosphor bronze cylinder with a diameter of 2.54 cm and a length of 15.24 cm ($L/d = 6$):

$L = 15.24$ cm $\quad D = 2.54$ cm $\quad \rho = 8.86$ g/cm^3

$C_p = 0.09$ cal/(g °C) $\quad k = 0.165$ cal/(cm^2 s °C)/cm $\quad \alpha = 0.2074$ cm^2/s.

Now we take $(hR/k) = 0.15$; we will later determine whether this is an appropriate choice. Using tabulated roots for eq. (6.51), we find that

n	λ_n	$\lambda_n R$	A_n
1	0.42	0.5376	1.0356
2	3.05	3.8706	−0.0492
3	5.54	7.0369	+0.0202
4	8.02	10.188	?
5	10.50	13.3349	?
6	12.98	16.4797	?

You may want to try to complete this table as an exercise. Now we will compute the centerline temperature of the phosphor bronze specimen 5 s after its immersion in the heated bath:

$$\text{First term of infinite series: } 0.8625$$
$$\text{Second term: } -3.221 \times 10^{-6}.$$

This is a desirable behavior in an infinite series solution and the result corresponds to a temperature $T(r=0,\ t=5\ \text{s})$ of 12.9°C. Did we select the correct value for the Biot modulus? We may be able to determine this by examining Figure 6.9.

Experimental data for two different cylindrical samples, phosphor bronze and acrylic plastic (Plexiglas®), are provided in Figures 6.9 and 6.10. The ratio of the thermal diffusivities for these two materials is

$$\frac{\alpha_{pb}}{\alpha_{acry}} = \frac{0.207}{0.0012} = 172.5.$$

Both samples initially were at a uniform temperature of 3°C; at $t = 0$, each was immersed in a heated bath with $T_b = 75°C$.

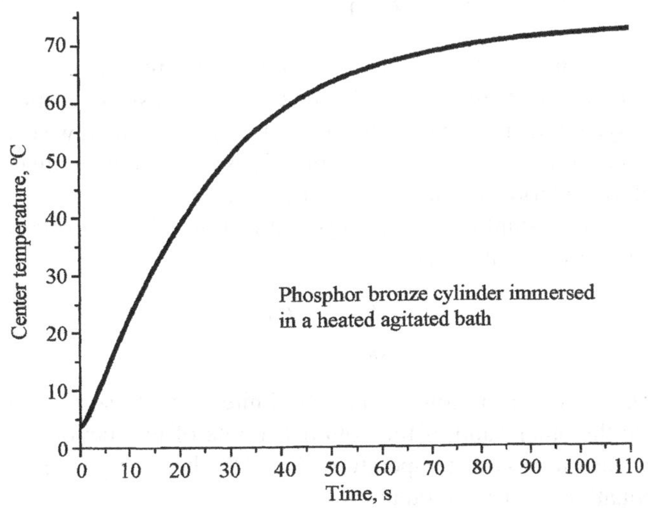

FIGURE 6.9. Center temperature of a phosphor bronze cylinder ($d = 2.54$ cm) after immersion in a heated bath maintained at 75°C, (the initial cylinder temperature was 3°C).

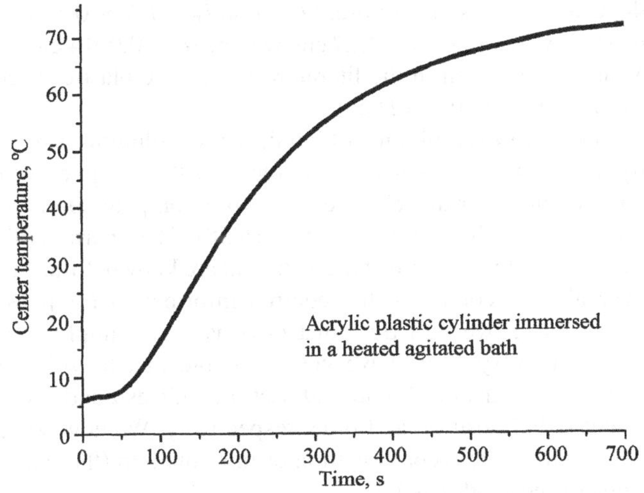

FIGURE 6.10. Center temperature of an acrylic plastic cylinder ($d = 2.54$ cm) after immersion in a heated bath maintained at 75°C, (the initial cylinder temperature was 3°C).

The main difference between these two cases is the location of the resistance to heat transfer. For the phosphor bronze cylinder, the principal resistance is *outside* the material ($r > R$); for the acrylic plastic, the main resistance is *inside*. So, for materials that are poor conductors, the surface temperature will very rapidly attain T_b and the analytic solution is found using eq. (6.50) with (6.45). The results for this case can be compiled in a very useful way for different values of the parameter, $\alpha t/R^2$.

We shall illustrate one use of Figure 6.11. The center temperature of the acrylic plastic cylinder (Figure 6.9) was about

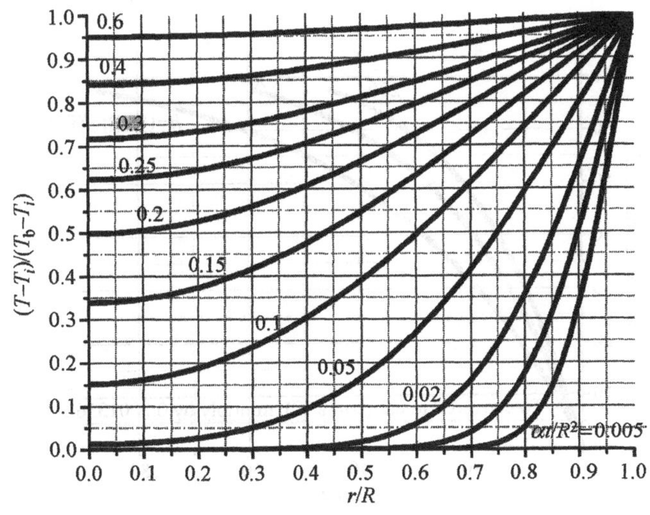

FIGURE 6.11. Temperature distributions for transient conduction in a long cylinder. The initial temperature of the material is T_i; at $t = 0$, the outer surface ($r = R$) is instantaneously heated to T_b. The curves represent values of $\alpha t/R^2$ ranging from 0.005 to 0.60 and the left-hand side of the figure corresponds to the center of the cylinder. The data appearing in this figure were computed numerically.

44°C at $t = 250$ s. Therefore, $(T - T_i)/(T_b - T_i) \approx 0.57$ and $\alpha t/R^2 \approx 0.22$. Since $R = 1.27$ cm, we find $\alpha \approx 0.0014$ cm²/s. Values for α given in the literature for acrylic plastic range from 0.00118 to 0.00121 cm²/s.

One common limitation of infinite series solutions is readily apparent. If t is small, *many* terms will be required for convergence. Fortunately, we can easily compute solutions for the partial differential equation (6.38) if the thermal diffusivity and the heat transfer coefficient are known. Since we have already compiled the required information for phosphor bronze, we will treat that case as our example. Our plan is to vary h until we get a suitable match with the experimental data in Figure 6.9. Let the indices i and j represent radial position and time, respectively. We now write a finite difference representation of this equation (the initial value of the i-index is 1):

$$\theta_{i,j+1} = \alpha\Delta t\left[\frac{\theta_{i+1,j} - 2\theta_{i,j} + \theta_{i-1,j}}{(\Delta r)^2} + \frac{1}{(i-1)\Delta r}\frac{\theta_{i+1,j} - \theta_{i,j}}{\Delta r}\right] + \theta_{i,j}. \tag{6.52}$$

Note how this equation allows us to compute the temperature on the new time-step row $(j + 1)$, using only known, old temperatures. This is another example of an *explicit* algorithm for solution of the parabolic partial differential equation. It does have the usual problem with respect to numerical stability; the quotient $\alpha\ \Delta t/(\Delta r)^2$ must be smaller than 0.5. Solutions for three values of the heat transfer coefficient are shown in Figure 6.12.

The computed results shown in Figure 6.12 can be compared with the experimental data for the phosphor bronze cylinder given in Figure 6.9; the comparison shows that choosing $h = 150$ Btu/(ft² h °F), or 0.02034 in cal/(cm² s °C), produces excellent agreement.

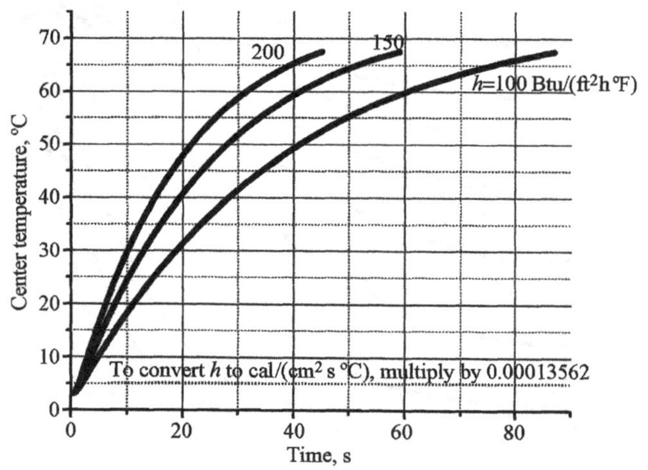

FIGURE 6.12. Center temperature histories for a phosphor bronze cylinder immersed in a heated bath maintained at 75°C. The initial temperature of the cylinder was 3°C. Curves are shown for heat transfer coefficients of 0.01356, 0.02034, and 0.02712 cal/(cm² s °C), corresponding to 100, 150, and 200 Btu/(ft² h °F), respectively.

6.6 STEADY-STATE CONDUCTION PROBLEMS IN SPHERICAL COORDINATES

Heat transfer problems in spherical coordinates are sometimes given minimal attention in engineering coursework. That may not be justifiable since there are many important nonisothermal processes occurring in spheres and sphere-like objects. Let us think of a few examples: catalyst pellets, combustion of granular solids, grain drying, fluidized bed reactors, ball bearing production and operation, ore reduction, grinding and milling, resin and bead production, spray drying, etc.

For radially directed conduction (and no production term), eq. (6.4) becomes

$$\frac{d}{dr}\left(r^2\frac{dT}{dr}\right) = 0. \tag{6.53}$$

Upon integration, we find

$$T = \frac{C_1}{r} + C_2. \tag{6.54}$$

For a spherical shell extending from R_1 to R_2, with surface temperatures T_1 and T_2, we find

$$C_1 = \frac{T_2 - T_1}{(1/R_2 - 1/R_1)} \tag{6.55}$$

and the corresponding flux is given by

$$q_r = \frac{k}{r^2}\left(\frac{T_2 - T_1}{1/R_2 - 1/R_1}\right). \tag{6.56}$$

Equation (6.56) indicates that a multilayered sphere, an "onion" for example, could be treated analogously to the multilayered cylinder. Since the product r^2q_r is constant, we can isolate the ΔT's and add the expressions together to eliminate all the interior interfacial temperatures.

If a constant thermal energy production S is occurring in a spherical entity, then

$$T = -\frac{S}{6k}r^2 - \frac{C_1}{r} + C_2. \tag{6.57}$$

This solution, of course, must be finite at $r = 0$, so $C_1 = 0$. On the other hand, if the volumetric rate of production is a linear function of temperature ($S = \beta T$), then the governing equation must be written:

$$\frac{d^2T}{dr^2} + \frac{2}{r}\frac{dT}{dr} + \frac{\beta}{k}T = 0. \tag{6.58}$$

It is convenient to define a new dependent variable $\theta = rT$; we can then rewrite eq. (6.58) as

$$\frac{d^2\theta}{dr^2} + \frac{\beta}{k}\theta = 0, \tag{6.59}$$

with the solution

$$T = \frac{A}{r}\sin\sqrt{\frac{\beta}{k}}r + \frac{B}{r}\cos\sqrt{\frac{\beta}{k}}r. \tag{6.60}$$

Again, the temperature must be finite at the center, so $B = 0$. If we assign a temperature T_s at the surface of the sphere ($r = R$), then the two solutions (for constant and linearly dependent production) can be written as

$$\frac{T}{T_s} = \frac{S}{6kT_s}(R^2 - r^2) + 1 \quad (S \text{ constant}) \tag{6.61}$$

and

$$\frac{T}{T_s} = \frac{R}{r}\frac{\sin\sqrt{(\beta/k)}r}{\sin\sqrt{(\beta/k)}R} \quad (S = \beta T). \tag{6.62}$$

The differences between the two temperature distributions are subtle if center temperatures are set equal. However, if the thermal energy fluxes at the surface ($r = R$) are forced to be equal, then the center temperature with eq. (6.62) will of course be higher.

6.7 TRANSIENT CONDUCTION PROBLEMS IN SPHERICAL COORDINATES

A number of problems of practical interest are governed by

$$\frac{\partial T}{\partial t} = \alpha\left[\frac{1}{r^2}\frac{\partial}{\partial r}\left(r^2\frac{\partial T}{\partial r}\right)\right]. \tag{6.63}$$

As we have already noted, the operator appearing on the right-hand side of eq. (6.63) suggests the substitution $\theta = rT$, which results in

$$\frac{\partial\theta}{\partial t} = \alpha\frac{\partial^2\theta}{\partial r^2}. \tag{6.64}$$

We begin with the Dirichlet problem in which the surface of the sphere is instantaneously heated (or cooled) to some new temperature T_s. Application of the product method results in

$$T = C_1\exp(-\alpha\lambda^2 t)\left[\frac{A}{r}\sin\lambda r + \frac{B}{r}\cos\lambda r\right], \tag{6.65}$$

and once again since T must be finite at $r = 0$, $B = 0$. We choose to rewrite eq. (6.65) as

$$T = T_s + \frac{A}{r}\exp(-\alpha\lambda^2 t)\sin\lambda r, \tag{6.66}$$

because of our surface boundary condition; consequently, $\sin(\lambda R) = 0$ and $\lambda = n\pi/R$. Equation (6.66) becomes

$$T - T_s = \sum_{n=1}^{\infty}\frac{A_n}{r}\exp(-\alpha\lambda_n^2 t)\sin\lambda_n r \tag{6.67}$$

and the initial condition is applied, at $t = 0$, $T = T_i$. Once again we see a half-range Fourier sine series and the A_n's can be immediately determined by integration, resulting in the solution:

$$T - T_s = \sum_{n=1}^{\infty}\frac{2(T_s - T_i)R\cos n\pi}{n\pi}\frac{\exp}{r}\left[-\frac{\alpha n^2\pi^2 t}{R^2}\right]\sin\frac{n\pi r}{R}. \tag{6.68}$$

We can look at the application of eq. (6.68) to a familiar situation. A watermelon with a diameter of 20 in. and a uniform temperature of 80°F is removed from the field and placed in ice water at 35°F. The melon is a poor conductor with a thermal diffusivity α of about 0.0055 ft^2/h; how long will it take for the temperature at $r = 0.25$ ft to fall to 45°F? You might want to use the series solution to show that the melon must be immersed for about 25 h. Alternatively, the results for transient conduction in a sphere can be compiled in a manner analogous to Figure 6.11 for cylinders; consult Figure 6.13.

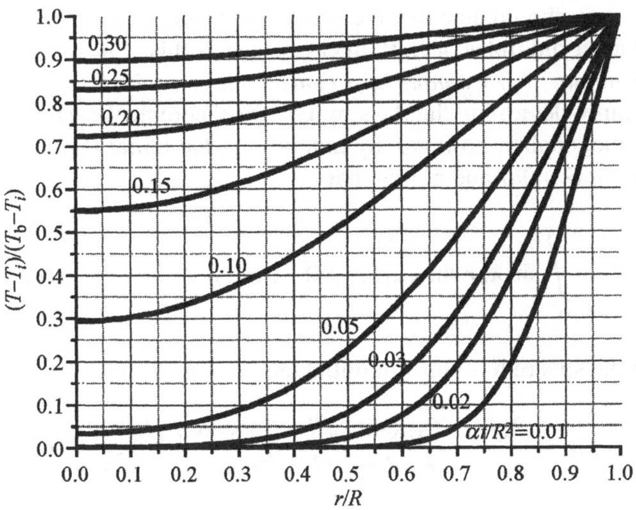

FIGURE 6.13. Temperature distributions for transient conduction in a sphere. The initial temperature of the object is T_i; at $t = 0$, the outer surface ($r = R$) is instantaneously heated to T_b. The curves represent values of $\alpha t/R^2$ ranging from 0.01 to 0.30 and the center of the sphere corresponds to the left-hand side of the figure. The data appearing in this figure were computed numerically.

TABLE 6.1. The First Seven Roots for the Transcendental Equation (6.70) for Four Values of the Biot Modulus

Biot	0.05	0.5	5.0	50
$\lambda_1 R$	0.3854	1.1656	2.5704	3.0788
$\lambda_2 R$	4.5045	4.6042	5.3540	6.1581
$\lambda_3 R$	7.7317	7.7899	8.3029	9.2384
$\lambda_4 R$	10.9088	10.9499	11.3348	12.3200
$\lambda_5 R$	14.0697	14.1017	14.4080	15.4034
$\lambda_6 R$	17.2237	17.2497	17.5034	18.4887
$\lambda_7 R$	20.3737	20.3958	20.6121	21.5763

Make use of these data for the watermelon cooling problem cited above and confirm the estimated time.

In contrast to the situation treated above, if the thermal conductivity of the sphere is large, the resistance to heat transfer may occur for $r > R$, that is, outside the sphere. For this case, just as we saw for metallic cylinders, we must use a Robin's-type boundary condition at $r = R$:

$$-k\left.\frac{\partial T}{\partial r}\right|_{r=R} = h(T_{r=R} - T_\infty). \quad (6.69)$$

When we apply eq. (6.69) to (6.65), we get the transcendental equation

$$\tan \lambda R = \frac{\lambda R}{1-(hR/k)}. \quad (6.70)$$

The values of λ that we need must come from the roots of this equation. Examine Table 6.1 for the Biot modulus values ranging 0.05–50.

You should note that the successive values of $\lambda_n R$ are not integer multiples of $\lambda_1 R$. In cases such as this, $\Sigma A_n \sin(\lambda_n r)$ *is not another example of a Fourier series problem*, and we cannot determine the A_n's by Fourier theorem. We can use orthogonality, however, by multiplying the initial condition by $\sin(\lambda_m r)dr$ and noting that

$$\int_0^R \sin \lambda_n r \sin \lambda_m r dr = 0 \quad \text{for} \quad n \neq m. \quad (6.71)$$

If the sphere has uniform initial temperature T_i, then

$$A_n = \frac{2(T_i - T_\infty)(\sin \lambda_n R - \lambda_n R \cos \lambda_n R)}{\lambda_n R - 1/2 \sin 2\lambda_n R}. \quad (6.72)$$

It is reasonable to ask when the result eq. (6.72) must be used, that is, when must we employ the Robin's-type boundary condition at the surface of the sphere? If we take a material that is a poor conductor, like acrylic plastic, and monitor its surface temperature following immersion in a heated fluid, we may be able to come to some conclusion. Figure 6.14 shows the approach of the sphere's surface temperature to the heated bath value. For these computed results, $R = 3.175$ cm and $\alpha = 0.0012$ cm^2/s. At the sphere's surface, 90% of the ultimate temperature change is accomplished in about 13 s. Though this is not instantaneous, it must be put into perspective: It will take several thousand seconds for this acrylic plastic sphere to come to (virtual) thermal equilibrium with the heated bath. Assuming that $T(r = R)$ acquires the bath value immediately following immersion is at least reasonably appropriate. The computed temperature distributions are shown in Figure 6.15, *using the Robin's-type boundary condition at the surface.* Compare these results with the idealized case described by Figure 6.13.

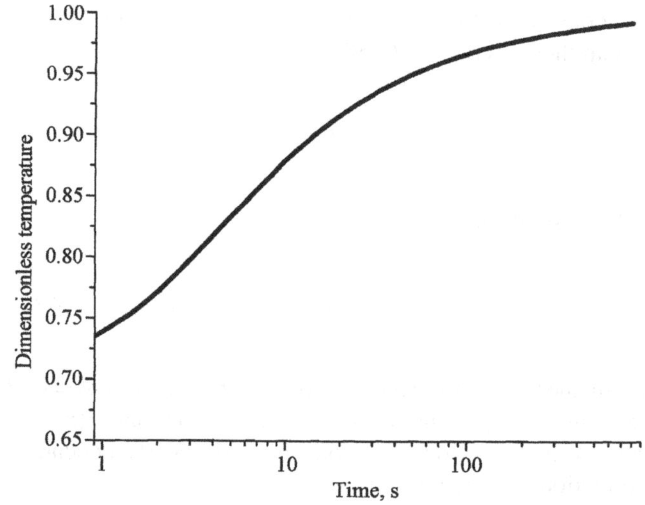

FIGURE 6.14. Approach of the surface temperature of an acrylic plastic sphere to the heated bath value following immersion. The process is not complete even at $t = 1000$ s. On the other hand, the process is 75% complete in about 1.4 s.

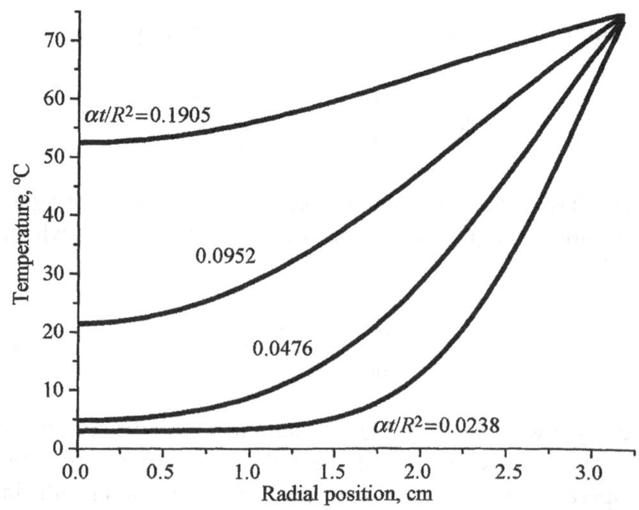

FIGURE 6.15. Computed temperature distributions for an acrylic plastic sphere immersed in a heated water bath maintained at 75°C. Curves are shown for $\alpha t/R^2$ ranging from 0.0238 to 0.1905.

6.8 KELVIN'S ESTIMATE OF THE AGE OF THE EARTH

It has occurred to many, including Fourier and Kelvin, that the age of the earth might be estimated from the known geothermal gradient at the surface. The earth is a composite sphere consisting of the crust (~10 km approximate thickness), the mantle (~2900 km), a liquid core (~2200 km), and a solid center. While the average density near the surface is about 2.8 g/cm^3, the core is much more dense, resulting in an average planetary specific gravity of about 5.5. As a result, the density, heat capacity, and thermal conductivity all change with depth and a descriptive equation for conduction in the interior must be written as

$$\frac{\partial}{\partial t}(\rho C_p T) = \left[\frac{1}{r^2}\frac{\partial}{\partial r}\left(r^2 k \frac{\partial T}{\partial r}\right)\right] + S_N. \qquad (6.73)$$

The source term S_N is added to account for the production of thermal energy by radioactive decay. Naturally, the production varies with rock type but a ballpark figure (per unit mass) is on the order of 2×10^{-6} cal/g per year. The thermal conductivity of the earth's crust is widely given as 0.004 cal/(cm s °C), whereas for solid nickel, k is about 0.14 cal/(cm s °C). The thermal conductivity of metals usually decreases a little for the molten state while the heat capacity changes only slightly. With the known inhomogeneities, solution of eq. (6.73) would not be easy; more important, it might not even be necessary.

Kelvin (1864) realized that only a small fraction of the earth's initial thermal energy has been lost. Consequently, if the cooling has been mainly confined to layers near the surface, then curvature can be neglected. By assuming that the surface temperature of the "young" earth instantaneously acquired a low value and neglecting the production of thermal energy, eq. (6.20) can be used to approximate T. Accordingly, we find at the surface

$$\left.\frac{\partial T}{\partial y}\right|_{y=0} = \frac{T_i}{\sqrt{\pi \alpha t}}. \qquad (6.74)$$

Measurements show that the geothermal gradient is on the order of 20°C per km, or roughly 2×10^{-4}°C per cm. If the initial temperature of the molten earth was 3800°C and the thermal diffusivity α taken to be 0.01 cm^2/s, then the required time for cooling would be about 3.65×10^8 years. In fact, Kelvin's original estimate was 94×10^6 years (see Carslaw and Jaeger, 1959, p. 85), which is of course contrary to all available geologic evidence. This analysis has three principal flaws: the earth (as noted above) is not homogeneous, the melting point of rock is affected by pressure, and heat is continuously being generated beneath the surface by radioactive species. Rather than simply adopting the error function solution, a more reasonable analysis might be made by numerical solution of

$$\frac{\partial}{\partial t}(\rho C_p T) = \frac{\partial}{\partial y}\left(k \frac{\partial T}{\partial y}\right) + S_N. \qquad (6.75)$$

The controversy engendered by Kelvin's estimate of 1864 persisted throughout the nineteenth century and the problem attracted many investigators, including Oliver Heaviside. In 1895, Heaviside used his operational method to solve the Kelvin problem for flow of heat in a body with spatially varying conductivity. His methods were largely discounted by mathematicians of the day; Heaviside lacked a formal education and his eccentricities contributed to biases against his work. Nevertheless, Kelvin himself expressed admiration for Heaviside (see Nahin, 1983). That may have been of little solace; Heaviside died impoverished in 1925 with his many contributions to the emerging field of electrical engineering unappreciated. The story of Oliver Heaviside is a sad footnote to the history of applied mathematics and it demonstrates how difficult it is for an unorthodox approach to find acceptance in the face of established authority.

6.9 SOME SPECIALIZED TOPICS IN CONDUCTION

6.9.1 Conduction in Extended Surface Heat Transfer

Extended surfaces, or fins, are used to cast off unwanted thermal energy to the surroundings; we can find specific applications in air-cooled engines, intercoolers for compressors, and heat sinks for electronic components and computer processors. Generally, such fins are constructed from high-conductivity metals like aluminium, copper, or brass, and they often have a large aspect ratio (thin relative to the length of projection into the fluid phase). Because they are made from materials with large conductivities, most of the resistance to heat transfer is in the fluid film surrounding the fin's surface. Under these conditions, we *may be able* to assume that the temperature in the fin is nearly constant with respect to transverse position, that is, the temperature is a function only of position along the major axis projecting away from the heated object. With these conditions in mind, we take the conduction equation and append a loss term using Newton's law of cooling. For example, consider a rectangular fin with width W and thickness b; it projects into the fluid a distance L in the $+y$-direction (Figure 6.16).

The governing equation for this steady-state case is

$$k\frac{d^2T}{dy^2} - \frac{2h}{b}(T - T_\infty) = 0. \qquad (6.76)$$

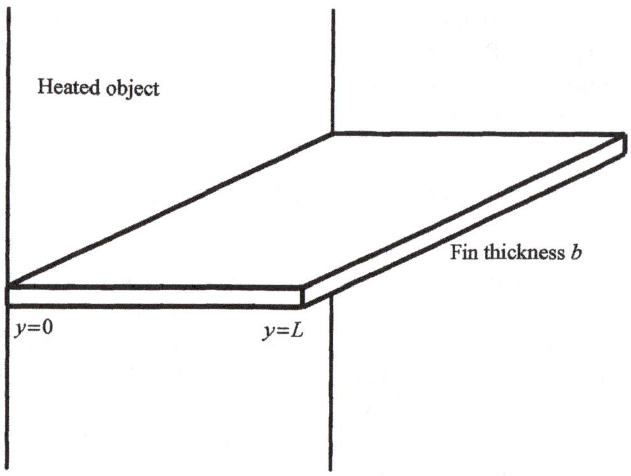

FIGURE 6.16. A rectangular fin of width W and thickness b. It projects into the fluid from the wall, from $y=0$ to $y=L$.

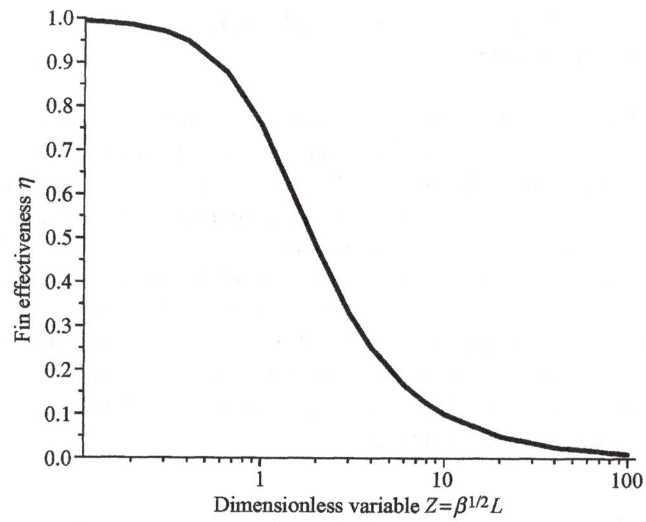

FIGURE 6.17. The effectiveness η of a rectangular fin as a function of dimensionless product, $Z = \sqrt{\beta L}$.

We set $\theta = (T - T_\infty)$ and let $\beta = 2h/bk$. At the wall we have an elevated temperature: at $y=0$, $T=T_0$. But what boundary condition shall we use at the end of the fin where $y=L$? There are at least three possibilities. If the fin is *very long*, we might take $T(y=L)=T_\infty$. If bW is only a small fraction of the surface area $2LW$, then we could assume that there is very little heat loss through the end of the fin: $\left.\frac{dT}{dy}\right|_{y=L} = 0$. If the loss through the end of the fin is significant, we must write a Robin's-type condition by equating the conductive flux with the Newton's law of cooling. If we employ the second option, the solution is

$$\frac{\theta}{\theta_0} = \frac{T - T_\infty}{T_0 - T_\infty} = \frac{e^{-\sqrt{\beta}L}e^{+\sqrt{\beta}y} + e^{+\sqrt{\beta}L}e^{-\sqrt{\beta}y}}{e^{-\sqrt{\beta}L} + e^{+\sqrt{\beta}L}}$$

$$= \cosh\sqrt{\beta}y - \left(\tanh\sqrt{\beta}L\right)\sinh\sqrt{\beta}y. \qquad (6.77)$$

The total heat loss from the fin is determined by integrating the flux $h(T - T_\infty)$ over the surface area (*both* sides). In 1923, Harper and Brown reported a study of the *effectiveness* of the rectangular fin; they formed a quotient comparing the total heat dissipated by the fin to the thermal energy that would be cast off if the entire fin were maintained at the wall temperature T_0.

$$\eta = \frac{2\int_0^L Wh(T - T_\infty)dy}{2\int_0^L Wh(T_0 - T_\infty)dy} = \frac{\tanh\sqrt{\beta}L}{\sqrt{\beta}L}. \qquad (6.78)$$

It is to be noted that the integrals in (6.78) are *over the surface* of the fin; since we have only a one-dimensional model, the integration with respect to z has been replaced by multiplication by the fin width W.

We should examine Figure 6.17 recalling that $\beta = 2h/bk$. We observe that the effectiveness of the fin is improved by an *increase in thermal conductivity* of the metal, an *increase in the thickness* of the fin, and a *decrease* in the magnitude of the heat transfer coefficient. One can perhaps imagine the difficulty faced by the heat transfer engineers as they struggled with these findings in the context of a demanding application such as an air-cooled aircraft engine illustrated in Figure 6.18. Finding the optimum fin length, spacing (pitch), and thickness for *all operating conditions* would be extremely challenging to say the least; in fact, it is clear from the historical record of the Boeing B-29 in World War II that satisfactory cooling was never achieved for the Wright 3350 engine (a two-row radial of about 2000 hp).

Next we consider a circular fin with thickness b mounted on a pipe or perhaps upon an air-cooled engine cylinder. The fin extends from the outer surface of the pipe ($r=R_1$) to the radial position $r=R_2$. The appropriate steady-state model is written as

$$k\frac{d}{dr}\left(r\frac{dT}{dr}\right) - \frac{2hr}{b}(T - T_\infty) = 0. \qquad (6.79)$$

We set $\beta = 2h/kb$ and let $\theta = T - T_\infty$. Thus,

$$\theta = AI_0\left(\sqrt{\beta}r\right) + BK_0\left(\sqrt{\beta}r\right). \qquad (6.80)$$

The boundary condition at $r=R_1$ is clear: $\theta = T_s - T_\infty$. But what about the edge of the fin at $r=R_2$? We have the same three possibilities as noted in the rectangular case above; we stipulate that the fin is quite thin relative to its length (projection), consequently,

$$A = B\frac{K_1\left(\sqrt{\beta}R_2\right)}{I_1\left(\sqrt{\beta}R_2\right)}. \qquad (6.81)$$

FIGURE 6.18. Close-up of a two-row radial engine that has been partially disassembled and sectioned for instructional purposes (photo courtesy of the author).

Once again we have determined the temperature distribution in the fin with relative ease. There are two aspects of these problems that the reader may wish to contemplate further. Is the temperature variation in the transverse (z-) direction really negligible, and under what circumstances will the heat transfer coefficient be independent of position/temperature?

Jakob (1949) reviewed results for other fin geometries, including triangular wedges and trapezoids. For the former, he shows that the governing equation is

$$\frac{d^2\theta}{dx^2} + \frac{1}{x}\frac{d\theta}{dx} - \frac{1}{x}\frac{hL}{ky_0}\theta = 0, \tag{6.82}$$

where x is measured from the point (vertex) of the fin toward the base (where the heated surface is located). The half-thickness of the wedge at the base is y_0 and the length of projection into the fluid phase is L. By defining a new independent variable ψ,

$$\psi = \frac{hL}{ky_0}x, \tag{6.83}$$

we find

$$\frac{d^2\theta}{d\psi^2} + \frac{1}{\psi}\frac{d\theta}{d\psi} - \frac{1}{\psi}\theta = 0, \quad \text{with the solution} \tag{6.84}$$

$$\theta = AI_0\left(2\sqrt{\psi}\right) + BK_0\left(2\sqrt{\psi}\right). \tag{6.85}$$

For boundary conditions, we have $d\theta/d\psi = 0$ at $\psi = x = 0$ and $\theta = T_s - T_\infty$ at $x = L$. It is appropriate for the reader to wonder whether wedge-shaped fins might violate one of our underlying assumptions—namely, that the temperature of the fin is essentially constant with respect to transverse position (perpendicular to the projection into the fluid phase). If the heat transfer coefficient (h) is unusually large (or if hL/k is large), then such a deviation can occur as illustrated by the temperature distribution in the triangular (wedge-shaped) fin shown in Figure 6.19.

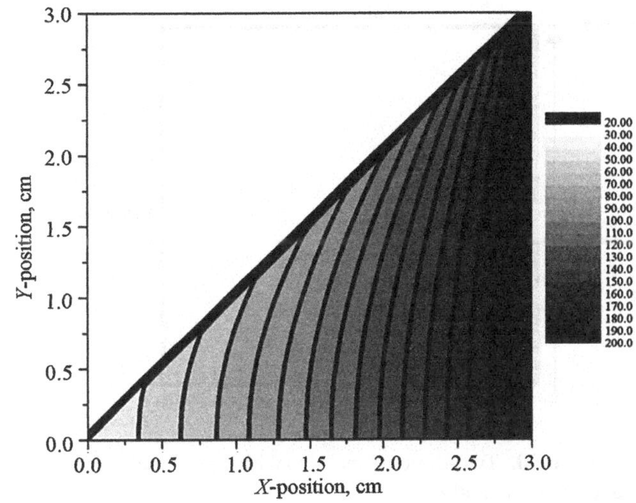

FIGURE 6.19. Example of a computed temperature distribution in the upper half of a wedge-shaped fin with a *very large* heat transfer coefficient. When h is small, the steep gradients are confined to regions very near the surface of the metal fin and the underlying assumptions of the analytic solution are satisfied.

6.9.2 Anisotropic Materials

We observed in the introduction that there are many materials with directional characteristics in their structures; familiar examples include carbon–fiber composites and wood. In the case of pine (wood), the thermal conductivities parallel and perpendicular to the board's face are reported to be

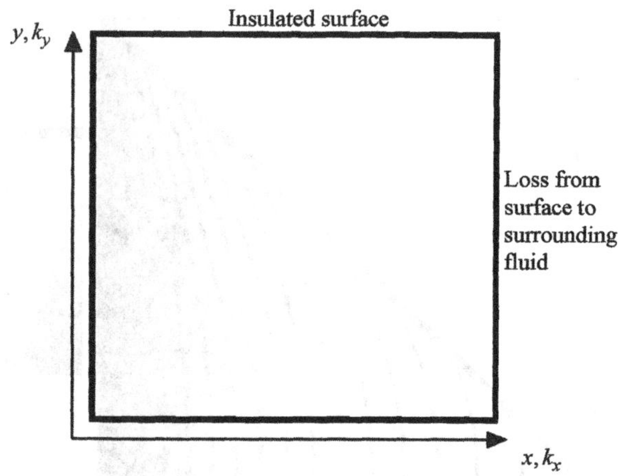

FIGURE 6.20. Two-dimensional slab with directionally dependent conductivities k_x and k_y.

0.000834 and 0.000361 cal/(cm s °C), respectively. Consequently, a transient conduction problem in a two-dimensional slab of such a material must begin with

$$\rho C_p \frac{\partial T}{\partial t} = \left[\frac{\partial}{\partial x}\left(k_x \frac{\partial T}{\partial x}\right) + \frac{\partial}{\partial y}\left(k_y \frac{\partial T}{\partial y}\right) \right]. \quad (6.86)$$

We will explore an example case in which a slab of pine has some initial temperature T_i. At $t = 0$, the temperatures of a couple of faces are instantaneously elevated to new (and possibly different) values (see Figure 6.20). Since the ratio of the conductivities k_x/k_y is about 2.31, we wonder if we can expect the developing temperature distribution in the slab to exhibit some interesting features.

Problems of this type are quite easily solved numerically (Figure 6.21)—the explicit algorithm for this problem can be rapidly coded in just about any high-level language as illustrated by the following example program (PBCC, PowerBASIC™ Console Compiler).

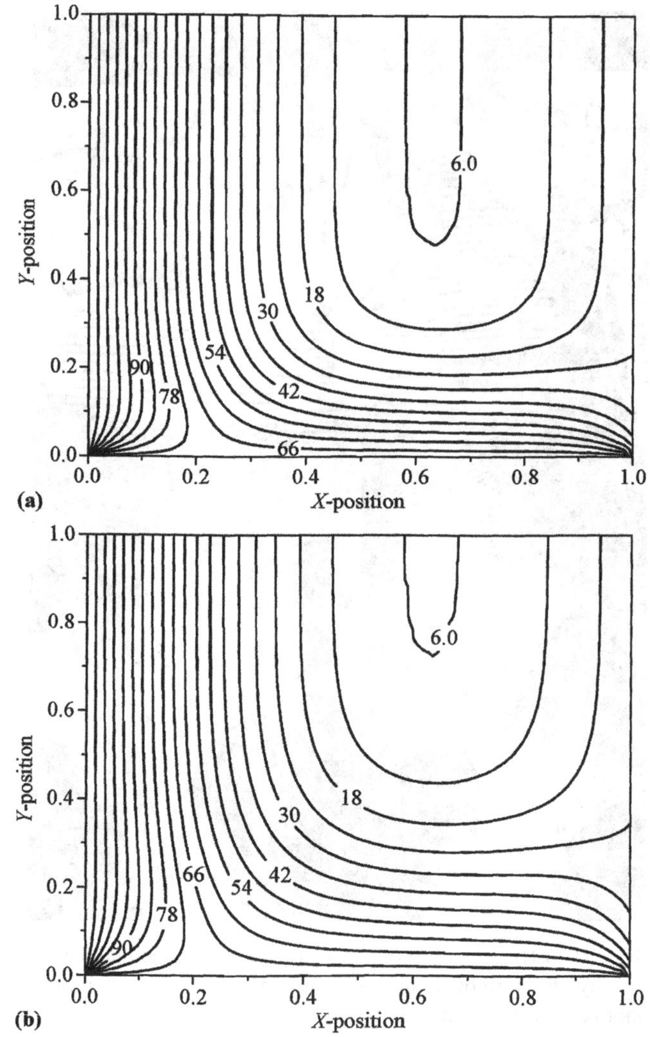

FIGURE 6.21. (a) and (b) Comparison of results with $k_x/k_y = 2.31$ (a) and $k_y = k_x$ (b). The contour plots are for $\alpha_x t/L^2 = 0.0166$. The differences become very subtle at larger t, with the main effect that thermal energy has been transported a little farther toward the top of the slab.

```
    #COMPILE EXE
#DIM ALL
  GLOBAL dx,dy,dt,kx,ky,rho,cp,ttime,tair,d2tdx2,d2tdy2,h,i,j AS SINGLE
FUNCTION PBMAIN
  DIM t(60,60,2) AS SINGLE
dx=0.0166667:dy=0.0166667:dt=0.01:kx=0.000834:ky=0.000361:rho=0.55:cp=0.42
  ttime=0:tair=25:h=0.02
REM *** initialize temp field
      FOR i=1 TO 59
          FOR j=1 TO 59
              t(i,j,1)=0
              NEXT j:NEXT i
          FOR j=0 TO 60
              t(0,j,1)=120:t(0,j,2)=120
```

```
            NEXT j
        FOR i=0 TO 60
            t(i,0,1)=70:t(i,0,2)=70
            NEXT i
REM *** perform interior computation
  100 FOR j=1 TO 59
      FOR i=1 TO 59
          d2tdx2=(t(i+1,j,1)-2*t(i,j,1)+t(i-1,j,1))/dx^2
          d2tdy2=(t(i,j+1,1)-2*t(i,j,1)+t(i,j-1,1))/dy^2
          t(i,j,2)=dt/(rho*cp)*(kx*d2tdx2+ky*d2tdy2)+t(i,j,1)
          NEXT i:NEXT j
REM *** top boundary
      FOR i=1 TO 59
          t(i,60,2)=(4*t(i,59,2)-t(i,58,2))/3
          NEXT i
REM *** far right boundary
      FOR j=1 TO 59
          t(60,j,2)=(h*dx/kx*tair+t(59,j,2))/(1+h*dx/kx)
          NEXT j
          t(60,60,2)=t(60,59,2):t(60,0,2)=t(60,1,2)
      ttime=ttime+dt
      PRINT ttime,t(30,30,2)
REM *** swap time values
          FOR i=0 TO 60
              FOR j=0 TO 60
                  t(i,j,1)=t(i,j,2)
                  NEXT j:NEXT i
      IF ttime>20 THEN 200 ELSE 100
REM *** write results to file
  200 OPEN ''c:tblock20.dat'' FOR OUTPUT AS #1
  FOR j=0 TO 60
      FOR i=0 TO 60
          WRITE#1,i*dx,j*dy,t(i,j,1)
          NEXT i:NEXT j
      CLOSE
END FUNCTION
```

6.9.3 Composite Spheres

As we saw previously, many problems of radially directed conduction in spheres can be transformed into simpler problems in slabs, we need only to set $\theta = rT$ and then adopt results from the equivalent problem in rectangular coordinates. However, there is a rather common exception. Consider a sphere comprised of multiple (two) layers, each with distinct thermal conductivity. Let material "1" extend from the center to $r = R_{12}$, and let material "2" extend from R_{12} to the surface at $r = R_s$. The governing equations are, of course,

$$\rho_1 C_{p1} \frac{\partial T_1}{\partial t} = k_1 \left[\frac{1}{r^2} \frac{\partial}{\partial r} \left(r^2 \frac{\partial T_1}{\partial r} \right) \right] \quad \text{and}$$

$$\rho_2 C_{p2} \frac{\partial T_2}{\partial t} = k_2 \left[\frac{1}{r^2} \frac{\partial}{\partial r} \left(r^2 \frac{\partial T_2}{\partial r} \right) \right]. \tag{6.87}$$

Clearly, both these equations can be readily transformed into "slab" versions. But for the boundary between the two materials, we must have

$$\text{at } r = R_{12}, \quad T_1 = T_2, \quad \text{and}$$

$$-k_1 \left. \frac{\partial T_1}{\partial r} \right|_{r=R_{12}} = -k_2 \left. \frac{\partial T_2}{\partial r} \right|_{r=R_{12}}. \tag{6.88}$$

It is the latter (equating the fluxes at the interface) that poses the problem; should we attempt the transformation, we find eq. (6.89) for the two temperature gradients:

$$\frac{\partial T_1}{\partial r} = \frac{1}{r} \frac{\partial \theta_1}{\partial r} - \frac{\theta_1}{r^2} \quad \text{and} \quad \frac{\partial T_2}{\partial r} = \frac{1}{r} \frac{\partial \theta_2}{\partial r} - \frac{\theta_2}{r^2}. \tag{6.89}$$

This is not a form that we have seen or employed in problems involving conduction in rectangular slabs. Carslaw and Jaeger (1959) observe that many problems involving conduction in composite materials can be solved by application of the Laplace transform, and they provide a solution for the composite sphere (see 13.9, VII, p. 351). We also note that this is the type of problem that confronted Kelvin in his attempt to estimate the age of the earth; Heaviside's operational method was later shown to be a subset of the Laplace transform technique.

We can find a familiar example of a composite sphere (and on a much smaller scale) in the golf ball. Modern golf balls have typical diameter and mass of about 42.68 mm and 45.63 g, respectively, producing a gross density of about 1.12 g/cm^3. In recent years, golf ball manufacturers have transitioned from rubber-wound, balata-covered balls with liquid centers to solid, multilayer balls with polybutadiene cores and Surlyn® (a copolymer of ethylene and methacrylic acid) or polyurethane covers. Depending upon the desired spin and flight characteristics, the ball may have two, three, or four layers. For golfers who play in cold weather, maintaining the desirable properties of the elastomer layers can be a challenge. Imagine, for example, that a ball starts out with an initial uniform temperature of 80°F (26.7°C). It might be put into play on a long hole and exposed continuously to an ambient temperature of 0°C for a period of 10–15 min. One can appreciate the importance of the temperature distribution in the ball; it would be necessary of course to evaluate the impact the cold might have upon the ball's coefficient of restitution (COR). We shall defer further exploration of this problem, saving it for a student exercise.

6.10 CONCLUSION

Most heat transfer processes in *fluids* utilize fluid motion, even if it is only inadvertent motion arising from localized buoyancy (natural convection). Indeed, in the chemical process industries, much effort is devoted to enhancing fluid motion to produce larger heat transfer coefficients and improve process efficiency. But in the solid phase, thermal energy is transferred molecule-to-molecule by conduction. Thus, it is not only an important transfer mechanism, it is often the *only significant mechanism* of heat transfer. Nowhere could one find a better contemporary (and critically important) example than in solid-state electronic devices; thermal energy is produced in such applications, and we typically have multilayer fabrication with different thermal conductivities in each layer. This is but one example of an application where the *conduction* of thermal energy may constrain both design and operation since power limitations are often imposed upon such devices by the rate of molecular transport of thermal energy.

REFERENCES

Carslaw, H. S. *An Introduction to the Theory of Fourier's Series and Integrals*, 3rd revised edition, Dover Publications, New York (1950).

Carslaw, H. S. and J. C. Jaeger. *Conduction of Heat in Solids*, 2nd edition, Oxford University Press, Oxford (1959).

Fourier, J. B. J. *On the Propagation of Heat in Solid Bodies*, Paris Institute (1807).

Harper, D. R. and W. R. Brown. Mathematical Equations for Heat Conduction in the Fins of Air-Cooled Engines. *NACA Report 158* (1923).

Herivel, J. *Joseph Fourier, the Man and the Physicist*, Clarendon Press, Oxford (1975).

Holman, J. P. *Heat Transfer*, 8th edition, McGraw-Hill, New York (1997).

Jakob, M. *Heat Transfer*, Vol. 1, Wiley, New York (1949).

Kelvin, Lord The Secular Cooling of the Earth. *Transactions of the Royal Society of Edinburgh*, 23:157 (1864).

Nahin, P. J. Oliver Heaviside: Genius and Curmudgeon. *IEEE Spectrum*, 20:63 (1983).

7

HEAT TRANSFER WITH LAMINAR FLUID MOTION

7.1 INTRODUCTION

Our consideration of heat transfer *with* fluid motion is initiated by extending equations (6.2) through (6.4) to include both the fluid velocity and the volumetric rate of thermal energy production (by unspecified mechanism):

$$\rho C_p\left(\frac{\partial T}{\partial t}+v_x\frac{\partial T}{\partial x}+v_y\frac{\partial T}{\partial y}+v_z\frac{\partial T}{\partial z}\right)=k\left[\frac{\partial^2 T}{\partial x^2}+\frac{\partial^2 T}{\partial y^2}+\frac{\partial^2 T}{\partial z^2}\right]+S, \tag{7.1}$$

$$\rho C_p\left(\frac{\partial T}{\partial t}+v_r\frac{\partial T}{\partial r}+\frac{v_\theta}{r}\frac{\partial T}{\partial \theta}+v_z\frac{\partial T}{\partial z}\right)=k\left[\frac{1}{r}\frac{\partial}{\partial r}\left(r\frac{\partial T}{\partial r}\right)+\frac{1}{r^2}\frac{\partial^2 T}{\partial \theta^2}+\frac{\partial^2 T}{\partial z^2}\right]+S, \tag{7.2}$$

$$\rho C_p\left(\frac{\partial T}{\partial t}+v_r\frac{\partial T}{\partial r}+\frac{v_\theta}{r}\frac{\partial T}{\partial \theta}+\frac{v_\phi}{r\sin\theta}\frac{\partial T}{\partial \phi}\right)=k\left[\frac{1}{r^2}\frac{\partial}{\partial r}\left(r^2\frac{\partial T}{\partial r}\right)+\frac{1}{r^2\sin\theta}\frac{\partial}{\partial\theta}\left(\sin\theta\frac{\partial T}{\partial\theta}\right)+\frac{1}{r^2\sin^2\theta}\frac{\partial^2 T}{\partial\phi^2}\right]+S. \tag{7.3}$$

You can see immediately that there has been a fundamental change in the level of complexity of the generalized problem. Consider (7.1) in three dimensions with an arbitrary flow field. The dependent variables are now T, v_x, v_y, v_z, and p. It will be necessary for us to solve the energy equation (7.1), all three components of the Navier–Stokes equation, and the equation of continuity, all simultaneously—a formidable task. Furthermore, the generalized production term S could be nonlinear in velocity (viscous dissipation S_v) or perhaps in temperature (chemical reaction S_c). For production by the viscous dissipation in rectangular coordinates, S_v is

$$S_v=2\mu\left[\left(\frac{\partial v_x}{\partial x}\right)^2+\left(\frac{\partial v_y}{\partial y}\right)^2+\left(\frac{\partial v_z}{\partial z}\right)^2\right]+\mu\left[\left(\frac{\partial v_x}{\partial y}+\frac{\partial v_y}{\partial x}\right)^2+\left(\frac{\partial v_x}{\partial z}+\frac{\partial v_z}{\partial x}\right)^2+\left(\frac{\partial v_y}{\partial z}+\frac{\partial v_z}{\partial y}\right)^2\right]. \tag{7.4}$$

We must be able to anticipate the circumstances for which production by (7.4) may become important. Consider a shaft 2 in. in diameter rotating at 2000 rpm in a journal bearing, and assume that the gap between the surfaces is 0.0015 in. At the shaft surface, the tangential velocity will be about 532 cm/s and the velocity gradient (neglecting curvature) will be 139,633 s^{-1}. If the viscosity of the lubricating oil is 2.9 cp, then $S_v \approx 13.5$ cal/(cm^3 s). Clearly, small clearances with large velocity differences will lead to significant production of thermal energy.

In the case of S_c, assuming a first-order, elementary, exothermic chemical reaction,

$$S_c=k_0\exp\left(-\frac{E}{RT}\right)C_A\left|\Delta H_{rxn}\right|. \tag{7.5}$$

Equation 7.5 indicates that rapid kinetics, combined with a strongly exothermic reaction, can make S_c very large indeed.

Transport Phenomena: An Introduction to Advanced Topics, By Larry A. Glasgow

What is the magnitude of a large ΔH_{rxn}? For the combustion of propane at 25°C, $\Delta H_C = -530.6$ kcal/gmol.

In addition to the possibility of thermal energy production, we encounter two other common problems in cases where viscous fluid flow is combined with heat transfer: buoyancy resulting from a localized reduction in density (gases and liquids), and viscosity reduction (liquids) resulting from elevated temperatures. With regard to buoyancy, if the change in ρ is not too great, then we can modify the equation of motion by adding the Boussinesq approximation; this consists of a term (force per unit volume) appended to an equation of motion such as

$$\rho \frac{Dv}{Dt} = \mu \left[\frac{\partial^2 v_z}{\partial y^2} \right] + \rho g \beta (T - T_\infty), \quad (7.6)$$

where β is the coefficient of volumetric expansion. The mean density is used in 7.6; note that ρ is *not included* in the substantial time derivative on the left-hand side of the equation. This cannot be correct. Nevertheless, the Boussinesq approximation works well for many free (or natural) convection problems when the driving force is not too large. We will study several examples later in this chapter.

The problem posed by viscous liquids with $\mu = \mu(T)$ is also familiar; we will look at four examples in Figure 7.1. Note how the viscosities of glycerol and castor oil decrease by *two* orders of magnitude over this temperature range. For the lower temperatures, the viscosity data for castor oil are roughly described by

$$\mu = \mu_0 \exp \left[-\frac{A(T - T_0)}{T_0} \right], \quad (7.7)$$

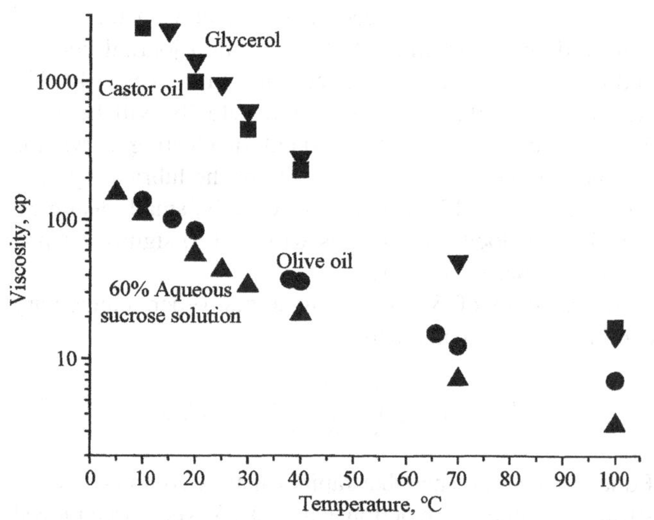

FIGURE 7.1. Viscosity in centipoises for glycerol, castor oil, olive oil, and a 60% aqueous sucrose solution between 10 and 100°C. These data were adapted from Lange (1961) and DOW™.com.

with $\mu_0 = 2420$ cp ($T_0 = 10$°C) and $A = 0.81$. In addition to oils, many organic liquids such as phenols, glycols, and alcohols exhibit pronounced $\mu(T)$. In these cases, we must expect coupling between the energy and momentum equations.

We conclude this introduction by looking at a familiar example that serves to underscore how effectively a fluid motion can be used to increase heat (or mass) transfer. Suppose we immerse a slightly heated spherical object in a quiescent fluid, such that heat transfer in the fluid occurs solely by conduction and at a very low rate. The fluid phase process will be approximately described by

$$0 = \frac{1}{r^2} \frac{d}{dr} \left(r^2 \frac{dT}{dr} \right). \quad (7.8)$$

Integrating this equation and using the boundary conditions

$$\text{at } r = R, T = T_S \quad \text{and} \quad \text{at } r \to \infty, T = T_\infty$$

allow us to find the first constant of integration: $C_1 = R(T_S - T_\infty)$. The flux of thermal energy away from the object is now written with both Fourier's law and Newton's "law" of cooling and the two expressions (both on the fluid side) are equated:

$$\frac{k}{R}(T_S - T_\infty) = h(T_S - T_\infty). \quad (7.9)$$

Obviously, the *limiting* Nusselt number *hd/k* for a sphere is 2. Now we have a convenient opportunity to assess the importance of fluid motion to the heat transfer process. Imagine that the fluid is moved past the sphere at such a velocity that analytic solution is no longer possible. Ranz and Marshall (1952) developed a correlation for this case:

$$Nu = \frac{hd}{k} = 2 + 0.6\, Re^{1/2}\, Pr^{1/3}. \quad (7.10)$$

Consequently, if we move water past a 10 cm diameter sphere at 300 cm/s with $T_S = 90$°C and $T_\infty = 20$°C, then $Nu \approx 660$, which is more than 300 times larger than the limiting value. Even *modest* fluid motions will greatly enhance heat and mass transfer.

7.2 PROBLEMS IN RECTANGULAR COORDINATES

Consider a pressure-driven flow occurring between two planar surfaces, separated by a distance of $2B$, with a constant heat flux at both surfaces. The arrangement is illustrated in Figure 7.2.

The velocity distribution in the duct is given by

$$v_z = \frac{1}{2\mu} \frac{dp}{dz} (y^2 - B^2), \quad (7.11)$$

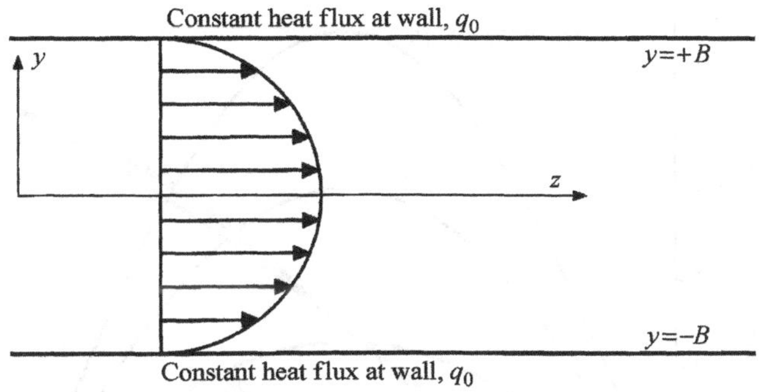

FIGURE 7.2. Poiseuille flow in a semi-infinite duct with constant heat flux at the walls.

and the governing equation for this situation is

$$\rho C_p v_z \frac{\partial T}{\partial z} = k\left[\frac{\partial^2 T}{\partial y^2} + \frac{\partial^2 T}{\partial z^2}\right]. \quad (7.12)$$

Note that axial conduction has been included in (7.12), although for this particular problem, $\partial^2 T/\partial z^2 = 0$. Why? We should also ask under what conditions may axial conduction be safely neglected in more general heat transfer problems? To help us answer this question, we shall put velocity (v_z) and position (y and z) into dimensionless forms:

$$v_z^* = v_z/\langle v_z\rangle, \qquad y^* = y/(2B), \qquad \text{and} \qquad z^* = z/(2B).$$

The result is (verify for yourself)

$$v_z^* \frac{\partial T}{\partial z^*} = \frac{1}{Re\,Pr}\left[\frac{\partial^2 T}{\partial y^{*2}} + \frac{\partial^2 T}{\partial z^{*2}}\right]. \quad (7.13)$$

For the tube flow with heat transfer into the fluid, Singh (1958) demonstrated that axial conduction is unimportant if the product $RePr$ is greater than 100. We can assess what this condition means with respect to Reynolds number by looking at Pr's for some familiar liquids. For water, n-butyl alcohol, and light lubricating oil (all at 60°F), we find the Prandtl numbers of 8.03, 46.6, and 1170, respectively. In these cases, the Reynolds number does not have to be very large for the condition to be satisfied. Turning our attention back to the problem at hand,

$$\frac{1}{2\mu}\frac{dp}{dz}(y^2 - B^2)\frac{dT_m}{dz} = \alpha\frac{d^2 T}{dy^2}. \quad (7.14)$$

Note that $\partial T/\partial z$ has been replaced by dT_m/dz; the latter is a constant (if the heat transfer coefficient h is fixed) and a simple energy balance will show that the bulk fluid temperature must increase linearly in the flow direction. The resulting temperature distribution is

$$T - T_s = \frac{1}{2\alpha\mu}\frac{dp}{dz}\frac{dT_m}{dz}\left(\frac{y^4}{12} - \frac{B^2 y^2}{2} + \frac{5B^4}{12}\right). \quad (7.15)$$

From an engineering perspective, we are likely to be interested in the Nusselt number (or heat transfer coefficient h). Since $q = h(T_s - T_m)$, we must evaluate the bulk fluid temperature from (7.15) and the velocity distribution. The result will be a seventh-degree polynomial in y, to be evaluated from 0 to B, yielding

$$T_m - T_s = -\frac{0.4857}{\alpha}\langle v_z\rangle\frac{dT_m}{dz}B^2. \quad (7.16)$$

The Nusselt number for this case, $2hB/k$, is 8.235.

7.2.1 Couette Flow with Thermal Energy Production

Production of thermal energy by viscous dissipation is expected in lubrication problems, as we saw previously. Consider a Couette flow in a rectangular geometry with the upper planar surface moving at a constant velocity V, as shown in Figure 7.3. The plates are separated by a small distance δ, so the velocity gradient is large.

For this case, we have

$$\rho C_p v_z \frac{\partial T}{\partial z} = k\left[\frac{\partial^2 T}{\partial y^2}\right] + \mu\left(\frac{\partial v_z}{\partial y}\right)^2. \quad (7.17)$$

If external cooling is used to maintain the surface temperatures at T_0 (both sides), then the problem is described by

$$\frac{d}{dy}\left(\frac{dT}{dy}\right) = -\frac{\mu}{k}\frac{V^2}{\delta^2}. \quad (7.18)$$

And then the temperature distribution in the fluid is given by

$$T - T_0 = \frac{\mu}{2k}\frac{V^2}{\delta^2}(\delta y - y^2). \quad (7.19)$$

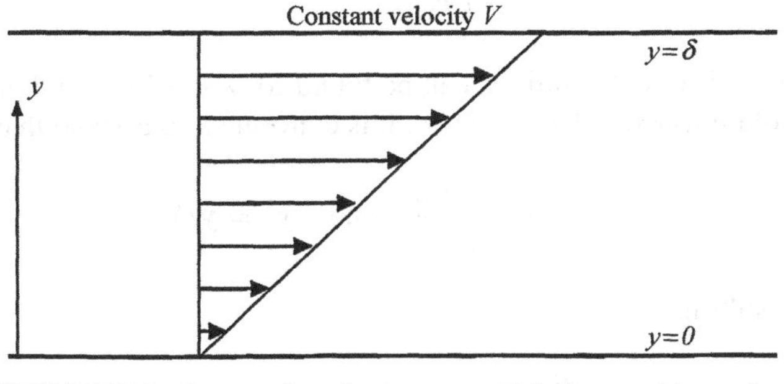

FIGURE 7.3. Couette flow between parallel planes with production of thermal energy by viscous dissipation.

It is apparent from (7.19) that the maximum temperature (at the center of the duct) is simply

$$T_{\max} - T_0 = \frac{\mu V^2}{8k}. \tag{7.20}$$

Selecting values for viscosity, plate velocity, and thermal conductivity, for example, 15 cp, 50 ft/s, and 0.00065 cal/(cm s °C), respectively, we find a centerline temperature rise of 1.6°C. Under more severe conditions, however, the temperature increase may be large enough to significantly affect viscosity. This will distort the velocity distribution and require solution of coupled differential equations. A classic illustration of this situation follows.

7.2.2 Viscous Heating with Temperature-Dependent Viscosity

The Gavis–Laurence problem is a modification of the previous example. Two planar surfaces are separated by a distance δ; the upper plate moves with velocity V and the lower surface is fixed. The viscosity of the liquid is taken to be a sensitive function of temperature, approximately described by

$$\mu = \mu_0 \exp\left[-\frac{A(T - T_0)}{T_0}\right]. \tag{7.21}$$

In this case, the momentum and energy equations are written as

$$\frac{d}{dy}\left(\mu(T)\frac{dv_z}{dy}\right) = 0 \tag{7.22}$$

and

$$k\frac{d^2T}{dy^2} + \mu\left(\frac{dv_z}{dy}\right)^2 = 0. \tag{7.23}$$

Gavis and Laurence (1968) demonstrated that a unique solution for the temperature profile exists *only* when

$$\lambda = \frac{A\tau_0^2\delta^2}{kT_0\mu_0} = 3.5138. \tag{7.24}$$

Two different solutions can be found for $\lambda < 3.5138$ and no solutions exist if $\lambda > 3.5138$. It is convenient to assume that

$$\theta = \frac{A(T - T_0)}{T_0} \quad \text{and} \quad y^* = y/\delta,$$

resulting in

$$\frac{d^2v^*}{dy^{*2}} - \left(\frac{d\theta}{dy^*}\right)\left(\frac{dv^*}{dy^*}\right) = 0 \tag{7.25}$$

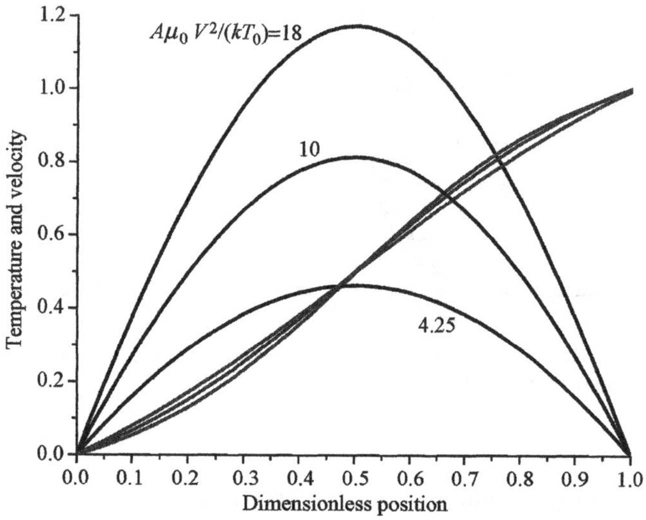

FIGURE 7.4. Characteristic results for the Gavis–Laurence problem. The velocity and the temperature distributions are shown (both dimensionless). The parameter $A\mu_0 V^2/(kT_0)$ was assigned the values 4.25, 10, and 18. The effect of $\mu(T)$ upon the velocity distribution is subtle.

and

$$\frac{d^2\theta}{dy^{*2}} + \frac{A\mu_0 V^2}{kT_0}\exp(-\theta)\left(\frac{dv^*}{dy^*}\right)^2 = 0. \tag{7.26}$$

The boundary conditions for this problem are

$$\begin{aligned} &\text{at } y^* = 0,\ \theta = 0 \text{ and } v^* = 0,\\ &\text{at } y^* = 1,\ \theta = 0 \text{ and } v^* = 1. \end{aligned} \tag{7.27}$$

The Gavis–Laurence problem is particularly interesting because of the existence of multiple solutions. One might ask whether this is merely another curious example of the behavior of nonlinear equations, or a direct result of the functional choice for $\mu(T)$. One should also think whether the nonunique temperature profiles would be physically realizable in such an apparatus. Some typical results for this problem are shown in Figure 7.4; note how the viscosity variation distorts the velocity profiles.

7.2.3 The Thermal Entrance Region in Rectangular Coordinates

We now wish to consider M. Andre Leveque's treatment of heat transfer from a flat surface (maintained at elevated temperature) to a fluid whose velocity distribution can, at least locally, be described by $v_x = cy$. The situation is as depicted in Figure 7.5.

Although Leveque is mentioned by name by Schlichting (1968) and Knudsen and Katz (1958), his work is often omitted from contemporary texts and monographs in heat

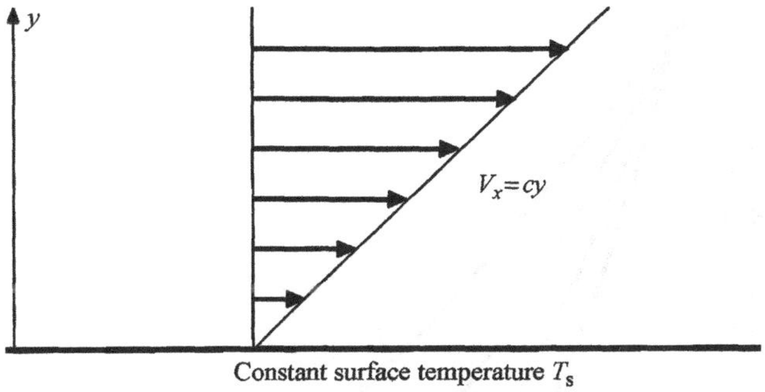

FIGURE 7.5. Heat transfer to a moving fluid from a plate maintained at T_s. The fluid motion (close to the wall) is approximately described by $v_x = cy$.

transfer. Niall McMahon (2004) of the Dublin City University observed that there is very little online information available about Leveque. McMahon notes that Leveque's dissertation entitled "Les Lois de la Transmission de Chaleur par Convection" was submitted in Paris in 1928. Some portions of it were also published in *Annales des Mines*, 13:210, 305, and 381 (Leveque, 1928). Leveque's development is practically useful in both heat and mass transfer; for a case in point, you may refer to pages 397 and 398 in Bird et al. (2002).

We shall assume that the appropriate form of the energy equation is

$$v_x \frac{\partial T}{\partial x} = \alpha \frac{\partial^2 T}{\partial y^2}. \tag{7.28}$$

Note that once again axial conduction has been neglected. Recall our earlier observation regarding the Peclet number *Pe* ($Pe = RePr$). Generally speaking, the local Nusselt number increases dramatically as the Peclet number exceeds about 100, and axial conduction becomes unimportant. However, the Leveque case offers us another line of reasoning. Consider the two second derivatives:

$$\frac{\partial^2 T}{\partial x^2} \quad \text{and} \quad \frac{\partial^2 T}{\partial y^2}.$$

Suppose we sought a crude dimensionally correct representation for these derivatives. We would need to select characteristic lengths in both the x- and y-directions. Since the thermal energy is just beginning to penetrate the moving fluid, an appropriate Δy can be many times smaller than an appropriate length in the flow direction (Δx). Furthermore, these widely disparate lengths must be squared, increasing the relative importance of transverse conduction.

Assuming $v_x = cy$ and defining a new independent variable η,

$$\eta = y\left(\frac{c}{9\alpha x}\right)^{1/3}, \tag{7.29}$$

we obtain an ordinary differential equation:

$$\frac{d^2 T}{d\eta^2} + 3\eta^2 \frac{dT}{d\eta} = 0. \tag{7.30}$$

We reduce the order of the equation (by letting $\phi = dT/d\eta$, for example) and integrate twice, resulting in

$$T = \int_0^{\eta} C_1 \exp(-\eta^3) d\eta + C_2. \tag{7.31}$$

You should verify that

$$C_1 = \frac{T_\infty - T_s}{\Gamma(4/3)} \quad \text{and} \quad C_2 = T_s. \tag{7.32}$$

The local Nusselt number is evaluated by equating both Fourier's law and Newton's law of cooling:

$$Nu_x = \frac{hx}{k} = \frac{(c/9\alpha)^{1/3} x^{2/3}}{\Gamma(4/3)}. \tag{7.33}$$

This is a significant result, of value to us for both heat and mass transfer in cases where the assumed linear velocity profile is a reasonable approximation. In entrance problems where the penetration of heat or mass from the wall into the moving fluid is just getting started, the Leveque solution is quite accurate. Results are provided in Figure 7.7 for dimensionless temperature θ as a function of η. We define the dimensionless temperature as $\theta = (T - T_s)/(T_\infty - T_s)$.

We can look at an example using these results; from Figure 7.6 we note that $\theta \approx 0.9$ for $\eta = 1$. Assume water is flowing past a heated plate with $c = 10\,\text{s}^{-1}$ and $\alpha = 0.00141\,\text{cm}^2/\text{s}$. If we set $x = 10$ cm, we find $Nu_x = 48$; the y-position corresponding to $\eta = 1$ is just 0.233 cm. If the water approaches

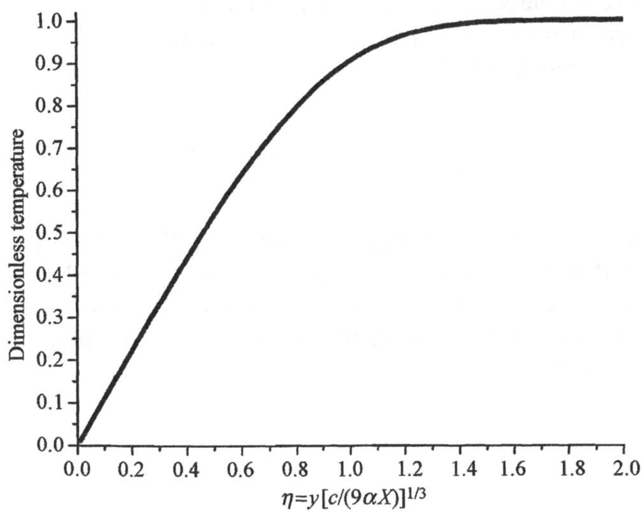

FIGURE 7.6. Results from the Leveque analysis of heat transfer to a moving fluid from a plate maintained at temperature T_s.

the heated plate at 55°F and if $T_s = 125$°F, then the temperature at the chosen location is 62°F. Under these conditions, the penetration of thermal energy into the flowing liquid is slight and the Leveque analysis gives excellent results.

7.2.4 Heat Transfer to Fluid Moving Past a Flat Plate

When a fluid at temperature T_∞ flows past a heated plate maintained at T_w, a thermal boundary layer will develop analogous to the momentum boundary layer that we discussed in Chapter 4. If we neglect buoyancy and the variation of viscosity with temperature, then the momentum transfer is decoupled from the energy equation and the flow field can be determined independently using the Prandtl equations:

$$\frac{\partial v_x}{\partial x} + \frac{\partial v_y}{\partial y} = 0 \tag{7.34}$$

and

$$v_x \frac{\partial v_x}{\partial x} + v_y \frac{\partial v_x}{\partial y} = \nu \frac{\partial^2 v_x}{\partial y^2}. \tag{7.35}$$

To include heat transfer, we must add the energy equation; if we allow the possibility of energy production by viscous dissipation, we obtain

$$v_x \frac{\partial T}{\partial x} + v_y \frac{\partial T}{\partial y} = \alpha \frac{\partial^2 T}{\partial y^2} + \frac{\mu}{\rho C_p}\left(\frac{\partial v_x}{\partial y}\right)^2. \tag{7.36}$$

You should be struck by the similarity between (7.35) and (7.36). In fact, if we omit thermal energy production and set $\nu = \alpha$ (i.e., $Pr = 1$), the two equations are the same and the dimensionless velocity distribution (which we determined previously) *is* the solution for the heat transfer problem as well. Thus, under these conditions,

$$\frac{T - T_w}{T_\infty - T_w} = \frac{v_x}{V_\infty} = f'(\eta). \tag{7.37}$$

Obviously, this is a special case and we will find soon that the Prandtl number will affect the temperature distribution significantly. We recall from Chapter 4 that Blasius defined a similarity variable η and incorporated the stream function ψ such that

$$\eta = y\left(\frac{V_\infty}{\nu x}\right)^{1/2}, \quad v_x = V_\infty f'(\eta), \quad \text{and}$$

$$v_y = \frac{1}{2}\left(\frac{\nu V_\infty}{x}\right)^{1/2}(\eta f' - f). \tag{7.38}$$

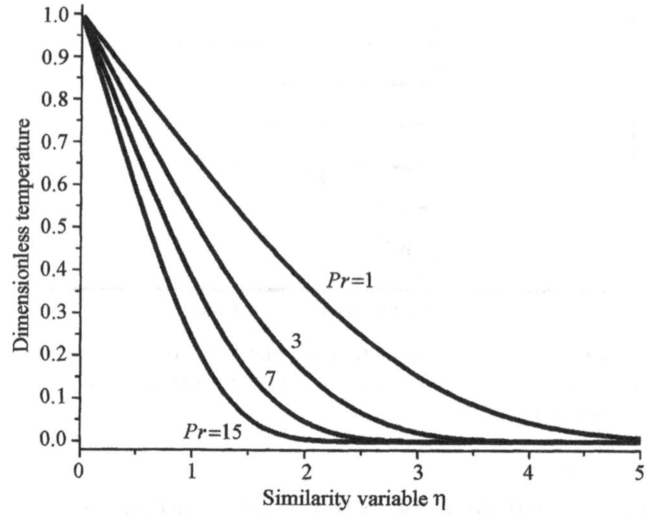

FIGURE 7.7. Dimensionless temperature distributions for the flow past a flat plate with heat transfer from the plate to the fluid for the Prandtl numbers of 1, 3, 7, and 15 *without* the production of thermal energy by viscous dissipation ($T_w = 65$°C and $T_\infty = 20$°C).

The momentum transport problem is then governed by the Blasius equation

$$f''' + \frac{1}{2} f f'' = 0, \tag{7.39}$$

and under the circumstances described here, we can solve the flow problem independently of (7.36). If we do not impose any restriction upon the Prandtl number and if we include production of thermal energy by viscous dissipation, then the energy equation (7.36) is transformed to the ordinary differential equation:

$$\frac{d^2T}{d\eta^2} + \frac{Pr}{2} f \frac{dT}{d\eta} = -Pr \frac{V_\infty{}^2}{2C_p}(f'')^2. \tag{7.40}$$

It is apparent from (7.40) that the Prandtl number will affect the temperature distribution; this is confirmed by the computational results shown in Figure 7.7. How significant will the *Pr* effect be? Consider the following abbreviated list of *Pr*'s (rough values for approximate ambient conditions)—these data show that even among the common fluids, we see variations in *Pr* over many orders of magnitude.

	Prandtl Number
Mercury (Hg)	4.6×10^{-6}
Air	0.7
Water	7
Ethylene glycol	200
Engine oil	10,000

Note the effect of *Pr* upon the temperature distributions in Figure 7.7: You can see that if the Prandtl number is large,

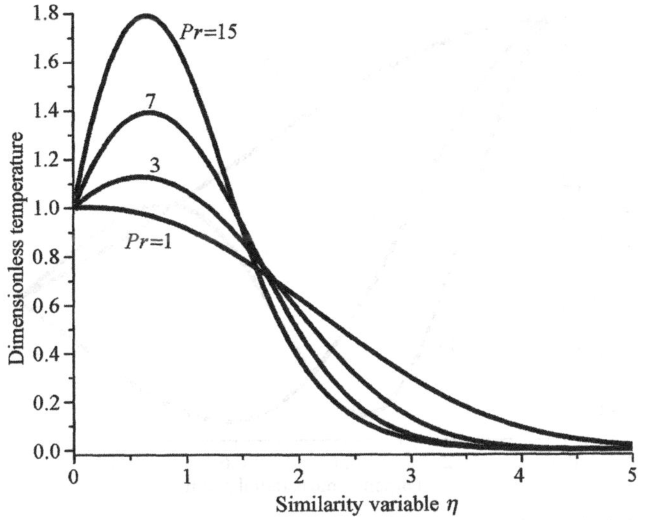

FIGURE 7.8. Dimensionless temperature distributions for the flow past a flat plate with heat transfer for the Prandtl numbers of 1, 3, 5, and 15 *including strong* production of thermal energy by viscous dissipation ($T_w = 65°C$ and $T_\infty = 20°C$).

the thermal penetration is limited as expected. The local heat flux is given by

$$q_y(x) = -k\left.\frac{\partial T}{\partial y}\right|_{y=0} = -k\left(\frac{V_\infty}{\nu x}\right)^{1/2}\left.\frac{dT}{d\eta}\right|_{\eta=0}. \quad (7.41)$$

For the results shown in Figure 7.7 ($T_w = 65°C$ and $T_\infty = 20°C$), the correct values for $dT/d\eta$ at $\eta = 0$ are -14.9425, -21.827, -29.066, and -37.5346 for the Prandtl numbers of 1, 3, 7, and 15, respectively. How might we expect the temperature distributions to change if we include production by viscous dissipation?

The data in Figure 7.8 show that the production of thermal energy by viscous dissipation will be especially significant at the larger Prandtl numbers (of course, since the viscosity is high relative to the thermal diffusivity). At $Pr = 15$, the maximum temperature occurs at $\eta \approx 0.6$; compare the curves here with the corresponding distributions shown in Figure 7.7.

7.3 PROBLEMS IN CYLINDRICAL COORDINATES

We begin with fully developed laminar flow in a tube with constant heat flux at the wall, as illustrated in Figure 7.9.

We assume that the heat transfer coefficient does not vary with axial position. This is equivalent to setting

$$\frac{\partial}{\partial z}\left(\frac{T_s - T}{T_s - T_m}\right) = 0. \quad (7.42)$$

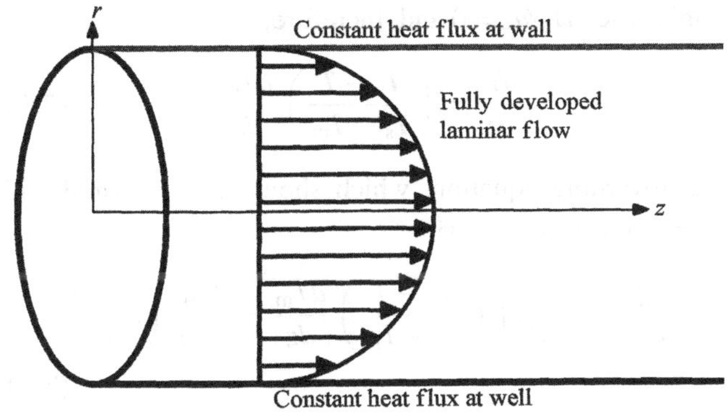

FIGURE 7.9. Heat transfer to fully developed laminar flow in a tube with constant heat flux q_s at the wall.

By the Newton's "law" of cooling, $q_s = h(T_s - T_m)$, and since both h and q_s are constants, we conclude

$$\frac{\partial T}{\partial z} = \frac{dT_m}{dz} = \frac{dT_s}{dz}. \quad (7.43)$$

It is important that the reader understand that both the bulk fluid and wall temperatures (T_m and T_s) increase linearly in the flow direction. Substitution of the parabolic velocity distribution into the energy equation results in

$$\frac{2\langle v_z\rangle}{\alpha}\left[1 - \frac{r^2}{R^2}\right]\frac{dT_m}{dz} = \frac{1}{r}\frac{d}{dr}\left(r\frac{dT}{dr}\right). \quad (7.44)$$

We integrate twice noting that $dT/dr = 0$ at the centerline and that $T = T_s$ at the wall. The result is

$$T - T_s = -\left(\frac{2\langle v_z\rangle}{\alpha}\right)\frac{dT_m}{dz}\left(\frac{r^4}{16R^2} - \frac{r^2}{4} + \frac{3R^2}{16}\right). \quad (7.45)$$

For engineering purposes, we may be more interested in either the heat transfer coefficient or the Nusselt number $Nu = hd/k$. This will require that we determine the bulk fluid temperature by integration:

$$T_m - T_s = -\frac{11}{96}\left(\frac{2\langle v_z\rangle}{\alpha}\right)\frac{dT_m}{dz}R^2. \quad (7.46)$$

We use the defining equation for h and an energy balance for the slope (rate of change of T in the flow direction) of the bulk fluid temperature to show

$$Nu = \frac{hd}{k} = \frac{194}{44} = 4.3636. \quad (7.47)$$

We should contrast this result with the case of constant wall temperature that might, for example, be achieved by the condensation of saturated steam on the outside of the tube.

In this case, $dT_s/dz = 0$ and, therefore,

$$\frac{\partial T}{\partial z} = \left(\frac{T_s - T}{T_s - T_m}\right)\frac{dT_m}{dz}. \tag{7.48}$$

The governing equation, which should be compared with (7.44), can be written as

$$\frac{2\langle v_z\rangle}{\alpha}\left[1 - \frac{r^2}{R^2}\right]\left(\frac{T_s - T}{T_s - T_m}\right)\frac{dT_m}{dz} = \frac{1}{r}\frac{d}{dr}\left(r\frac{dT}{dr}\right). \tag{7.49}$$

This is clearly a more complicated situation than the constant heat flux case. The solution can be found by successive approximation; $T(r)$ for the constant q_s case is substituted into the left-hand side of (7.49) and a new $T_1(r)$ is found. Of course, the bulk fluid temperature T_m must be found by integration as well. The process is repeated until the Nusselt number attains its ultimate value 3.658. Note that this value is about 16% lower than the constant heat flux case. We can understand this difference by intuiting the shapes of the temperature profiles for the two cases. What effect will the constant wall temperature have upon $T(r)$ for $r \to R$?

7.3.1 Thermal Entrance Length in a Tube: The Graetz Problem

Suppose that the velocity distribution in a tube is fully developed prior to the contact with a heated section of a tube wall. At this point, say $z = 0$, the fluid has a uniform temperature of T_∞. It is convenient to let $r^* = r/R$, $z^* = z/R$, and $\theta = (T - T_s)/(T_\infty - T_s)$. Since the velocity distribution is given by

$$v_z = 2\langle v_z\rangle\left[1 - \frac{r^2}{R^2}\right],$$

the appropriate energy equation can be written as

$$[1 - r^{*2}]\frac{\partial\theta}{\partial z^*} = \frac{1}{Re\,Pr}\frac{1}{r^*}\frac{\partial}{\partial r^*}\left(r^*\frac{\partial\theta}{\partial r^*}\right). \tag{7.50}$$

This equation is a candidate for separation; we let $\theta = f(r^*)g(z^*)$. The resulting differential equation for g is elementary, yielding

$$g = C_1\exp\left(-\frac{\lambda^2}{Re\,Pr}z^*\right). \tag{7.51}$$

However, the equation for f is of the Sturm–Liouville type:

$$\frac{d^2 f}{dr^{*2}} + \frac{1}{r^*}\frac{df}{dr^*} + \lambda^2(1 - r^{*2})f = 0. \tag{7.52}$$

Despite appearances, the solution of (7.52) cannot be expressed in terms of Bessel functions. Equation (7.52) can be

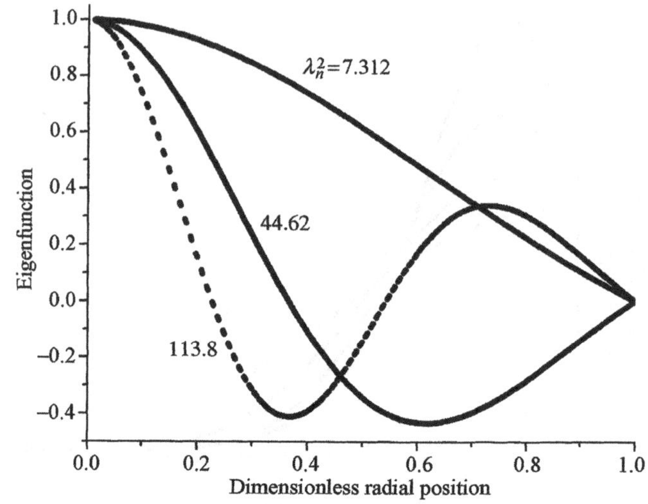

FIGURE 7.10. The first three eigenfunctions for the Graetz problem with λ_n^2: 7.312, 44.62, and 113.8.

solved numerically as a characteristic value problem and, of course, there are an infinite number of λ's that produce valid solutions. A Runge–Kutta scheme can be used to identify values for λ_n:

n	= 1	2	3	4	5
λ_n^2	= 7.312	44.62	113.8	215.2	348.5

The eigenfunctions obtained with the first three of these parametric values are shown in Figure 7.10. As one might expect, the series begins to converge rapidly as z^* increases. It is common practice to write the solution as

$$\theta = \sum_{n=1}^{\infty} C_n f_n(r^*)\exp\left(-\frac{\lambda_n^2}{Re\,Pr}z^*\right). \tag{7.53}$$

Note that for $z = 0$, $\theta = 1$; this suggests the use of orthogonality for determination of the C_n's. Jakob (1949) and Sellars et al. (1956) summarize the procedure (which was developed by Graetz, 1885). The eigenfunctions are orthogonal on the interval 0–1 using the weighting function $r^*(1 - r^{*2})$. The resulting coefficients are

$$+1.480 - 0.8035 + 0.5873 - 0.4750 + 0.4044 - 0.3553$$
$$+0.3189 - 0.2905 + 0.2677 - 0.2489 \text{ etc.}$$

The temperature distribution itself may be of less interest in engineering applications than the rate of heat flow, but we can differentiate and set $r^* = 1$ to find the heat flux at the wall; for the Graetz problem, the result is

$$q_w = -\frac{k}{R}\sum_{n=1}^{\infty} C_n f'_n(1)\exp\left(-\frac{\lambda_n^2}{Re\,Pr}z^*\right)(T_w - T_\infty). \tag{7.54}$$

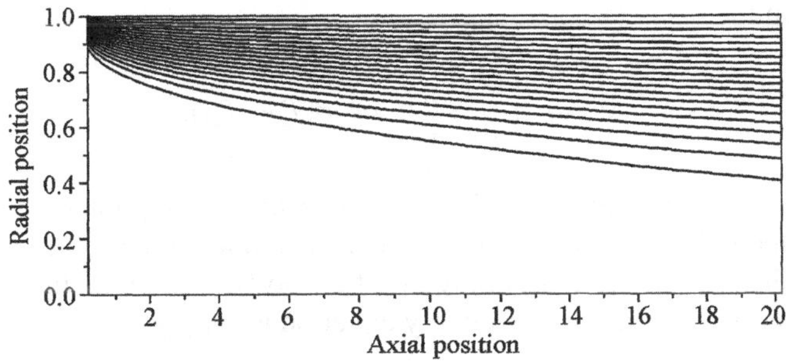

FIGURE 7.11. Development of the thermal boundary layer for the classical Graetz problem with $RePr = 1000$. The scale for the radial direction has been greatly expanded and the computation carried out to an axial (z-) position of 20 radii.

And the local Nusselt number can be written as

$$Nu = \frac{1/2\sum_{n=1}^{\infty} C_n f'_n(1)\exp(-(\lambda_n^2/Re\ Pr)z^*)}{\sum_{n=1}^{\infty}(C_n f'_n(1)/\lambda_n^2)\exp(-(\lambda_n^2/Re\ Pr)z^*)}. \quad (7.55)$$

The Graetz problem has continued to attract attention in recent years. For example, Gupta and Balakotaiah (2001) extended the analysis to the case where an exothermic catalytic reaction is occurring at the tube wall. They demonstrated that the Graetz problem with surface reaction has mutiple solutions for certain parametric choices. Coelho et al. (2003) considered variations of the Graetz problem for a viscoelastic fluid with constant wall temperature, constant heat flux, and thermal energy production by viscous dissipation. It should also be pointed out that the Graetz problem (7.50) is extremely easy to solve numerically, one can simply forward march in the z-direction, computing new temperatures for all interior r-positions (making use of symmetry at the center). An illustration of such a computation is shown in Figure 7.11 for $RePr = 1000$. You should be able to anticipate the effects of changing $RePr$ upon the development of $T(r,z)$.

The Graetz analysis described above is appropriate for large values of Pr (ν/α) where the velocity distribution is fully developed. In many heat exchange applications, however, we can expect simultaneous development in both the momentum and thermal transport problems. When the Prandtl number is less than or comparable to 1, it will be necessary to write the energy equation as

$$\rho C_p\left(v_r\frac{\partial T}{\partial r} + v_z\frac{\partial T}{\partial z}\right) = k\left[\frac{1}{r}\frac{\partial}{\partial r}\left(r\frac{\partial T}{\partial r}\right)\right]. \quad (7.56)$$

An approximate solution for this problem can be obtained by omitting the convective transport of thermal energy in the radial direction (v_r is likely to be important only for very small z's). We can then make use of Langhaar's (1942) analysis of laminar flow in the entrance of a cylindrical tube. His solution of the linearized equation of motion (see the previous discussion in Chapter 3) resulted in the relation

$$\frac{v_z}{\langle v_z\rangle} = \frac{I_0(\phi(z) - I_0(\phi(z)\cdot(r/R))}{I_2(\phi(z))}. \quad (7.57)$$

The function $\phi(z)$ has the following numerical values for specific combinations of $z/d/Re$:

$\phi(z)$	$\frac{z/d}{Re}$
20	0.000205
11	0.00083
8	0.00181
6	0.00358
5	0.00535
4	0.00838
3	0.01373
2	0.02368
1	0.04488
0.4	0.0760

Kays (1955) and Heaton et al. (1964) used this approach to find an approximate numerical solution for the combined entrance region problem. Heaton et al. extended Langhaar's method to include developing flow in an annulus and they obtained results for the annulus, flow between parallel plates, and flow through a cylindrical tube, all with constant heat flux at the wall. Their data for the tube are presented graphically in Figure 7.12 for the Prandtl numbers of 0.01, 0.7, and 10. Note that for the Prandtl numbers ranging from 0.7 to 10, the Nusselt number has roughly approached the expected value of 4.36 for $(z/d)/(RePr)$ of about 0.1. Accordingly, if $RePr = 1000$, about 100 tube diameters will be required to complete profile development in the entrance region.

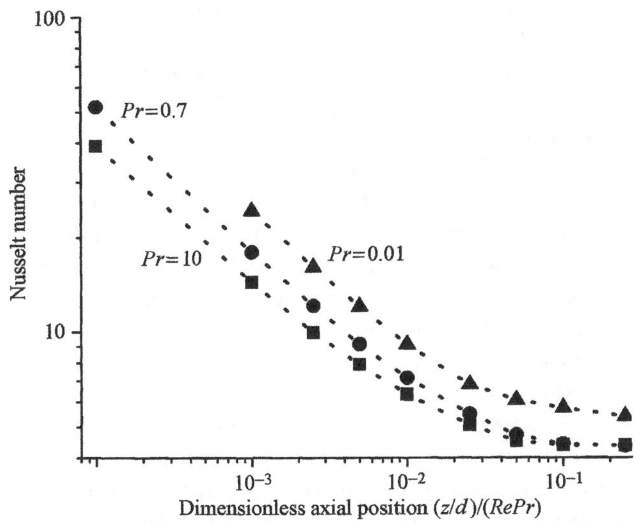

FIGURE 7.12. The Nusselt number as a function of $(z/d)/(RePr)$ for the combined entrance problem in a cylindrical tube with constant heat flux at the wall. These data (for the Prandtl numbers of 0.01, 0.7, and 10) were adapted from Heaton et al. (1964).

7.4 NATURAL CONVECTION: BUOYANCY-INDUCED FLUID MOTION

Consider the following table of liquid densities for temperatures ranging from 0 to 30°C:

Density (g/cm³) T (°C)	Water	Ethanol	Mercury
0	0.99987	0.80625	13.5955
10	0.99973	0.79788	13.5708
20	0.99823	0.78945	13.5462
30	0.99568	0.78097	13.5217

Note that the densities of water, ethanol, and mercury, decrease by 0.42%, 3.14%, and 0.54%, respectively, as the temperature increases from 0 to 30°C. Clearly, localized transfer of thermal energy can result in a fluid of reduced density being overlain by a higher density fluid. This common occurrence can result in a buoyancy-driven flow; we refer to such a phenomenon as free or natural convection. For a confined fluid, localized heating can produce regions of recirculation (commonly called convection rolls).

In such cases, the energy and momentum equations are coupled since $\rho=\rho(T)$. However, it is common practice to add an external force term to the equation of motion employing the volumetric coefficient of expansion (β), for example,

$$\rho\beta g_z(T-T_\infty) \quad \text{where} \quad \beta=-\frac{1}{\rho}\left(\frac{\partial\rho}{\partial T}\right)_p. \tag{7.58}$$

This is referred to in the literature as the Boussinesq approximation, as we saw in the introduction to this chapter. We should recognize that any solutions obtained in this fashion will be restricted to modest thermal driving forces. The reason that this often works well is because the volumetric coefficient of expansion (β) is usually quite small; if ΔT is modest, the effect on density may be 1% or less. We also note that natural convection can result in a velocity distribution that contains a point of inflection. Recall from our earlier discussions that this is a clear indication of a marginally stable laminar flow. We should not expect laminar flow to persist in free convection in cases where the thermal driving force is large. Indeed, the transition from laminar to turbulent flow is easily visualized in the plumes from candles or cigarettes.

Consider two infinite vertical parallel planes, spaced $2b$ apart: one surface is heated slightly and the other is cooled. We expect upwardly directed flow on the heated side and downward motion on the cooled side. With the Boussinesq approximation, the governing equations take the form

$$\rho C_p v_z\frac{\partial T}{\partial z}=k\left[\frac{\partial^2 T}{\partial y^2}+\frac{\partial^2 T}{\partial z^2}\right] \tag{7.59}$$

and

$$0=\mu\left[\frac{\partial^2 v_z}{\partial y^2}\right]+\rho g_z\beta(T-T_m). \tag{7.60}$$

Suppose we decide to impose some major simplifications upon this problem. Let us neglect conduction in the z-direction and omit the convective transport as well. With these severe restrictions, the energy equation is simply

$$\frac{d^2T}{dy^2}=0, \quad \text{with the solution} \quad T=C_1y+C_2.$$

Since one surface is maintained at T_h and the other at T_c, the constants of integration are $C_1=(T_h-T_c)/2b$ and $C_2=(T_h+T_c)/2$. Note that the latter is just the mean fluid temperature T_m. Therefore, equation (7.60) can be integrated directly to yield

$$v_z=-\frac{\rho g_z\beta}{\mu}\left(\frac{T_h-T_c}{2b}\right)\left[\frac{y^3}{6}-\frac{b^2y}{6}\right]. \tag{7.61}$$

What does the velocity distribution look like? You can see immediately that v_z is zero at the center and at both walls. For positive y less than b, the velocity is positive; for negative y greater than $-b$, the velocity is negative. Note also that there is a point of inflection at $y=0$.

7.4.1 Vertical Heated Plate: The Pohlhausen Problem

Consider an infinite vertical plate maintained at an elevated temperature T_s that is immersed in a fluid. The fluid in proximity to the plate is warmed and fluid motion ensues. By the no-slip condition, the velocity at the plate surface is zero and at large transverse distances, the thermal driving force disappears and the velocity asymptotically approaches zero. Therefore, we can anticipate a velocity profile with a point of inflection.

The governing equations for this case will be

$$v_y\frac{\partial T}{\partial y}+v_z\frac{\partial T}{\partial z}=\alpha\frac{\partial^2 T}{\partial y^2} \tag{7.62}$$

and

$$v_y\frac{\partial v_z}{\partial y}+v_z\frac{\partial v_z}{\partial z}=\nu\frac{\partial^2 v_z}{\partial y^2}+g\beta(T-T_\infty). \tag{7.63}$$

You may recognize the similarity to Prandtl's boundary-layer equation. This has not occurred by chance; the same argument has been made, namely, the characteristic length in the transverse direction (δ) is very much smaller than the characteristic vertical length scale (L). It seems likely that a similarity transformation might be appropriate here and, indeed, this is exactly the approach that Pohlhausen (1921) and Schmidt and Beckmann (1930) took. Let

$$\eta = \frac{Cy}{z^{1/4}}, \quad F(\eta) = \frac{\psi}{4\nu C z^{3/4}},$$
$$C = \left[\frac{g\beta(T_s - T_\infty)}{4\nu^2}\right]^{1/4}, \quad \text{and} \quad \theta = \frac{T - T_\infty}{T_s - T_\infty}. \tag{7.64}$$

Remember that introduction of the stream function will result in the increase of order of the momentum equation (7.62) from 2 to 3. The resulting coupled ordinary differential equations are

$$\frac{d^2\theta}{d\eta^2} + 3Pr\,F\frac{d\theta}{d\eta} = 0 \tag{7.65}$$

and

$$\frac{d^3F}{d\eta^3} + 3F\frac{d^2F}{d\eta^2} - 2\left(\frac{dF}{d\eta}\right)^2 + \theta = 0. \tag{7.66}$$

This is a fifth-order system and we note that the Prandtl number Pr occurs as a parameter in eq. (7.65). Accordingly, a separate solution will be required for each fluid of interest, subject to the following boundary conditions:

$$\text{at } \eta = 0, \quad v_y = v_z = 0, \text{ which means}$$
$$F = F' = 0 \text{ and } \theta = 1, \text{ and}$$
$$\text{as } \eta \to \infty, \quad v_y = v_z = 0, \text{ so } F' = 0 \text{ and } \theta = 0.$$

Typical results (obtained with the fourth-order Runge–Kutta algorithm) for the vertical heated plate are shown in Figures 7.13 and 7.14 for Pr's ranging from 0.1 to 100.

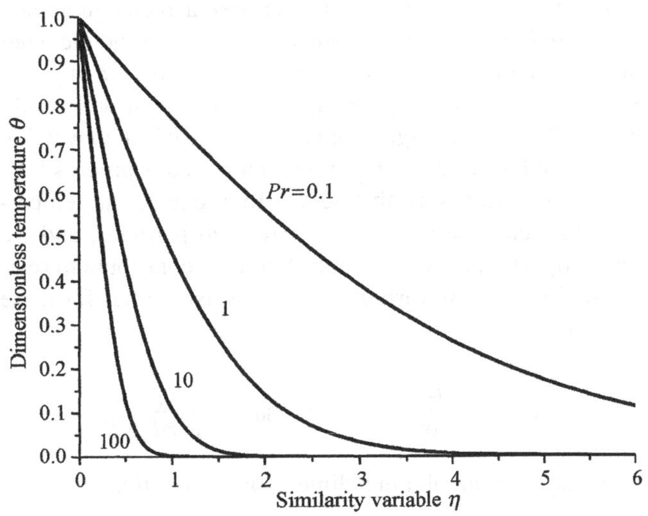

FIGURE 7.13. Dimensionless temperature distributions for natural convection from a vertical heated plate with $Pr = 0.1, 1, 10,$ and 100.

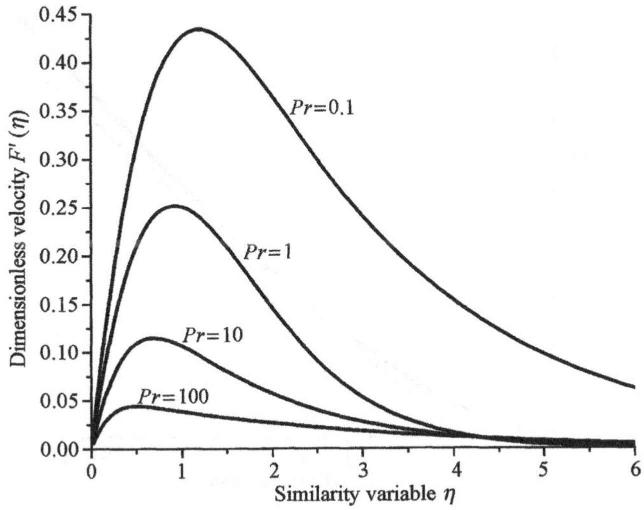

FIGURE 7.14. Dimensionless velocity distributions for the Pohlhausen problem with $Pr = 0.1, 1, 10,$ and 100.

Note that the quotient identified as C in (7.64) is related to the Grashof number Gr. Normally, we take

$$Gr = \frac{g\beta(T_s - T_\infty)L^3}{\nu^2}. \tag{7.67}$$

What is the physical significance of this grouping? One *might* suggest that Gr is the ratio of buoyancy and viscous forces—but note that there is no characteristic velocity. What we really have is

$$\frac{\text{(buoyancy forces)(inertial forces)}}{\text{(viscous forces)}^2}.$$

We conclude that Gr is an extremely useful parameter because it serves as an indicator of heat transfer regime, namely, if $Gr \gg Re^2 \Rightarrow$ natural convection and if $Gr \ll Re^2 \Rightarrow$ forced convection. Note that Eckert and Jackson (1951), in a study of free convection with a vertical isothermal plate, concluded that transition occurs for $Ra_z = Gr_z Pr \approx 10^9$. This is an important limitation of the similarity solution.

Many experimental measurements have been made for the vertical heated plate and a comparison with the model is provided in Figure 7.15. Note that agreement is generally good in the intermediate region of Rayleigh numbers. At large Ra, the flow becomes turbulent as noted above. At small Ra, Ede (1967) suggested that the boundary layer becomes so thick that the usual Prandtl assumptions no longer apply.

7.4.2 The Heated Horizontal Cylinder

The long horizontal cylinder is an extremely important heat transfer geometry because of its common use in process engineering applications. The first successful treatment of

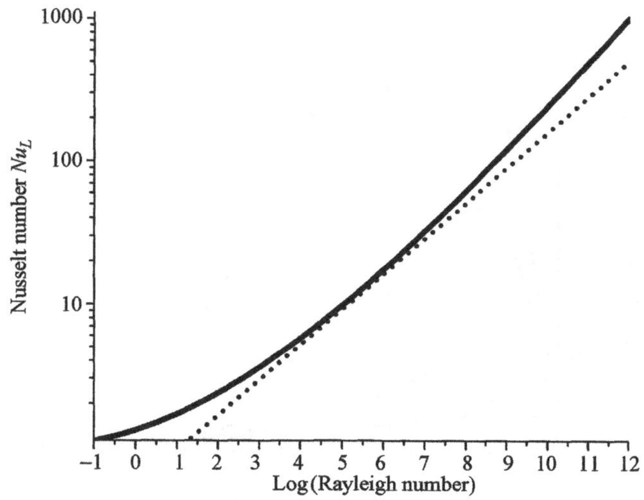

FIGURE 7.15. Comparison of the model (dashed line) with the approximate locus of experimental data (heavy, solid curve) for air. The Nusselt number Nu_L is plotted as a function of the $\log_{10}$ of the Rayleigh number $Ra_L = Gr_L Pr$.

this problem was carried out by R. Hermann (1936). His approach was an extension of Pohlhausen's analysis of the vertical heated plate, though we should note that no similarity solution is possible for the horizontal cylinder. The equations employed (excluding continuity) are

$$v_x \frac{\partial v_x}{\partial x} + v_y \frac{\partial v_x}{\partial y} = \nu \frac{\partial^2 v_x}{\partial y^2} + g\beta(T - T_\infty)\sin\left(\frac{x}{R}\right) \quad (7.68)$$

and

$$v_x \frac{\partial T}{\partial x} + v_y \frac{\partial T}{\partial y} = \alpha \frac{\partial^2 T}{\partial y^2}. \quad (7.69)$$

In usual boundary-layer fashion, the x-coordinate represents distance along the surface of the cylinder and y is normal to the surface, extending into the fluid. White (1991) notes that Hermann's calculations are in good agreement with experimental data; Hermann found the mean Nusselt number for this case was:

$$Nu_{\mathrm{m}} = 0.402(Gr\ Pr)^{1/4}. \quad (7.70)$$

The characteristic length for the Grashof number is the cylinder diameter. Hermann was able to transform the governing partial differential equations into a system of ordinary equations that corresponded with Pohlhausen's development for the vertical heated plate. This was accomplished by defining a new independent variable q such that

$$q = y \cdot g(x), \quad \psi(x, y) = p(q) \cdot f(x), \quad \text{and} \quad T(x, y) = \theta(q).$$

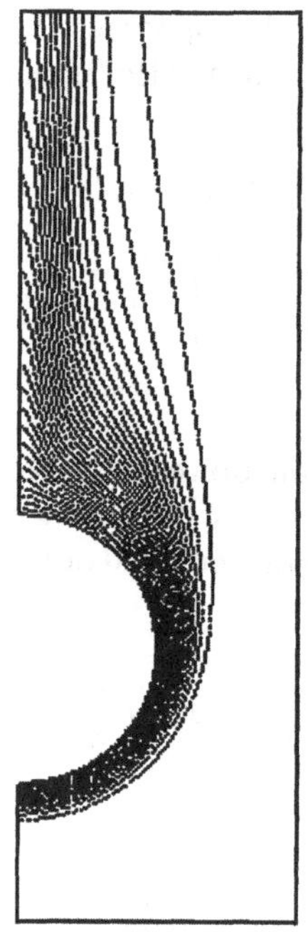

FIGURE 7.16. Characteristic thermal plume (in air) resulting from a slightly heated horizontal pipe. The isotherms shown range from 303°C at the pipe surface to 293°C. This example was computed with COMSOL^{TM}.

Thus, for a given Pr, he was able to directly use Pohlhausen's existing numerical results. Additional details for Hermann's solution procedure can be found in *NACA Technical Memorandum 1366*. An illustration of a typical thermal plume from a heated horizontal pipe is shown in Figure 7.16.

7.4.3 Natural Convection in Enclosures

Heating a surface of a fluid-filled enclosure can result in buoyancy-induced circulation; consider a rectangular box filled with fluid with the bottom slightly heated and the other walls maintained at some temperature T_s. If the ΔT imposed upon the bottom is very small, no fluid motion will result. But if ΔT is a little larger, we can expect natural convection to occur. What are the competing factors in this process? We have thermal diffusion that serves to attenuate the temperature difference between proximate fluid particles, and we have buoyant and viscous forces that may contribute to relative motion. We can formulate characteristic times for these processes:

$$\tau_{\text{thermal}} = \frac{L^2}{\alpha} \quad \text{and} \quad \tau_{\text{motion}} = \frac{\mu}{\rho g \beta L \Delta T}.$$

Obviously, we can obtain a dimensionless quotient:

$$\frac{\tau_{\text{thermal}}}{\tau_{\text{motion}}} = \frac{\rho g \beta L^3}{\alpha \mu} \Delta T = Ra. \quad (7.71)$$

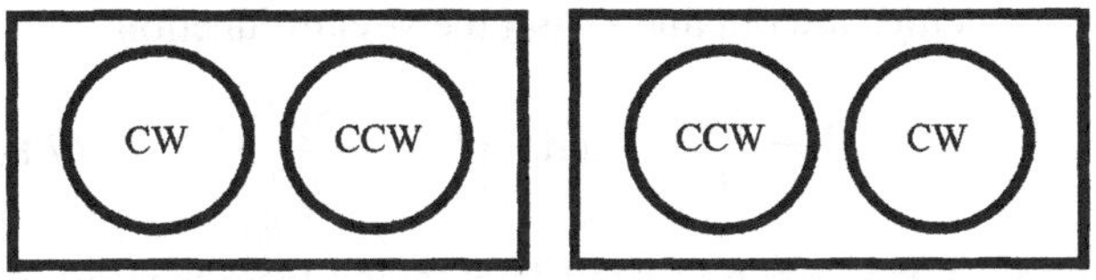

FIGURE 7.17. Typical patterns of circulation in a rectangular enclosure, but with opposite rotations (CW: clockwise, CCW: counter-clockwise).

This is known as the Rayleigh number in honor of Lord Rayleigh (John William Strutt, 1842–1919, winner of the Nobel Prize in Physics in 1904). You will recognize, as we noted previously, that $Ra = GrPr$. In cases where buoyancy is dominant (the timescale for relative fluid motion is small), the molecular transport of thermal energy cannot suppress local temperature differences and buoyancy-driven fluid motion ensues. The onset of this condition is marked by a critical value of the Rayleigh number Ra_c. In the fluid-filled enclosures, if the Rayleigh number is *slightly* higher than Ra_c, then the resulting flow is highly ordered, consisting of a series of closed circulations (sometimes referred to as convection rolls). Adjacent vortical structures necessarily rotate in opposite directions, but an interesting question arises: What are the factors that cause a particular structure (or roll) to rotate clockwise? Or counterclockwise? In fact, in a well-designed and carefully executed Rayleigh–Benard experiment, the situations depicted in Figure 7.17 are equally probable.

Berge et al. (1984) have pointed out that this means the transition that occurs at Ra_c is a bifurcation between stationary states. Naturally, as Ra continues to increase, we can expect to see additional instabilities, resulting ultimately in a bifurcation diagram not unlike the logistic map we discussed much earlier in this text. This in turn suggests that the Rayleigh–Benard convection *might* serve as a useful analogue for study of the onset of turbulence.

It may have occurred to you that there are similarities between the stability of the Rayleigh–Benard phenomenon and the stability of the Couette flow between concentric cylinders. The analogy is particularly appropriate in the case of the latter when the rotational motion is dominated by the angular velocity of the inner cylinder. You may recall that in this case, the initial instability predicted by Taylor's analysis leads to a succession of stable secondary flows. When this occurs, we say that the "principle of exchange of stabilities" is valid, which simply means that the frequency parameter ($\sigma = \omega L^2/\nu$) is real and the marginal states are characterized by $\sigma = 0$.

The discussion above leads us to the foundation of a linearized stability analysis of the Rayleigh–Benard convection. Consider a layer of fluid with no motion, but upon which a steady adverse temperature gradient (warm at the bottom and cool at the top) is maintained. Under these conditions, the hydrodynamic equations merely describe a state of constant stress; all the velocity vector components are zero. Since the imposed temperature gradient is fixed, the appropriate energy equation (assuming constant k) appears simply as $\nabla^2 T = 0$. An appropriate solution is $T = T_s - \lambda z$, and the corresponding density and pressure distributions must be linear functions of z.

We now assume that a small disturbance is imposed upon the static fluid in the form of velocity and temperature fluctuations; these must be described with the Navier–Stokes and energy equations. However, we neglect all terms that are nonlinear with respect to the perturbations. Thus, the inertial terms are dropped from the equation of motion and the convective transport terms are omitted from the equation of energy. Excluding continuity, we then have the following equations:

$$\frac{\partial v_i}{\partial t} = -\frac{\partial}{\partial x_i}\left(\frac{p'}{\rho_s}\right) + \nu\nabla^2 v_i + g\beta\theta_i, \tag{7.72}$$

$$\frac{\partial \theta}{\partial t} = \lambda v_i + \alpha\nabla^2\theta. \tag{7.73}$$

Chandrasekhar (1961) shows how these equations can be written in terms of the z-components of vorticity and velocity; a specific functional dependence is assumed for the perturbations (as usual with the method of small disturbances). It is possible to eliminate θ between these equations resulting in a disturbance equation (with $W(z)$ as the amplitude function):

$$\left(D^2 - a^2\right)^3 W = -a^2 Ra\, W, \tag{7.74}$$

where Ra is the Rayleigh number and D represents d/dz. Written out, the differential equation is

$$\frac{d^6 W}{dz^6} - 3a^2\frac{d^4 W}{dz^4} + 3a^4\frac{d^2 W}{dz^2} - a^6 W = -a^2 Ra\, W. \tag{7.75}$$

The origin is placed at the center, so a solution is sought from $z = -1/2$ to $z = +1/2$. The boundary conditions are $W = dW/dz = \left(D^2 - a^2\right)^2 W = 0$ for $z = \pm 1/2$. The problem thus posed is a sixth-order characteristic value problem. Reid and Harris (1958) determined the exact eigenvalues for the first even mode of instability; they found that the lowest value of Ra_c occurred with (dimensionless wave number) $a = 3.117$. This critical Rayleigh number was found to be 1707.762 for a fluid layer contained between two horizontal walls. The classical view is that this critical value Ra_c is independent of the Prandtl number. However, there is evidence that this presumption is incorrect, and a brief discussion of this point will be given at the end of the next section.

7.4.4 Two-Dimensional Rayleigh–Benard Problem

Consider a viscous fluid initially at rest contained within a two-dimensional rectangular enclosure; at $t=0$, the bottom surface is heated such that the dimensionless temperature at that surface is 1:

$$\theta = \frac{T - T_i}{T_s - T_i} = 1.$$

For all other surfaces, $\theta = 0$ for all t. Naturally, the buoyancy-driven fluid motion will ensue, and depending upon the *W/h* ratio of the enclosure, we can expect to see convection roll(s) develop in response to the temperature difference. This is an example of the Benard (1900) problem first treated theoretically by Lord Rayleigh. Chow (1979) has provided a detailed illustration of a practical method for solving this type of problem, and we follow his example with a few modifications here.

The equations that must be solved are

$$\rho\left(\frac{\partial v_x}{\partial t} + v_x\frac{\partial v_x}{\partial x} + v_y\frac{\partial v_x}{\partial y}\right) = -\frac{\partial p}{\partial x} + \mu\left[\frac{\partial^2 v_x}{\partial x^2} + \frac{\partial^2 v_x}{\partial y^2}\right], \tag{7.76}$$

$$\rho\left(\frac{\partial v_y}{\partial t} + v_x\frac{\partial v_y}{\partial x} + v_y\frac{\partial v_y}{\partial y}\right) = -\frac{\partial p}{\partial y} + \mu\left[\frac{\partial^2 v_y}{\partial x^2} + \frac{\partial^2 v_y}{\partial y^2}\right] + \rho g\beta\Delta T, \tag{7.77}$$

and

$$\rho C_p\left(\frac{\partial T}{\partial t} + v_x\frac{\partial T}{\partial x} + v_y\frac{\partial T}{\partial y}\right) = k\left[\frac{\partial^2 T}{\partial x^2} + \frac{\partial^2 T}{\partial y^2}\right]. \tag{7.78}$$

It is convenient to eliminate pressure by cross-differentiating (7.76) and (7.77) and subtracting the former from the latter. Since the z-component of the vorticity vector is defined by

$$\omega_z = \left(\frac{\partial v_y}{\partial x} - \frac{\partial v_x}{\partial y}\right), \tag{7.79}$$

or $[\nabla \times v]_z$, the problem can be recast in terms of the vorticity transport equation and the energy equation. This provides us with a straightforward solution procedure.

The lower surface is located at $y=0$ and the upper surface is at $y=H$. We define the other dimensionless quantities as follows:

$$x^* = x/H, \quad y^* = y/H, \quad t^* = \frac{\mu t}{\rho_0 H^2}, \quad v_x^* = \frac{\rho_0 H v_x}{\mu},$$

$$v_y^* = \frac{\rho_0 H v_y}{\mu}, \quad \psi^* = \frac{\rho_0 \psi}{\mu}, \quad \text{and} \quad \Omega = \frac{\rho_0 H^2 \omega}{\mu}.$$

The velocities are obtained from the stream function

$$v_x^* = \frac{\partial \psi^*}{\partial y^*} \quad \text{and} \quad v_y^* = -\frac{\partial \psi^*}{\partial x^*}, \tag{7.80}$$

and the stream function itself is obtained from the vorticity distribution:

$$\Omega = -\left(\frac{\partial^2 \psi^*}{\partial x^{*2}} + \frac{\partial^2 \psi^*}{\partial y^{*2}}\right). \tag{7.81}$$

In a dimensionless form, the governing equations (energy and vorticity) can be written as

$$\frac{\partial \theta}{\partial t^*} + \frac{\partial (v_x^*\theta)}{\partial x^*} + \frac{\partial (v_y^*\theta)}{\partial y^*} = \frac{1}{Pr}\left[\frac{\partial^2 \theta}{\partial x^{*2}} + \frac{\partial^2 \theta}{\partial y^{*2}}\right] \tag{7.82}$$

and

$$\frac{\partial \Omega}{\partial t^*} + \frac{\partial (v_x^*\Omega)}{\partial x^*} + \frac{\partial (v_y^*\Omega)}{\partial y^*} = Gr\frac{\partial \theta}{\partial x^*} + \frac{\partial^2 \Omega}{\partial x^{*2}} + \frac{\partial^2 \Omega}{\partial y^{*2}}. \tag{7.83}$$

Note the similarities between the two equations; of course, the implication is that we can use the same procedure to solve both. We must use a stable differencing scheme for the convective terms, and the method developed by Torrance (1968) is known to work well for both natural convection and rotating flow problems. The generalized solution procedure follows:

1. Calculate the stream function from the vorticity distribution using SOR.
2. Find the velocity vector components from the stream function.
3. Compute vorticity on the new time-step row explicitly.
4. Calculate temperature on the new time-step row explicitly.

Depending upon the desired spatial resolution, the optimal relaxation parameter will generally fall in the range $1.7 < \omega < 1.9$. In the case of the example appearing here, $\omega \cong 1.75$ seems to work well. We select the parametric values:

$$Pr = 6.75, \quad Gr = 1000, \quad \Delta x^* = 0.0667, \quad \text{and}$$
$$\Delta t^* = 0.0005.$$

Since the box is much wider than it is deep, the right-hand boundary (at the center of the enclosure) is a plane of symmetry where $\partial\theta/\partial x^* = 0$. Conveniently, we can also take the stream function ψ to be zero everywhere on the computational boundary. In the sequence shown in Figure 7.18, the evolution of the recirculation patterns is illustrated.

The Benard flow described above has been the object of some disagreement in the limiting cases of very small *Pr*. Lage et al. (1991) carried out an extensive test of a finding put

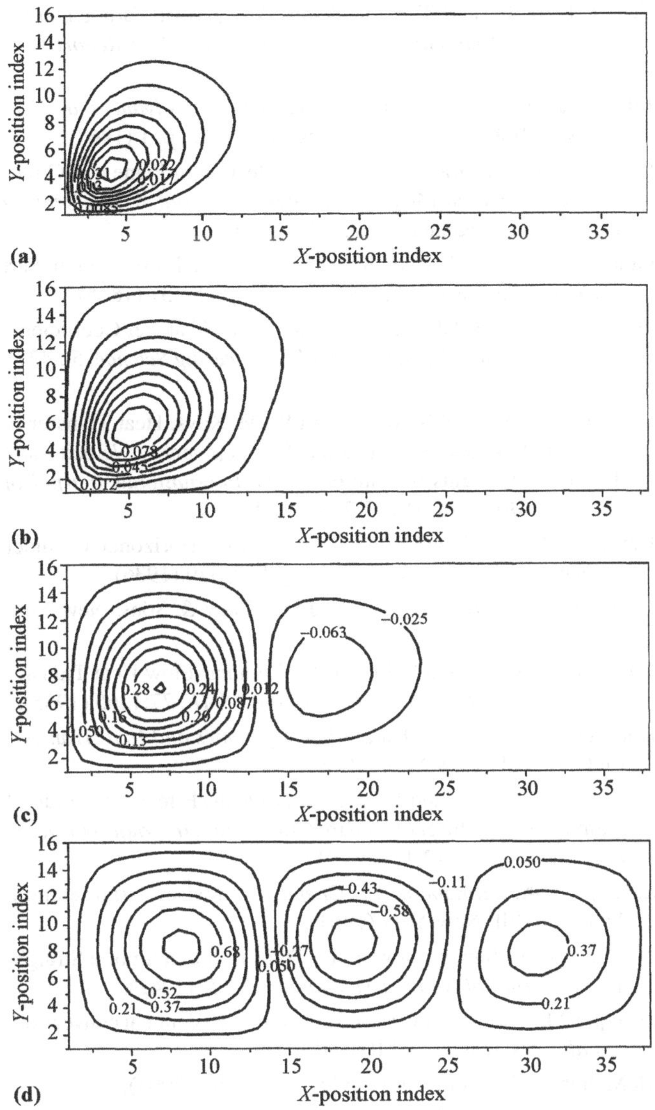

FIGURE 7.18. Evolution of convection rolls in a rectangular enclosure at dimensionless times of 0.1, 0.2, 0.4, and 0.8.

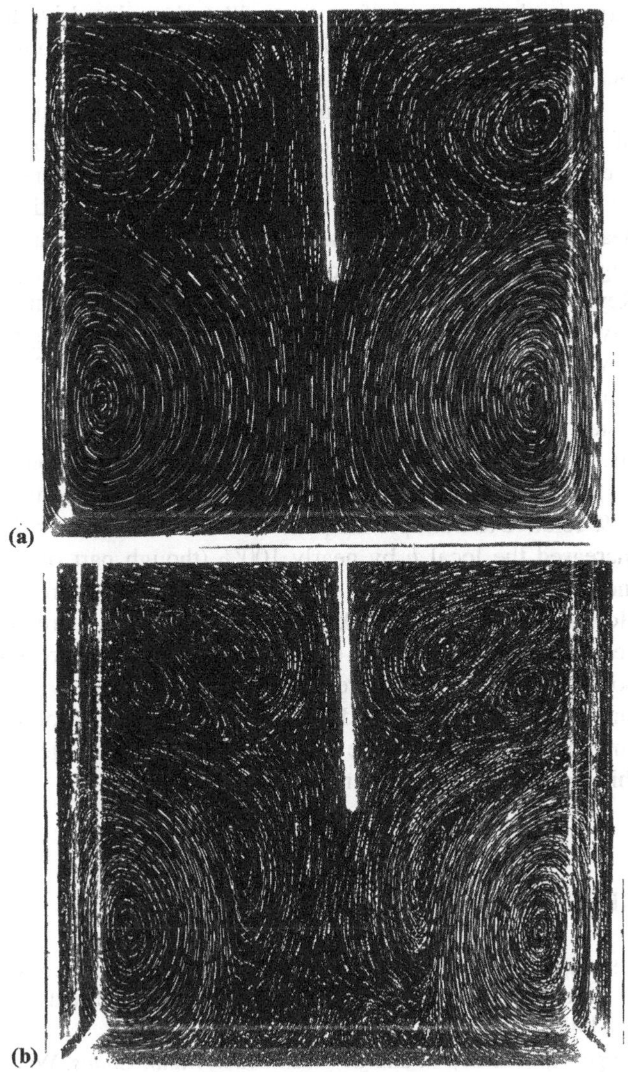

FIGURE 7.19. Convection patterns in a 3 in. (7.62 cm) glass cube filled with water, heated on all surfaces by immersion in a heated bath. The temperature of the bath is increased linearly but the mean driving force is constant. Note that there are four convection rolls in the top image and eight for the bottom. These remarkable images are shown through the courtesy of Dr. Richard G. Akins, who carried out extensive studies of natural convection for liquids in enclosures.

forward by Chao et al. (1982) and Bertin and Ozoe (1986) that the critical Rayleigh number increases significantly as the Prandtl number decreases. Lage et al. confirmed that Ra_c increases sharply as Pr drops below about 0.1; in fact, they found that the critical Rayleigh number was about 3000 at $Pr = 6 \times 10^{-4}$ (as opposed to 1707.8). They also discovered that the natural shape for *near-critical* convection rolls at low Pr was approximately square. Furthermore, their results were shown to be independent of the aspect ratio of the enclosure.

Transient natural convection in enclosures can present a rich panoply of behaviors as noted above. In the sequence of experimental visualizations shown in Figure 7.19 (courtesy of Dr. Richard G. Akins), striking differences are seen in the number and location of convection rolls. These experiments were conducted using a glass cube (3 in. on each side) filled with water. The cube was immersed in a heated bath in which the bath temperature was increased linearly with time; this resulted in a constant thermal driving force between the bath and the fluid in the cube.

7.5 CONCLUSION

In this chapter, we noted the importance of the Prandtl number several times. The Prandtl number also plays a very important role in the Rayleigh–Benard problems. Consider eq. (7.82); if Pr is large, then the convective transport terms such as $\partial/\partial x^*(v_x^*\theta)$ will drive secondary instabilities. If, on the other hand, the Prandtl number is small, then the secondary instabilities will be of hydrodynamic character. That is, the inertial terms in the equation of motion will (primarily) drive the secondary instabilities. The interested reader should consult Berge et al. (1984) for additional detail.

Finally, some general comments regarding the influence of fluid motion upon the rate of heat transfer are in order. We have seen that even modest fluid velocities will increase heat transfer. Confronted with the need to extract additional heat duty from an existing piece of equipment, a heat transfer engineer will immediately consider higher flow rate (larger Reynolds numbers). However, one can also increase the intensity of fluid motions *normal* to the surface by changing the flow direction, or by promoting turbulence. In a study of heat transfer with air flowing past a surface, Boelter et al. (1951) tested plates with small vertical strips installed with 1 in. spacing. They found that 0.125 in. strips (turbulence promoters) increased the local heat transfer coefficient by roughly 73% relative to a simple flat plate. The use of 0.375 in. strips increased the local h by nearly 100% (though part of that increase was attributed to extended surface heat transfer). However, Boelter et al. (1951) also found that the increased heat transfer was almost exactly offset by the increased power consumption required to maintain the same average air velocity. When coupled with likely increases in fouling and possibly corrosion, the value of altering the flow field in this manner may not be very great.

REFERENCES

Benard, H. Les Tourbillons cellulaires dans une nappe liquide. *Revue generale des Sciences pures et appliquees*, 11: 1261 and 1309 (1900).

Berge, P., Pomeau, Y., and C. Vidal. *Order Within Chaos*, Wiley-Interscience, New York (1984).

Bertin, H. and H. Ozoe. Numerical Study of Two-Dimensional Convection in a Horizontal Fluid Layer Heated from Below by Finite Element Method. *International Journal of Heat and Mass Transfer*, 29:439 (1986).

Bird, R. B., Stewart, W. E., and E. N. Lightfoot. *Transport Phenomena*, 2nd edition, Wiley, New York (2002).

Boelter, L. M. K., Young, G., Greenfield, M. L., Sanders, V. D., and M. Morgan. An Investigation of Aircraft Heaters, XXXVII: Experimental Determination of Thermal and Hydrodynamical Behavior of Air Flowing Along a Flat Plate Containing Turbulence Promoters. *NACA Technical Note* 2517 (1951).

Chandrasekhar, S. *Hydrodynamic and Hydromagnetic Stability*, Dover Publications, New York (1961).

Chao, P., Churchill, S. W., and H. Ozoe. The Dependence of the Critical Rayleigh Number on the Prandtl Number. *Convection Transport and Instability Phenomena*, Braun, Karlsruhe (1982).

Chow, C. Y. *An Introduction to Computational Fluid Mechanics*, Seminole Publishing (1979).

Coelho, P. M., Pinho, F. T., and P. J. Oliveira. Thermal Entry Flow for a Viscoelastic Fluid: The Graetz Problem for the PTT Model. *International Journal of Heat and Mass Transfer*, 46: 3865 (2003).

Eckert, E. R. G. and T. W. Jackson. Analysis of Turbulent Free Convection Boundary Layer on a Flat Plate. *NACA Report* 1015 (1951).

Ede, A. J. Advances in Free Convection. In: *Advances in Heat Transfer*, Vol. 4, Academic Press, New York, p. 1 (1967).

Gavis, J. and R. L. Laurence. Viscous Heating in Plane and Circular Flow Between Moving Surfaces. *Industrial & Engineering Chemistry Fundamentals*, 7:232 (1968).

Graetz, L. Uber die Warmeleitungsfahigkeit von Flussigkeiten, Part 2. *Annual Review of Physical Chemistry*, 25:337 (1885).

Gupta, N. and V. Balakotaiah. Heat and Mass Transfer Coefficients in Catalytic Monoliths. *Chemical Engineering Science*, 56:4771 (2001).

Heaton, H. S., Reynolds, W. C., and W. M. Kays. Heat Transfer in Annular Passages: Simultaneous Development of Velocity and Temperature Fields in Laminar Flow. *International Journal of Heat and Mass Transfer*, 7:763 (1964).

Hermann, R. Free Convection and Flow Near a Horizontal Cylinder in Diatomic Gases. *VDI Forschungsheft*, 379:(1936).

Jakob, M. *Heat Transfer*, Vol. 1: John Wiley & Sons, New York (1949).

Kays, W. M. Numerical Solutions for Laminar-Flow Heat Transfer in Circular Tubes. *Transactions of the ASME*, 77:1265 (1955).

Knudsen, J. G. and D. L. Katz. *Fluid Dynamics and Heat Transfer*, McGraw-Hill, New York (1958).

Lage, J. L., Bejan, A., and J. Georgiadis. On the Effect of the Prandtl Number on the Onset of Benard Convection. *International Journal of Heat Flow*, 12:184 (1991).

Lange, N. A. *Handbook of Chemistry*, revised 10th edition, McGraw-Hill, New York (1961).

Langhaar, H. L. Steady Flow in the Transition Length of a Straight Tube. *Journal of Applied Mechanics*, A-55: (1942).

Leveque, M. A. Les lois de la transmission de chaleur par convection. *Annales des Mines*, 13:210 (1928).

McMahon, N. Website, Dublin City University (2004).

Pohlhausen, E. Der Warmeaustausch zwischen festen Kopern und Flussigkeiten mit kleiner Reibung und kleiner Warmeleitung. *ZAMM*, 1:115 (1921).

Ranz, W. E. and W. R. Marshall, Jr. Evaporation from Drops. *Chemical Engineering Progress*, 48:141 (1952).

Reid, W. H. and D. L. Harris. Some Further Results on the Benard Problem. *Physics of Fluids*, 1:102 (1958).

Schlichting, H. *Boundary-Layer Theory*, 6th edition, McGraw-Hill, New York (1968).

Schmidt, E. and W. Beckmann. Das Temperatur- und Geschwindigkeitsfeld von einer Warme abegbenden senkrechten Platte bei naturlicher Konvektion. *Forsch Ing-Wes*, 1:391 (1930).

Sellars, J. R., Tribus, M., and J. S. Klein. Heat Transfer to Laminar Flow in a Round Tube or Flat Conduit: The Graetz Problem Extended. *Transactions of the ASME*, 78:441 (1956).

Singh, S. N. Heat Transfer by Laminar Flow in a Cylindrical Tube. *Applied Scientific Research*, Section A, 7:325 (1958).

Torrance, K. E. Comparison of Finite-Difference Computations of Natural Convection. *Journal of Research of the National Bureau of Standards*, 72B:281 (1968).

White, F. M. *Viscous Fluid Flow*, 2nd edition, McGraw-Hill, Boston (1991).

8

DIFFUSIONAL MASS TRANSFER

8.1 INTRODUCTION

When a student of transport phenomena is asked to write a description of a molar (*molar or molal*—see Skelland (1974) for the absolute last word on the difference) flux in mass transfer, the response is generally

$$N_{Ay} = -D_{AB}\frac{\partial C_A}{\partial y} \quad \text{or} \quad N_{Ay} = -CD_{AB}\frac{\partial x_A}{\partial y}. \tag{8.1}$$

This expression, Fick's first law, is correct only under very particular conditions, so we should take a moment to consider the migration of a species i more broadly. In a system with n-components, we could define both mass-average and molar-average velocities:

$$v = \frac{1}{\rho}\sum_{i=1}^{n}\rho_i v_i \quad \text{and} \quad V^* = \frac{1}{C}\sum_{i=1}^{n} C_i V_i^*. \tag{8.2}$$

In a binary system, if the solute concentration is very low, we see $v \cong V^*$. We also note emphatically that we must not regard these quantities as the velocities of individual molecules—this is continuum mechanics! It is apparent that the motion of component i can be defined in three ways: relative to stationary coordinates, relative to the mass-average velocity, and relative to the molar-average velocity. Accordingly, given the different velocities, we can define the flux for component "A" relative to either of the pair defined in eq. (8.2). We should make two observations: First, in many engineering applications, the physical frame of reference is tied to an interface, boundary, reactive surface, etc. In such cases, a moving frame of reference would not assist the analyst. Second, many of the problems that are of interest to us involve a fairly small amount of solute in a large volume of solvent, that is, we can frequently assume a dilute solution.

Adolf Fick proposed eq. (8.1) in 1855 through analogy with Fourier's law; we can follow his reasoning through the following translation of Fick's own words: "It was quite natural to suppose this law of diffusion of a salt in its solvent must be identical with that according to which the diffusion of heat in a conducting body takes place." This is an appealing assumption because when eq. (8.1) is applied to transient molecular transport in rectangular coordinates, we obtain Fick's second law (or the diffusion equation):

$$\frac{\partial C_A}{\partial t} = D_{AB}\left[\frac{\partial^2 C_A}{\partial x^2} + \frac{\partial^2 C_A}{\partial y^2} + \frac{\partial^2 C_A}{\partial z^2}\right]. \tag{8.3}$$

The analogous relations in cylindrical and spherical coordinates are

$$\frac{\partial C_A}{\partial t} = D_{AB}\left[\frac{1}{r}\frac{\partial}{\partial r}\left(r\frac{\partial C_A}{\partial r}\right) + \frac{1}{r^2}\frac{\partial^2 C_A}{\partial \theta^2} + \frac{\partial^2 C_A}{\partial z^2}\right] \tag{8.4}$$

and

$$\frac{\partial C_A}{\partial t} = D_{AB}\left[\frac{1}{r^2}\frac{\partial}{\partial r}\left(r^2\frac{\partial C_A}{\partial r}\right) + \frac{1}{r^2\sin\theta}\frac{\partial}{\partial \theta}\left(\sin\theta\frac{\partial C_A}{\partial \theta}\right) + \frac{1}{r^2\sin^2\theta}\frac{\partial^2 C_A}{\partial \phi^2}\right]. \tag{8.5}$$

Transport Phenomena: An Introduction to Advanced Topics, By Larry A. Glasgow

Of course, these equations are the same as the conduction equation(s) for molecular heat transfer; solutions developed for transient conduction problems can be directly utilized for certain unsteady diffusion problems. This is undeniably attractive, but it is essential that we understand the limitations of equations (8.3–8.5).

Fick's second law can be applied to diffusion problems in solids and in stationary liquids. It can also be applied to *equimolar counterdiffusion* in binary systems, where, for example, every molecule of "A" moving in the $+y$ direction is *countered* by a molecule of "B" moving in the $-y$ direction. Therefore,

$$N_{Ay} + N_{By} = 0. \tag{8.6}$$

This is critically important because we need to represent the combined flux of "A" with respect to fixed coordinates as

$$N_{Ay} = -CD_{AB}\frac{\partial x_A}{\partial y} + x_A(N_{Ay} + N_{By}). \tag{8.7}$$

Let us examine the right-hand side of eq. (8.7): the first part accounts for random molecular motions of species "A." Though we cannot say with certainty where any single molecule of "A" will be located at a given time, we recognize that there will be a net movement of "A" from the regions of higher concentration to those where "A" is less prevalent. Thus, the molecular mass transport occurs "downhill" (in the direction of decreasing concentration) just as heat transfer by conduction occurs in the direction of decreasing temperature. However, there is also an obvious difference between heat and mass transfer: Suppose species "A" is moving through a medium consisting of mainly "B" at high(er) rate. Under such circumstances, molecular transport and the resulting motion of the fluid work in concert producing a convective flux that must be added to Fick's first law. This is the reason why the product $x_A(N_{Ay} + N_{By})$ appears on the right-hand side of eq. (8.7). It is to be noted that for the multicomponent diffusion problems in gases, the concentration gradient for a particular species must be written in terms of the fluxes of *all* species. This is accomplished with the Stefan–Maxwell equations, which will be discussed in Chapter 11.

We can easily illustrate the problem that arises in binary systems at higher mass transfer rates. Suppose we have a spill of a volatile organic compound such as methanol in a plant or processing environment; we begin by examining the vapor pressure as a function of temperature, shown in Figure 8.1.

Furthermore, suppose that the temperature is 35°C and that the liquid methanol pool has been in place for some time. Under these circumstances,

$$x_{A0} \cong \frac{200}{760} = 0.263, \tag{8.8}$$

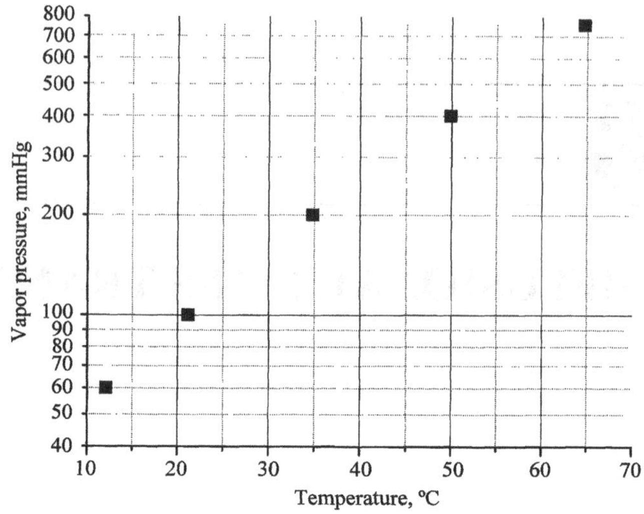

FIGURE 8.1. The vapor pressure of methanol (mmHg) as a function of temperature (°C).

and the flux at the liquid–vapor interface is

$$N_{Az} = -\frac{cD_{AB}}{1 - x_{A0}}\left.\frac{\partial x_A}{\partial z}\right|_{z=0}. \tag{8.9}$$

Note that the flux of methanol at the interface has been increased by about 35% over eq. (8.1), *assuming* that $\left.\frac{\partial x_A}{\partial z}\right|_{z=0}$ remains unchanged. We will consider this problem in greater detail later.

8.1.1 Diffusivities in Gases

In our previous discussions we have said little about actual determination of the molecular diffusivities: ν (momentum), α (thermal energy), and D_{AB} (binary diffusivity). One might conclude from this omission that data are available in the literature to provide the needed values. This is not entirely true, especially in the case of D_{AB}. Measurement of the binary diffusivity poses challenges that we do not see with either ν or α . In the case of kinematic viscosity of liquids, for example, one can use a simple device such as a Cannon–Fenske (pipette-type) viscometer, and measure ν's for liquids in a manner of minutes. No similar elementary technique is available for the measurement of D_{AB}. Philibert's (2006) account of the history of diffusion underscores this point; although Thomas Graham worked on problems of diffusion in gases around 1830, nearly 40 years elapsed before Maxwell was able to calculate D_{AB} using Graham's data (for carbon dioxide in air). Remarkably, Maxwell's diffusivity is within about 5% of the modern value.

For monatomic gases in which the density is low enough to guarantee two-body collisions, the transport properties can be determined from first principles. Reed and Gubbins (1973) provide a readable summary of the procedure. In the specific case of D_{AB}, the theory (using the Lennard-Jones 6–12 potential function) results in

$$D_{AB} = \frac{3}{16} \frac{[2\pi kT(M_A + M_B)]^{1/2}}{\sqrt{M_A M_B}} \frac{f_D}{(n\pi\sigma_{AB}^2 \Omega_D)}. \quad (8.10)$$

The M_i's are the formula weights, n is the number density of the mixture, σ_{AB} is the Lennard-Jones force constant (which must be estimated by a combining rule from the pure constituents), and Ω_D is the collision integral. Equation (8.10) is quite useful and it has been subjected to a large number of tests. Reid and Sherwood (1966) provide comparisons with experimental values for more than 100 gaseous systems. This method provides particularly good results for spherical nonpolar molecules. By inserting appropriate numerical values for the constants and assuming that the number density is adequately represented by the ideal gas law, eq. (8.10) can be written as follows:

$$D_{AB} = 0.001858 \frac{[M_A + M_B]^{1/2}}{\sqrt{M_A M_B}} \frac{T^{3/2}}{p\sigma_{AB}^2 \Omega_D}. \quad (8.11)$$

In eq. (8.11), p is in atmospheres, T in Kelvin, and D_{AB} in cm^2/s.

We will carry out a test of eq. (8.11) for air and helium at 300K.

	Air	Helium
σ, force constant	3.711 Å	2.551 Å
ε_0/k, depth of potential well	78.6	10.22

Now we employ the combining rules to obtain the necessary values for the mixture:

$$\sigma_{AB} = \frac{\sigma_A + \sigma_B}{2} = 3.131\ \text{Å}$$

and

$$\frac{\varepsilon_{0AB}}{k} = \sqrt{\left(\frac{\varepsilon_0}{k}\right)_A \left(\frac{\varepsilon_0}{k}\right)_B} = 28.34\ \text{K}.$$

We compute the quotient $kT/\varepsilon_{0\,AB} = 10.59$. An approximate value of the collision integral can now be obtained from one of the many available tabulations: $\Omega_D \cong 0.738$. The diffusivity resulting from this calculation is $D_{AB} = 0.711\ cm^2/s$ at 300K. Let us examine how this value compares with the available experimental data in Figure 8.2.

For many gases at ambient pressures, binary diffusivities range roughly from 0.1 to 1 cm^2/s. Since the Schmidt number is the ratio

$$Sc = \frac{\nu}{D_{AB}}, \quad (8.12)$$

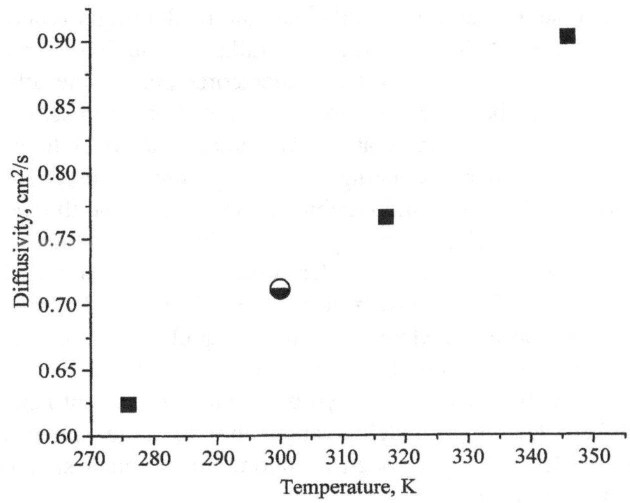

FIGURE 8.2. Comparison of experimental diffusivities (filled squares) for the air–helium system compared with the value calculated (half-filled circle) using eq. 8.10. The agreement is excellent in this case.

we can expect (for a variety of gases in air) to see the Schmidt numbers of about 1. This is illustrated in Table 8.1 (recall that for air at 0°C and 1 atm pressure, $\nu = 0.133\ cm^2/s$).

8.1.2 Diffusivities in Liquids

For a pure liquid, a central molecule can have about 10 nearest-neighbors. Contrast this with the coordination number (n_c) in solids; ice, for example, has $n_c = 4$. Though better, this is not as attractive from a modeling perspective as the low-pressure gas; when a molecule has a single nearest-neighbor, we can employ pairwise additivity and construct an effective model from first principles. The implication for liquids,

TABLE 8.1. Schmidt numbers at 1 atm pressure and 0°C for a variety of gases in air.

System	Schmidt Number, ν/D_{AB}
Air–acetone	1.60
Air–ammonia	0.61
Air–benzene	1.71
Air–chlorine	1.42
Air–ethane	1.22
Air–hydrogen	0.22
Air–methanol	1.00
Air–naphthalene	2.57
Air–oxygen	0.74
Air–propane	1.51
Air–toluene	1.86
Air–water (vapor)	0.60

Source: These data were excerpted from Sherwood and Pigford (1952).

of course, is that a physically accurate model might require solution of a "10-body" problem. Muller and Gubbins (2001) provided a nice graphic that underscores part of the difficulty: The "bond" energy for Ne–Ne is about 0.14 kJ/mol; for water this value is about 21 kJ/mol (due to hydrogen bonding). Other associating fluids range upward to perhaps 100 kJ/mol. Muller and Gubbins point out that the thermodynamic behaviors of "simple" fluids (those for which the interactions are mainly van der Waals attractions and weak electrostatic forces) have been successfully modeled over the past few decades. Many associating liquids, unfortunately, continue to elude fundamentally sound description.

Einstein proposed a "hydrodynamic" theory utilizing Stokes' law; the model is applicable to large spherical molecules moving through a continuum of much smaller solvent molecules:

$$\frac{D_{AB}\mu_B}{T} = \frac{k_B}{6\pi R_A}. \tag{8.13}$$

R_A is the radius of molecule "A" and k_B is 1.38×10^{-16} dyn cm/K. The Stokes–Einstein model is easily tested, we need only to prepare a plot of diffusivities against a range of solvent viscosities. Hayduk and Cheng (1971) have done this for carbon tetrachloride in solvents ranging from hexane to decalin, with very good results. Suppose that we try to apply this to an arbitrary system, say benzene in water. At 25°C, the experimentally measured diffusivity is about 1.09×10^{-5} cm²/s. Applying eq. (8.13), we find

$$D_{AB} = \frac{(1.38 \times 10^{-16})(298)}{(6)(3.1416)(0.01)(2.65 \times 10^{-8})} = 8.2 \times 10^{-6}\,\text{cm}^2/\text{s}.$$

The estimate is about 24% low. This would not be adequate for most engineering purposes.

Numerous investigators have proposed empirical correlations for diffusivities in *dilute* solutions; the Wilke–Chang (1955) equation is a commonly cited example:

$$D_{AB} = 7.4 \times 10^{-8} \frac{(\phi M_B)^{1/2} T}{\mu_B V_A^{0.6}}, \tag{8.14}$$

where M_B is the molecular weight of the solvent, T is the absolute temperature (K), μ is the viscosity of the solvent (cp), and V_A is the molal volume of the solute (cm³/g mol) at its boiling point. Note that the temperature and viscosity dependencies are exactly the same as those of the Stokes–Einstein model. The difference is that the Wilke–Chang correlation accounts for the association tendency of the solvent (through the parameter ϕ) and the size of the solute molecule (through V_A). Generally speaking, the Wilke–Chang correlation performs adequately for many aqueous systems, but it seems to struggle especially with systems where alcohols are the solvents. In such cases, 40% (or larger) errors are routine.

The state of affairs for the liquid phase is quite unsatisfactory. We do not have a comprehensive, molecular-based theory available that can be used universally to predict transport properties (such as diffusivity) from first principles. However, this may be changing; Muller and Gubbins note that SAFT (statistical associating fluid theory) may offer the prospect of success in modeling nonideal liquids. Indeed, they provide the interested reader with a good starting point for an exploration of the thermodynamics of complicated (or what thermodynamicists call nonregular) fluids and solutions.

8.2 UNSTEADY EVAPORATION OF VOLATILE LIQUIDS: THE ARNOLD PROBLEM

In the introduction, we described a scenario in which liquid methanol was evaporating; we want to revisit this type of problem and provide greater detail. Again, suppose we have a spill of a volatile liquid hydrocarbon that results in a large liquid pool overlain by still air. In particular, let the hydrocarbon be the very volatile *n*-pentane at 18.5°C such that the vapor pressure p^* is about 400 mmHg. The interfacial equilibrium mole fraction will be $x_{A0} \cong 400/760 = 0.526$; the diffusivity for these conditions is about 0.081 cm²/s. Our concern is the rate of mass transfer from the liquid pool into the vapor phase (the $+z$-direction). If we choose to write

$$\frac{\partial C_A}{\partial t} = D_{AB}\frac{\partial^2 C_A}{\partial z^2} \quad \text{or} \quad \frac{\partial x_A}{\partial t} = D_{AB}\frac{\partial^2 x_A}{\partial z^2}, \tag{8.15}$$

we have the familiar solution (assuming the total molar concentration is constant):

$$\frac{x_A}{x_{A0}} = \text{erfc}\left(\frac{z}{\sqrt{4D_{AB}t}}\right). \tag{8.16}$$

We can evaluate the molar flux at the interface by differentiation:

$$N_{A0} = C_{A0}\sqrt{\frac{D_{AB}}{\pi t}}. \tag{8.17}$$

This analysis produces the following results for the cited example:

Time (s)	N_{A0}, g mol/(cm² s)
0.001	1.12×10^{-4}
0.01	3.53×10^{-5}
0.1	1.12×10^{-5}
1	3.53×10^{-6}
10	1.12×10^{-6}
100	3.53×10^{-7}

We now want to correct the above results for the evaporation of n-pentane into air by adding the convective flux, that is, we recognize that this is *not a system with zero velocity.* Since the pentane will evaporate rapidly, we write continuity equations for each species:

$$\frac{\partial C_A}{\partial t} + \frac{\partial N_{Az}}{\partial z} = 0 \quad \text{and} \quad \frac{\partial C_B}{\partial t} + \frac{\partial N_{Bz}}{\partial z} = 0. \tag{8.18}$$

We add the equations together and note that the total molar concentration is constant. Therefore,

$$\frac{\partial}{\partial z}(N_{Az} + N_{Bz}) = 0. \tag{8.19}$$

Clearly, the sum of the fluxes is independent of z; if we can determine that sum at *any* z location, then we know it everywhere. If "B" is insoluble in the liquid "A", then at the interface

$$N_{Az} + N_{Bz} = N_{Az0} = -\frac{CD_{AB}}{1 - x_{A0}} \frac{\partial x_A}{\partial z}\bigg|_{z=0}. \tag{8.20}$$

It is clear that the correct form for the continuity equation must be written as

$$\frac{\partial x_A}{\partial t} = D_{AB}\frac{\partial^2 x_A}{\partial z^2} + \left[\frac{D_{AB}}{1 - x_{A0}} \frac{\partial x_A}{\partial z}\bigg|_{z=0}\right] \frac{\partial x_A}{\partial z}. \tag{8.21}$$

Compare this equation with eq. (8.15). J. H. Arnold solved this problem in 1944 and it is worthwhile for us to outline a few of the important steps in the analysis. We define a new variable using the Boltzmann transformation

$$\eta = \frac{z}{\sqrt{4D_{AB}t}} \tag{8.22}$$

and introduce it in (8.21). This substitution results in

$$-2\eta x'_A = x''_A + x'_A\left[\frac{1}{1 - x_{A0}} x'_A\big|_{\eta=0}\right]. \tag{8.23}$$

If we let $\psi = x_A/x_{A0}$ and $\phi_0 = -\frac{1}{2}\frac{x_{A0}}{1-x_{A0}}\psi'\big|_{\eta=0}$, then

$$\psi'' + 2(\eta - \phi_0)\psi' = 0. \tag{8.24}$$

Please note that ϕ_0 does *not* depend upon η. We can reduce the order of eq. (8.24) and integrate immediately yielding

$$\frac{d\psi}{d\eta} = C_1 \exp(-\eta^2 - 2\phi_0\eta). \tag{8.25}$$

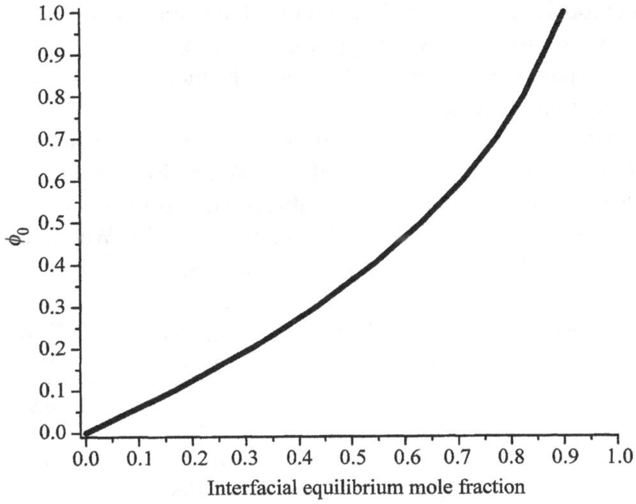

FIGURE 8.3. Illustration of the variation of ϕ_0 with x_{A0} for the Arnold problem.

We need only to complete the square and integrate to find the solution

$$\frac{x_A}{x_{A0}} = \frac{1 - \text{erf}(\eta - \phi_0)}{1 + \text{erf}(\phi_0)}. \tag{8.26}$$

The initial condition must be used to find the relationship between interfacial equilibrium mole fraction and ϕ_0:

$$\frac{x_{A0}}{1 - x_{A0}} = \sqrt{\pi}\phi_0 \exp(\phi_0^2)(1 + \text{erf}\,\phi_0). \tag{8.27}$$

A table of corresponding numerical values and a more useful graph (Figure 8.3) follow:

x_{A0}	ϕ_0
0	0
0.1	0.0586
0.2	0.1222
0.3	0.1920
0.4	0.2697
0.6	0.4608
0.8	0.7506
0.9	1.0063
1.0	∞

We are now in a position to return to our n-pentane example. The molar flux at the interface using the Arnold correction is

$$N_{A0} = C\phi\sqrt{\frac{D_{AB}}{t}}. \tag{8.28}$$

At $t = 1$ s, the corresponding flux is 4.581×10^{-6} g mol/(cm^2 s); this is 30% larger than the value we calculated previously using Fick's second law. One can easily imagine circumstances involving approach to the flammability limit (or perhaps toxicity threshold) where the increased flux could be absolutely critical!

How much difference will this correction make with regard to the concentration profiles? We will look at an example using diethyl ether (*very* volatile) evaporating into air. For $T = 18°C$, $D_{AB} = 0.089\ cm^2/s$ and $x_{A0} = 0.526$. We choose $t = 40$ s and calculate the following results:

Z-Position (cm)	Transformation Variable η	X_A/X_{A0} Fick	X_A/X_{A0} Arnold
0.5	0.1325	0.86	0.90
1	0.265	0.72	0.80
2	0.53	0.47	0.58
4	1.06	0.135	0.24
8	2.12	0.002	0.008

It is evident that the Arnold correction is *very important* in the unsteady evaporation of volatile liquids; both the flux at the interface and the concentration profile will be significantly different from those obtained from Fick's second law whenever x_{A0} is large.

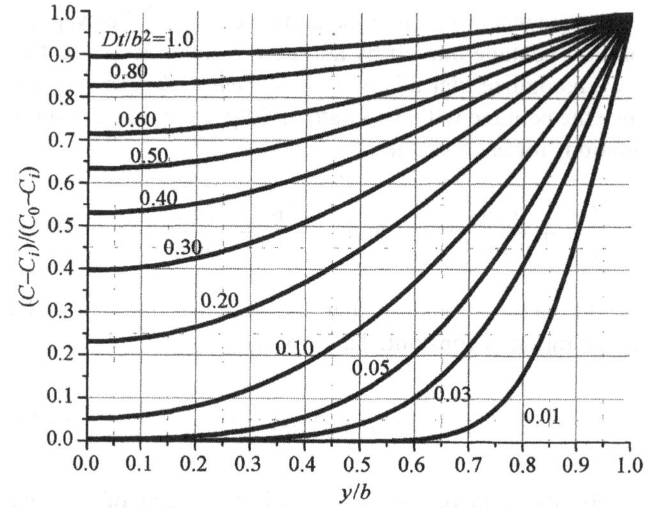

FIGURE 8.4. Transient diffusion in a plane sheet of thickness $2b$. The initial concentration in the sheet is C_i and the surface concentration (for all t) is C_0. Concentration distributions are provided for values of the parameter Dt/b^2 of 0.01, 0.03, 0.05, 0.1, 0.2, 0.3, 0.4, 0.5, 0.6, 0.8, and 1.0. The left-hand side of the figure corresponds to the center of the sheet. These concentration distributions were determined by computation.

8.3 DIFFUSION IN RECTANGULAR GEOMETRIES

The starting point for these problems is eq. (8.3). We begin with an example illustrating the similarities between conduction problems that we explored in Chapter 6 and certain diffusion problems. Consider a plane sheet or slab of thickness $2b$. The initial concentration of "A" in the interior is C_{Ai}; at $t = 0$, the surface concentration is changed to a new value C_{A0}. We place the origin ($y = 0$) on the sheet's centerline and write the governing equation:

$$\frac{\partial C_A}{\partial t} = D_{AB}\frac{\partial^2 C_A}{\partial y^2}. \tag{8.29}$$

This, of course, is a prime candidate for application of the product method. We define a dimensionless concentration as

$$C = \frac{C_A - C_{Ai}}{C_{A0} - C_{Ai}}, \tag{8.30}$$

such that $C = 0$ initially and $C \to 1$ as $t \to \infty$. The reader may wish to show that

$$C = 1 + \sum_{n=1}^{\infty} B_n \exp(-D_{AB}\lambda_n^2 t)\cos\lambda_n y,$$

where

$$\lambda_n = \frac{(2n-1)\pi}{2b}. \tag{8.31}$$

The result for this problem may be conveniently represented graphically as shown in Figure 8.4.

To illustrate the use of Figure 8.4, let $b = 0.1$ cm, $t = 2000$ s, and $D = 1 \times 10^{-6}\ cm^2/s$, therefore, $Dt/b^2 = 0.2$. At $y = 0.05$ cm, $(C - C_i)/(C_0 - C_i) \approx 0.45$. The flux at the surface can also be obtained from this figure (using the same parametric values) since for $y/b = 1$,

$$\frac{b}{C_0 - C_i}\frac{dC}{dy} \cong -1.32.$$

8.3.1 Diffusion into Quiescent Liquids: Absorption

Consider a gas–liquid interface located at $y = 0$; the liquid extends in the y-direction and is either infinitely deep or very deep relative to the expected penetration of species "A". An impermeable barrier separates the two phases up to $t = 0$. When it is removed, "A" enters the liquid phase and mass transfer by diffusion in the y-direction ensues. The governing equation is

$$\frac{\partial C_A}{\partial t} = D_{AB}\frac{\partial^2 C_A}{\partial y^2}. \tag{8.32}$$

We assume that equilibrium at the interface is established rapidly, which is generally true unless a surfactant is present to hinder transport across the interface. It is convenient to define a dimensionless concentration $C = C_A/C_{As}$, where C_{As} is determined by the solubility of "A" in the liquid phase. You may immediately recognize that this problem is fully analogous to Stokes' first problem (viscous flow near a wall suddenly set in motion) and also to the conduction of thermal energy into a (semi-) infinite slab. If we again employ the Boltzmann transformation

$$\eta = \frac{y}{\sqrt{4D_{AB}t}}, \tag{8.33}$$

then it is a simple matter to show

$$\frac{C_A}{C_{As}} = \text{erfc}\left(\frac{y}{\sqrt{4D_{AB}t}}\right). \tag{8.34}$$

We can illustrate the rate at which a transport process like this occurs with an example. Carbon dioxide is to be absorbed into (initially pure) water; at 25°C, the diffusivity is about 2×10^{-5} cm^2/s. We construct the following table for the fixed y-position, 10 cm:

Time (s)	$\sqrt{4D_{AB}t}$	η, for $y = 10$ cm	C_A/C_{As}
100	0.089	112.4	0
1000	0.283	35.34	0
10,000	0.894	11.18	0
100,000	2.828	3.54	0
1,000,000	8.944	1.12	0.11
10,000,000	28.284	0.354	0.62

We note that it is going to take about 10 or 11 days for appreciable carbon dioxide to show up at a y-position just 10 cm below the water surface: Diffusion in liquids is slow! This particular example also has important implications with respect to climate change. The solubility of carbon dioxide in seawater is about 0.09 g per kg, though this value is affected by both temperature and pressure. It is recognized that the world's oceans constitute a very large sink for CO_2 and numerous investigations are underway to explore possibilities of sequestration in seawater. But it is also clear that the current rate of anthropic generation of CO_2 is considerably larger than the rate of absorption; consequently, the concentration of carbon dioxide in the atmosphere continues to rise (in fact, we are rapidly approaching 400 ppm). We will not be able to rely upon absorption at the gas–liquid interface (to lessen the impact of burning fossil fuels) as it is too slow; therefore, there is much current emphasis upon carbon capture from power plant flue gases. A recent report in *Chemical and Engineering News* (Thayer, 2009) notes that scrubbing processes using alkanolamines or ammonia are being tested successfully. Yet the carbon dioxide, once captured, still has to go *somewhere* for long-term storage. This is why companies like Norway's *Statoil* have been injecting CO_2 into sediments at the bottom of the North Sea. Though very expensive, the scheme might be made viable by taxes upon CO_2 emissions.

8.3.2 Absorption with Chemical Reaction

We want to extend the previous example by adding chemical reaction. Once again, there is initially no "A" present in the liquid phase. At $t = 0$, the gas and liquid are brought into contact; species "A" diffuses into the liquid where it undergoes an irreversible first-order chemical reaction:

$$\frac{\partial C_A}{\partial t} = D_{AB}\frac{\partial^2 C_A}{\partial y^2} - k_1 C_A. \tag{8.35}$$

The reader may note the similarity to certain heat transfer problems, for example, conduction in a metal rod or pin with loss from the surface to the surrounding fluid. This is a very well-known problem treated successfully by P. V. Danckwerts in 1950. It holds a prominent place in the chemical engineering literature and presents a couple of features that are of special interest to us. The first of those concerns an alternative solution procedure. We will use the Laplace transform and reduce eq. (8.35) to an ordinary differential equation:

$$s\overline{C}_A = D_{AB}\frac{d^2\overline{C}_A}{dz^2} - k_1\overline{C}_A. \tag{8.36}$$

Recall that with the Laplace transform, the time derivative is replaced by multiplication by "s" and that the initial value for C_A must be subtracted. In our case, of course, that concentration is zero. Accordingly,

$$\frac{d^2\overline{C}_A}{dz^2} - \left(\frac{k_1 + s}{D_{AB}}\right)\overline{C}_A = 0, \tag{8.37}$$

which leads us directly to the subsidiary equation:

$$\overline{C}_A = c_1 \exp\left(-\sqrt{\beta}z\right) + c_2 \exp\left(+\sqrt{\beta}z\right). \tag{8.38}$$

The transform must remain finite as $z \to \infty$, so $c_2 = 0$. At the interface ($z = 0$), the concentration is determined from the solubility of "A" in the liquid. For convenience, we assume that the concentration is written in dimensionless form such that

$$C_A(z = 0) = 1 \text{ and, consequently, } c_1 = \frac{1}{s}.$$

It remains for us to invert the transform; referring to an appropriate table, we find

$$\frac{C_A}{C_{A0}} = \frac{1}{2}\exp\left(-\sqrt{\frac{k_1}{D_{AB}}}z\right)\text{erfc}\left[\frac{z}{\sqrt{4D_{AB}t}} - \sqrt{k_1 t}\right] + \frac{1}{2}\exp\left(+\sqrt{\frac{k_1}{D_{AB}}}z\right)\text{erfc}\left[\frac{z}{\sqrt{4D_{AB}t}} + \sqrt{k_1 t}\right]. \tag{8.39}$$

We are now in a position to assess the impact of reaction upon the mass transfer rate in absorption and the effects are illustrated in Figure 8.5.

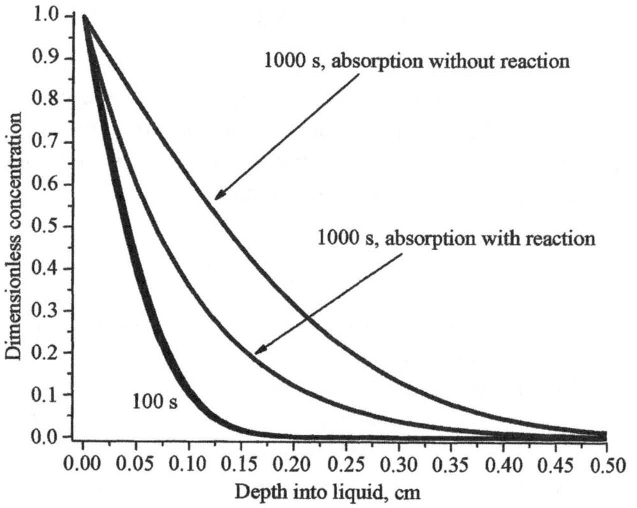

FIGURE 8.5. Comparison of concentration profiles for absorption into a quiescent liquid at 100 and 1000 s with comparable curves for absorption with reaction. The two curves at $t = 100$ s are virtually coincident.

Note how the chemical reaction has steepened the concentration gradient at the surface. This is referred to as *enhancement*; the chemical reaction has enhanced the rate of absorption and diminished the penetration of the solute species "A" into the liquid phase. The *enhancement factor E* is used to assess the impact of the chemical reaction upon mass transfer; it is the ratio of the amount of "A" absorbed into a reacting liquid in time t to the amount that would be absorbed over time t in the absence of reaction.

8.3.3 Concentration-Dependent Diffusivity

There are many real systems for which the diffusivity depends upon concentration, and one of the more interesting studies of this situation was carried out by Wagner (1950) who set

$$D_{AB} = D_0 \frac{C_A}{C_{A0}}. \tag{8.40}$$

Suppose we have diffusion into a semi-infinite medium with the interface located at $x = 0$. We define a dimensionless concentration

$$C = \frac{C_A}{C_{A0}} \tag{8.41}$$

such that

$$\frac{\partial C}{\partial t} = \frac{\partial}{\partial x}\left[D_0 C \frac{\partial C}{\partial x}\right]. \tag{8.42}$$

We again apply the familiar transformation $\eta = x/\sqrt{4D_0 t}$, which produces a second-order nonlinear ODE:

$$C\frac{d^2C}{d\eta^2} + \left(\frac{dC}{d\eta}\right)^2 + 2\eta\frac{dC}{d\eta} = 0. \tag{8.43}$$

No closed-form solution is known for this equation. But we can carry out a numerical exploration of this model and compare it with the result we obtained previously from the unsteady transport into a semi-infinite medium, where $\partial C/\partial t = D_{AB}(\partial^2 C/\partial x^2)$; we have already observed that the Boltzmann transformation yields the ordinary differential equation:

$$\frac{d^2C}{d\eta^2} + 2\eta\frac{dC}{d\eta} = 0. \tag{8.44}$$

We know that at $\eta = 0$, $C = 1$ and we also know that for eq. (8.44), as $\eta \to \infty$, $C \to 0$. Using a Runge–Kutta algorithm, we can obtain the comparison. We set $C(0) = 1$ and use the definition of the error function to show that $\left.\frac{dC}{d\eta}\right|_{\eta=0} = -\frac{2}{\sqrt{\pi}} = -1.128379$. We can solve eq. (8.44) numerically and then try the same procedure with eq. (8.43) (see Figure 8.6). We should probably expect some difficulties in the latter case as the concentration C decreases, since

$$\frac{d^2C}{d\eta^2} = \frac{-(dC/d\eta)^2 - 2\eta(dC/d\eta)}{C}. \tag{8.45}$$

The difference between the two models evident in Figure 8.6 is remarkable. In the case of Wagner's model, the advancing velocity of the diffusing component is *strictly definable*. We

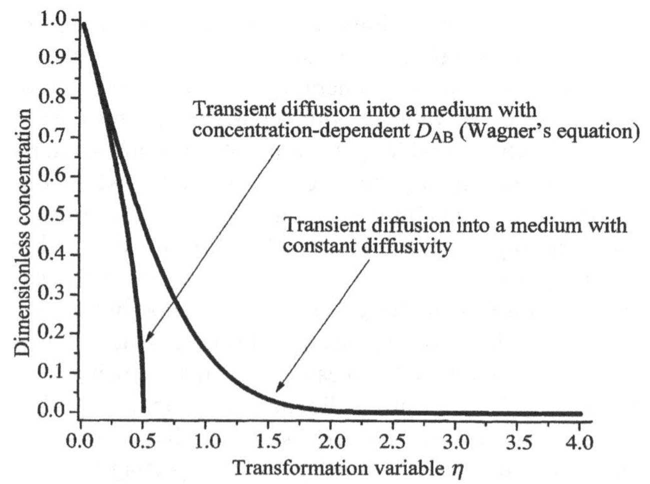

FIGURE 8.6. Comparison of the erfc solution for transient diffusion in an infinite medium with Wagner's (1950) model incorporating a concentration-dependent diffusivity.

note that $C=0$ at about $\eta=0.51$. Consequently,

$$0.51 \approx \frac{x}{\sqrt{4D_0 t}}, \text{ and accordingly, } x \approx 0.51\sqrt{4D_0 t}.$$

We differentiate

$$\left(\frac{dx}{dt}\right)\Big|_{C=0} = 0.51\sqrt{\frac{D_0}{t}}. \tag{8.46}$$

Therefore, if $D_0 = 1\times 10^{-5}$ cm^2/s, then $dx/dt =$ 0.00161 cm/s at $t=1$ s.

8.3.4 Diffusion Through a Membrane

A membrane is a semipermeable barrier that allows a solute (or permeate) to pass through. Membranes are employed for many separation processes, including water treatment, desalination, drug delivery and controlled release, artificial kidneys (dialysis), etc. They are made from a wide range of materials such as cellulose acetate, ethyl cellulose, and spun polysulfone. We tend to think of membrane-based separation as a "new" process, but as Philibert (2006) notes, the Scottish chemist Thomas Graham described the technique in 1854. Perhaps even more intriguing is the experiment carried out by Jean-Antoine Nollet in the eighteenth century. Nollet demonstrated that water would pass through a membrane (a pig's bladder), diluting an ethanol solution by osmosis.

We want to examine transient diffusion through a membrane in which the dimensionless solute concentration is instantaneously elevated on one side of the membrane and maintained at zero on the other. Let the membrane extend from $x=0$ to $x=b$; the governing equation is

$$\frac{\partial C_A}{\partial t} = D\frac{\partial^2 C_A}{\partial x^2}. \tag{8.47}$$

We have omitted subscripts on D here because the diffusion coefficient in this equation must be determined empirically. Ultimately, the dimensionless concentration profile across the membrane must take the form $C=(1-x/b)$. The product method can be used to show (and the reader should verify) that

$$C = \left(1-\frac{x}{b}\right) + \sum_{n=1}^{\infty} A_n \exp(-D\lambda_n^2 t)\sin\lambda_n x, \tag{8.48}$$

where $\lambda_n = n\pi/b$. Application of the initial condition produces the expected half-range Fourier sine series and the coefficients (the A_n's) are determined by the Fourier theorem:

$$A_n = \frac{2}{b}\int_0^b \left(\frac{x}{b}-1\right)\sin\lambda_n x\,dx. \tag{8.49}$$

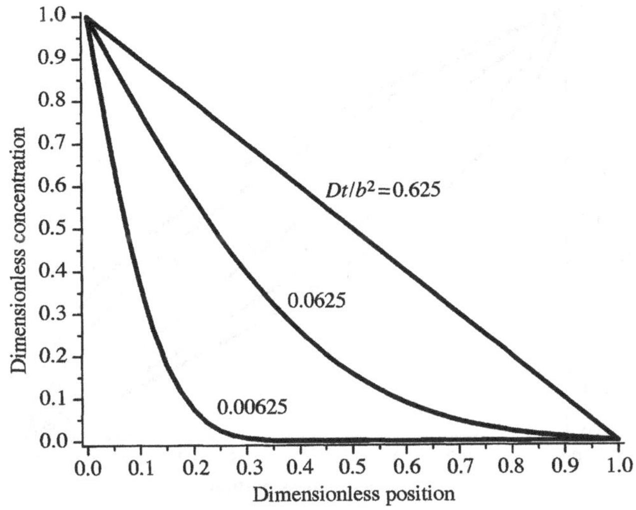

FIGURE 8.7. Concentration profiles across a membrane for values of the parameter Dt/b^2 of 0.00625, 0.0625, and 0.625. For the latter, the steady-state condition is virtually attained.

The analytic solution can be used to determine how rapidly the ultimate (linear) profile is established across the membrane, and some results are shown in Figure 8.7.

8.3.5 Diffusion Through a Membrane with Variable *D*

It is worthwhile to consider what happens to the mass transfer process examined in the previous section if the diffusion coefficient is a function of concentration. Our starting point is eq. (8.47) but with D taken into the operator:

$$\frac{\partial C_A}{\partial t} = \frac{\partial}{\partial x}\left(D\frac{\partial C_A}{\partial x}\right). \tag{8.50}$$

We now set $D = D_0(1+aC_A)$ and assume a steady-state operation. The resulting equation is

$$\frac{d^2 C_A}{dx^2} = -\frac{a}{1+aC_A}\left(\frac{dC_A}{dx}\right)^2. \tag{8.51}$$

The transport process and the shape of the concentration distribution across the membrane will be significantly affected by the constant a. If the diffusion coefficient decreases with concentration (a is negative), then the gradient must be larger (more negative) where the permeate concentration is high. You can see in Figure 8.8 that for $a=-0.9$, $C(x)$ is very steep at $x=0$. Conversely, if a is large, the concentration profile will be concave down (and very steep at dimensionless positions approaching 1).

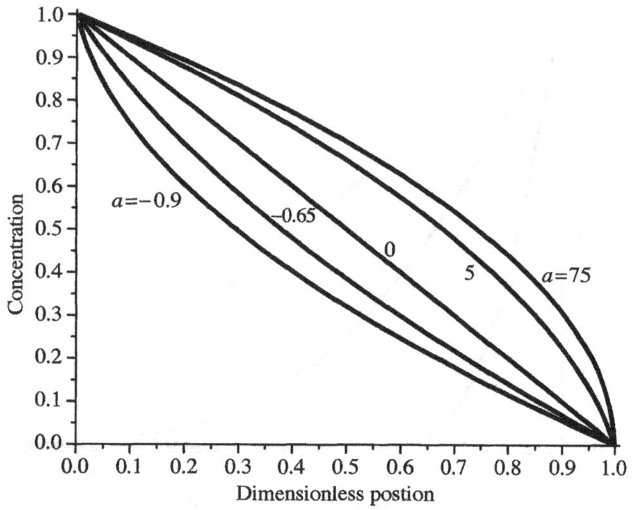

FIGURE 8.8. Steady-state concentration distributions across a membrane with variable diffusion coefficient: $D = D_0(1 + aC)$. Curves are shown for values of the parameter a of −0.9, −0.65, 0.0, 5.0, and 75.

8.4 DIFFUSION IN CYLINDRICAL SYSTEMS

The general equation for this class of problem, assuming angular symmetry, is

$$\frac{\partial C_A}{\partial t} = D_{AB}\left[\frac{1}{r}\frac{\partial}{\partial r}\left(r\frac{\partial C_A}{\partial r}\right) + \frac{\partial^2 C_A}{\partial z^2}\right] + R_A. \quad (8.52)$$

For the steady-state problems in long cylinders with no chemical reaction, we find

$$0 = \frac{1}{r}\frac{d}{dr}\left(r\frac{dC_A}{dr}\right), \text{ which yields } C_A = C_1 \ln r + C_2. \quad (8.53)$$

Suppose species "A" is diffusing through a permeable annular solid with $R_1 < r < R_2$. At $r = R_1$, the concentration is C_{A1}, and at $r = R_2$, the permeate is carried away by the solvent phase such that $C_{A2} = 0$. Consequently,

$$C_1 = \frac{C_{A1}}{\ln(R_1/R_2)}, \text{ and the flux at } r \text{ is } -D\frac{dC_A}{dr} = -D\frac{C_1}{r}. \quad (8.54)$$

Once again, the subscript has been dropped from the diffusion coefficient since we are no longer talking about a molecular property. This new "D" is determined by the characteristics of the pores in the permeable annulus as well as the size and shape of the permeate species.

8.4.1 The Porous Cylinder in Solution

Now imagine a porous cylinder, initially saturated with "A" that is placed in a nearly infinite liquid bath containing little (or even no) solute. If there is no resistance to mass transfer between the surface of the cylinder and the solvent phase, then the concentration at $r = R$ can be set to a constant value C_{As} or perhaps zero if the solvent volume is large. This situation is described by the equation

$$\frac{\partial C_A}{\partial t} = D\left[\frac{1}{r}\frac{\partial}{\partial r}\left(r\frac{\partial C_A}{\partial r}\right)\right]. \quad (8.55)$$

As we noted in the preceding section, D is an "effective" diffusivity that must be determined empirically. We can apply the product method by letting $C_A = f(r)g(t)$; two ordinary differential equations are obtained:

$$\frac{dg}{dt} = -D\lambda^2 g \quad \text{and} \quad f'' + \frac{1}{r}f' + \lambda^2 f = 0. \quad (8.56)$$

By our hypothesis, the solution must then have the form

$$C_A = C_1 \exp(-D\lambda^2 t)[AJ_0(\lambda r) + BY_0(\lambda r)]. \quad (8.57)$$

The concentration of "A" must be finite at the center of the cylinder, so $B = 0$. It is convenient to define a dimensionless concentration

$$C = \frac{C_A - C_{As}}{C_{Ai} - C_{As}} \text{ such that } C = 0 \text{ at } r = R. \quad (8.58)$$

This, of course, requires that $J_0(\lambda R) = 0$, and consequently,

$$\frac{C_A - C_{As}}{C_{Ai} - C_{As}} = \sum_{n=1}^{\infty} A_n \exp(-D\lambda_n^2 t) J_0(\lambda_n r). \quad (8.59)$$

The cylinder is initially saturated with "A"—the corresponding concentration is C_{Ai}; thus at $t = 0$, we have

$$1 = \sum_{n=1}^{\infty} A_n J_0(\lambda_n r). \quad (8.60)$$

The reader should use orthogonality to show

$$A_n = \frac{2/(\lambda_n R)}{J_1(\lambda_n R)}. \quad (8.61)$$

Now, suppose we have a porous cylinder saturated with benzene; assume $R = 1$ cm and $D \approx 0.5 \times 10^{-5}$ cm²/s. At $t = 0$, the cylinder is immersed in a large agitated reservoir of pure water. How long will it take for the dimensionless concentration to fall to 0.94 at $r = 1/2$ cm? We can use the infinite series solution to show that $t_{req} \approx 6000$ s. The reader may wish to check to see how many terms are needed for reasonable

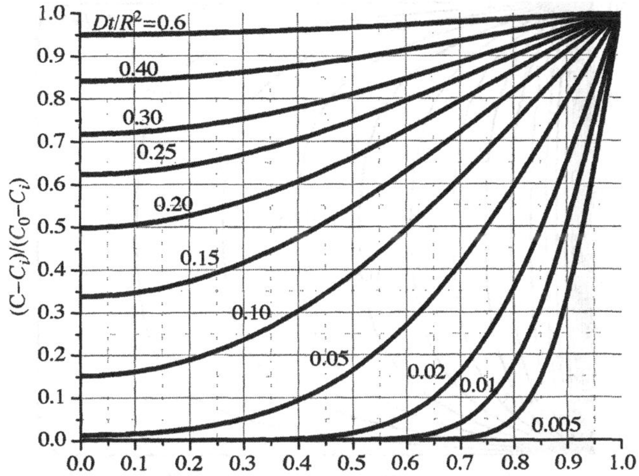

FIGURE 8.9. Transient diffusion in a long cylinder of radius R. The initial concentration in the cylinder is C_i and the surface concentration for all t is C_0. Concentration distributions are provided for values of the parameter Dt/R^2 of 0.005, 0.01, 0.02, 0.05, 0.10, 0.15, 0.20, 0.25, 0.30, 0.40, and 0.60. The left-hand side of the figure corresponds to the center of the long cylinder. These concentration profiles were determined by computation.

convergence if t is only 1000 s. The solution for this problem can be conveniently represented graphically as shown by Figure 8.9.

Let us illustrate the use of Figure 8.9 with an example. Suppose we have a cylinder with a diameter of 1 cm; if $t = 1500$ s and $D = 2.5 \times 10^{-5}$ cm^2/s, then $Dt/R^2 = 0.15$ and at the center of the cylinder,

$$\frac{C - C_i}{C_0 - C_i} \cong 0.34.$$

At the same time t, the flux at $r = R$ will be proportional to the slope of the 0.15 curve at the right-hand side of the figure:

$$\frac{R}{C_0 - C_i}\frac{dC}{dr} \cong 0.96.$$

8.4.2 The Isothermal Cylindrical Catalyst Pellet

You may recall that there are seven steps in heterogeneous catalysis:

1. Transport of reactant from the fluid phase to the pellet's surface
2. Transport of reactant to the interior of the pellet
3. Adsorption of reactant at an active site
4. Reaction
5. Desorption of product from the reactive site
6. Transport of the product back to the surface of the pellet
7. Transport of product from the surface to the bulk fluid phase

We will now develop a *homogeneous* model for a "long" cylindrical pellet that accounts for steps (2) and (4) from this list. Naturally, we must employ an effective diffusivity D, and we expect its value to be (*very roughly*) an order of magnitude smaller than the corresponding binary diffusivity D_{AB}. The precise value for D depends upon pore diameter and tortuousity, molecular shape and size, and so on; experimental measurement will be required for its determination. We assume that the rate of reaction is adequately described by the relation $k_1 a C_A$, where a is the available surface area per unit volume. Our starting point is the steady-state model,

$$\frac{d^2 C_A}{dr^2} + \frac{1}{r}\frac{dC_A}{dr} - \frac{k_1 a}{D} C_A = 0. \qquad (8.62)$$

Assuming $\beta = k_1 a/D$, we find that this example of Bessel's differential equation has the solution

$$C_A = C_1 I_0\left(\sqrt{\beta} r\right) + C_2 K_0\left(\sqrt{\beta} r\right). \qquad (8.63)$$

Since the concentration of reactant must be finite at the center of the pellet, $C_2 = 0$. At the surface, the concentration of "A" is C_{As}, consequently,

$$C_A = C_{As}\frac{I_0\left(\sqrt{\beta} r\right)}{I_0\left(\sqrt{\beta} R\right)}. \qquad (8.64)$$

While the concentration distribution in the interior of the pellet is certainly interesting, it does not tell us much about the actual operation of the catalytic process. In particular, suppose we wanted to know something about how the structure of the pellet (the configuration of the substrate) was affecting the conversion of reactant. In such cases, we might wish to examine the effectiveness factor η, which is defined as the total molar flow at the pellet's surface (taking into account both transport in the interior and the reaction) divided by the total molar flow at the surface *if all reactive sites are exposed to the surface concentration.* Therefore,

$$\eta = \frac{2\pi R L\left(-D\left.\frac{dC_A}{dr}\right|_{r=R}\right)}{-\pi R^2 L k_1 a C_{As}} = \frac{2}{\sqrt{\beta} R}\frac{I_1\left(\sqrt{\beta} R\right)}{I_0\left(\sqrt{\beta} R\right)}. \qquad (8.65)$$

Under isothermal conditions, the effectiveness factor must lie between 0 and 1; obviously, if $\eta \approx 1$, then the conversion of the reactant species is not significantly hindered by pore structure (mass transfer to the interior).

This example raises several important questions, for example, how long is long? What value of the ratio L/d is required to guarantee the validity of eq. (8.64)? If end effects must be included, how will $(\partial^2 C_A/\partial z^2)$ in squat cylinders affect η?

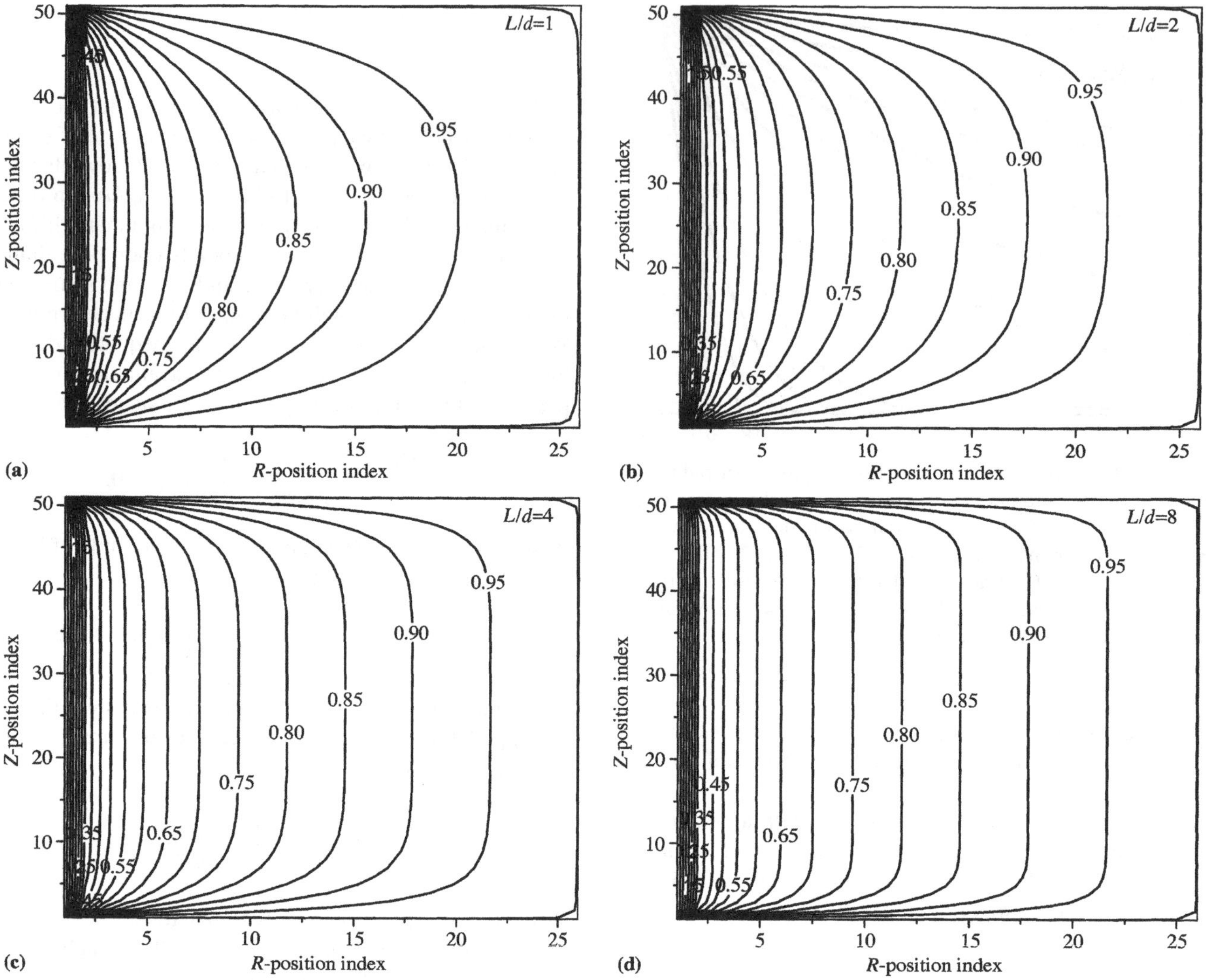

FIGURE 8.10. Diffusion in a cylinder with end effects. Across the top, left to right, $L/d = 1$ and 2 and across the bottom, $L/d = 4$ and 8. For these calculations, $Dt/R^2 = 0.45$.

And perhaps most important, what happens if a cylindrical catalyst pellet is operated nonisothermally? This last question will be the focus of a student exercise.

8.4.3 Diffusion in Squat (Small L/d) Cylinders

We implied above that if L/d is small, that is, less than perhaps 4 or 5, then diffusion in the axial direction will become important in cylinders. We should now give some definite form to this discussion. Suppose we have a diffusional transport into the interior of a "short" porous cylinder (perhaps a catalyst pellet). The governing equation must be written as

$$\frac{\partial C_A}{\partial t} = D\left[\frac{1}{r}\frac{\partial}{\partial r}\left(r\frac{\partial C_A}{\partial r}\right) + \frac{\partial^2 C_A}{\partial z^2}\right]. \tag{8.66}$$

We shall examine solutions for this equation for various values of L/d in the absence of reaction. We let L/d assume values of 1, 2, 4, and 8, and we fix the parameter Dt/R^2 at 0.45. The results are shown in Figure 8.10 for easy comparison. Note that the differences between the concentration distributions for L/d's of 4 and 8 are slight; indeed, at $L/d = 8$, transport through the ends of the cylinder is of little significance. At $L/d = 2$, however, transport in the z-direction is quite important.

8.4.4 Diffusion Through a Membrane with Edge Effects

Membranes usually have hardware supports and these supports can affect transport of the permeate. Suppose, for example, that a circular membrane is supported at the edges by an impermeable barrier (a clamping bracket). If the effective diameter of the membrane is only a small multiple of its thickness, then the governing equation must be rewritten as

$$\frac{\partial C_A}{\partial t} = D\left[\frac{\partial^2 C_A}{\partial r^2} + \frac{1}{r}\frac{\partial C_A}{\partial r} + \frac{\partial^2 C_A}{\partial z^2}\right]. \qquad (8.67)$$

Some computed results are shown in Figure 8.11. Obviously, the flux of the permeate will be reduced near the edges where the supporting hardware obstructs transport in the z-direction.

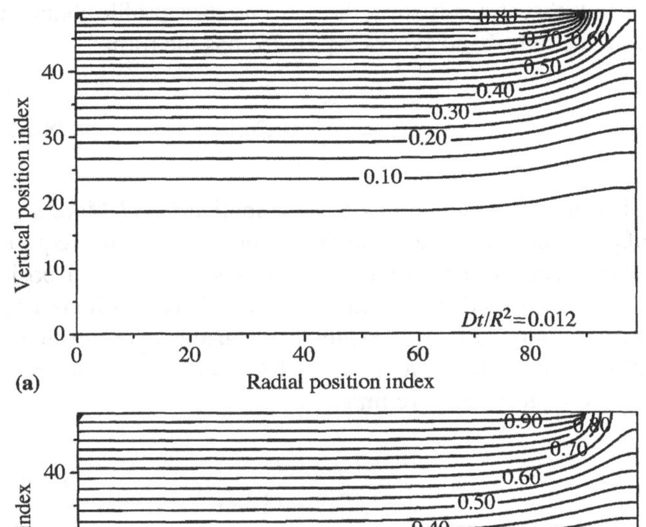

FIGURE 8.11. The evolution of edge effects in diffusional transport through a membrane. The three contour plots correspond to values of the parameter Dt/R^2 of 0.012, 0.024, and 0.048. The ratio of membrane thickness to diameter $h/2R$ is 1/4. The center of the membrane corresponds to the left-hand side of the figure, and the clamping bracket blocks 5% (of the top and bottom based upon the diameter) at the right-hand side of each figure.

We can assess the magnitude of this effect through solution of eq. (8.67). Assume that the membrane extends in the z-direction from 0 to h. Furthermore, set $C_A(z=0)=1$ and assume that transport into the fluid phase at $z=h$ occurs so rapidly that the concentration is effectively zero (there is no resistance to mass transfer in the fluid phase at $z=h$). Under these conditions, the interesting dynamics occur mainly over values of the parameter Dt/R^2 between 0 and about 0.05. Obviously, we could solve this problem for several different values of h/R, and possibly acquire a better understanding of the importance of the effect. A rule of thumb for transport through membranes is that edge effects are probably negligible if $h/R \leq 0.2$.

The impact of the supporting bracket upon the rate of permeate transport is apparent in Figure 8.11; however, we can quantify it by determining the value of the integral

$$\int_0^R 2\pi r\left(N_{Az}|_{z=h}\right) dr \qquad (8.68)$$

and forming a quotient using (8.68) twice, the numerator with edge effects taken into account and the denominator with no interference in the z-direction.

8.4.5 Diffusion with Autocatalytic Reaction in a Cylinder

Acetylene (C_2H_2) is used as a raw material in the production of some elastomers and plastics. It is also used for metal cutting because the oxy-acetylene flame has a theoretical temperature of about 3100°C. Acetylene also has the unfortunate tendency to decompose explosively (to oxygen and hydrogen) by a free-radical mechanism. It is because of this problem that acetylene is generally not compressed to pressures over 2 atm. It can be stored at higher pressure by dissolution in acetone, however, and this is usually done for commercial transport and storage. Acetylene decomposition presents some interesting features for our consideration; suppose we store acetylene in a bare steel cylinder. Because the free radicals are destroyed by contact with an iron surface, a concentration gradient is set up and mass transfer by diffusion will occur. But this process can be thwarted if the cylinder is large enough; the available surface area may no longer be adequate to control the population of free radicals and a runaway decomposition may ensue. A balance upon the free radical "A" results in

$$\frac{\partial C_A}{\partial t} = D_{AB}\left[\frac{1}{r}\frac{\partial}{\partial r}\left(r\frac{\partial C_A}{\partial r}\right)\right] + k_1 C_A. \qquad (8.69)$$

For the moment we will consider the steady-state problem, where

$$\frac{d^2 C_A}{dr^2} + \frac{1}{r}\frac{dC_A}{dr} + \frac{k_1}{D_{AB}}C_A = 0. \qquad (8.70)$$

The solution is familiar to us:

$$C_A = AJ_0\left(\sqrt{\frac{k_1}{D_{AB}}}r\right) + BY_0\left(\sqrt{\frac{k_1}{D_{AB}}}r\right). \qquad (8.71)$$

B must be zero to ensure a finite concentration at the center. At the steel wall the free radicals are destroyed and their concentration is effectively zero, thus,

$$J_0\left(\sqrt{\frac{k_1}{D_{AB}}}R\right) = 0. \qquad (8.72)$$

As we have seen previously, the first zero occurs at 2.404826. Consequently, a critical size for the steel cylinder can be specified:

$$R_{crit} = 2.404826\sqrt{\frac{D_{AB}}{k_1}}. \qquad (8.73)$$

Now we can return to eq. (8.69) for a very interesting study of the transient problem; we arbitrarily choose $R_{crit} = 10$ cm, so that $D_{AB}/k_1 = 17.2915$ cm, and we pick a convenient initial distribution of species "A" in the cylinder:

$$(1)\left[1-\left(\frac{r}{R}\right)^2\right]^2. \qquad (8.74)$$

By varying the actual cylinder radius a little above and a little below the critical value, we can get a sense of the dynamics of the process. Some computed results are shown in Figure 8.12.

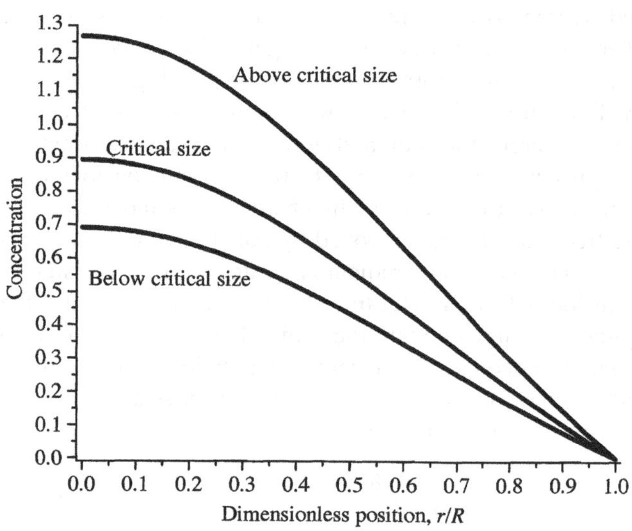

FIGURE 8.12. Concentration distributions for the autocatalytic process in a cylinder after 10 s. The three curves (top to bottom) represent above critical size, critically sized, and below critical size. Note that for the critically sized reactor, diffusion results in a rearrangement of the profile, with reduction in concentration at the center and an increase at larger r.

8.5 DIFFUSION IN SPHERICAL SYSTEMS

The starting point for this part of our discussion is eq. (8.5); with angular symmetry invoked and chemical reaction excluded, we have

$$\frac{\partial C_A}{\partial t} = D\left[\frac{\partial^2 C_A}{\partial r^2} + \frac{2}{r}\frac{\partial C_A}{\partial r}\right]. \qquad (8.75)$$

We note that at steady state, the concentration profile obtained from the right-hand side of eq. (8.75) has the form

$$C_A = \frac{C_1}{r} + C_2. \qquad (8.76)$$

What boundary conditions can be applied here? More significant, should we be concerned about $r=0$? If we require concentration to be symmetric (with respect to center position), what does that say about flux of "A" in the r-direction? We are going to press forward by focusing upon a spherical shell of thickness $R_2 - R_1$: Let $C_A = C_{A1}$ at $r = R_1$, and $C_A = C_{A2}$ at $r = R_2$. We find

$$C_1 = \frac{C_{A1} - C_{A2}}{(1/R_1) - (1/R_2)}. \qquad (8.77)$$

Consequently, the flux at any position r is

$$-D\frac{dC_A}{dr} = -D\frac{C_{A1} - C_{A2}}{(1/R_1) - (1/R_2)}(1/r^2). \qquad (8.78)$$

For transient problems to which eq. (8.75) applies, the transformation

$$C_A = \frac{\phi}{r} \text{ results in}$$

$$\frac{\partial \phi}{\partial t} = D\frac{\partial^2 \phi}{\partial r^2}. \qquad (8.79)$$

Of course, this parabolic partial differential equation has exactly the same form that we saw for a number of problems involving a slab. To illustrate, consider a sphere, initially at a uniform composition C_{Ai}, with the surface maintained at the constant value C_{As} for all t. We define a dimensionless concentration

$$C = \frac{C_A - C_{Ai}}{C_{As} - C_{Ai}}. \qquad (8.80)$$

We can use the product method to show

$$\frac{\phi}{r} = C = \frac{A}{r}\exp(-D\lambda^2 t)\sin\lambda r. \qquad (8.81)$$

Note that $\cos(\lambda r)$ has been dropped; the concentration must be finite at the center of the sphere. If the fluid phase offers no resistance to mass transfer, then $C=1$ at $r=R$ and we write

$$C = 1 + \frac{A}{r}\exp(-D\lambda^2 t)\sin\lambda r. \tag{8.82}$$

This requires that $\lambda = n\pi/R$, so the solution is simply

$$C = 1 + \sum_{n=1}^{\infty}\frac{A_n}{r}\exp(-D\lambda_n^2 t)\sin\lambda_n r, \tag{8.83}$$

with

$$A_n = \frac{2R}{n\pi}\cos n\pi. \tag{8.84}$$

A useful compilation of these results is provided in Figure 8.13.

Suppose that porous sorbent spheres were to be loaded with a solute species carried by an aqueous solution (such that $C_0 = 0.01$ g mol/cm^3). At $t=0$, the spheres are placed into the solution. Given $d=2$ cm and $D=2\times10^{-6}$ cm^2/s, what is the rate of uptake (per sphere) when $t=25{,}000$ s? The reader may wish to use Figure 8.13 to confirm that the answer is about 4.19×10^{-7} g mol/s per sphere.

Now we modify the previous case by adding a resistance to mass transfer offered by the fluid surrounding the spherical entity; all the preliminary steps are the same, but the boundary condition at the surface is changed to a Robin's-type relation:

$$-D\frac{\partial C_A}{\partial r}\bigg|_{r=R} = K\left(C_A|_{r=R} - C_{A\infty}\right). \tag{8.85}$$

Since

$$C = \frac{C_A - C_{Ai}}{C_{A\infty} - C_{Ai}}, \tag{8.86}$$

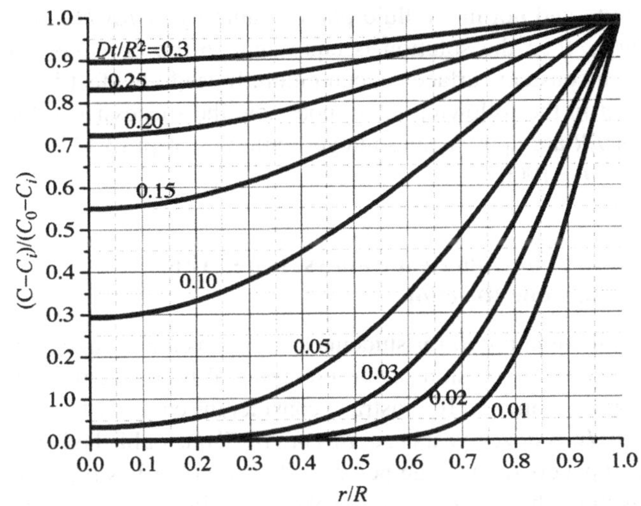

FIGURE 8.13. Transient diffusion in a sphere of radius R. The initial concentration in the sphere is C_i and the surface concentration for all t is C_0. Concentration distributions are provided for values of the parameter Dt/R^2 of 0.01, 0.02, 0.03, 0.05, 0.10, 0.15, 0.20, 0.25, and 0.3. The left-hand side of the figure corresponds to the center of the sphere. These concentration profiles were determined by computation.

the application of this boundary condition at the surface results in the transcendental equation (and the reader should verify this result):

$$-1 + \frac{\lambda R}{\tan\lambda R} = -\frac{KR}{D}. \tag{8.87}$$

The reader may recognize the similarity between the parameter *KR/D* and the Biot modulus discussed in Chapter 6. Once again we are comparing resistances (but this time with respect to mass transfer). If *KR/D* is small, then the fluid phase is significantly hindering the mass transfer process. If *KR/D* is large, then the principal resistance is in the spherical

First 12 Values for λR *for KR/D*'s from 0.01 to 1000.

KR/D	0.01	0.1	1.0	10.0	100	1000
$n=1$	0.17303	0.54228	1.57080	2.83630	3.11019	3.13845
$n=2$	4.49563	4.51566	4.71239	5.71725	6.22044	6.27690
$n=3$	7.72655	7.73820	7.85398	8.65870	9.33081	9.41535
$n=4$	10.90504	10.91329	10.99557	11.65321	12.44136	12.55380
$n=5$	14.06690	14.07330	14.13717	14.68694	15.55214	15.69226
$n=6$	17.22134	17.22656	17.27876	17.74807	18.66323	18.83071
$n=7$	20.37179	20.37621	20.42035	20.82823	21.77465	21.96916
$n=8$	23.51988	23.52370	23.56194	23.92179	24.88647	25.10761
$n=9$	26.66643	26.66980	26.70354	27.02501	27.99872	28.24607
$n=10$	29.81193	29.81495	29.84513	30.13535	31.11144	31.38452
$n=11$	32.95669	32.95942	32.98672	33.25106	34.22468	34.52298
$n=12$	36.10090	36.10339	36.12832	36.37089	37.33845	37.66143

entity and not in the fluid phase. Naturally, if *KR/D* is *very* large, then the solution is equivalent to the previous case with constant surface concentration. Indeed, this fact is evident in the following table, note how the successive values for λR are approaching integer multiples of pi (3.1416) for $KR/D = 1000$.

8.5.1 The Spherical Catalyst Pellet with Exothermic Reaction

A dilemma posed for students and professionals alike is the incredible explosion of the professional literature in transport phenomena. To illustrate, consider the case of *Physics of Fluids*. A dozen years ago, *Physics of Fluids* published about 350 papers on average per year. This number has increased by more than 40% in recent years (see Kim and Leal, 2008). Fortunately, the truly consequential developments in our field are much fewer in number, and the underlying principles of transport phenomena are fixed. Thus, a student can still be reasonably well informed by focused effort. An example: One of the classic problems in the chemical engineering literature is the spherical catalyst pellet operated nonisothermally; the student is encouraged to read the paper by Weisz and Hicks (1962). For the steady-state operation, the governing equations are

$$\frac{d^2C}{dr^2} + \frac{2}{r}\frac{dC}{dr} - \frac{k_1 aC}{D_{\text{eff}}} = 0 \tag{8.88}$$

and

$$\frac{d^2T}{dr^2} + \frac{2}{r}\frac{dT}{dr} - \frac{k_1 aC\Delta H_{\text{rxn}}}{k_{\text{eff}}} = 0. \tag{8.89}$$

For an exothermic reaction, ΔH_{rxn} is negative, and furthermore,

$$k_1 = k_0 \exp(-E/RT). \tag{8.90}$$

It is apparent that the two ordinary differential equations are coupled. There are three key dimensionless parameters associated with this problem:

$$\phi = R\sqrt{\frac{k_1 a}{D_{\text{eff}}}} \quad \text{(Thiele modulus)} \qquad \gamma = \frac{E}{RT_s} \quad \text{(Arrhenius number)}$$

$$\beta = \frac{-(\Delta H_{\text{rxn}})D_{\text{eff}}C_s}{k_{\text{eff}}T_s}. \quad \text{(Heat generation parameter)} \tag{8.91}$$

By making concentration, temperature, and radial position all dimensionless, it is possible to rewrite the governing equations as

$$\frac{d^2C}{dr^2} + \frac{2}{r}\frac{dC}{dr} - \phi^2 C = 0 \tag{8.92}$$

and

$$\frac{d^2T}{dr^2} + \frac{2}{r}\frac{dT}{dr} - \phi^2 \beta C. \tag{8.93}$$

The effectiveness factor for the modified equations is simply

$$\eta = -\frac{3}{\phi^2}\left(\frac{dC/dr|_{r=1}}{C|_{r=1}}\right). \tag{8.94}$$

We note that at steady state, the total heat flow at the surface of the sphere is equal to the heat generated in the interior by reaction. In turn, the total flow of reactant into the sphere must be equal to that consumed by the reaction. Consequently, we can write (for any r-position)

$$-4\pi R^2 k_{\text{eff}}\frac{dT}{dr} = -\Delta H_{\text{rxn}} 4\pi R^2 D_{\text{eff}}\frac{dC}{dr}. \tag{8.95}$$

We can integrate from an arbitrary r-position to the surface of the sphere and obtain the Damköhler relationship:

$$T - T_0 = \frac{\Delta H_{\text{rxn}} D_{\text{eff}}}{k_{\text{eff}}}(C - C_0). \tag{8.96}$$

This equation is of great value for two reasons: (1) It allows us to decouple the governing differential equations. (2) We can use it to estimate the maximum temperature difference for a particular catalytic reaction. As an example of the latter, let

$$\Delta H_{\text{rxn}} = -80,000\,\text{J/mol} \quad D_{\text{eff}} = 10^{-1}\,\text{cm}^2/\text{s}$$
$$C_0 = 4 \times 10^{-5}\,\text{mol/cm}^3 \quad k_{\text{eff}} = 16 \times 10^{-4}\,\text{J/(cm s °C)}$$

Accordingly,

$$T - T_0 = \frac{(80,000)(10^{-1})(4 \times 10^{-5})}{(16 \times 10^{-4})} = 200°\text{C},$$

assuming that the reactant concentration goes to zero at the pellet center.

A remarkable feature of the spherical nonisothermal catalyst pellet is the possibility of steady-state multiplicity; if the heat generation parameter is sufficiently large, one can find three distinct values of the effectiveness factor for a single Thiele modulus (with three valid concentration profiles). For strongly exothermic conditions, the effectiveness factor can be much larger than 1, though we generally try to avoid this condition to minimize risk of damage to the catalyst. What is the simplest change one could make to ensure that the operation does not enter the region of steady-state multiplicity?

8.5.2 Sorption into a Sphere from a Solution of Limited Volume

Consider a porous sorbent sphere placed in a well-agitated solution of limited volume; for example, an activated carbon "particle" immersed in a beaker of water containing an organic contaminant. The contaminant (or solute) species ("A") is taken up by the sphere and the concentration of "A" in the liquid phase is depleted. The governing equation for transport in the sphere's interior is

$$\frac{\partial C_A}{\partial t} = D\left[\frac{\partial^2 C_A}{\partial r^2} + \frac{2}{r}\frac{\partial C_A}{\partial r}\right]. \tag{8.97}$$

As we have seen previously, this equation can be transformed into an equivalent problem in a "slab" by setting $\phi = C_A r$. The total amount of "A" in solution initially is VC_{A0} and the rate at which "A" is removed from solution can be described by

$$4\pi R^2 D_{AB} \left.\frac{\partial C_A}{\partial r}\right|_{r=R}, \tag{8.98}$$

therefore, the total amount removed over a time t can be obtained by integration of eq. (8.98). The transformation of eq. (8.97) leads to

$$\frac{\partial \phi}{\partial t} = D\frac{\partial^2 \phi}{\partial r^2}, \tag{8.99}$$

which is a (familiar) candidate for separation of variables:

$$C_A = \frac{A}{r}\exp(-D\lambda^2 t)\sin\lambda r. \tag{8.100}$$

The cosine term has disappeared because the concentration of solute at the sphere's center must be finite. It is convenient to switch to dimensionless concentration, where

$$C = \frac{C_A - C_{Ai}}{C_{As} - C_{Ai}}. \tag{8.101}$$

It is likely that the sphere contains no solute initially, so $C_{Ai} = 0$. If the solution volume is unlimited, then

$$C = 1 + \sum_{n=1}^{\infty} \frac{A_n}{r}\exp(-D\lambda_n^2 t)\sin\lambda_n r, \tag{8.102}$$

where $\lambda_n = n\pi/R$. This solution provides the *lower limit* for the family of curves shown in Figure 8.14; if the solution volume is unlimited, then the fractional uptake by the particle (compared to the solute in the liquid phase) is effectively zero.

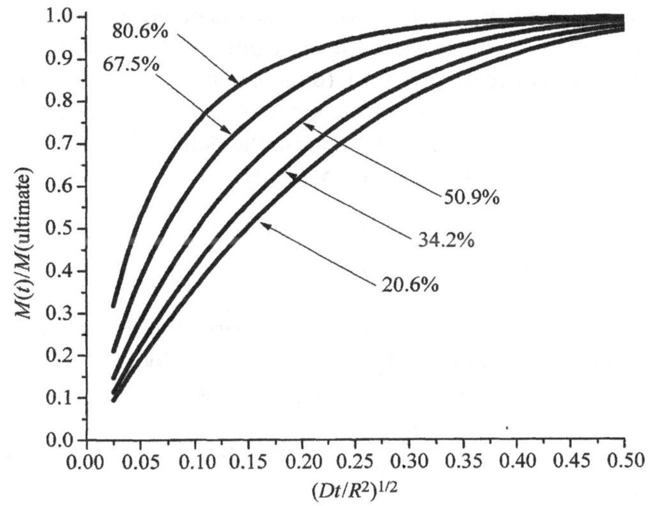

FIGURE 8.14. Sorption from a well-agitated solution of limited volume. The fractional uptake of the spherical particle, $M(t)/M(t\to\infty)$, is shown as a function of $\sqrt{(Dt/R^2)}$. The curves represent the portion of solute present in the solvent that is transferred to the sphere (80.6%, 67.5%, 50.9%, 34.2%, and 20.6%, from top to bottom). These data were obtained by computation.

8.6 SOME SPECIALIZED TOPICS IN DIFFUSION

8.6.1 Diffusion with Moving Boundaries

There are a number of important phenomena in diffusional mass transfer for which a moving boundary arises; generally this situation results from (1) a discontinuous change in diffusivity, (2) immobilization of the diffusing species (perhaps by phase change), or (3) chemical reaction where a constituent at the interface is consumed. We will consider the following two examples:

We will begin by considering problems of type (1)—specifically, let diffusion in a slab occur where the diffusion coefficient changes abruptly from D_1 to D_2 at a particular "boundary" concentration. Let the concentration in the slab be initially uniform; at $t = 0$, the concentration at one face is changed such that $C = 0$. This problem is described by two equations:

$$\frac{\partial C_{A1}}{\partial t} = D_1\frac{\partial^2 C_{A1}}{\partial y^2} \quad\text{and}\quad \frac{\partial C_{A2}}{\partial t} = D_2\frac{\partial^2 C_{A2}}{\partial y^2}. \tag{8.103}$$

At the moving boundary (the interface where the diffusivity changes abruptly), we have

$$C_{A1} = C_{A2} \quad\text{and}\quad D_1\frac{\partial C_{A1}}{\partial y} = D_2\frac{\partial C_{A2}}{\partial y}. \tag{8.104}$$

Crank (1975) points out that if the medium is infinite, then each region has an error function solution and the spatial position of the boundary must be proportional to $\sqrt{t}$. For a finite medium, such problems are easily handled numerically. Consider a medium that extends from $y=0$ to $y=b$ with an initial concentration of 1 (dimensionless). For all $t>0$, $C(y=b)=0$. The edge of the medium at $y=0$ is impermeable such that $\partial C/\partial y = 0$. Suppose the delineation between diffusivities occurs at $C=0.55$, and let

$$\frac{D_1}{D_2} = 60.$$

We solve the governing equations numerically (the behavior is shown in Figure 8.15) and find that the location of the "boundary" moves with time; in particular, we find

$$y_{\text{boundary}} \cong 0.002\sqrt{t} \tag{8.105}$$

until $D_1 t/b^2 > 0.3$. At this point, the finite character of the medium begins to be felt and the movement of the boundary deviates from the square-root dependence shown in eq. (8.105).

Another common type of moving boundary problem arises when a material is consumed by chemical reaction at an interface. For example, when a catalyst pellet becomes fouled by carbon deposition and loses its effectiveness, it may be regenerated by contact with oxygen at elevated temperatures. The carbon is converted to CO_2 quickly resulting in equimolar counterdiffusion in the matrix: Every O_2 coming in is balanced by CO_2 coming out. This problem is often referred to as the "shrinking core" model since the carbon interface retreats into the interior of the pellet as CO_2 is generated by the combustion.

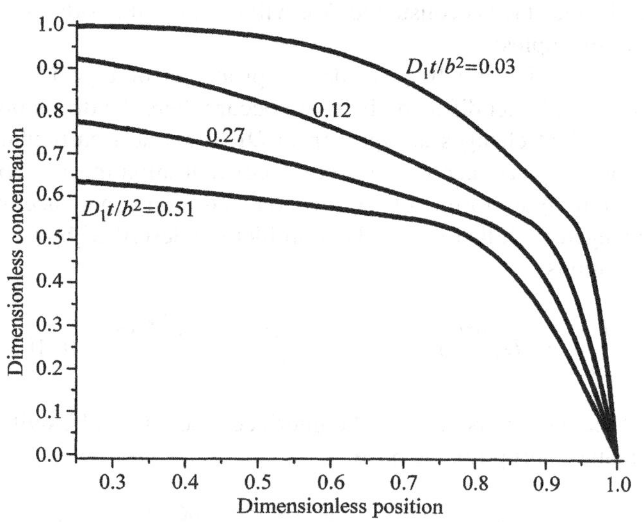

FIGURE 8.15. Dynamic behavior for a system in which the diffusivity changes abruptly at a concentration of 0.55. It is to be noted that the horizontal axis (position) has been truncated on the left to emphasize the motion of the "boundary." Curves are provided for values of the parameter $D_1 t/b^2$ of 0.03, 0.12, 0.27, and 0.51.

One approach to this problem is to assume that mass transfer process is nearly steady state (the carbon interface does not retreat rapidly). Consequently,

$$\frac{d}{dr}\left(r^2 D_{\text{eff}} \frac{dC_{\text{A}}}{dr}\right) = 0. \tag{8.106}$$

We can use this equation to determine the concentration distribution in the interior of the pellet; we assume that the effective diffusivity is constant and that appropriate boundary conditions are

$$\text{at } r = R,\ C_{\text{A}} = C_{\text{A}s}, \quad \text{and} \quad \text{at } r = R_{\text{C}},\ C_{\text{A}} = 0.$$

The latter implies that oxygen is consumed very quickly at the retreating carbon interface. The result is

$$C_{\text{A}} = \frac{C_{\text{A}s}}{((1/R_{\text{C}}) - (1/R))}\left(\frac{1}{R_{\text{C}}} - \frac{1}{r}\right). \tag{8.107}$$

We use this concentration profile to find the molar flux of oxygen at the carbon interface:

$$N_{\text{A}}|_{r=R_{\text{C}}} = -D_{\text{eff}} \left.\frac{dC_{\text{A}}}{dr}\right|_{r=R_{\text{C}}} = \frac{-D_{\text{eff}} C_{\text{A}s}}{(R_{\text{C}} - (R_{\text{C}}^2/R))}. \tag{8.108}$$

If the reaction occurs quickly, then the rate at which carbon is consumed must be directly related to the flux of oxygen at the interface. A balance on carbon leads us to

$$\frac{dR_{\text{C}}}{dt} = -\frac{D_{\text{eff}} C_{\text{A}s}/(\rho_{\text{C}}\phi)}{(R_{\text{C}} - (R_{\text{C}}^2/R))}. \tag{8.109}$$

ϕ is the volume fraction of carbon and ρ_{C} is the carbon molar density. We can use this differential equation to estimate the time required for regeneration:

$$t_{\text{req}} = \frac{\rho_{\text{C}}\phi R^2}{6 D_{\text{eff}} C_{\text{A}s}}. \tag{8.110}$$

This example of a moving boundary problem can be made considerably more interesting by considering transient diffusion in a catalytic cylinder with a small L/d ratio. The distribution of oxygen in the pellet will now be governed by

$$\frac{\partial C_{\text{A}}}{\partial t} = D_{\text{eff}}\left[\frac{\partial^2 C_{\text{A}}}{\partial r^2} + \frac{1}{r}\frac{\partial C_{\text{A}}}{\partial r} + \frac{\partial^2 C_{\text{A}}}{\partial z^2}\right]. \tag{8.111}$$

An interested student might explore the shape that the retreating carbon interface assumes in this truncated cylinder; what would you expect to see?

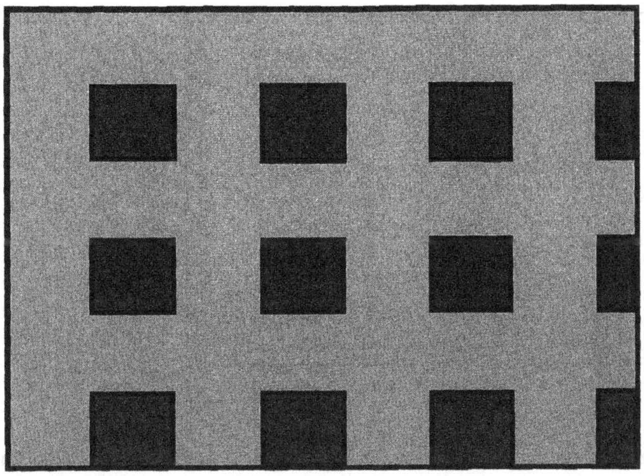

FIGURE 8.16. Upper left-hand corner of a model medium with impermeable blocks placed on a square lattice. About one-quarter of the medium is occluded by inserted bodies.

8.6.2 Diffusion with Impermeable Obstructions

One approach to the modeling of diffusional mass transfer in heterogeneous media is to place impermeable obstructions in the continuous phase either randomly or on a regular lattice. For example, one might place rectangular blocks into a fluid region in the manner indicated in Figure 8.16.

Bell and Crank (1974) demonstrated that steady-state problems in (repeating) media of the type illustrated above could be treated by subset, that is, it is only necessary to consider a portion of the domain (a rectangular region with a re-entrant corner for the case illustrated above). The method holds for both staggered and square arrays of blocks. Of greater interest perhaps is the study of the transient diffusion problem for this model, where

$$\frac{\partial C_A}{\partial t} = D_{AB}\left[\frac{\partial^2 C_A}{\partial x^2} + \frac{\partial^2 C_A}{\partial y^2}\right]. \tag{8.112}$$

This approach makes it possible for the analyst to see how the migration of the solute species is affected both by the impermeable regions and by different boundary conditions applied at the edges of the domain. For example, consider a case in which the impermeable blocks are placed on a square lattice; component "A" enters the medium through the left-hand boundary. Figure 8.17 shows how the blocks affect the transient migration of the solute species.

The preceding example is particularly significant in connection with contaminant transport in porous media. Naturally, the number and size of the impermeable regions will alter the development of the contaminant plume; these quantities could be adjusted to simulate a contamination event if one had an estimate of the void fraction (or structure) of the medium of interest.

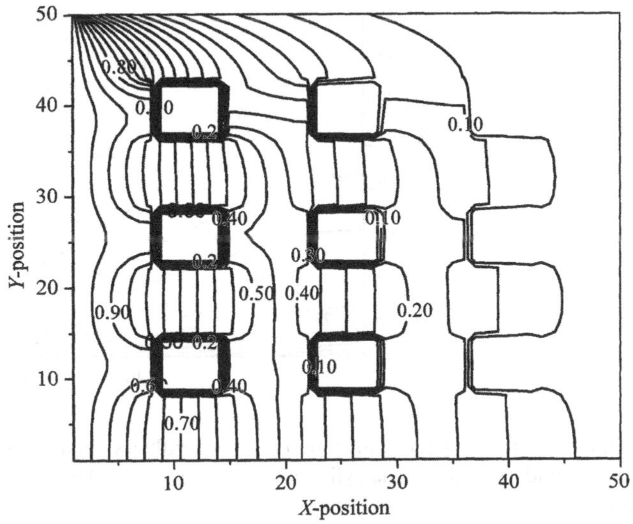

FIGURE 8.17. Concentration contours for diffusion through a rectangular region with impermeable blocks inserted on a square lattice. The solute species enters on the left-hand side of the figure. The bottom boundary is impermeable to the solute, and there is loss at both the top and right-hand side by Robin's-type boundary conditions. Note the effect of the blocks upon transport of the solute. There was no solute present in the rectangular region initially.

8.6.3 Diffusion in Biological Systems

Biological systems could not function without diffusional mass transfer through phospholipid bilayers (cell membranes) and tissue. We will look at one specific example below, but the reader is cautioned that this is a complicated field and a good starting point for background information would be one of the many specialized references such as Fournier (1999) or Truskey et al. (2004).

Consider the supply of oxygen to tissue surrounding a capillary; the Krogh model for this phenomenon utilizes concentric cylinders and two partial differential equations: one for oxygen concentration in the capillary and one for concentration in the tissue. For our purposes, we will focus upon transport in the tissue only:

$$\frac{\partial C}{\partial t} = \frac{D_t}{\varepsilon_t}\left[\frac{\partial^2 C}{\partial r^2} + \frac{1}{r}\frac{\partial C}{\partial r} + \frac{\partial^2 C}{\partial z^2}\right] - \frac{\mathrm{MR}_0 H}{\varepsilon_t}. \tag{8.113}$$

In eq. (8.113), ε_t is the void volume fraction, H is the Henry's law constant, and MR_0 is the metabolic requirement (the rate of consumption). The capillary wall does not offer much resistance to oxygen transfer, so appropriate boundary conditions for this problem are

$$r = R_t, \quad \frac{\partial C}{\partial r} = 0$$

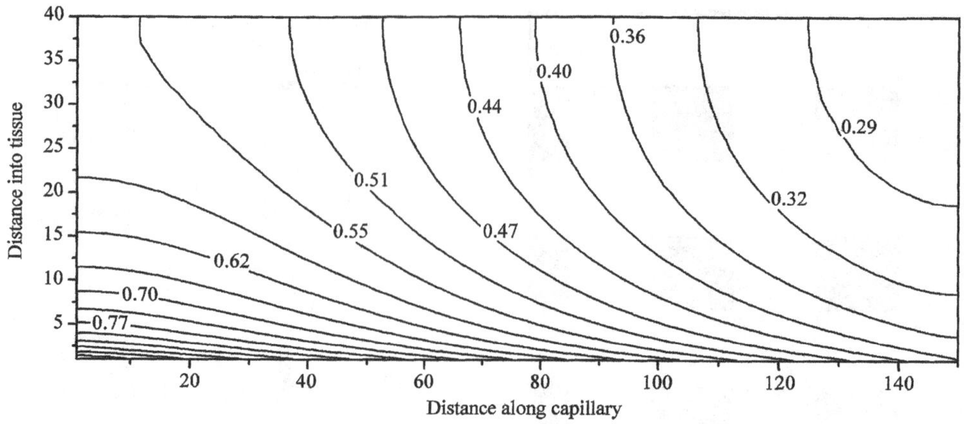

FIGURE 8.18. Oxygen distribution in tissue with assumed linearly decreasing concentration in the capillary. There is no oxygen flux at the outer boundary of the tissue cylinder (top of the figure). This is a computed result for an intermediate time that illustrates the change from the initial distribution.

and

$$z = 0 \quad \text{and} \quad z = L, \quad \frac{\partial C}{\partial z} = 0.$$

We will assume for this example that the oxygen concentration in the capillary decreases linearly in the direction of flow (we will consider the convective aspect of this problem in the next chapter), and some characteristic results are shown in Figure 8.18.

8.6.4 Controlled Release

There are many cases where an active agent (drug, pesticide, fertilizer, etc.) must be dispersed or introduced into a system at a controlled rate. In the case of drug delivery, for example, simple oral ingestion of a tablet or capsule may result in a rapid rise of drug concentration followed by a lengthy period of decay as the agent is metabolized or purged from the system. This points directly to our objective: We want the drug concentration to quickly rise above the minimum threshold for effectiveness, but remain below the level of toxicity. And typically, we would like this condition to persist for some time. Hence, the need for an effective method of controlled release (or delivery).

Fan and Singh (1989) summarized many of the techniques that have been employed for this purpose. For example, we might consider encapsulation (the drug is surrounded by a polymeric barrier) where the release is limited by diffusion through the wall. Or alternatively, the active agent might be dispersed in a polymer matrix such that the rate of release is controlled by either diffusion through, or the erosion of, the polymer material. The latter arrangement is often referred to as the "monolithic" device. There are other options as well, and Fan and Singh note that it is possible to classify them according to the nature of the rate-controlling process: these groupings include diffusion, reaction, swelling, and osmosis.

For our purposes, it will be sufficient to focus upon diffusion-controlled release in which the active agent is surrounded by a polymeric shell. We shall assume that Fick's law is capable of describing the transport of the active agent through the capsule material. Peterlin (1983) reviewed this aspect of controlled release and Crank (1975) described the "time-lag" method for the determination of the needed diffusivities. We will now illustrate the latter for a *long* cylindrical membrane in the form of a tube. The radii of the inner and outer surfaces are R_1 and R_2, respectively, and constant concentration of the penetrant species is maintained for all time such that $C(r=R_1)=1$. We also assume that the penetrant is continuously removed from the outer surface such that $C(r=R_2)=0$. The governing equation is

$$\frac{\partial C}{\partial t} = D\left[\frac{\partial^2 C}{\partial r^2} + \frac{1}{r}\frac{\partial C}{\partial r}\right]. \tag{8.114}$$

A solution is easily obtained by application of the product method and this is left to the student as an exercise. Our immediate interest is determining the value of D for transport through the encapsulating polymer. We do this by calculating the amount of the penetrant species that has passed through the membrane after time t. Of course, this will vary with the thickness of the polymer layer, $R_2 - R_1$. We let the ratio R_2/R_1 assume several values ranging from 1.2 to 2 and compare the results as shown in Figure 8.19.

An estimate for the diffusivity can be obtained from Figure 8.19, as indicated by the following example: We take the curve for $R_2/R_1 = 1.35$, fit a straight line to it (at larger t), and then extrapolate to the point of intersection with the x-axis. This will occur at a value of about 0.16. Crank (1975) notes that the intercept should occur at a *lag* value of $D\tau/(R_2 - R_1)^2$

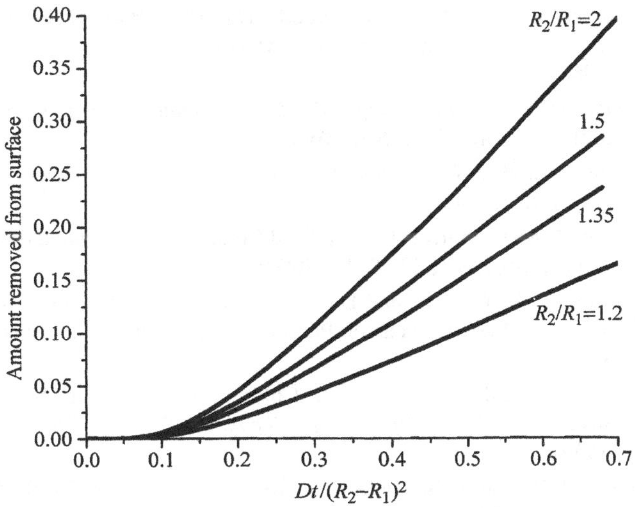

FIGURE 8.19. Amount of penetrant species removed from the outer surface of the cylindrical polymer capsule after time t. The four curves are for values of the ratio R_2/R_1 of 1.2, 1.35, 1.5, and 2. These results were obtained by numerical solution of eq. (8.114).

corresponding to

$$\frac{R_1^2 - R_2^2 + (R_1^2 + R_2^2)\ln(R_2/R_1)}{4\ln(R_2/R_1)}.$$

For the conditions chosen for the calculations, this quotient is about 0.0018. Using radii of 0.3 and 0.405 cm and a diffusivity of 2×10^{-6} cm²/s, the lag is found to be about 900 s. Peterlin states that the steady flow of permeant is established in about 5τ but the calculations presented in Figure 8.19 show that about 3τ is probably sufficient for most purposes.

You will also note from the figure that at large t, the amount of permeant that has passed through the polymer encapsulation increases linearly with time; that is, the release rate is constant. This is the desirable behavior from the standpoint of drug delivery, but the reader is cautioned that these results are predicated upon a constant concentration at the inner surface (R_1) and zero concentration at $r = R_2$. The latter, of course, means that in order for the results to be applicable *in vivo*, the permeant must be continuously swept away from the outer surface of the delivery device.

The application of eq. (8.114) is limited because it pertains to cases for which *L/d* is large. Fu et al. (1976) recognized the obvious advantages of a more general theory that could accommodate the continuum of shapes ranging from the long cylinder (capsule) to the flat disk (tablet). The starting point for such an analysis must be

$$\frac{\partial C}{\partial t} = D\left[\frac{\partial^2 C}{\partial r^2} + \frac{1}{r}\frac{\partial C}{\partial r} + \frac{\partial^2 C}{\partial z^2}\right]. \tag{8.115}$$

Fu et al. considered the case in which the active agent is distributed uniformly throughout a polymer matrix and they compared the model with experimental data for the fractional drug release. Their trials were conducted with pyrimethamine dispersed in silicone rubber and they reported a diffusivity for this system of 1.10×10^{-10} cm²/s.

8.7 CONCLUSION

Diffusional mass transfer is ubiquitous, and many of the mass transfer processes that are crucial to life, and particularly those occurring in aqueous systems, are diffusion limited. That is, the overall process rate is controlled by molecular mass transfer. Consider characteristic timescales formulated for molecular transport of momentum, heat, and mass in a tube ($R = 1$ cm) with an aqueous fluid:

$$\frac{R^2}{\nu} \approx 100, \quad \frac{R^2}{\alpha} \approx 700, \quad \text{and} \quad \frac{R^2}{D_{AB}} \approx 100,000.$$

Thus, the timescales are roughly in the ratio of 1:7:1000. Obviously, mass transfer by molecular diffusion is *very* slow; from an engineering perspective, anything we can do to enhance the rate of mass transfer is certain to be valuable. But what are our options? Of course, we recognize that we can increase the temperature or energetically move (or agitate) the fluid phase. But there may be other opportunities as well. For example, we might think about combining driving forces, possibly by adding an electric field (electrophoresis), or we might use a large temperature difference (Soret effect) to augment diffusion (which does occur in chemical vapor deposition). Certainly, we are well advised to keep such processes in mind, but just as we saw in the case of heat transfer, for many practical circumstances, fluid motion is the key to effective mass transfer. This realization leads us directly to Chapter 9.

REFERENCES

Arnold, J. H. Studies in Diffusion III: Unsteady-State Vaporization and Absorption. *Transactions of the American Institute of Chemical Engineers*, 40:361 (1944).

Bell, G. E. and J. Crank . Influence of Imbedded Particles on Steady-State Diffusion. *Journal of the Chemical Society, Faraday Transaction 2*, 70:1259 (1974).

Crank, J. *The Mathematics of Diffusion*, 2nd edition, Oxford University Press, London (1975).

Danckwerts, P. V. Absorption by Simultaneous Diffusion and Chemical Reaction. *Transactions of the Faraday Society*, 46:300 (1950).

Fan, L. T. and S. K. Singh. *Controlled Release: A Quantitative Treatment*, Springer-Verlag, Berlin (1989).

Fournier, R. L. *Basic Transport Phenomena in Biomedical Engineering*, Taylor&Francis, Philadelphia (1999).

Fu, C., Hagemeir, C., Moyer, D., and E. W. Ng. A Unified Mathematical Model for Diffusion from Drug–Polymer Composite Tablets. *Journal of Biomedical Materials Research*, 10:743 (1976).

Hayduk, W. and S. C. Cheng. Review of Relation Between Diffusivity and Solvent Viscosity in Dilute Liquid Solutions. *Chemical Engineering Science*, 26:635 (1971).

Kim, J. and L. G. Leal. Editorial: Fifty Years of *Physics of Fluid. Physics of Fluids*, 20:1 (2008).

Muller, E. A. and K. E. Gubbins. Molecular-Based Equations of State for Associating Fluids: A Review of SAFT and Related Approaches. *Industrial & Engineering Chemistry Research*, 40:2193 (2001).

Peterlin, A. Transport of Small Molecules in Polymers. In: *Controlled Drug Delivery* (S. D. Bruck, editor), CRC Press, Boca Raton (1983).

Philibert, J. One and a Half Century of Diffusion: Fick, Einstein, Before and Beyond. *Diffusion Fundamentals*, 4:6.1 (2006).

Reed, T. M. and K. E. Gubbins. *Applied Statistical Mechanics: Thermodynamic and Transport Properties of Fluids*, McGraw-Hill, New York (1973).

Reid, R. C. and T. K. Sherwood. *The Properties of Gases and Liquids*, 2nd edition, McGraw-Hill, New York (1966).

Sherwood, T. K. and R. L. Pigford. *Absorption and Extraction*, 2nd edition, McGraw-Hill, New York (1952).

Skelland, A. H. P. *Diffusional Mass Transfer*, Wiley-Interscience, New York (1974).

Thayer, A. M. Chemicals to Help Coal Come Clean. *Chemical and Engineering News*, 28, 87:18 (2009).

Truskey, G. A. Yuan, F., and D. F. Katz. *Transport Phenomena in Biological Systems*, Pearson Prentice Hall, Upper Saddle River, NJ (2004).

Wagner, C. Diffusion of Lead Chloride Dissolved in Solid Silver Chloride. *Journal of Chemical Physics*, 18:1227 (1950).

Weisz, P. B. and J. S. Hicks. The Behavior of Porous Catalyst Particles in View of Internal Mass and Heat Diffusion Effects. *Chemical Engineering Science*, 17:265 (1962).

Wilke, C. R. and P. Chang. Correlation of Diffusion Coefficients in Dilute Solutions. *AIChE Journal*, 1:264 (1955).

9

MASS TRANSFER IN WELL-CHARACTERIZED FLOWS

9.1 INTRODUCTION

We noted at the end of Chapter 8 that fluid motion was crucial to effective mass transfer in fluids and between fluids and solids. Based upon our previous exposure to heat transfer where we encountered the product *RePr*, we recognize that the product of the Reynolds number and the Schmidt number *ReSc* must provide important information about the rate of convective mass transfer. Indeed, consider the following correlations developed for very specific situations:

Mass transfer between a sphere and moving gases (Froessling equation):

$$Sh = \frac{Kd}{D_{AB}} = 2 + 0.552\, Re^{1/2} Sc^{1/3}. \tag{9.1}$$

Mass transfer in a wetted-wall column (Gilliland–Sherwood correlation):

$$Sh = 0.023\, Re^{0.83}\, Sc^{0.44}. \tag{9.2}$$

Mass transfer between a plate of length L and a moving fluid:

$$Sh_m = 0.66\, Re_L^{1/2} Sc^{1/3}. \tag{9.3}$$

In each of these cases, an increase in fluid velocity increases the mass transfer coefficient. If all other parameters of the given problem are held constant, then the rate of mass transfer must be increased by the motion. For spheres, for example, the available data (see Steinberger and Treybal, 1960) show that increasing the Reynolds number from 10 to 10,000 results in a 20-fold increase in the Sherwood number. Clearly, for mass transfer involving fluids, we can, and we must, exploit velocity. In this chapter, we will mainly confine ourselves to highly ordered flows where the variation of velocity with position is well characterized. For the most part, we will assume that the transport of species "A" is being superimposed upon an established laminar flow; the rate of mass transfer is taken to be small enough so that the velocity field is little affected. We will also assume that the system is a binary one, consisting of "A" and "B", although as a practical matter, many multicomponent systems can be treated as if they were effectively binary.

The starting points for our analyses are the equations of change (continuity equations); for the general case in rectangular, cylindrical, and spherical coordinates, they can be written as

$$\frac{\partial C_A}{\partial t} + v_x\frac{\partial C_A}{\partial x} + v_y\frac{\partial C_A}{\partial y} + v_z\frac{\partial C_A}{\partial z} = D_{AB}\left[\frac{\partial^2 C_A}{\partial x^2} + \frac{\partial^2 C_A}{\partial y^2} + \frac{\partial^2 C_A}{\partial z^2}\right] + R_A, \tag{9.4}$$

$$\frac{\partial C_A}{\partial t} + v_r\frac{\partial C_A}{\partial r} + \frac{v_\theta}{r}\frac{\partial C_A}{\partial \theta} + v_z\frac{\partial C_A}{\partial z} = D_{AB}\left[\frac{1}{r}\frac{\partial}{\partial r}\left(r\frac{\partial C_A}{\partial r}\right) + \frac{1}{r^2}\frac{\partial^2 C_A}{\partial \theta^2} + \frac{\partial^2 C_A}{\partial z^2}\right] + R_A, \tag{9.5}$$

Transport Phenomena: An Introduction to Advanced Topics, By Larry A. Glasgow

and

$$\frac{\partial C_A}{\partial t} + v_r \frac{\partial C_A}{\partial r} + \frac{v_\theta}{r}\frac{\partial C_A}{\partial \theta} + \frac{v_\phi}{r \sin\theta}\frac{\partial C_A}{\partial \phi} = D_{AB}\left[\frac{1}{r^2}\frac{\partial}{\partial r}\left(r^2 \frac{\partial C_A}{\partial r}\right) + \frac{1}{r^2 \sin\theta}\frac{\partial}{\partial \theta}\left(\sin\theta \frac{\partial C_A}{\partial \theta}\right) + \frac{1}{r^2 \sin^2\theta}\frac{\partial^2 C_A}{\partial \phi^2}\right] + R_A. \tag{9.6}$$

Note the similarity between these equations (9.4–9.6) and the corresponding energy and Navier–Stokes equations. These common features will allow us to adapt and make direct use of some solutions from heat transfer. Moreover, solution procedures we used previously should be applicable here as well.

9.2 CONVECTIVE MASS TRANSFER IN RECTANGULAR COORDINATES

9.2.1 Thin Film on a Vertical Wall

Consider a thin liquid film (extending from $y=0$ to the free surface at $y=\delta$) flowing down a flat, soluble wall, as illustrated in Figure 9.1. Species "A" dissolves, entering the fluid phase, and is then carried in the z-direction by the fluid motion:

Our starting point for this case is a suitably simplified eq. (9.4):

$$v_z \frac{\partial C_A}{\partial z} = D_{AB}\left[\frac{\partial^2 C_A}{\partial y^2} + \frac{\partial^2 C_A}{\partial z^2}\right]. \tag{9.7}$$

The velocity distribution in the flowing film, if it is thin and if the motion is slow enough to prevent ripple formation, is

$$v_z = \frac{\rho g}{\mu}\left[\delta y - \frac{y^2}{2}\right]. \tag{9.8}$$

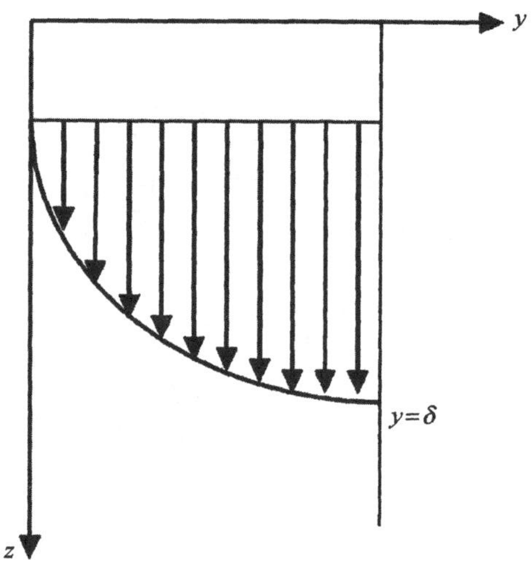

FIGURE 9.1. Thin liquid film flowing down a slightly soluble vertical wall.

Furthermore, our previous experience with heat transfer suggests that *molecular* transport in the flow (z-) direction might be negligible, particularly if the soluble species does not penetrate very far into the flowing liquid. Alternatively, we might suggest that the characteristic length for the z-direction should be much larger than that for the y-direction:

$$l_z \gg \delta.$$

Therefore,

$$\frac{\rho g}{\mu}\left[\delta y - \frac{y^2}{2}\right]\frac{\partial C_A}{\partial z} = D_{AB}\frac{\partial^2 C_A}{\partial y^2}. \tag{9.9}$$

Further simplification is possible if we allow the velocity distribution to be approximated by the linear form

$$v_z \cong \alpha y, \tag{9.10}$$

which is appropriate if y is very small. At this point, you should recognize our intent; we will now apply the Leveque analysis by setting

$$\frac{C_A}{C_{As}} = f(\eta) \quad \text{and} \quad \eta = y\left(\frac{\alpha}{9 D_{AB} z}\right)^{1/3}. \tag{9.11}$$

The transformation results in the ordinary differential equation $f'' + 3\eta^2 f' = 0$. You may also recall that the solution for this problem can be written as

$$\frac{C_A}{C_{As}} = 1 - \frac{\int_0^\eta \exp(-\eta^3)d\eta}{\Gamma(4/3)}. \tag{9.12}$$

Now we will explore a specific situation in which a water film flows down a wall made of cast benzoic acid; we want to see how well this approximate solution works. For benzoic acid in water at 14°C, we have

$$C_{As} = 1.96 \times 10^{-5}\,\text{g mol/cm}^3 \quad \text{and}$$
$$D_{AB} = 5.41 \times 10^{-6}\,\text{cm}^2/\text{s}.$$

We fix z at 20 cm, choose $\delta = 0.15$ cm (thick!), and let $\alpha = 14{,}700$ L/s, which means that $\eta = 247y$. We want to determine the concentration of benzoic acid in water at y-positions ranging from 10^{-4} cm to 10^{-2} cm. The resulting profile is shown in Figure 9.2.

It is essential that we understand the limitations of this solution. To achieve this, we will explore the problem treated above, but we will select a thinner film and a larger z-position. We set $z = 500$ cm and we let $\delta = 0.08$ cm. For the Leveque profile, we select $\alpha = 7840$ L/s, while in the case of the corrected analysis, we will solve

$$\frac{\partial C_A}{\partial z} = \frac{D_{AB}}{(\rho g/\mu)[\delta y - (y^2/2)]}\frac{\partial^2 C_A}{\partial y^2} \tag{9.13}$$

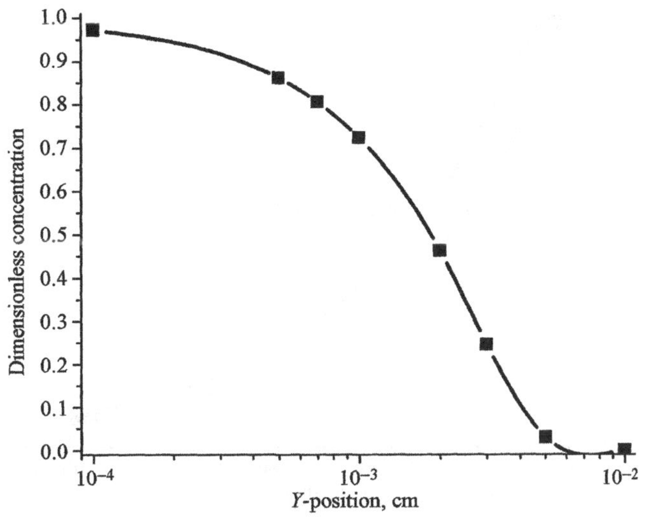

FIGURE 9.2. Concentration profile for benzoic acid in flowing water film. Note that the penetration of the soluble species at this z-position (20 cm) only amounts to about 7% of the film thickness.

numerically by forward marching in the z-direction. The results are compared in Figure 9.3.

These results show that the Leveque approximation works surprisingly well, even when the penetration of the solute species corresponds to a quarter of the film thickness. More important, note that the flux at the wall will be *very similar* for these two solutions.

9.2.2 Convective Transport with Reaction at the Wall

We now turn our attention to a case in which species "A" is transported by a flowing fluid in the z-direction, one of the walls of the rectangular channel is catalytic (Figure 9.4). At

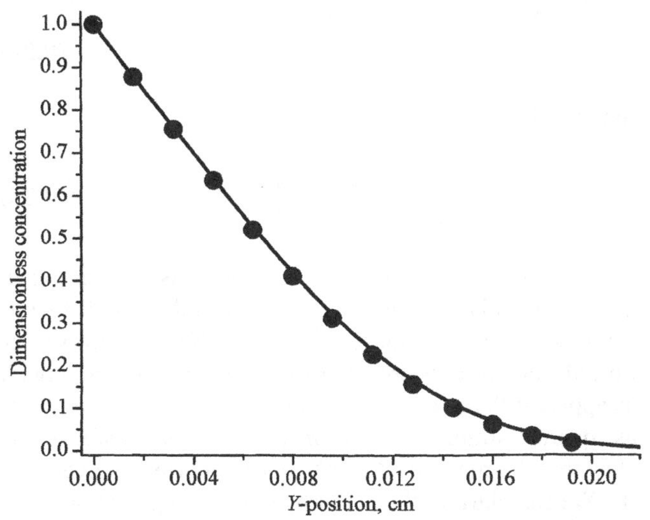

FIGURE 9.3. Comparison of the Leveque approximation (filled circles) with the correct solution (solid line) for the dissolution of benzoic acid into a flowing water film. Significant deviation appears only for y-positions larger than about 0.018 cm.

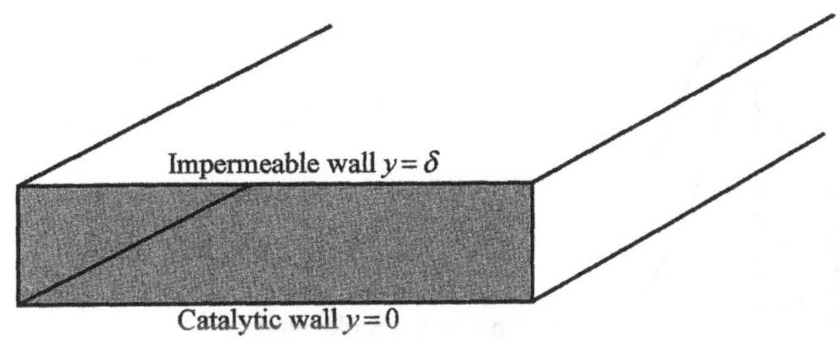

FIGURE 9.4. Rectangular duct with $W \gg \delta$ and a catalytic wall at $y = 0$.

that surface, "A" disappears rapidly and the opposing wall is nonreactive and impermeable. The reaction enters the picture as a boundary condition since it occurs at the wall only. The governing equation is

$$v_z \frac{\partial C_A}{\partial z} = D_{AB} \frac{\partial^2 C_A}{\partial y^2}. \tag{9.14}$$

For the first case of interest here, we incorporate a dimensionless concentration and assume plug flow in the duct:

$$\frac{\partial C}{\partial z} = \frac{D}{V} \frac{\partial^2 C}{\partial y^2}. \tag{9.15}$$

This is a candidate for separation of variables; we assume $C = f(y)g(z)$, resulting in the two ordinary differential equations that are solved to yield

$$g = C_1 \exp\left(-\frac{D}{V}\lambda^2 z\right) \quad \text{and} \quad f = A \sin \lambda y + B \cos \lambda y. \tag{9.16}$$

The catalytic surface is located at $y = 0$ and the impermeable surface at $y = \delta$. The reader should verify that

$$C = \sum_{n=1}^{\infty} A_n \exp\left(-\frac{D}{V}\lambda_n^2 z\right) \sin \lambda_n y, \tag{9.17}$$

where

$$\lambda_n = \frac{(2n-1)\pi}{2\delta}. \tag{9.18}$$

As we may expect with problems of this type, the leading coefficients are determined by applying the "initial" (actually entrance) condition ($C = 1$ for all y) and the Fourier theorem, resulting in:

$$A_n = \frac{4}{(2n-1)\pi}. \tag{9.19}$$

We shall fix $V\delta/D = 40$, set the channel height $\delta = 2$ cm, and explore the behavior of the plug flow solution, which is illustrated in Figure 9.5.

This brings us to the critical question with respect to this example: How different will the results be if we account

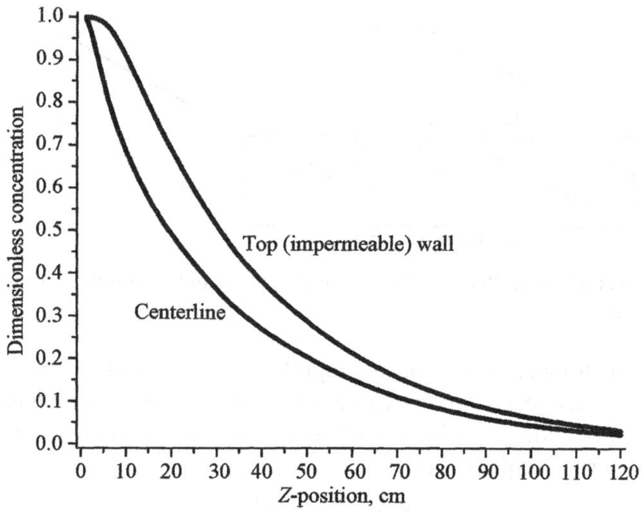

FIGURE 9.5. Evolution of concentration in a duct with one catalytic wall located at $y = 0$ for the plug flow case. The curves show the concentration at the upper (impermeable) wall and at the channel centerline.

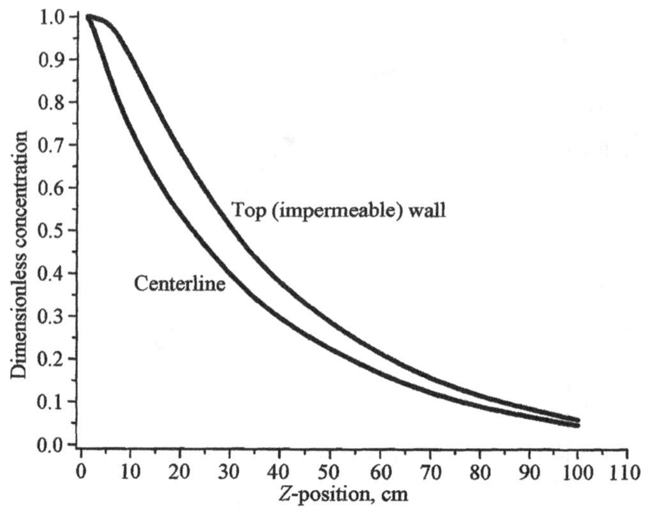

FIGURE 9.6. Evolution of concentration in a duct with *laminar* flow and one catalytic wall located at $y = 0$. The curves show the concentration at the upper (impermeable) wall and at the channel centerline. Note that the differences between these results and those for plug flow (Figure 9.5) are subtle. The centerline concentrations are slightly higher in this (the laminar flow) case.

for the variation of velocity with respect to y-position? That is, what impact will the no-slip conditions applied at $y = 0$ and $y = \delta$ have upon the change in concentration in the z-direction? This is important, because similar situations will arise when we will discuss the significance of dispersion in chemical reactors.

We start by noting that the velocity distribution will have the form

$$v_z = \frac{1}{2\mu}\frac{dp}{dz}(y^2 - \delta y). \tag{9.20}$$

The maximum velocity occurs at the centerline ($y = \delta/2$), so

$$v_z = \frac{4V_{\max}}{\delta^2}(\delta y - y^2). \tag{9.21}$$

The governing equation is now

$$\frac{4V_{\max}}{\delta^2}(\delta y - y^2)\frac{\partial C}{\partial z} = D\frac{\partial^2 C}{\partial y^2}. \tag{9.22}$$

This equation can be attacked using the very same method we employed for the "corrected" Leveque analysis. Once again, we set $\langle v_z\rangle\delta/D = 40$; for this flow, $V_{\max} = 3/2\langle v_z\rangle$. Typical results are shown in Figure 9.6.

9.2.3 Mass Transfer Between a Flowing Fluid and a Flat Plate

We assume species "A" is transferred either from the plate to the fluid, or from the fluid to the plate. Let the plate's surface correspond to $y = 0$ and place the origin at the leading edge. The governing equation is

$$v_x\frac{\partial C_A}{\partial x} + v_y\frac{\partial C_A}{\partial y} = D_{AB}\frac{\partial^2 C_A}{\partial y^2}. \tag{9.23}$$

The similarity to Prandtl's equation for the laminar boundary layer on a flat plate is to be noted. In a familiar process, we set

$$\eta = y\sqrt{\frac{V_\infty}{\nu x}}, \quad \phi = \frac{C_A - C_{A0}}{C_{A\infty} - C_{A0}}, \quad \text{and} \quad \psi = \sqrt{\nu x V_\infty} f(\eta), \tag{9.24}$$

which results in

$$\frac{d^2\phi}{d\eta^2} + \frac{1}{2}Scf\frac{d\phi}{d\eta} = 0. \tag{9.25}$$

We see that the Schmidt number Sc appears as a parameter in eq. (9.25); recall that Sc is the ratio of the molecular diffusivities for momentum and mass (ν and D_{AB}). If $Sc = 1$, the velocity profile and the concentration distribution will be identical. It is apparent that we must solve this eq. (9.25) and the Blasius equation simultaneously *unless* the mass transfer rate is so low that the movement of "A" does not affect the velocity field. We can clarify this matter by considering the boundary conditions that must be applied to solve this problem:

$$\begin{aligned} &\text{At } \eta = 0, \; C_A = C_{A0}, \text{ so } \phi = 0. \\ &\text{As } \eta \to \infty, \; C_A = C_{A\infty}, \text{ so } \phi \to 1. \end{aligned} \tag{9.26}$$

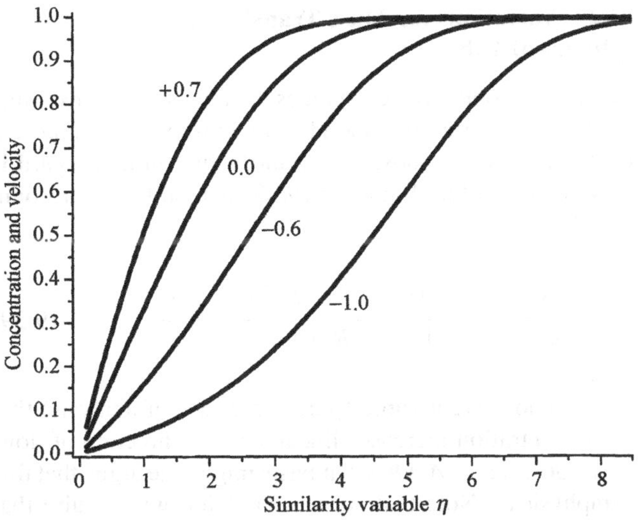

FIGURE 9.7. The effects of mass transfer between a flat plate and a flowing fluid upon the laminar boundary layer for $Sc = 1$. The dimensionless velocity and concentration profiles are shown and the Blasius profile is labeled 0.0, that is, $f(0) = 0$.

We also know from Chapter 4 that f' (which is v_x/V_∞) must be 0 at the plate's surface and must approach 1 as y becomes large. Therefore, $f'(0) = 0$ and $f'(\infty) = 1$. However, the system we have described is of fifth order—we need one more boundary condition. *If the rate of mass transfer is low, then* $v_y(\eta = 0) = 0$, *so* $f(0) = 0$. *If the rate of mass transfer is large, we note*

$$v_{y0} = -\frac{1}{2}\sqrt{\frac{\nu V_\infty}{x}}\, f(0). \tag{9.27}$$

By defining $Re_x = xV_\infty/\nu$, we find

$$f(0) = -2\sqrt{Re_x}\,\frac{v_{y0}}{V_\infty}. \tag{9.28}$$

Some interesting features of this problem are now clear; see the computed results in Figure 9.7. If the rate of mass transfer from the plate to the fluid is large, the boundary layer will be pushed away from the surface (which is referred to as blowing). Furthermore, this situation can result in a velocity profile with a point of inflection suggesting that the flow is destabilized by the mass transfer process. On the other hand, if we have a high rate of mass transfer from the fluid to the plate surface (referred to as suction), the boundary layer will be drawn down toward the plate. Such a scenario could be (and has been) used to reduce drag and even delay or prevent separation.

The molar flux of "A" at the plate surface is given by

$$N_{Ay}\big|_{y=0} = -D_{AB}(C_{A\infty} - C_{A0})\sqrt{\frac{V_\infty}{\nu x}}\,\phi'\big|_{\eta=0}. \tag{9.29}$$

9.3 MASS TRANSFER WITH LAMINAR FLOW IN CYLINDRICAL SYSTEMS

9.3.1 Fully Developed Flow in a Tube

We turn our attention to the case in which mass transfer occurs between a fluid flowing through a cylindrical tube and the tube wall. The process we are describing is a common one and it could involve sublimation, dissolution, condensation, or perhaps a reactive wall in which a species "A" is consumed. We could also envision a solute diffusing through a porous wall, possibly a transpiration process. Our first concern in such problems should be the Schmidt number Sc. Recall that we discovered that for many gases in air, Sc is on the order of 1. This of course means that the molecular diffusivities for momentum and mass have the same magnitude. If such a fluid enters the cylindrical tube, the velocity and concentration profiles will develop simultaneously, and at about the same rate. On the other hand, if we consider a similar process but with a solute species transported through a *liquid* phase, we might find much larger Sc. For example, for a variety of solutes in water, the Schmidt number ranges from 500 to about 1500 (Arnold, 1930). In these cases, we can usually assume the process is *fully developed hydrodynamically*; we only need to concern ourselves with the mass transfer portion of the problem. For the most general case under steady-state conditions, we have

$$v_r\frac{\partial C_A}{\partial r} + v_z\frac{\partial C_A}{\partial z} = D_{AB}\left[\frac{1}{r}\frac{\partial}{\partial r}\left(r\frac{\partial C_A}{\partial r}\right) + \frac{\partial^2 C_A}{\partial z^2}\right] + R_A. \tag{9.30}$$

Now we assume that there is no chemical reaction in the fluid phase, that we are far enough downstream from the entrance to assume the velocity distribution is fully developed, and that we can neglect axial diffusion:

$$v_z\frac{\partial C_A}{\partial z} = D_{AB}\left[\frac{\partial^2 C_A}{\partial r^2} + \frac{1}{r}\frac{\partial C_A}{\partial r}\right]. \tag{9.31}$$

We will consider the case in which we have mass transfer from the wall into the fluid phase; the interfacial equilibrium concentration (at $r = R$) is C_{As}. If, in addition, we assume "plug" flow and define a dimensionless concentration as

$$\phi = \frac{C_{As} - C_A}{C_{As} - C_{Ai}}, \tag{9.32}$$

then

$$\frac{\partial \phi}{\partial z} = \frac{D_{AB}}{V}\left[\frac{\partial^2 \phi}{\partial r^2} + \frac{1}{r}\frac{\partial \phi}{\partial r}\right]. \tag{9.33}$$

This is a good candidate for the product method, so we let $\phi = f(r)g(z)$, which yields

$$f = AJ_0(\lambda r) + BY_0(\lambda r) \quad \text{and} \quad g = C_1 \exp\left(-\frac{D}{V}\lambda^2 z\right). \tag{9.34}$$

The concentration must be finite at the center and equal to C_{As} at the wall (so $\phi(R) = 0$). Consequently, we obtain

$$\phi = \sum_{n=1}^{\infty} A_n \exp\left(-\frac{D}{V}\lambda_n^2 z\right) J_0(\lambda_n r). \tag{9.35}$$

We can find the leading coefficients in the usual fashion through orthogonality; note that at $z=0$, we have the inlet concentration C_{Ai}. Therefore, $1 = \sum_{n=1}^{\infty} A_n J_0(\lambda_n r)$, and we multiply both sides by $rJ_0(\lambda_m r)dr$ and integrate from 0 to R. The reader may wish to show that

$$\frac{C_{As} - C_A}{C_{As} - C_{Ai}} = \sum_{n=1}^{\infty} \frac{2}{\lambda_n R J_1(\lambda_n R)} \exp\left(-\frac{D}{V}\lambda_n^2 z\right) J_0(\lambda_n r). \tag{9.36}$$

Since we are interested in how this infinite series behaves, we select some parametric values: $D/V = 8 \times 10^{-6}$ cm, $z = 18{,}000$ cm, $R = 4$ cm, and we choose a particular radial position $r = 3$ cm. The first six terms of the series solution have the values 0.5124, 0.3110, 0.1119, 0.0106, −0.0118, and −0.0066. Therefore, $\phi \cong 0.927$.

Although obtaining the concentration distribution is important, in many practical cases, the *rate of mass transfer* is critical. This suggests that we should focus on the determination of the Sherwood number *Sh*, where $Sh = Kd/D_{AB}$; accordingly,

$$-D_{AB} \left.\frac{\partial C_A}{\partial r}\right|_{r=R} = K(C_{Am} - C_{As}), \tag{9.37}$$

where C_{Am} is the mean concentration that must be determined by integration across the cross section. Since we have plug flow, we need only to integrate $C_A(r)$, not the product of $v_z(r)C_A(r)$. We obtain the Fickian flux by differentiation:

$$-D_{AB} \left.\frac{\partial C_A}{\partial r}\right|_{r=R} = -\frac{2D_{AB}(C_{As} - C_{Ai})}{R} \sum_{n=1}^{\infty} \exp\left(-\frac{D}{V}\lambda_n^2 z\right). \tag{9.38}$$

The reader can gain valuable practice by completing this example with the determination of *Sh*.

9.3.2 Variations for Mass Transfer in a Cylindrical Tube

We should contemplate changes to the previous example that might make it correspond more closely to the physical reality; clearly, the most important feature in that regard is the velocity profile. Equation (9.33) is modified to account for $v_z(r)$:

$$\frac{\partial \phi}{\partial z} = \frac{D_{AB}}{V_{max}(1 - r^2/R^2)} \left[\frac{\partial^2 \phi}{\partial r^2} + \frac{1}{r}\frac{\partial \phi}{\partial r}\right]. \tag{9.39}$$

Now, suppose we assume (purely for ease of analysis) that the concentration increases linearly in the direction of flow, that is, $\partial C_A/\partial z = A$. On what basis might one argue that this is unphysical? Note that such a condition will require that the interfacial equilibrium concentration (C_{As}) also increase linearly in the z-direction (if the mass transfer coefficient is constant). If we press forward, ignoring the obvious objection,

$$C_A - C_{As} = \frac{V_{max}A}{D_{AB}} \left[\frac{r^2}{4} - \frac{r^4}{16R^2} - \frac{3R^2}{16}\right]. \tag{9.40}$$

This should be familiar to you; it is identical to the constant heat flux (at the wall of a tube) problem that we explored in Chapter 7. One might ask whether this result could ever be useful (perhaps for small z)?

Of course, eq. (9.31), with constant concentration at the wall, is precisely the same as the Graetz problem we examined in Chapter 7. You may recall that in that case, application of the product method results in a Sturm–Liouville problem for which eigenvalues and eigenfunctions must be determined. Many investigators have computed results for this problem and Brown (1960) provides an interesting comparison of the eigenvalues that have been obtained, beginning with Graetz in 1883 and 1885 and Nusselt in 1910. Lawal and Mujumdar (1985) point out that the classical approach to the Graetz problem suffers from poor convergence near the entrance (which is not surprising). We can easily circumvent this problem; it should be immediately apparent to you that eq. (9.39) can be solved numerically by merely forward marching in the z-direction. If we employ a sufficiently small Δz, we can obtain very accurate results. For this example, we will let

$$D_{AB} = 2 \times 10^{-5}\ \text{cm}^2/\text{s}, \quad R = 4\ \text{cm}, \quad \text{and}$$
$$v_z(r = 0) = V_{max} = 5\ \text{cm/s}$$

and compute concentration distributions at z-positions corresponding to values of $z/(1000d)$ of 0.25, 0.75, 1.75, 3.75, 7.8, 15.8, and 32. The results are shown in Figure 9.8.

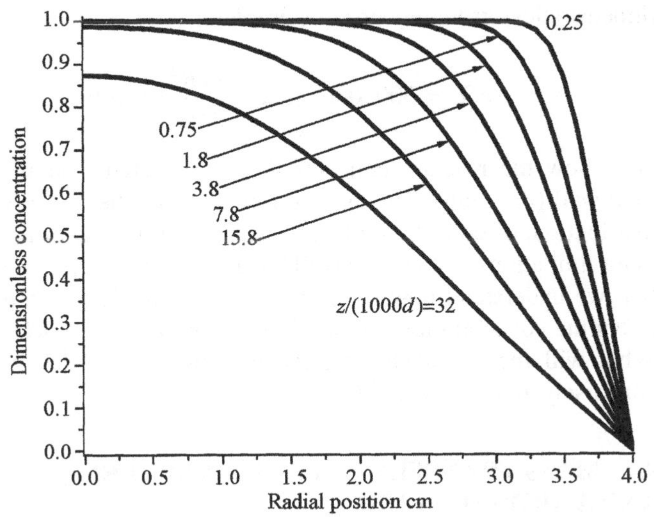

FIGURE 9.8. Evolution of the concentration distribution for the Graetz problem in mass transfer. These results were computed for values of $z/(1000d)$ of 0.25, 0.75, 1.8, 3.8, 7.8, 15.8, and 32.

Now we reconsider eq. (9.40); suppose we rearrange it as follows:

$$\frac{C_{As} - C_A}{V_{max}A/D_{AB}} = \frac{1}{3}\left[\frac{3R^2}{16} + \frac{r^4}{16R^2} - \frac{r^2}{4}\right]. \tag{9.41}$$

The reader may wish to explore (9.41) to see if this function corresponds to any of the distributions shown in Figure 9.8. Should it?

9.3.3 Mass Transfer in an Annulus with Laminar Flow

We discovered previously that the velocity distribution for fully developed laminar flow in an annulus is

$$v_z = \frac{1}{4\mu}\frac{dp}{dz}r^2 + C_1 \ln r + C_2, \tag{9.42}$$

with

$$C_1 = -\frac{(1/4\mu)(dp/dz)(R_2^2 - R_1^2)}{\ln(R_2/R_1)}. \tag{9.43}$$

The second constant of integration is found by applying the no-slip condition at either R_1 or R_2. As we noted in Chapter 3, the location of maximum velocity corresponds to

$$R_{max} = \sqrt{\frac{(R_2^2 - R_1^2)}{2\ln(R_2/R_1)}}. \tag{9.44}$$

For mass transfer occurring between the fluid and the wall(s) of the annulus with a sufficiently large product $ReSc$, we have

$$\left(\frac{1}{4\mu}\frac{dp}{dz}r^2 + C_1 \ln r + C_2\right)\frac{\partial C_A}{\partial z} = D_{AB}\left[\frac{1}{r}\frac{\partial}{\partial r}\left(r\frac{\partial C_A}{\partial r}\right)\right]. \tag{9.45}$$

If $\partial C_A/\partial z$ were approximately *constant*, eq. (9.45) could be immediately integrated to produce an analytic solution. However, we really need to start by determining how realistic this simplification would be. Suppose an aqueous fluid containing the reactant species "A" enters an annulus with one reactive wall (at $r = R_2$), where "A" is rapidly consumed. Let $Re = 1000$ and $Sc = 500$. We can compute the changes in concentration with z-position, and find the average concentration (C_{Am}) by integration:

$$C_{Am} = \frac{\int_{R_1}^{R_2} 2\pi r C_A(r) v_z(r) dr}{\pi(R_2^2 - R_1^2)\langle v_z\rangle}. \tag{9.46}$$

The results show that for this case of laminar flow in an annulus with one reactive wall, the average concentration does not decrease linearly except for perhaps $\frac{z}{(d_2 - d_1)} < 125$. The results also indicate that the Sherwood number $Sh = K(d_2 - d_1)/D_{AB}$, which is computed from

$$Sh = \frac{(d_2 - d_1)\left.\frac{\partial C_A}{\partial r}\right|_{r=R_2}}{(C_{A2} - C_{Am})}, \tag{9.47}$$

decreases rapidly in the z-direction, as shown in Figure 9.9.

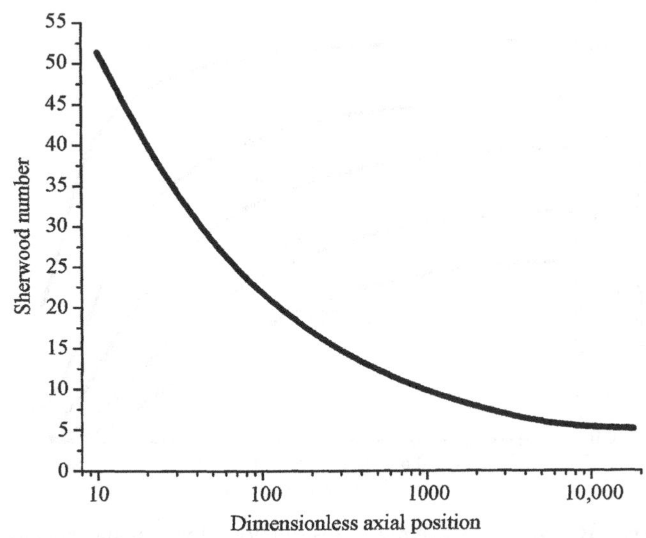

FIGURE 9.9. Sherwood number for laminar flow through an annulus with one reactive wall (located at $r = R_2$). The reaction at the surface is very rapid. The horizontal axis is dimensionless, $z/(R_2 - R_1)$.

9.3.4 Homogeneous Reaction in Fully Developed Laminar Flow

We would like to investigate a steady (fully developed) laminar flow in a tube accompanied by a homogeneous first-order chemical reaction (disappearance of the reactant species "A"). In particular, we would like to explore the effects of the radial variation of velocity upon the concentration distribution. The governing equation for this case, neglecting axial diffusion, is

$$v_z \frac{\partial C_A}{\partial z} = D_{AB}\left[\frac{\partial^2 C_A}{\partial r^2} + \frac{1}{r}\frac{\partial C_A}{\partial r}\right] - k_1 C_A. \quad (9.48)$$

It is convenient for us to rewrite the equation as

$$\frac{\partial C_A}{\partial z} = \frac{R[(\partial^2 C_A/\partial r^2 + (1/r)(\partial C_A/\partial r)] - (k_1 R/D_{AB})C_A}{ReSc[1-(r^2/R^2)]}. \quad (9.49)$$

Given an initial concentration or an initial concentration distribution, we can adapt eq. (9.49) to explicitly compute the concentration downstream $C_A(r,z)$. Our boundary conditions are as follows: for $r=0$, $\partial C_A/\partial r = 0$ (symmetry); at $r=R$, $\partial C_A/\partial r = 0$ (impermeable wall); and for $z=0$, $C_A=1$ (uniform concentration at the inlet). For this example, we set $Re=1000$, $Sc=500$, $R=2$ cm, and $k_1=0.002\ \mathrm{s}^{-1}$ and merely forward march in the z-direction computing the new concentration distributions as we go. Some results are shown in Figure 9.10. Note that there are two important dimensionless groupings for this problem:

$$Pe = ReSc = 500,000 \quad \text{and} \quad \frac{k_1 R^2}{D_{AB}} = 800.$$

Note how the reactant concentration is depleted near the tube wall ($r=2$ cm). This is a consequence of the velocity distribution, of course, and these results point to one of the main limitations of the (ideal) PFTR model. It is important that we understand how the parameters Pe and k_1R^2/D_{AB} affect the concentration distributions shown in Figure 9.10. What will the effects be if $ReSc$ is increased to 10^6, or conversely, reduced to 10^4?

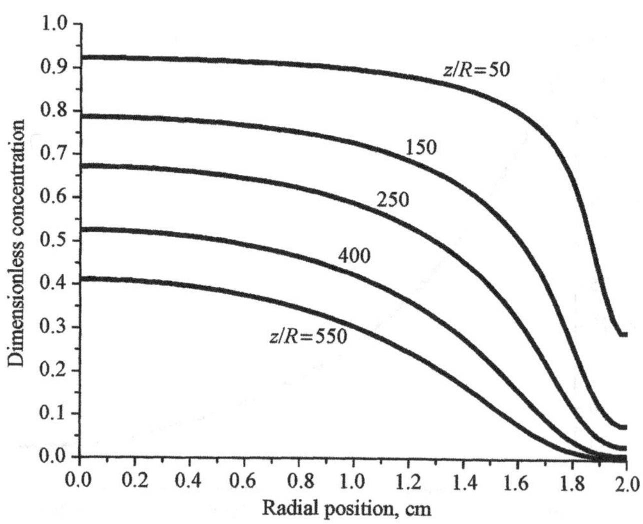

FIGURE 9.10. Concentration distributions for a homogeneous first-order reaction in fully developed laminar flow in a tube. The wall is impermeable and the Reynolds and Schmidt numbers are 1000 and 500, respectively. The curves represent dimensionless axial positions (*z/R*) of 50, 150, 250, 400, and 550.

9.4 MASS TRANSFER BETWEEN A SPHERE AND A MOVING FLUID

The sphere immersed in a flowing fluid presents some difficulties; if the Reynolds number is *very* small (creeping fluid motion) such that the inertial forces can be disregarded, then the flow field can be determined as shown by Bird et al. (2002):

$$v_r = V_\infty\left[1 - \frac{3}{2}\left(\frac{R}{r}\right) + \frac{1}{2}\left(\frac{R}{r}\right)^3\right]\cos\theta \quad (9.50)$$

and

$$v_\theta = V_\infty\left[-1 + \frac{3}{4}\left(\frac{R}{r}\right) + \frac{1}{4}\left(\frac{R}{r}\right)^3\right]\sin\theta. \quad (9.51)$$

However, these velocity vector components are limited to Reynolds numbers less than 0.1. The source of the problem, of course, is the adverse pressure gradient that results from the flow around any bluff body; the boundary layer gets pushed away from the surface (separation) and a region of recirculation is established in the wake. Investigators have explored several alternative approaches to the problem of mass transfer between a flowing fluid and a sphere as a result. Examples of these methods include application of boundary-layer theory near the stagnation point (Spalding, 1954), matched perturbation expansions (Acrivos and Taylor, 1962), transformation to a parabolic-type partial differential equation through introduction of the stream function and new independent variables (Gupalo and Ryazantsev), and numerical solution (Conner and Elghobashi). Of course, throughout the history of engineering practice, we have relied upon correlations for problems of this type as we indicated in the introduction to this chapter.

The challenges presented by flow around spheres are well known. Stokes' solution for creeping fluid motion indicates that the flow around a sphere is symmetric, fore and aft. This is not really correct, even at the low Reynolds numbers. Many attempts have been made to improve the analysis, beginning with Oseen (1910), who recognized that the neglected inertial forces might be important at significant distances from the object's surface. His approach involved inclusion of linearized inertial terms; we accomplish this, for example, by proposing

$$v_x \frac{\partial v_x}{\partial x} \approx V_\infty \frac{\partial v_x}{\partial x}. \tag{9.52}$$

Earlier, Whitehead (1889) had discovered that a simple perturbation correction for Stokes' velocity field failed at large r (the interested reader should explore Whitehead's *paradox*). Such difficulties precluded progress on analytic solutions until the technique of matched asymptotic expansions was employed in mid-twentieth century.

We begin by considering the steady case in which the relative velocity between a fluid sphere and the moving immiscible fluid is constant; the Reynolds number is relatively small but the Peclet number may be large. The governing equation is

$$v_r \frac{\partial C}{\partial r} + \frac{v_\theta}{r}\frac{\partial C}{\partial \theta} = D_{AB}\left[\frac{1}{r^2}\frac{\partial}{\partial r}\left(r^2 \frac{\partial C}{\partial r}\right)\right]. \tag{9.53}$$

Gupalo and Ryazantsev (1972) solved this problem in an approximate way by introducing the stream function

$$v_r = \frac{1}{r^2 \sin\theta}\frac{\partial \psi}{\partial \theta} \quad \text{and} \quad v_\theta = -\frac{1}{r\sin\theta}\frac{\partial \psi}{\partial r}, \tag{9.54}$$

and by changing the independent variables, resulting in the parabolic partial differential equation:

$$\frac{\partial C}{\partial \tau} = \frac{\partial^2 C}{\partial \psi^2}. \tag{9.55}$$

This is of course attractive because the familiar error function solution can be utilized directly if the boundary conditions are written as

$$\psi = 0,\ C = 0 \quad \text{and} \quad \psi \to \infty,\ C = C_0$$

The problem with this technique is that of limited applicability, as the solution is valid for small Reynolds numbers only.

For larger Re, numerical solution will be required. Conner and Elghobashi (1987) solved this problem for the Reynolds numbers up to 130 by using a variation of the technique devised by Patankar and Spalding (Patankar, 1980). Obviously, it is critical that the computed flow field accurately portray the wake region if the mass transfer is to be properly characterized. Conner and Elghobashi compared their computed results for both the size of the standing vortex and the point of separation against available experimental data and the agreement was very good. An adaptation of their results for the local $Sh(Re)$ is given in Figure 9.11 for $Re = 48$ and $Sc = 2.5$.

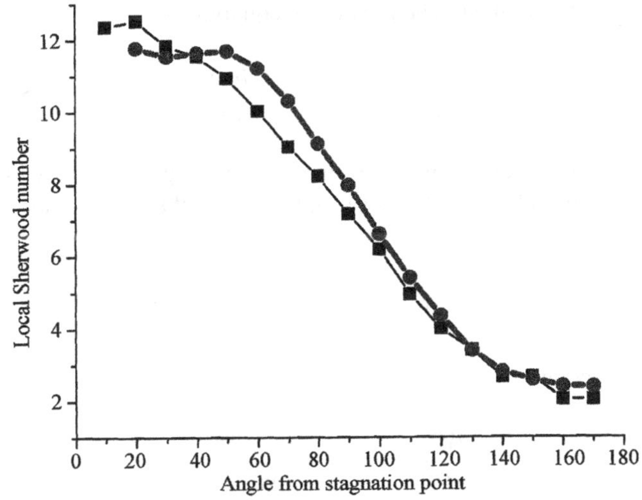

FIGURE 9.11. Local Sherwood number on a sphere immersed in a moving fluid with $Re = 48$ and $Sc = 2.5$. The curve represented by the filled circles was computed by Conner and Elghobashi (1987) and it is compared to Froessling's experimental data (filled squares).

9.5 SOME SPECIALIZED TOPICS IN CONVECTIVE MASS TRANSFER

9.5.1 Using Oscillatory Flows to Enhance Interphase Transport

Drummond and Lyman (1990) note that oscillating flows can be used to increase interphase heat and mass transport; among applications appearing in the literature are drying, combustion, and gas dispersion. In the case of spherical entities dispersed in an oscillating fluid, there is an important threshold: If the amplitude of the fluid oscillations is much smaller than the diameter of the sphere, then the transport processes are controlled by acoustic streaming (motion induced by sound, or pressure, waves). Drummond and Lyman computed mass fraction contours for a spherical particle immersed in a zero-mean oscillating fluid (for which the free-stream velocity was $V_\infty = V_1 \sin \omega t$). Their results are useful in the effort to understand how oscillations might enhance transport in multiphase systems. Our immediate interest is a little different, however, because we want to consider a nonzero mean flow in a duct or passageway.

We examined an oscillatory flow in a physiological context in Chapter 3 (periodic flow in the femoral artery of a dog). We now want to look at an oscillatory flow in a rectangular duct with the intent to examine possible enhancement of interphase transport. Consider a rectangular duct with height $2h$; the origin is located at the center of the duct and flow occurs in the x-direction in response to a periodically applied pressure gradient. The governing equation is

$$\frac{\partial v_x}{\partial t} = -\frac{1}{\rho}\frac{\partial p}{\partial x} + \nu\frac{\partial^2 v_x}{\partial y^2}. \tag{9.56}$$

We represent the driving force, pressure, with $-(1/\rho)(\partial p/\partial x) = P_0 \cos \omega t$, and we define the dimensionless variables:

$$t^* = \omega t, \quad y^* = \frac{y}{h}, \quad \text{and} \quad V^* = \frac{v_x}{(P_0/\omega)}.$$

Consequently,

$$\frac{\partial V^*}{\partial t^*} = \cos t^* + \frac{\nu}{\omega h^2}\frac{\partial^2 V^*}{\partial y^{*2}}. \tag{9.57}$$

Karagoz (2002) used a transformation approach and solved this problem analytically. Our ultimate goal is different, so we seek a numerical solution; we want to see what impact the oscillations will have upon interphase transport, particularly mass transfer enhancement. Note that the driving force in eq. (9.56) is symmetric and no *net* flow will occur under these conditions. However, we can solve this problem and check our results against Karagoz before moving on to the more realistic conditions that are of interest to us. We let the parameter $\nu/(\omega h^2)$ be 1/16 and show some results for V^* in Figure 9.12.

Now that our method for the flow computation has been verified, we move to the real issue: Can we use such a flow to our advantage in mass transfer? We will change the pressure term to produce net flow in the positive x-direction; let $\cos(t^*)$ be replaced by $1/2 + \cos(t^*)$. The reader may wish to verify

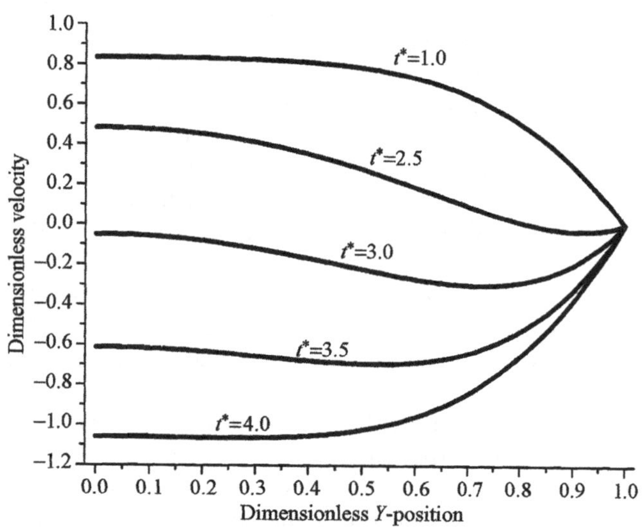

FIGURE 9.12. Velocity distributions for the oscillatory pressure-driven flow in a rectangular duct. The curves correspond to dimensionless times of 1.0, 2.5, 3.0, 3.5, and 4.0.

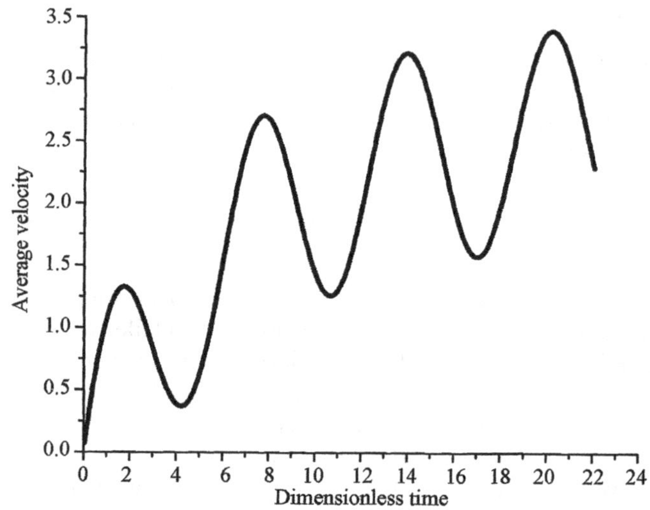

FIGURE 9.13. Spatial average velocity in the duct with the fluid subjected to an oscillatory pressure gradient. The fluid was initially at rest.

that the average velocity in the x-direction will oscillate and increase with time, as shown in Figure 9.13.

Now we are in a position to consider the possible mass transfer enhancement. For the same rectangular duct with a *locally* soluble wall, we have (neglecting axial diffusion)

$$\frac{\partial C_A}{\partial t} + v_x\frac{\partial C_A}{\partial x} = D_{AB}\frac{\partial^2 C_A}{\partial y^2}. \tag{9.58}$$

By defining

$$C^* = \frac{C_A}{C_{As}} \quad \text{and} \quad x^* = \frac{x}{h},$$

we obtain

$$\frac{\partial C^*}{\partial t^*} = \frac{D_{AB}}{\omega h^2}\frac{\partial^2 C^*}{\partial y^{*2}} - \frac{P_0}{h\omega^2}V^*\frac{\partial C^*}{\partial x^*}. \tag{9.59}$$

We compute the concentration field and note its evolution in Figure 9.14 as flow is initiated.

The contour plots shown in Figure 9.14 illustrate how the concentration profile(s) is distorted by the flow oscillations. A number of studies have appeared in the literature that focused upon significant heat transfer enhancement that can occur for what are called "zero-mean" oscillatory flows between parallel planes. Li and Yang (2000) point out that the exact mechanism by which this augmentation arises is uncertain. One possibility, of course, is the laminar–turbulent transition, if such occurred in the reported experiments. There is evidence in the literature, for example, Hino et al. (1976), that pulsatile conditions in a pipe may actually provide greater stability than that seen in the normal Hagen–Poiseuille flow.

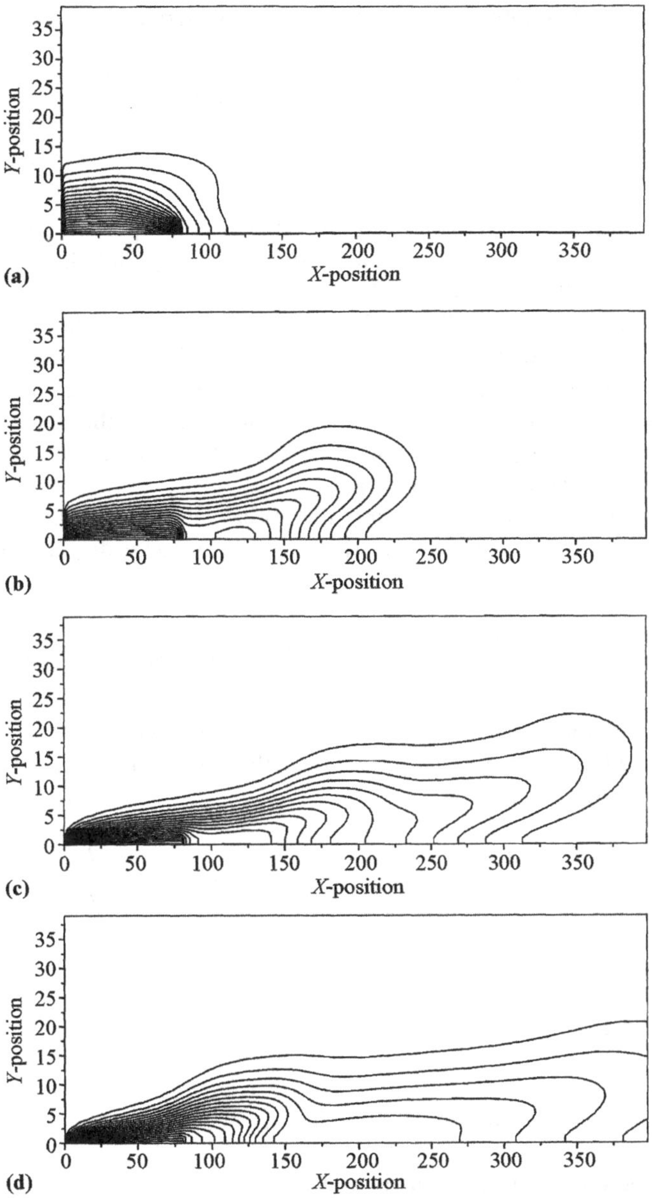

FIGURE 9.14. Concentration contours computed for the oscillatory start-up flow in a rectangular duct with a soluble wall at the lower left corner. These results are computed for dimensionless times (t^*) of 5, 10, 15, and 20.

9.5.2 Chemical Vapor Deposition in Horizontal Reactors

Organometallic chemical vapor deposition (or OMCVD) is a process by which semiconductor and microelectronic devices are fabricated. For example, gallium arsenide films are grown on a heated substrate (or susceptor) by the combination of gaseous species trimethylgallium and arsine (AsH_3). The chemical reaction takes place on the surface and if it is rapid, the limiting step in the process may be mass transfer. Hydrogen is widely used as a carrier gas in CVD processes and since the actual film growth rate in such processes is fairly small, the gas velocities are small as well (often 10 cm/s). Thus, the Reynolds numbers for many CVD processes are small enough (often 10–100) to consider the flow to be highly ordered. We assume for this example case that the chamber over the susceptor is rectangular in cross section, extending from $y = 0$ to $y = h$. We will also assume that the channel height h is much less than the channel width W and that the flow occurs in the x-direction, across the heated surface. Consequently, we start with a tentative model with a fully developed velocity distribution:

$$\frac{\partial C_A}{\partial t} + \frac{6\langle V\rangle}{h}\left[y - \frac{y^2}{h}\right]\frac{\partial C_A}{\partial x} = D_{AB}\left[\frac{\partial^2 C_A}{\partial x^2} + \frac{\partial^2 C_A}{\partial y^2}\right], \tag{9.60}$$

with the following conditions:

at $x = 0$, $C_A = C_0$, for all y,
$y = 0$, $C_A = 0$ (rapid surface reaction), and
$y = h$, $\frac{\partial C_A}{\partial y} = 0$ (impermeable upper boundary).

This gives us a starting point that we must regard as semi-quantitative. Although the simple model (9.60) will reveal one of the unpleasant truths of CVD (that film deposition is not spatially uniform), we note that it is likely that neither the velocity nor the concentration distributions would be fully developed. In addition, the temperature difference in such reactors can be quite large. Often the susceptor will be maintained at 600–1000K, depending upon the process, and the temperature of the upper wall of the chamber may be several hundred K lower. This large temperature difference may give rise to the Soret effect (thermal diffusion) and if the gases are light, the phenomenon may not be negligible. We now consider the case where concentration and temperature gradients coexist; the combined mass flux in the y-direction can be written as

$$J_{Ay} = -\rho D_{AB}\frac{d\omega_A}{dy} - \rho D_T\omega_0(1 - \omega_0)\frac{dT}{dy}, \tag{9.61}$$

where D_T is the thermodiffusion coefficient. Platten (2006) notes that the Soret coefficient, defined as D_T/D_{AB}, can be either positive or negative depending upon the sign of D_T. For the system consisting of water and ethanol (0.6088 and 0.3912, by mass), Platten cites a number of experimental studies indicating that the Soret coefficient is about $3.2 \times 10^{-3}K^{-1}$. An interesting comparison can be made utilizing eq. (9.61); we set the mass flux equal to zero, resulting in

$$\frac{d\omega_A}{dy} = -\frac{D_T}{D_{AB}}\omega_0(1 - \omega_0)\frac{dT}{dy}. \tag{9.62}$$

We assume $D_T/D_{AB} = 0.003\text{K}^{-1}$ and use the mass fractions for the water–ethanol system cited above, resulting in

$$\frac{dT}{dy} \approx 1400\frac{d\omega_A}{dy}. \tag{9.63}$$

That is, the temperature gradient (K per unit length) would need to be about 1400 times larger than the concentration (mass fraction) gradient in order for the flux of "A" to be canceled out by the Soret effect. Since the temperature gradients in CVD reactors can be very large, it is clear that the Soret effect may be important.

Tran and Scroggs (1992) used a commercial CFD code to model the performance of a CVD reactor with two-dimensional axisymmetric flow and they concluded that the Soret effect could not be discounted. They added thermodiffusion to their continuity equation. Furthermore, the large temperature difference between the susceptor and the upper boundary (confining wall) suggests that a buoyancy-driven fluid motion should be added to the pressure-driven flow through the reactor. Recall that the Rayleigh number $Ra = GrPr$ can be used to assess whether the buoyancy-driven fluid motions may arise; on a vertical wall, the threshold value of Ra is approximately 10^9. Jensen (1989) points out that with such large ΔT's common in CVD (perhaps 400K), the usual Boussinesq approximation $\rho g\beta(T_H - T_C)$ would not be an appropriate fix for the equation of motion. An equation of state must be used in such cases to represent the changes in gas density. Furthermore, the convection rolls that develop in horizontal CVD reactors require that an accurate model of the resulting flow be three dimensional.

9.5.3 Dispersion Effects in Chemical Reactors

When we speak of dispersion in chemical reactors, we are referring to processes by which a component is distributed or scattered in one or more directions. Usually this scattering is the result of relative fluid motions and diffusion, working in concert. Clearly, if the local reactant concentration is diminished as a result of these phenomena, then the local rate of reaction will be reduced. The end result is that the conversion that could be (or might have been) obtained according to the idealized reactor models cannot be achieved. Our purpose in this section is to examine the dispersion models so that we might be better prepared to analyze the mass transfer phenomena occurring in flow reactors; we would also like to be able to explain why real reactors may not perform as indicated by the usual simplified models. A very readable introduction to this field has been provided by Himmelblau and Bischoff (1968) and a more complete coverage can be found in Wen and Fan (1975).

We begin by considering a tubular reactor and acknowledging the possibility of dispersion in both the radial and axial directions. For this general case, we write

$$\frac{\partial C_A}{\partial t} + v_z\frac{\partial C_A}{\partial z} = D_R\left[\frac{1}{r}\frac{\partial}{\partial r}\left(r\frac{\partial C_A}{\partial r}\right)\right] + D_L\frac{\partial^2 C_A}{\partial z^2} + S. \tag{9.64}$$

You can see that we have employed different dispersion coefficients for the radial and axial directions (D_R and D_L); we should think about the physical conditions that might dictate a difference. The reader should also make special note of the fact that *we are assuming that the mixing phenomena occurring in flow reactors can be represented as though they are diffusional processes*. We will not question the underlying validity of such modeling—contenting ourselves with successes where they occur.

Suppose we now *assume* that radial dispersion is unimportant; this will reduce eq. (9.64) to an *axial* dispersion model:

$$\frac{\partial C_A}{\partial t} + v_z\frac{\partial C_A}{\partial z} = D_L\frac{\partial^2 C_A}{\partial z^2} + S. \tag{9.65}$$

Equation (9.65) is usually the appropriate choice if $L/d \gg 1$ and the flow is turbulent. We make this equation dimensionless by setting

$$t^* = \frac{v_z t}{L}, \quad z^* = \frac{z}{L}, \quad Pe_L = Re_L Sc_L = \frac{v_z L}{D_L},$$
$$\text{and} \quad C^* = \frac{C_A}{C_{A0}}.$$

The result is

$$\frac{\partial C^*}{\partial t^*} + \frac{\partial C^*}{\partial z^*} = \frac{1}{Pe_L}\frac{\partial^2 C^*}{\partial z^{*2}} + S^*. \tag{9.66}$$

The task confronting us is to use experimental data to identify the best possible value for Pe_L, that is, the value of the dispersion coefficient that most nearly describes the observed behavior for our reactor. For select cases (such as δ-function input and a "doubly infinite" reactor), the analytic solution is known, for example,

$$C^* = \frac{1}{2}\left(\frac{Pe_L}{\pi t^*}\right)^{1/2}\exp\left[-\frac{Pe_L(1-t^*)^2}{4t^*}\right]. \tag{9.67}$$

Some results for this model are given in Figure 9.15.

The results shown in Figure 9.15 may, however, be only minimally useful for us and the difficulty is twofold: It is not easy to extract the optimal value of the dispersion coefficient from eq. (9.67), and it may be difficult to obtain a close physical approximation to a δ-function input. Estimates for the Peclet number can be obtained from the tracer distribution(s),

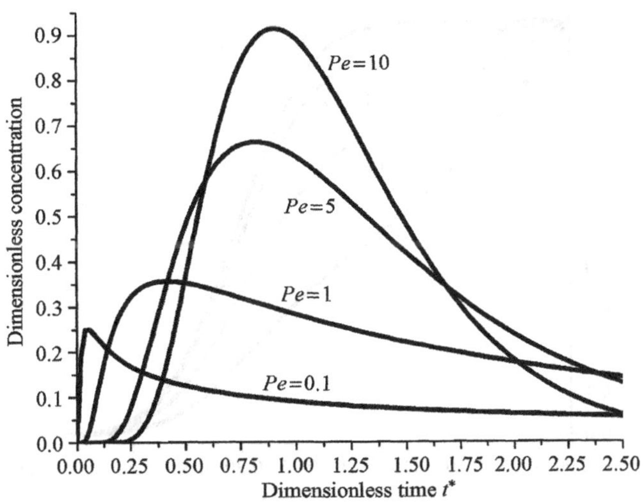

FIGURE 9.15. Response curves for the axial dispersion model, eq. (9.67), subjected to a δ-function input for the Peclet numbers ranging from 0.1 to 10.

since

$$\mu = \frac{\int_0^\infty t^* C^* dt^*}{\int_0^\infty C^* dt^*} \quad \text{and} \quad \mu = 1 + \frac{2}{Pe_L}. \qquad (9.68)$$

The second moment about the mean (variance) can also be used to estimate the Peclet number and it has been shown that $\sigma^2 = (2/Pe_L) + (8/Pe_L^2)$. This estimate is generally more reliable than that obtained from the mean. Equation (9.66) is a candidate for solution by the method of Laplace transform if S^* has an appropriate mathematical form, for example, a delta function. This is particularly convenient since the mean and the standard deviation can be obtained by differentiation of the transform (with respect to s). For a more comprehensive treatment of flow situations (including different values for the dispersion coefficient on either side of the test section), see Van der Laan (1958) and Aris (1958).

We observed above that it is physically difficult to apply a delta function input to a real reactor. Generally, it will be necessary for the analyst to approximate a real input with some numerical facsimile. Since eq. (9.66) is readily solved numerically, this is not at all formidable. The results from such calculations are shown in Figure 9.16 to better illustrate the effects of Pe_L. Two sets of results are provided; each shows the evolution of the tracer "spike" as it is transported downstream. It is to be noted that the Reynolds number is based on the diameter of the reactor and not on the length in the flow direction. The first curve is for $Pe = 12$ and the second is for $Pe = 4$.

We now examine actual tracer data (see Figure 9.17) from a prototype flow reactor. In this case, the reactor is a rectangular flume with four vortex-producing segments. It was designed specifically to produce circulation and retention in

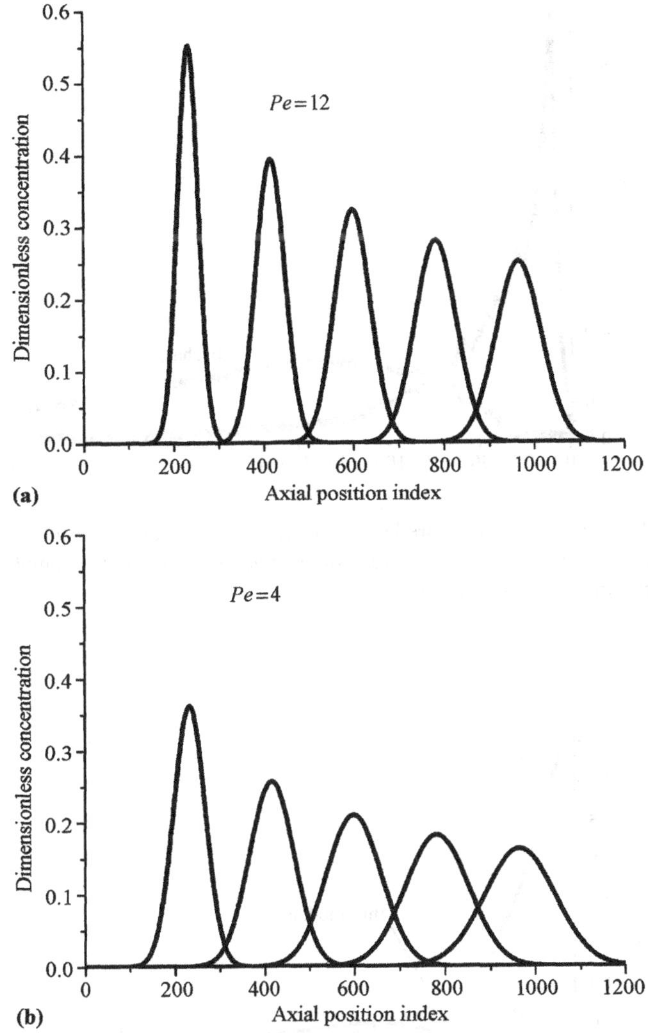

FIGURE 9.16. Comparison of the evolution of a tracer plume as it is transported downstream in a flow reactor. For (a), $Pe = 12$ and for (b), $Pe = 4$.

each of the segments, even at relatively low velocities. We would like to determine whether the simple axial dispersion model can adequately represent these results.

For this simulation, the fluid velocity is fairly low but the dispersion coefficient will need to be very large. Consequently, the Peclet number will be small (Figure 9.18). We would like to see if this rather simple axial dispersion model can mimic the behavior seen in Figure 9.17. For this case, the average velocity in the device is about 5 cm/s.

9.5.4 Transient Operation of a Tubular Reactor

Let us now consider the transient operation of an isothermal tubular reactor with a first-order homogeneous reaction and the possibility of axial dispersion (in the literature, such situations are often referred to as TRAM problems). The

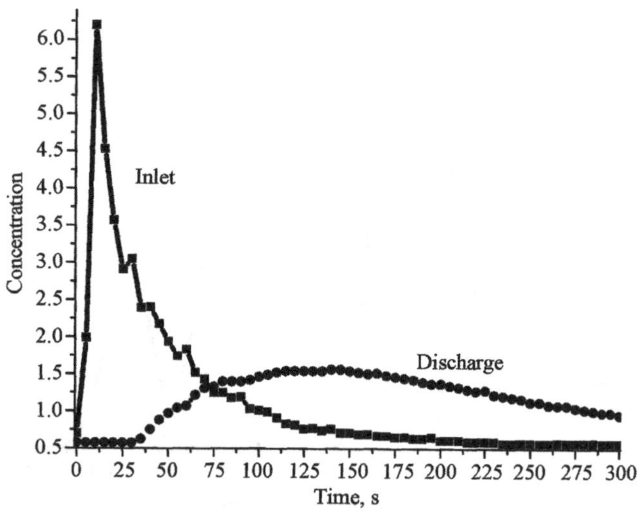

FIGURE 9.17. Measured tracer concentrations at the inlet and discharge of a prototype flow reactor designed specifically to promote mixing at low velocities.

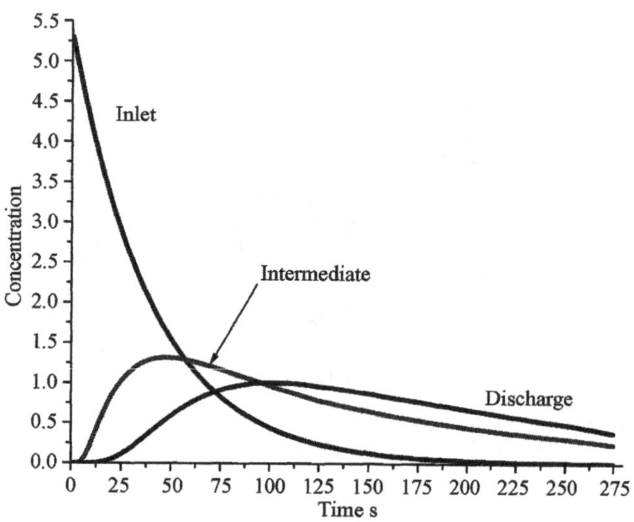

FIGURE 9.18. Computed concentrations at three points within the prototype device: inlet, intermediate, and discharge. The Peclet number for these results is about 0.002. Compare these data with those shown in Figure 9.17 (which include an offset of 0.5).

appropriate equation is

$$\frac{\partial C_A}{\partial t} + v_z \frac{\partial C_A}{\partial z} = D \frac{\partial^2 C_A}{\partial z^2} - k_1 C_A. \tag{9.69}$$

We render the problem dimensionless by setting

$$C = \frac{C_A}{C_{A\,in}}, \quad z^* = \frac{z}{L}, \quad \text{and} \quad t^* = \frac{t}{\tau},$$

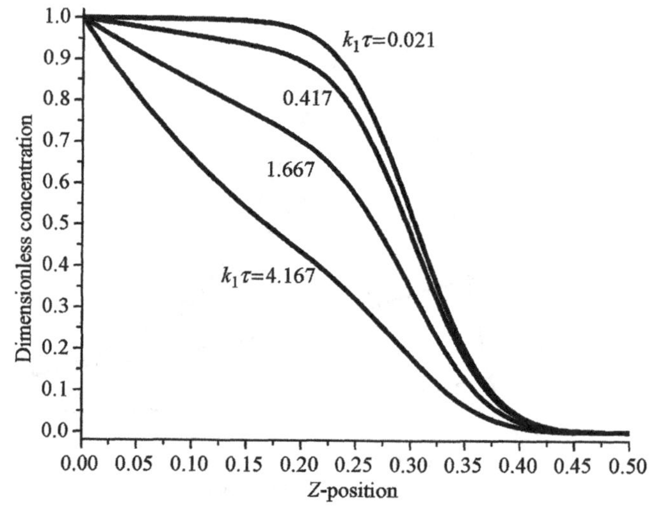

FIGURE 9.19. Distribution of reactant in an isothermal tubular reactor with dispersion for values of $k_1\tau$ of 0.021, 0.417, 1.667, and 4.167. The curves shown here are for an intermediate time (0.3).

therefore,

$$\frac{\partial C}{\partial t^*} = \frac{D\tau}{L^2}\frac{\partial^2 C}{\partial z^{*2}} - \frac{V\tau}{L}\frac{\partial C}{\partial z^*} - k_1 \tau C. \tag{9.70}$$

Problems of this type are easily solved numerically, and to demonstrate this we will choose

$$\frac{D\tau}{L^2} = 0.004 \quad \text{and} \quad \frac{V\tau}{L} = 1.$$

We will vary the rate constant k_1 to better examine the effects of reaction rate upon the development of the concentration profile. We assume that the reactor initially contains no reactant species; at $t = 0$, the feed of "A" commences. For values of $k_1\tau$ of 0.021, 0.417, 1.667, and 4.167, we obtain the results shown in Figure 9.19 at an intermediate time (0.3).

Note the effect of the axial dispersion upon the reactant front as it is transported down the reactor; there is a considerable "smoothing" at the corners and the slope one would expect to see with the plug flow operation is significantly reduced.

Now we would like to modify the previous example by the inclusion of thermal effects. In particular, suppose the homogeneous reaction is strongly exothermic. We presume that control of the process will be maintained by the removal of heat at the reactor wall. This suggests, of course, that T may vary substantially in the r-direction; we neglect this possibility for the time being. For this case, the model must be written using both continuity and energy equations and they are coupled through the reaction term:

$$\frac{\partial C_A}{\partial t} = D\frac{\partial^2 C_A}{\partial z^2} - V\frac{\partial C_A}{\partial z} - k_0 \exp\left(-\frac{E}{RT}\right) C_A \tag{9.71}$$

and

$$\frac{\partial T}{\partial t} = \alpha\frac{\partial^2 T}{\partial z^2} - V\frac{\partial T}{\partial z} + \frac{|\Delta H|\, k_0}{\rho C_p} \times \exp\left(-\frac{E}{RT}\right) C_A - \frac{2h}{\rho C_p R}(T - T_c). \tag{9.72}$$

Please note the similarity between the two equations. The parallel is really apparent if we define a reduced temperature for the energy equation by letting

$$\theta = \frac{\rho C_p T}{|\Delta H|}. \tag{9.73}$$

The reader should carry this out and then add the continuity and energy equations together; the reaction terms cancel of course. In fact, if we restrict our attention to the steady-state operation with adiabatic conditions, the equations can be decoupled producing an unexpectedly simple ordinary differential equation (as long as the Lewis number $Le = \alpha/D$ is equal to 1). The last stipulation is often at least approximately true and the reader is referred to Perlmutter (1972) for more details.

We now solve eqs. (9.71) and (9.72), using the parametric choices common to the previously considered isothermal case, but with a strongly exothermic first-order reaction. For these computed results, $E/(RT_{in}) = 18.25$ and the dimensionless production term $\frac{C_{A\,in}|\Delta H_{rxn}|}{\rho C_p T_{in}} = 21{,}053$. Once again we select an intermediate time for this transient problem. Although the distribution is little changed from the previous results, the parametric sensitivity is revealed (through variation of the heat transfer coefficient) in the dimensionless temperature distributions illustrated in Figure 9.20.

For the model illustrated by Figure 9.20, a slightly smaller heat transfer coefficient results in an unstable situation; the threshold lies between $h = 0.175$ and $h = 0.15$. Bilous and Amundson (1956) point out that this kind of parametric sensitivity can manifest itself in a real reactor in different ways. Of course, a "run-away" hot spot could be catastrophic, but it could also promote a side reaction that would adversely affect yield and/or product quality. It is the task of the reactor designer to make sure that regions of parametric sensitivity are avoided. The easiest way to do this is to make certain that the heat generated by the chemical reaction can *never exceed* the rate of heat removal. If one uses the *feed concentration* of the reactant and the *maximum temperature* (as shown in Figure 9.20) in the thermal energy production term

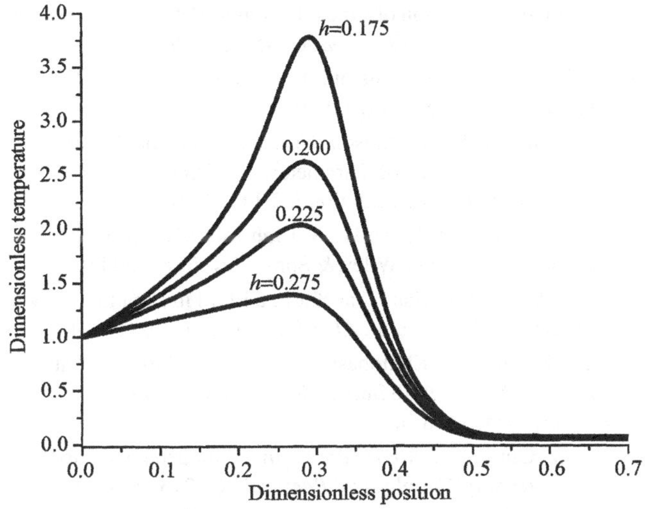

FIGURE 9.20. Illustration of the effects of heat transfer coefficient upon the dimensionless temperature distribution at fixed (intermediate) time. The numerical value of the heat transfer coefficient ranges from 0.175 to 0.275.

and then selects the heat transfer coefficient (or heat removal rate) accordingly, a conservative design will result.

9.6 CONCLUSION

In this chapter we have seen the importance of fluid motion to mass transfer. Many problems of interest for the laminar and other well-characterized flows can be solved readily through analytic and elementary numerical techniques. However, for most industrial-scale mass transfer processes, turbulence is the usual state of fluid motion. The reason for this is easy to understand by considering a central "blob" of "A" placed in continuous phase of "B": In turbulence, eddies distort the fluid region containing species "A" producing numerous projections (like tentacles or arms) of elevated concentration. Consequently, the "surface" over which the mass transfer occurs is increased and the local differences in concentration are enhanced. This combination increases the effectiveness of molecular diffusion and speeds up the dispersion process. A useful interpretive schematic of this phenomenon was developed by Corrsin (1959) and was reproduced by Monin and Yaglom (1971) (see Section 10.2, pp. 591–592). We will discuss this phenomenon in greater detail in Chapter 10 in connection with the Fokker–Planck equation and its application to (the modeling of) the turbulent molecular mixing.

REFERENCES

Acrivos, A. and T. D. Taylor. Heat and Mass Transfer from Single Spheres in Stokes Flow. *Physics of Fluids*, 5:387 (1962).

Aris, R. On the Dispersion of Linear Kinematic Waves. *Proceedings of the Royal Society of London A*, 245:268 (1958).

Arnold, J. H. Studies in Diffusion. II. *A Kinetic Theory of Diffusion in Liquid Systems*, 52:3937 (1930).

Bilous, O. and N. R. Amundson. Chemical Reactor Stability and Sensitivity, II. Effect of Parameters on Sensitivity of Empty Tubular Reactors. *AIChE Journal*, 2:117 (1956).

Bird, R. B., Stewart, W. E., and E. N. Lightfoot. *Transport Phenomena*, 2nd edition, John Wiley & Sons, New York (2002).

Brown, G. M. Heat or Mass Transfer in a Fluid in Laminar Flow in a Circular or Flat Conduit. *AIChE Journal*, 6:179 (1960).

Conner, J. M. and S. E. Elghobashi. Numerical Solution of Laminar Flow Past a Sphere with Surface Mass Transfer. *Numerical Heat Transfer*, 12:57 (1987).

Corrsin, S. Outline of Some Topics in Homogeneous Turbulent Flow. *Journal of Geophysical Research*, 64:2134 (1959).

Drummond, C. K. and F. A. Lyman. Mass Transfer from a Sphere in an Oscillating Flow with Zero Mean Velocity. *NASA Technical Memorandum* 102566 (1990).

Gupalo, Y. P. and Y. S. Ryazantsev. Mass and Heat Transfer from a Sphere in Laminar Flow. *Chemical Engineering Science*, 27:61 (1972).

Himmelblau, D. M. and K. B. Bischoff. *Process Analysis and Simulation: Deterministic Systems*, John Wiley & Sons, New York (1968).

Hino, M., Sawamoto, M., and S. Takasu. Experiments on Transition to Turbulence in an Oscillatory Pipe Flow. *Journal of Fluid Mechanics*, 75:193 (1976).

Jensen, K. F. Transport Phenomena and Chemical Reaction Issues in OMVPE of Compound Semiconductors. *Journal of Crystal Growth*, 98:148 (1989).

Karagoz, I. Similarity Solution of the Oscillatory Pressure Driven Fully Developed Flow in a Channel. *Uludag Universitesi MMFD*, 7:161 (2002).

Lawal, A. and A. S. Mujumdar. Extended Graetz Problem: A Comparison of Various Solution Techniques. *Chemical Engineering Communications*, 39:91 (1985).

Li, P. and K. T. Yang. Mechanisms for the Heat Transfer Enhancement in Zero-Mean Oscillatory Flows in Short Channels. *International Journal of Heat and Mass Transfer*, 43:3551 (2000).

Monin, A. S. and A. M. Yaglom *Statistical Fluid Mechanics*, MIT Press, Cambridge, MA (1971).

Oseen, C. W. Uber die Stokessche Formel und uber die verwandte Aufgabe in der Hydrodynamik. Arkiv for Mathematik, Astronomi och Fysik, 6:75 (1910).

Patankar, S. V. *Numerical Heat Transfer and Fluid Flow*, Hemisphere Publishing, Washington (1980).

Perlmutter, D. D. *Stability of Chemical Reactors*, Prentice-Hall, Englewood Cliffs (1972).

Platten, J. K. The Soret Effect: A Review of Recent Experimental Results. *Journal of Applied Mechanics*, 73:5 (2006).

Spalding, D. B. Mass Transfer in Laminar Flow. *Proceedings of the Royal Society of London A*, 221:78 (1954).

Steinberger, R. L. and R. E. Treybal. Mass Transfer from a Solid Soluble Sphere to a Flowing Liquid Stream. *AIChE Journal*, 6:227 (1960).

Tran, H. T. and J. S. Scroggs. Modeling and Optimal Design of a Chemical Vapor Deposition Reactor. *Proceedings of the 31st Conference on Decision and Control* (1992).

Van der Laan, E. T. Notes on the Diffusion Type Modeling for the Longitudinal Mixing in Flow. *Chemical Engineering Science*, 7:187 (1958).

Wen, C. Y. and L. T. Fan. *Models for Flow Systems and Chemical Reactors*, Marcel Dekker, New York (1975).

Whitehead, A. N. Second Approximations to Viscous Fluid Motion. *Quarterly Journal of Mathematics*, 23:143 (1889).

10

HEAT AND MASS TRANSFER IN TURBULENCE

10.1 INTRODUCTION

Suppose we take a container of cold water and supply heat to the bottom. We measure the temperature at a single point in the container to see how T varies with time. Because the thermal energy is supplied at a sufficiently high rate, we will get buoyancy-driven turbulence in the liquid. Clearly, this is a special kind of turbulence—not very energetic with low-frequency fluctuations. Since our measurements are made with a small thermocouple, this is entirely appropriate; we want the process dynamics to conform to the response time of the instrument. An excerpt from the resulting time-series data is provided in Figure 10.1. The fluctuations seen here result from the scalar quantity (T) being carried past the measurement point by the buoyancy-driven eddies.

It is apparent that the "mean" fluid temperature is increasing in an expected manner. In fact, if we use a macroscopic thermal energy balance (say, $mC_p(dT/dt) = hA\Delta T$) to model this transient heating process, we could obtain an approximate match to the gross behavior shown here. Naturally, we could not reproduce the fluctuations apparent in Figure 10.1. While this kind of macroscopic model is useful for engineering applications, it may strike a dissonant chord with students of transport phenomena; we would like to have a better understanding of how the scalar quantities (temperature and concentration) are transported by turbulence.

We will initiate this part of our discussion by writing the energy equation in rectangular coordinates, omitting the production mechanisms:

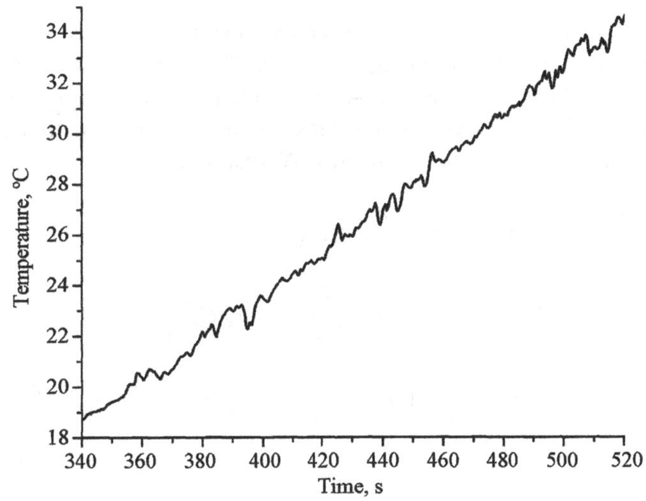

FIGURE 10.1. Point temperature measured in a container of water (640 g with an initial temperature of 6°C) heated from the bottom.

$$\rho C_p\left(\frac{\partial T}{\partial t} + v_x\frac{\partial T}{\partial x} + v_y\frac{\partial T}{\partial y} + v_z\frac{\partial T}{\partial z}\right) = k\left[\frac{\partial^2 T}{\partial x^2} + \frac{\partial^2 T}{\partial y^2} + \frac{\partial^2 T}{\partial z^2}\right]. \quad (10.1)$$

The level of complexity is now obvious; even in our beaker of heated water, the turbulence is three dimensional and time

Transport Phenomena: An Introduction to Advanced Topics, By Larry A. Glasgow

dependent. We cannot solve eq. (10.1) without the detailed knowledge of all the three velocity vector components. This is a formidable problem and it is appropriate for us to look for possible simplifications. We take one of the convective transport terms for illustration and rewrite it for an incompressible fluid using the Reynolds decomposition:

$$\frac{\partial}{\partial x}(v_x T) \Rightarrow \frac{\partial}{\partial x}\left[(V_x + v_x')\left(\overline{T} + T'\right)\right]. \tag{10.2}$$

We time average the result, remembering that this process automatically entails a loss of information. Since the quantities that are first order with respect to the fluctuations disappear (for a statistically stationary process), we are left with

$$\frac{\partial T}{\partial t} + \frac{\partial}{\partial x}(V_x T) + \frac{\partial}{\partial y}(V_y T) + \frac{\partial}{\partial z}(V_z T) = \frac{\partial}{\partial x}\left[\alpha\frac{\partial T}{\partial x} - \overline{v_x'T'}\right]$$
$$+\frac{\partial}{\partial y}\left[\alpha\frac{\partial T}{\partial y} - \overline{v_y'T'}\right] + \frac{\partial}{\partial z}\left[\alpha\frac{\partial T}{\partial z} - \overline{v_z'T'}\right]. \tag{10.3}$$

The procedure carried out above resulted in three new terms, the turbulent energy fluxes $v_i'T'$ (such quantities are referred to as the velocity–scalar covariance, or simply the scalar flux). The very same steps can be carried out with the continuity equation for species "A" resulting in

$$\frac{\partial C_A}{\partial t} + \frac{\partial}{\partial x}(V_x C_A) + \frac{\partial}{\partial y}(V_y C_A)$$
$$+\frac{\partial}{\partial z}(V_z C_A) = \frac{\partial}{\partial x}\left[D_{AB}\frac{\partial C_A}{\partial x} - \overline{v_x'C_A'}\right]$$
$$+\frac{\partial}{\partial y}\left[D_{AB}\frac{\partial C_A}{\partial y} + \overline{v_y'C_A'}\right] + \frac{\partial}{\partial z}\left[D_{AB}\frac{\partial C_A}{\partial z} + \overline{v_z'C_A'}\right]. \tag{10.4}$$

It is to be noted that the development above, if applied to the thermal energy production by viscous dissipation or to the production of "A" by chemical reaction, could result in additional new quantities being generated. For example, consider a bimolecular kinetic description like $k_2 C_A C_A$ that would result in $k_2\overline{(C_A + C_A')(C_A + C_A')}$. In this example, the turbulent fluctuations would affect the rate of reaction. We will return to this point later, but for the time being we will omit such complications.

There is also an important distinction between turbulent transport processes occurring in internal and external turbulent flows. For example, consider a turbulent wake or a free jet; the flow near the edges is intermittently turbulent. Since turbulence is only present for a fraction of the time, models based upon differential equations that are continuous in time are clearly inappropriate. Hinze (1975) notes that for the free turbulent flows, it is not possible to draw upon parallels between the transport processes, a topic to be discussed in more detail in the following section.

10.2 SOLUTION THROUGH ANALOGY

You may have noticed that the turbulent energy fluxes were taken to the right-hand side of eq. (10.3) and combined with the molecular transport (conduction) terms. This has been done in anticipation that a gradient transport model might be used to achieve closure. Recall our previous observation that the mean flow and the turbulence are only weakly coupled; we should not expect a gradient transport model to work well except under special circumstances. In particular, we know from experience that such an approach is likely to apply only to cases with a single dominant length scale and a single dominant velocity scale. Heat transfer to turbulent flows in ducts is such a case, and because of its practical importance, some additional exploration is warranted. For the steady turbulent flow in a cylindrical tube, we have

$$\frac{\partial}{\partial z}(V_z T) = \frac{1}{r}\frac{\partial}{\partial r}\left[\alpha r\frac{\partial T}{\partial r} - \left(r\overline{v_r'T'}\right)\right]. \tag{10.5}$$

Assuming that the turbulent energy flux can be replaced by a gradient transport model with an eddy (or turbulent) diffusivity, we obtain

$$\frac{\partial}{\partial z}(V_z T) = \frac{1}{r}\frac{\partial}{\partial r}\left[r(\alpha + \varepsilon_H)\frac{\partial T}{\partial r}\right]. \tag{10.6}$$

Using the same approach for momentum transport, we find

$$\frac{1}{\rho}\frac{\partial P}{\partial z} = \frac{1}{r}\frac{\partial}{\partial r}\left[r(\nu + \varepsilon_M)\frac{\partial V_z}{\partial r}\right]. \tag{10.7}$$

Note the similarity between the two equations; this is certainly suggestive. For now, we observe that if the functional forms for ε_H and ε_M were known, this pair of equations could be solved to obtain the velocity and temperature distributions.

The problem, of course, is that the functionality of these eddy diffusivities is likely to be different for *every* turbulent transport problem. We have obtained a relatively simple model, but the scope of application is limited. Nevertheless, if we want to argue that the mechanisms for momentum and heat (or mass) transfer are the same, we ought to have an appreciation of how the diffusivities vary with position in appropriate flow situations. Some data obtained by Page et al. (1952) for turbulent flow of air through a rectangular duct are shown in Figure 10.2.

These data are important to us for a couple of reasons. In the nineteenth century, Reynolds proposed that the laws governing momentum and heat transfer were the same. Certainly, the similarity in form for (10.6) and (10.7) suggests why this hypothesis is so attractive. If the rate of momentum transport was either known or measured, then the dimensionless temperature or the rate of heat transfer would also be known. Indeed, the equivalence would make it possible to express

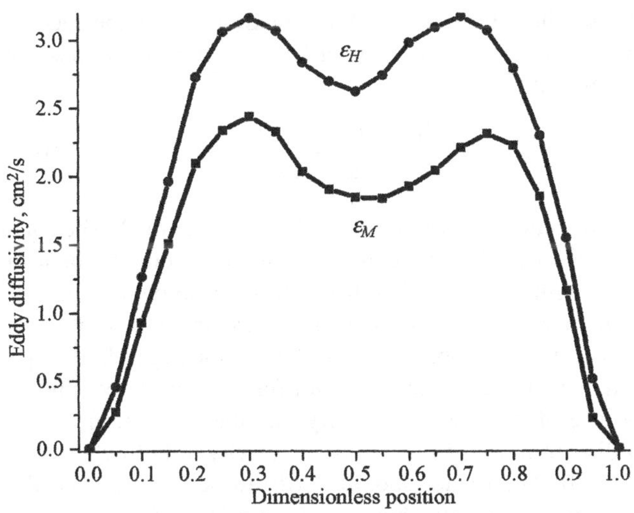

FIGURE 10.2. Eddy diffusivities (cm^2/s) for thermal energy ε_H and for momentum ε_M measured for the flow of air in a rectangular duct at $Re = 9370$ by Page et al. (1952).

the Reynolds analogy through relation of the friction factor to either the dimensionless temperature or the heat transfer coefficient. Two of the more common forms for the analogy (heat transfer to a fluid flowing through a tube with constant wall temperature) are

$$\ln\left[\frac{T_0 - T_1}{T_0 - T_2}\right] = \frac{fL_{12}}{R} \quad \text{and} \quad St = \frac{f}{2}, \tag{10.8}$$

where St is the Stanton number, $St = h/\rho C_p V_z$, and f is the friction factor defined by the equation $F = AKf$.

If only it were that easy. Unfortunately, Stanton's (1897) experimental data failed to substantiate Reynolds analogy and it became apparent that the Reynolds hypothesis was not entirely correct. Rayleigh (1917) pointed out that if consideration was limited to a steady laminar shear flow between parallel planes, the analogy was sound and the dimensionless velocity and temperature profiles would be identical. However, if the Reynolds number is large enough such that the motion becomes turbulent, then pressure $p = f(x,y,z,t)$; Rayleigh noted that the governing equation for momentum transport is changed in such cases and the analogy then fails. He speculated that if one considered only the time-averaged values (for the turbulent flow case), then the analogy would still fail. Stanton disagreed and he presented some data obtained by J. R. Pannell (air passed through a heated 2 in. diameter brass pipe) that indicated the only discrepancies between the time-averaged temperature and velocity profiles occurred very near the tube axis. Stanton attributed the difference to the fact that the thermal entrance length was not achieved in Pannell's apparatus. We now know, of course, that Reynolds' analogy is correct only under very special circumstances.

We should also note that for the data shown in Figure 10.2, ε_H is roughly 30–40% larger than ε_M. This makes it difficult to see how equating the eddy diffusivities is a good idea. Furthermore, the reader is urged to carefully study the shape of these curves, as this will be particularly significant for us in a moment.

In the first half of the twentieth century, much effort was devoted to fixing the Reynolds analogy by accounting for the variation of velocity near the wall. Prandtl (1910), for example, included the "laminar" sublayer and found

$$Nu = \frac{(f/2)Re\,Pr}{1 + 5\sqrt{f/2}(Pr - 1)} \tag{10.9}$$

for heat transfer and

$$Sh = \frac{(f/2)Re\,Sc}{1 + 5\sqrt{f/2}(Sc - 1)} \tag{10.10}$$

for mass transfer. Stanton's data agree reasonably well with Prandtl's modification. von Karman (1939) took the analogy process an additional step by including the complete "universal" velocity distribution, resulting in

$$Nu = \frac{(f/2)Re\,Pr}{1 + 5\sqrt{(f/2)}\{Pr - 1 + \ln[1 + 5/6(Pr - 1)]\}}. \tag{10.11}$$

One might think that the analogy idea, which is more than 130 years old, should have disappeared into the sunset. However, it continues to attract occasional attention; for an example, see Lin's (1994) work on laminar forced convection on a flat plate. This system was chosen because of the ease with which comparisons could be made against computed (similarity) solutions. The older analogies worked well for constant wall temperature as long as $Pr \approx 1$. They were not satisfactory for the case of constant heat flux, nor did they perform well for the small Prandtl numbers ($Pr \ll 1$).

There are other limits to the applicability of the "analogy" approach. For example, it is necessary that neither heat nor mass transfer affects the velocity distribution, a stipulation we know will be violated if the rate of either heat or mass transport is large. It is also necessary that both the (pairs of) molecular *and* eddy diffusivities be equal; we need $\nu = \alpha$ (or $\nu = D_{AB}$) and $\varepsilon_M = \varepsilon_H$. Recall that for air at typical ambient temperatures, $Pr = 0.72$ and for carbon dioxide in air, $Sc = 0.96$. The data shown in Figure 10.2 make it very clear that although the eddy diffusivities may be comparable, $\varepsilon_M \neq \varepsilon_H$. Finally, Reynolds' analogy will certainly fail for external flows where boundary-layer separation occurs.

10.3 ELEMENTARY CLOSURE PROCESSES

The analogy approach to turbulent transport has been made to work adequately for a few cases. Let us presume, however, that we need more detail, that we are not only interested in the Nusselt number but also in the actual temperature distribution in the duct. We begin with eq. (10.6) but assume that we have a constant heat flux at the wall; this means that the bulk fluid (or mixing cup) temperature increases linearly in the direction of flow and accordingly,

$$\frac{\partial T}{\partial z} = \frac{dT_m}{dz} = \text{const.} \tag{10.12}$$

We define our position variable as $s = R - r$, such that

$$V_z \frac{dT_m}{dz} = \frac{1}{R-s}\frac{\partial}{\partial s}\left[(R-s)(\alpha+\varepsilon_H)\frac{\partial T}{\partial s}\right]; \tag{10.13}$$

we then integrate

$$\int_0^s (R-s)V_z \frac{dT_m}{dz} ds = (R-s)(\alpha+\varepsilon_H)\frac{\partial T}{\partial s} + C_1. \tag{10.14}$$

We can integrate the left-hand side either analytically or numerically, depending upon our choice for $V_z(s)$. If we take the velocity to be *constant* and note that at $s = R$, $\partial T/\partial s = 0$, then we find

$$\frac{\partial T}{\partial s} = -\frac{dT_m}{dz}\frac{V_z}{2}\frac{(R-s)}{(\alpha+\varepsilon_H)}. \tag{10.15}$$

By confining our attention to a region very close to the wall (very small s), where the eddy diffusivity is effectively zero,

$$T - T_0 = -\frac{dT_m}{dz}\frac{V_z R}{2\alpha}s. \tag{10.16}$$

Finally, we use an energy balance to relate the increase in bulk fluid temperature to the heat flux at the wall and introduce the dimensionless position s^+ (recall that $s^+ = sv^*/\nu$):

$$T - T_0 = -\frac{q_0}{\rho C_p} Pr \frac{s^+}{v^*}. \tag{10.17}$$

The flux has been taken to be positive for heat transfer *from the wall to the fluid.* Since the "laminar" sublayer extends to about $s^+ = 5$,

$$T(s^+ = 5) - T_0 = -\frac{5q_0}{\rho C_p}\frac{Pr}{v^*}. \tag{10.18}$$

The process illustrated above can be carried out for both the "buffer region" ($5 < s^+ < 30$) and the turbulent core ($s^+ > 30$). However, this requires that we obtain a functional form for the eddy diffusivity. Kays (1966) and Bird et al. (1960) show how this is done making use of the fact that the shear stress varies linearly with transverse position, accordingly,

$$\frac{\varepsilon_M}{\nu} = \frac{(1 - s/R)}{dv^+/ds^+} - 1. \tag{10.19}$$

And of course, the eddy diffusivities are assumed to be equal, $\varepsilon_H/\nu = \varepsilon_M/\nu$. So, if the dimensionless velocity gradient can be determined, the eddy diffusivities are "known." There is a problem here, as Kays points out: If we, in an uncomfortably circular process, obtain dv^+/ds^+ from the logarithmic equation, then the centerline behavior for ε_M is incorrect. Indeed, the use of "universal" velocity distribution will also result in discontinuities in the eddy diffusivity at both $y^+ = 5$ and $y^+ = 30$. The reader should verify these features and then re-examine the data shown in Figure 10.2. There are other possibilities, of course. Cebeci and Bradshaw (1984) note that a popular approach to determine the functionality of ε_M is to combine the mixing length expression developed by Nikuradse

$$l = R\left[0.14 - 0.08\left(1 - \frac{s}{R}\right)^2 - 0.06\left(1 - \frac{s}{R}\right)^4\right] \tag{10.20}$$

with Van Driest's damping factor, resulting in

$$\varepsilon_M = l^2\left(1 - e^{-s/A}\right)^2 \frac{dV_z}{ds}. \tag{10.21}$$

The constant A appearing in (10.21) is the damping length. The reader is urged to compare the shape of the resulting eddy diffusivity with the data in Figure 10.2.

Waving off the obvious objections and proceeding, we find

$$T - T(s^+ = 5) = -\frac{q_0}{\rho C_p}\frac{5}{v^*}\ln\left(\frac{s^+ Pr}{5} - Pr + 1\right) \tag{10.22}$$

for the "buffer" region ($5 < s^+ < 30$) and

$$T - T(s^+ = 30) = -\frac{q_0}{\rho C_p}\frac{2.5}{v^*}\ln\left(\frac{s^+}{30}\right) \tag{10.23}$$

for the turbulent core ($s^+ > 30$). It is to be kept in mind that these results are strictly valid only for large Prandtl numbers. Kays notes that for liquid metals, Pr can be very small; for example, for sodium at 700°F, $Pr = 0.005$. Under such circumstances, molecular conduction in the turbulent core cannot be neglected.

The results shown above can be used to calculate the temperature distribution for turbulent flow through a tube with constant heat flux at the wall. Observe that there is a significant difference between this case and the comparable heat transfer problem occurring in a laminar flow. For heat transfer in turbulent flow, the time-averaged temperature distribution functionally depends upon both the flow

$$\frac{T_0 - T}{T_0 - T_C} = \frac{(\varepsilon_H/\varepsilon_M)Pr + \ln(1+5(\varepsilon_H/\varepsilon_M)Pr) + 0.5\,F_1 \ln[(Re/60)\sqrt{(f/2)}(s/R)]}{(\varepsilon_H/\varepsilon_M)Pr + \ln(1+5(\varepsilon_H/\varepsilon_M)Pr) + 0.5\,F_1 \ln[(Re/60)\sqrt{(f/2)}]}. \tag{10.24}$$

rate (Reynolds number) and the Prandtl number. The analysis presented above can be improved in a number of ways, and, in fact, Martinelli (1947) changed the procedure to make it applicable to all fluids, including liquid metals. There is, however, little difference between the two analyses for $Pr > 1$. The dimensionless temperature in the turbulent core by Martinelli's analysis is shown in eq. (10.24). See above.

The parameter F_1 is a function of Re and Pr; for $Re = 100{,}000$ and $Pr = 0.1$, $F_1 = 0.83$. If $Re = 10{,}000$ and $Pr = 1$, $F_1 = 0.92$. An illustration of computed temperature profiles is given in Figure 10.3 for the Reynolds numbers of 10,000 and 100,000. The effect of changing Pr upon the shape of the profiles is noteworthy.

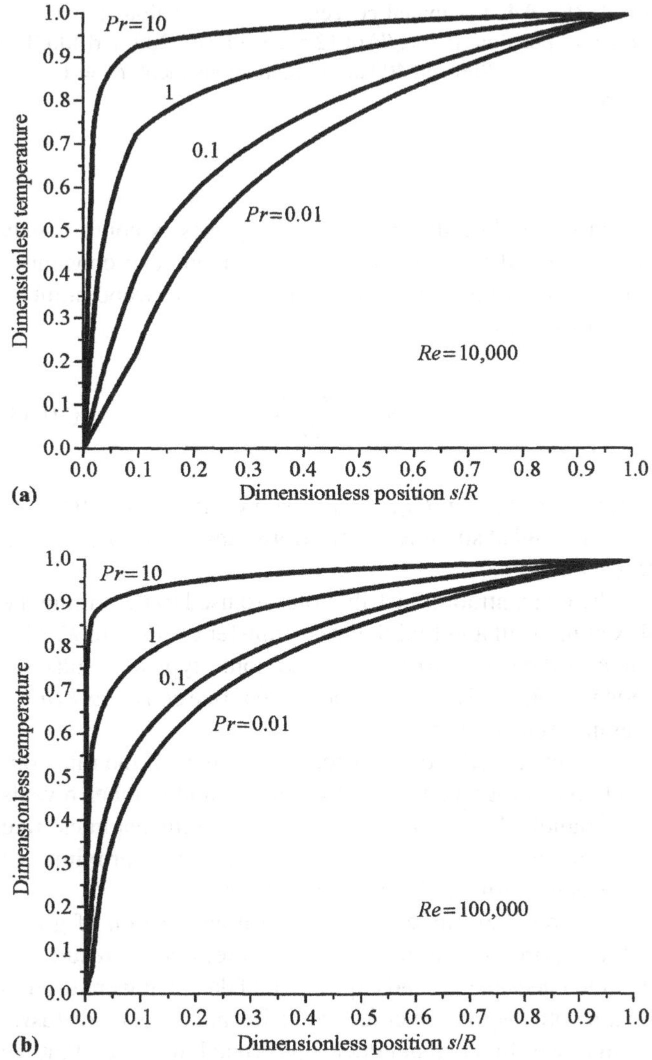

FIGURE 10.3. (a and b) Martinelli analogy: dimensionless temperature profiles $(T_0 - T)/(T_0 - T_{s=R})$ for $Re = 10{,}000$ and 100,000 and Prandtl numbers 10, 1, 0.1, and 0.01. For the lower figure, $Pr = 0.01$ and $Pr = 0.001$ are virtually indistinguishable. For these computed profiles, $\varepsilon_H = \varepsilon_M$.

Let us again draw attention to the significance of the Prandtl number in these two figures; a larger Pr moves the principal resistance closer to the wall. This is particularly evident at the lower Reynolds number, as in the case of Figure 10.3a.

We can also formulate a gradient transport model using Prandtl's mixing length hypothesis. For this case, we consider turbulent mass transport:

$$j_A^T = -l l_C \left|\frac{dV_x}{dy}\right| \frac{dC_A}{dy}, \tag{10.25}$$

where V_x and C_A are time-averaged velocity and concentration, respectively. Note that there are two mixing lengths in this expression, l_C is the mixing length for turbulent transport of a scalar. If the turbulent Schmidt number is equal to one (the eddy diffusivities for momentum and mass are equal, $\varepsilon_M = \varepsilon_D$), then the two mixing lengths are equal as well. Baldyga and Bourne (1999) observe that the mixing length model applied to the turbulent mass transport of species "A" may be more rational (than in the case of momentum transport) because of better conservation. If we take

$$l = \kappa y \quad \text{and} \quad l_C = \kappa_C y,$$

then

$$j_A^T = -\kappa\kappa_C y^2 \left|\frac{dV_x}{dy}\right| \frac{dC_A}{dy}. \tag{10.26}$$

Since

$$\kappa y \left|\frac{dV_x}{dy}\right| = \sqrt{\frac{\tau_W}{\rho}} = v^*, \tag{10.27}$$

we write

$$j_A^T = -\kappa_C y v^* \frac{dC_A}{dy}. \tag{10.28}$$

Baldyga and Bourne show that one can obtain a logarithmic profile for concentration through introduction of a suitable dimensionless concentration. Naturally, this process raises the very same concerns we encountered in Chapter 5; we know that a piecemeal approach to the time-averaged velocity (or time-averaged temperature/concentration) is unphysical. It is appropriate for us to take a moment to reconsider the circumstances for which this may be satisfactory.

Recall that closure achieved through gradient transport modeling will be useful only for cases in which we have a single dominant length scale and a single dominant velocity. Thus, we may be able to get a practical result for the turbulent transport processes occurring in duct or tube flows. Generally speaking, this type of modeling will not work nearly as well—or even at all—for the free (or external) turbulent flows. Suppose we write the time-averaged continuity equation for the transport of species "A" through a tube including first-order disappearance of the reactant (upper case letters are being used to represent time-averaged quantities):

$$V_z \frac{\partial C_A}{\partial z} = \frac{1}{r}\frac{\partial}{\partial r}\left(r(D_{AB}+\varepsilon_D)\frac{\partial C_A}{\partial r}\right) - k_1 C_A. \quad (10.29)$$

Please note that axial transport is being neglected and that ε_D is the turbulent (or eddy) diffusivity for mass transport. We can attack problems of this type successfully if we have accurate representations for both $V_z(r)$ and $\varepsilon_D(r)$. Indeed, Bird et al. 2002 provide a detailed example of such a calculation in Section 21.4; the results presented there show how the first-order disappearance of "A" results in mass-transfer enhancement (increased Sherwood number). We wish to examine a related problem, but with a little different approach.

Consider a turbulent flow through a rectangular duct for which the width (W) is significantly greater than the height ($2h$). The appropriate time-averaged continuity equation is

$$V_z \frac{\partial C_A}{\partial z} = \frac{\partial}{\partial y}\left((D_{AB}+\varepsilon_D)\frac{\partial C_A}{\partial y}\right) - k_1 C_A. \quad (10.30)$$

We will assume that the velocity distribution can be represented with the experimental data provided by Page et al. (1952); an approximate fit can be obtained with a variation of Prandtl's $1/n$ power law:

$$V_z = 552.7\left[1-\frac{y}{h}\right]^{0.152}, \quad (10.31)$$

where the maximum (centerline) velocity is 552.7 cm/s. We also approximate the variation of ε_D with a polynomial expression using three terms with different powers of $((1/2)-(y/2h)$ and assume that $\varepsilon_D \approx \varepsilon_H$. This choice for the polynomial guarantees that $\varepsilon_D = 0$ at the duct wall. Our computational algorithm is then obtained from

$$\frac{\partial C_A}{\partial z} = \frac{(D_{AB}+\varepsilon_D)(\partial^2 C_A/\partial y^2) + (\partial\varepsilon_D/\partial y)(\partial C_A/\partial y) - k_1 C_A}{V_z(y)}. \quad (10.32)$$

We shall compute concentration profiles as they evolve in the z-direction. We assume that "A" enters the duct with a *uniform distribution* with respect to the transverse (y-) direction.

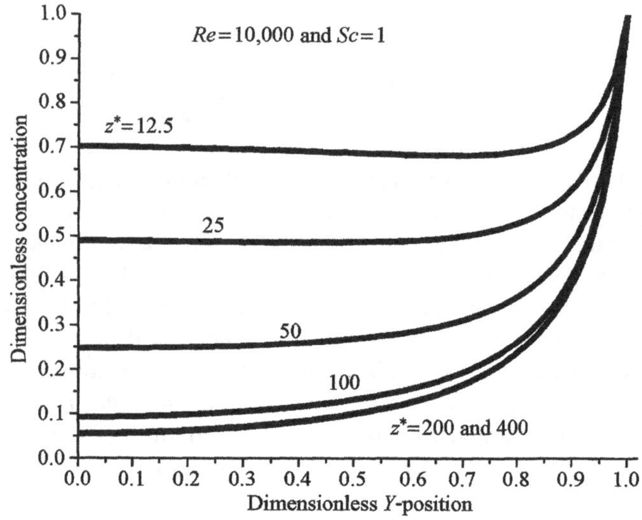

FIGURE 10.4. Computed concentration distributions for dimensionless z-positions ($z^* = z/h$) of 12.5, 25, 50, 100, 200, and 400. The profiles for $z^* = 200$ and 400 are virtually coincident. $Re = 1 \times 10^4$ and $Sc = 1$.

We also stipulate that the reactant species is continuously replenished at the walls as it is consumed. The concentration profile(s) can be used to compute the Sherwood number, which we define as

$$Sh = \frac{K(2h)}{D_{AB}}. \quad (10.33)$$

Typical results for $C_A(y,z)$ are shown in Figure 10.4 for $k_1 h^2/\nu = 89$ at streamwise positions ranging from $z^* = 12.5$ to $z^* = 400$.

The concentration distributions are used to determine the flux at the wall and find the mass transfer coefficient K. This value is then used to find Sh and some typical results are shown in Figure 10.5 for dimensionless rate constants $k_1 h^2/\nu$ ranging from 4.45 to 445.

The reader will note that for large values of the rate constant, asymptotic behavior of the Sherwood number reveals itself rapidly. Furthermore, an increase in dimensionless rate constant by a factor of 100 (from 4.45 to 445) approximately doubles the ultimate Sherwood number.

The above example is a successful application of gradient transport modeling; in this case, a reasonable result was obtained because we had experimental data that were used to obtain both the time-averaged velocity and the eddy diffusivity in a two-dimensional duct. It is crucial, however, that we again emphasize the problem with gradient transport models; as Leslie (1983) observes, "These expressions fail with monotonous regularity when they are applied to situations outside the range of the original experiments."

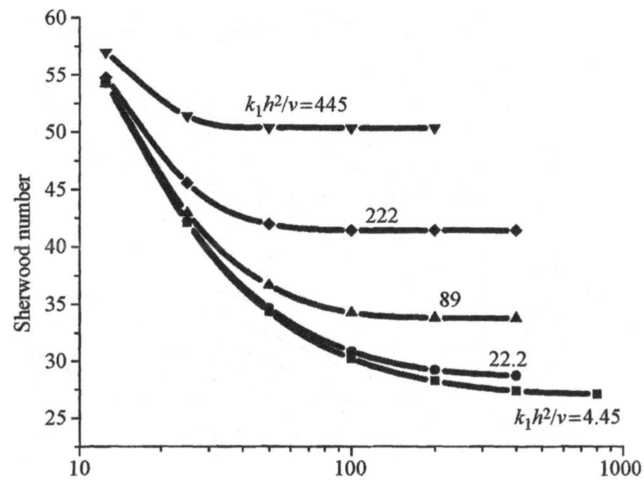

FIGURE 10.5. Sherwood numbers computed for the turbulent flow between parallel planar walls with $Re = 1 \times 10^4$. The curves show the enhancement effect produced by the homogeneous chemical reaction; it is apparent that the increasing rate constant lessens the decay of the Sherwood number with z^* (z/h). The dimensionless rate constant (k_1h^2/ν) ranges from 4.45 to 445 for the five curves.

10.4 SCALAR TRANSPORT WITH TWO-EQUATION MODELS OF TURBULENCE

We begin this part of our discussion by writing a transport equation for the scalar (concentration) in terms of the time-averaged concentration (C) and velocity (V) as

$$\frac{\partial C}{\partial t} + V_i \frac{\partial C}{\partial x_i} = \frac{\partial}{\partial x_i}\left[(D_{AB} + \varepsilon_D)\frac{\partial C}{\partial x_i}\right]. \tag{10.34}$$

Of course, the subscript i assumes values of 1 through 3 corresponding to the three principal directions. We can think of the eddy diffusivity as the product of velocity and length scales, $\varepsilon_M \propto vl$. Since $k = 1/2 v_i v_i$, we can obtain an appropriate velocity scale from $\sqrt{k}$. We also recall Taylor's inviscid estimate for the dissipation rate, $\varepsilon \approx v^3/l$; consequently, $k \sim \varepsilon^{3/2}$. Thus, we can represent the product of velocity and length scales in terms of the turbulent kinetic energy and the dissipation rate: $\sqrt{k}l = k^2/\varepsilon$. Therefore, eddy diffusivities are related to the distributions of k and ε, so typically

$$\varepsilon_D = 0.1(k^2/\varepsilon). \tag{10.35}$$

Now we need to pause for a moment and think about what might be required for solution of this hypothetical problem. We obviously need concentration, velocity, turbulent kinetic energy, and dissipation (C, V_i, k, ε). Of course, the velocity field is accompanied by variation in pressure (P), so we must add continuity; for an incompressible fluid, this means that

$$\nabla \cdot V = 0. \tag{10.36}$$

If the velocity field is three dimensional, we must solve this continuity equation, three components of the Reynolds-averaged Navier–Stokes equation, the scalar transport equation, and the energy (k) and dissipation (ε) equations, for a total of seven. This is a significant undertaking, the one that we would probably try to avoid if there were viable alternatives.

We observed previously that $k - \varepsilon$ modeling has become common, and indeed, it is used widely throughout the industry and academia. And although most workers in CFD acknowledge that such efforts are unlikely to yield fundamental progress in fluid mechanics, there are pressing requirements to find solutions to practical engineering problems. Consequently, $k - \varepsilon$ models are being used everywhere and for every purpose imaginable. A few recent examples appearing in the literature include pollutant dispersal in turbulent flows, heat transfer in coolant passages, heat transfer on a flat plate with high free-stream turbulence, turbulent natural convection in a fluid-saturated porous medium, and so on. We will examine just one scenario here, based upon the recent work of Kim and Baik (1999). Suppose we are concerned with heat and mass transport in an urban setting. In particular, consider mean flow across the top of a street "canyon" as illustrated in Figure 10.6.

The prevailing airflow moves across the top of the "canyon" in the x-direction and the surfaces are maintained at different temperatures (solar radiation heats the vertical wall on the right-hand side of the "canyon"). The objective is to develop a plausible model for heat and mass transfer in this urban space, with emphasis upon the buoyancy created by the solar heating of the (right-hand side) vertical surface. The mean flow is two dimensional, so Kim and Baik started by writing five equations:

$$\frac{\partial V_x}{\partial t} + V_x\frac{\partial V_x}{\partial x} + V_z\frac{\partial V_x}{\partial z} = -\frac{1}{\rho}\frac{\partial P}{\partial x} + \frac{\partial}{\partial x}\left(\varepsilon_M\frac{\partial V_x}{\partial x}\right) + \frac{\partial}{\partial z}\left(\varepsilon_M\frac{\partial V_x}{\partial z}\right), \tag{10.37}$$

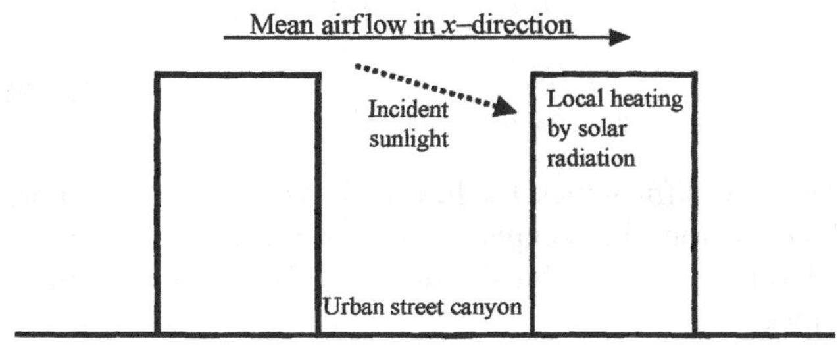

FIGURE 10.6. Urban street "canyon" in which the downstream building surface is heated by solar radiation.

$$\frac{\partial V_z}{\partial t} + V_x \frac{\partial V_z}{\partial x} + V_z \frac{\partial V_z}{\partial z} = -\frac{1}{\rho}\frac{\partial P}{\partial z} + g\frac{T - T_0}{T_0} + \frac{\partial}{\partial x}\left(\varepsilon_M \frac{\partial V_z}{\partial x}\right) + \frac{\partial}{\partial z}\left(\varepsilon_M \frac{\partial V_z}{\partial z}\right), \tag{10.38}$$

$$\frac{\partial V_x}{\partial x} + \frac{\partial V_z}{\partial z} = 0, \tag{10.39}$$

$$\frac{\partial T}{\partial t} + V_x \frac{\partial T}{\partial x} + V_z \frac{\partial T}{\partial z} = \frac{\partial}{\partial x}\left(\varepsilon_H \frac{\partial T}{\partial x}\right) + \frac{\partial}{\partial z}\left(\varepsilon_H \frac{\partial T}{\partial z}\right) + S_{\mathrm{T}} \tag{10.40}$$

$$\frac{\partial C}{\partial t} + V_x \frac{\partial C}{\partial x} + V_z \frac{\partial C}{\partial z} = \frac{\partial}{\partial x}\left(\varepsilon_C \frac{\partial C}{\partial x}\right) + \frac{\partial}{\partial z}\left(\varepsilon_C \frac{\partial C}{\partial z}\right) + S_{\mathrm{C}}. \tag{10.41}$$

Note that the fluid is taken to be incompressible, the Boussinesq approximation is used to account for buoyancy, and that source terms have been included in the energy and (mass transfer) continuity equations. Obviously, one must also model the eddy diffusivities in order to achieve closure. Since $\varepsilon_M = C_\mu(k^2/\varepsilon)$, we must include the equations for turbulent kinetic energy (k) and dissipation rate (ε):

$$\frac{\partial k}{\partial t} + V_x \frac{\partial k}{\partial x} + V_z \frac{\partial k}{\partial z} = \varepsilon_M \left\{ 2\left[\left(\frac{\partial V_x}{\partial x}\right)^2 + \left(\frac{\partial V_z}{\partial z}\right)^2\right] + \left(\frac{\partial V_x}{\partial z} + \frac{\partial V_z}{\partial x}\right)^2 \right\} - \varepsilon_H \frac{g}{T_a}\frac{\partial T}{\partial z} + \frac{\partial}{\partial x}\left(\frac{\varepsilon_M}{\sigma_k}\frac{\partial k}{\partial x}\right) \tag{10.42}$$

and

$$\frac{\partial \varepsilon}{\partial t} + V_x \frac{\partial \varepsilon}{\partial x} + V_z \frac{\partial \varepsilon}{\partial z} = C_1 \frac{\varepsilon}{k}\varepsilon_M \left\{ 2\left[\left(\frac{\partial V_x}{\partial x}\right)^2 + \left(\frac{\partial V_z}{\partial z}\right)^2\right] + \left(\frac{\partial V_x}{\partial z} + \frac{\partial V_z}{\partial x}\right)^2 \right\} - C_1 \varepsilon_H \frac{\varepsilon}{k}\frac{g}{T_a}\frac{\partial T}{\partial z} + \frac{\partial}{\partial x}\left(\frac{\varepsilon_M}{\sigma_\varepsilon}\frac{\partial \varepsilon}{\partial x}\right) + \frac{\partial}{\partial z}\left(\frac{\varepsilon_M}{\sigma_\varepsilon}\frac{\partial \varepsilon}{\partial z}\right) - C_2 \frac{\varepsilon^2}{k}. \tag{10.43}$$

The eddy diffusivities for heat and mass (ε_H and ε_C) are obtained from the computed value of ε_M using the numerical values assumed for the turbulent Prandtl and Schmidt numbers:

$$Pr^{\mathrm{T}} = \frac{\varepsilon_M}{\varepsilon_H} = 0.7 \quad \text{and} \quad Sc^{\mathrm{T}} = \frac{\varepsilon_M}{\varepsilon_C} = 0.9. \tag{10.44}$$

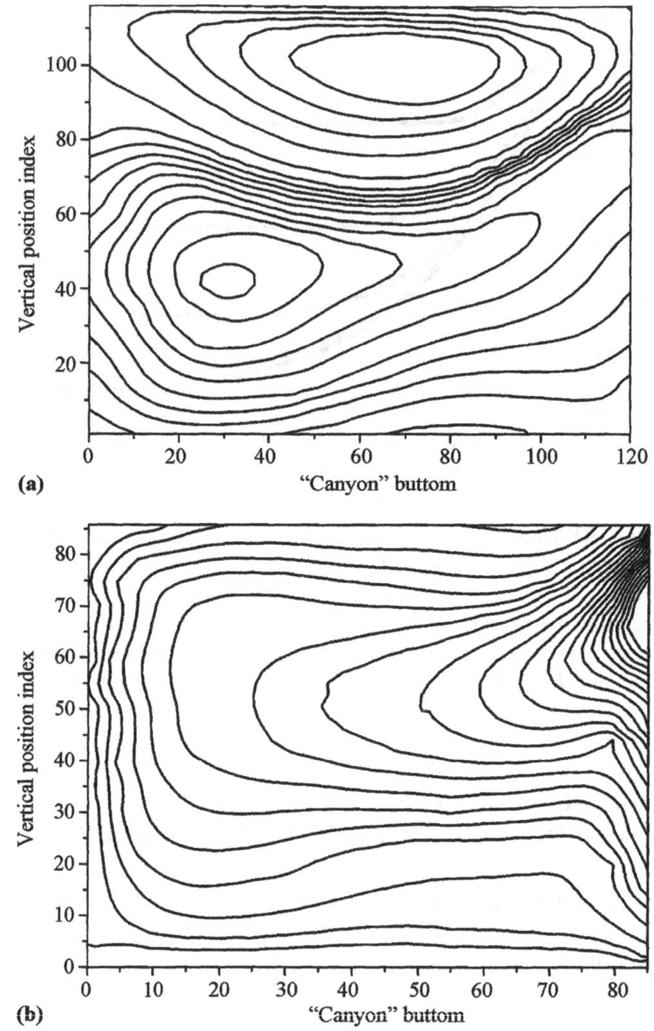

FIGURE 10.7. Approximate streamlines (a) and isotherms (b) for the case of a square "canyon" with solar heating of the downwind (right-hand) wall. These data have been reconstructed from an adaptation of the Kim–Baik computed results at $t = 1$ h.

The constants needed for this model are C_μ, σ_k, σ_ε, C_1, C_2, Pr^{T}, and Sc^{T}. Kim and Baik selected the corresponding numerical values 0.09, 1, 1.3, 1.44, 1.92, 0.7, and 0.9 and employed the SIMPLE (Patankar, 1980) algorithm for solution. Adapted excerpts from their results (streamlines and isotherms at $t = 1$ h) for the case of airflow across the top coupled with a heated wall (by solar radiation) on the right-hand side are shown in Figure 10.7.

10.5 TURBULENT FLOWS WITH CHEMICAL REACTIONS

Bear in mind that what we can provide here is merely an introduction; any reader with deeper interests in this field should turn to some of the specialized resources that are available.

I particularly recommend Fox (2003), Baldyga and Bourne (1999), and Libby and Williams (1994). At the beginning of this chapter, we noted that the reacting turbulent flows presented additional challenges. Let us revisit this issue and look at some of the complications. We begin by considering the limiting conditions for the reaction between species "A" and "B." In terms of the initial distributions of reactants, we have

Fully segregated ↔ Completely mixed.

For the chemical kinetics, the reaction may be

Very slow ↔ Very fast.

And for the flow field itself, we have

Highly ordered ↔ Enegetically turbulent.

For a chemical reaction occurring in a fluid, we could have *any* combination of these characteristics. Of course, we have familiar methods available to solve problems with a flow field characterized by (*Highly ordered*). But what about combinations involving turbulent flows? For example, if the reaction is very fast, then the controlling step is turbulent mixing. To facilitate this introductory discussion, we will need to spend a little effort considering characteristic times for mixing and reaction.

From an initially segregated state, we visualize a process in which large eddies transport material, producing a gross distribution but one that remains highly segregated. Smaller eddies continue this process, producing a structure with finer "grain." In some types of processes, such as stirred tank reactors, the time required for the gross convective mixing can be estimated from the circulation time (obtained from tracer studies). At dissipative scales, diffusion acts in concert with small distances and sharp concentration gradients to yield homogeneity. Let us focus our attention on this last step in the process, when the distributive, or convective, mixing is virtually complete. We presume that the volume elements of the material (or reactant) are of the size of the Kolmogorov microscale (ν is the kinematic viscosity):

$$\eta = \left(\frac{\nu^3}{\varepsilon}\right)^{1/4}. \tag{10.45}$$

Bourne (1992) notes that if this small volume element is taken to be roughly spherical, then the diffusional mixing time can be estimated:

$$t_{\mathrm{dm}} \approx \frac{\eta^2}{8D}. \tag{10.46}$$

In an aqueous medium with energetic turbulence, we might have $\eta \approx 50\ \mu\mathrm{m}$ and $D \approx 1\times10^{-5}\ \mathrm{cm^2/s}$; therefore, $t_{\mathrm{dm}} \approx 0.31$ s. In a very weak turbulent field, this time (t_{dm}) might be on the order of several hundred seconds or more.

The characteristic time for reaction, or chemical time, can be written for an elementary nth order reaction:

$$t_{\mathrm{cr}} = \left(k_n \overline{C_{\mathrm{A0}}}^{\,n-1}\right)^{-1} \tag{10.47}$$

Naturally, a very fast reaction results in a very small t_{cr}. For more complicated kinetic schemes, the chemical timescales can be obtained from the eigenvalues of the Jacobian of the chemical production (source) term; see pages 150–153 in Fox (2003). Exactly how the production term is closed depends upon how the timescales for mixing and chemical reaction compare. We can expect difficulties in developing suitable models when they are similar. The ratio of the mixing and chemical timescales forms a Damköhler number Da and the size of this dimensionless number can be used to guide selection of a closure procedure. For example, if the reaction is fast relative to the mixing rate (of components "A" and "B"), then the components will remain segregated.

Now, suppose we have a second-order kinetic description involving species "A" and "B." Employing the Reynolds decomposition and time averaging for an *isothermal process*, we find

$$-k_2\overline{(C_{\mathrm{A}}+C_{\mathrm{A}}')(C_{\mathrm{B}}+C_{\mathrm{B}}')} = -k_2\left(C_{\mathrm{A}}C_{\mathrm{B}}+\overline{C_{\mathrm{A}}'C_{\mathrm{B}}'}\right). \tag{10.48}$$

We see that a correlation has appeared relating the concentration fluctuations of the two species $\overline{C_{\mathrm{A}}'C_{\mathrm{B}}'}$ (the concentration covariance). Here, of course, is the closure problem; a first-order closure would be achieved if we were able to relate this correlation of fluctuations to the mean concentration(s). It seems likely that the time-averaged fluctuations $\overline{C_{\mathrm{A}}'C_{\mathrm{B}}'}$ may be affected by both the turbulent flow and the chemical reaction itself. Unfortunately, the real situation is often much worse than that indicated by eq. (10.48). Consider the case of a chemical reaction accompanied by a large temperature change—such is the case with combustion processes, for example. Under these circumstances, eq. (10.48) should be written in terms of mass fraction w:

$$k_m \exp(-E/RT)\rho^2 w_i w_j. \tag{10.49}$$

Applying the decomposition,

$$k_m \exp\left(-\frac{E}{R(\overline{T}+T')}\right)(\overline{\rho}+\rho')(\overline{\rho}+\rho')(\overline{w}_i+w_i')(\overline{w}_j+w_j'), \tag{10.50}$$

we see that products involving fluctuating quantities will include temperature, density, and mass fraction. Carrying out the indicated products *just* for density and mass fraction, we find (dropping the overbar for the average quantities):

$$\begin{aligned}&\rho^2\left(w_iw_j+w_iw_j'+w_i'w_j+w_i'w_j'\right)\\&+2\rho\rho'\left(w_iw_j+w_iw_j'+w_i'w_j+w_i'w_j'\right)\\&+\rho'\rho'\left(w_iw_j+w_iw_j'+w_i'w_j+w_i'w_j'\right).\end{aligned}\quad(10.51)$$

Now we rewrite the exponential portion of (10.49):

$$\begin{aligned}\exp\left(-\frac{E}{R\left(\overline{T}-T'\right)}\right)&=\exp\left(-\frac{T_A}{\left(\overline{T}-T'\right)}\right)=\exp(-\phi)\\&=1-\phi+\frac{\phi^2}{2!}-\frac{\phi^3}{3!}+\cdots.\end{aligned}\quad(10.52)$$

This process yields correlations (moments) involving density, mass fraction, and temperature of every (and all) order(s). Moreover, O'Brien (1980) notes that for a rapid reaction occurring in not-very-energetic turbulence, there may be no legitimate way to truncate the expansion (the fluctuating terms may be larger than the means). Hence, it is effectively impossible to achieve closure for this type of problem using conventional time averaging (this is an example where mass or *Favre* averaging becomes useful).

It is clear that we should expect a very high level of complexity due to the couplings between the physicochemical processes occurring in turbulent reactive flows. Consider that

- Exothermic reactions may produce changes in temperature, affecting density, viscosity, and pressure.
- Rapid reactions may result in length scales associated with the concentration fluctuations that are even smaller than the microscales of the turbulence itself.
- The concentration fluctuations may either enhance or diminish the overall rate of reaction; the correlation between "A" and "B" may be positive or negative depending upon the initial premixing or segregation of the reacting species.

As Leonard and Hill (1988) observed, "understanding the interaction of these processes presents a formidable challenge." Fortunately, there is a way around some of these difficulties, through use of the *transported pdf* (probability density function) method. A full exposition of this technique is beyond the scope of this book, however, we can lay a little groundwork for further exploration. Before we do that, we will review some of the older, elementary closure methods.

10.5.1 Simple Closure Schemes

Toor (1962, 1969) proposed a first-order closure scheme based upon the idea that the correlation of concentration fluctuations might depend solely upon hydrodynamics of the reactor; in particular, for the second-order irreversible reaction given by eq. (10.48),

$$\overline{C_A'C_B'}=-I_S C_{A0}C_{B0},\quad(10.53)$$

where I_S is the intensity of segregation and the concentrations are at the inlet. For an idealized plug flow reactor (PFR), the intensity of segregation is a function of axial position only, $I_S=f(z)$, and it can be determined from the decay of concentration fluctuations in a nonreactive system. For an infinitely fast reaction, Toor's hypothesis is based upon the assumptions that the reactants are fed in stoichiometric proportions, there is no premixing, and that their diffusivities are equal. Under these stipulations, the rate expression (10.48) can be written as

$$-k_2(C_AC_B-I_SC_{A0}C_{B0}),\quad(10.54)$$

where the zero subscripts refer to inlet concentrations assuming the two species are mixed without reaction.

Patterson (1981) described an "interdiffusion" model for the case of two *nearly segregated* components (by nearly segregated, we imply a three-spike distribution, with probability corresponding to the two pure components and one intermediate composition). The interdiffusion model resulted in

$$\overline{C_A'C_B'}=-\overline{C_A'^2}(1-\gamma)/(\beta(1+\gamma)),\quad\text{where}\quad(10.55)$$

$$\beta=C_{A0}/C_{B0}\ \text{and}\ \gamma=\left(\beta C_AC_B-\overline{C_A'^2}\right)/\left(\beta C_AC_B+\overline{C_A'^2}\right).$$

Leonard and Hill (1988) simulated a second-order irreversible chemical reaction in a decaying, homogeneous turbulent flow and compared Toor's closure scheme with Patterson's (1981). They found that Toor's model gave better results for their numerical simulation. They also discovered that regions of the flow with the largest reaction rates were correlated with the location of high strain rates. Leonard and Hill noted the implication: Relatively infrequent events in the turbulent flow field might have a significant effect upon the overall rate of conversion. This is a point we will return to later.

There is evidence in the literature that more complicated reaction schemes are less amenable to simple first-order closure schemes. Dutta and Tarbell (1989) examined the irreversible reactions

$$A+B\rightarrow C\quad\text{and}\quad C+B\rightarrow D$$

and found that neither the Bourne–Toor (1977) nor the Brodkey–Lewalle (1985) closure was able to correlate with available experimental data. They provided evaluations for four other closure schemes as well.

Dutta and Tarbell (1989) also cite an exponential decay for the intensity of segregation in a plug flow reactor:

$$I_S = \exp\left(-\frac{t}{\tau_M}\right), \quad (10.56)$$

where the timescale for turbulent micromixing τ_M is

$$\tau_M \cong \frac{2}{(3 - Sc^2)}\left(\frac{5}{\pi}\right)^{2/3}\left(\frac{l_c}{\varepsilon}\right)^{1/3}. \quad (10.57)$$

This result is valid for $Sc < 1$; it was developed by Corrsin (1964) who formulated a model for the decay of concentration fluctuations in a decaying isotropic turbulence. By Corrsin's analysis,

$$\frac{d}{dt}\left(\overline{C_i'C_i'}\right) = -2D_{AB}\overline{\left(\frac{\partial C'}{\partial x_i}\right)\left(\frac{\partial C'}{\partial x_i}\right)} \approx -12D_{AB}\frac{\overline{C'C'}}{\lambda_C^2}, \quad (10.58)$$

where λ_C is a concentration microscale analogous to the Taylor microscale introduced in Chapter 5.

10.6 AN INTRODUCTION TO pdf MODELING

Consider a scalar quantity, perhaps temperature, measured in a turbulent flow. This scalar will have a mean value and a fluctuation, which we will denote in the following way: $\langle\phi\rangle + \phi'$. The fluctuations will have a variance, which we will write as $\langle\phi'^2\rangle$. As we saw previously, coupling occurs between the velocity field and the scalar, resulting in a scalar flux: $\langle v_i\phi\rangle$. A transport equation for the scalar variance can be developed from the scalar flux equation, as shown by Fox (2003):

$$\frac{\partial\langle\phi'^2\rangle}{\partial t} + \langle V_j\rangle\frac{\partial\langle\phi'^2\rangle}{\partial x_j} = D\nabla^2\langle\phi'^2\rangle - \frac{\partial\langle v_j\phi'^2\rangle}{\partial x_j} + P_\phi - \varepsilon_\phi. \quad (10.59)$$

The first term on the right-hand side is the molecular transport of the scalar variance, which is unimportant in energetic turbulent flows. The last two terms on the right-hand side of this equation represent production and dissipation of the scalar variance, respectively. Production occurs as a result of the interaction between the scalar flux and the (mean) scalar gradient. Consequently, production is zero in a homogeneous scalar field. The dissipation term, as the name implies, represents the attenuation (or destruction) of the scalar variance. Physically, we can think about this by drawing an analogy with the decay of grid-generated turbulence in a wind tunnel. As we move farther downstream from the grid, we expect the mean square fluctuations $\overline{v_i v_i}$ to diminish. This is, however, not necessarily the case with a passive scalar variance (such as temperature). Jayesh and Warhaft (1992) studied the behavior of temperature fluctuations in grid-generated turbulence in a wind tunnel, for which a cross-stream temperature gradient was maintained (unchanging with respect to x, the flow direction). Their data show that the scalar variance $\langle\phi'^2\rangle$ *increases* in the x-direction under these conditions. Furthermore, their data also show that the scalar probability density function is not Gaussian for the higher turbulence Reynolds numbers in cases where the cross-stream temperature gradient is imposed. The non-Gaussian pdf's appear to be created by large infrequent temperature fluctuations, which are accompanied by enhanced scalar dissipation. The significance of this point will become clear in the next section: If ϕ and ε_ϕ are independent, then the conditional expectation of the scalar dissipation rate is constant with respect to the scalar field and the scalar pdf will be Gaussian. Since the small-scale mixing term in pdf modeling is expressed by the scalar dissipation rate (as Wang and Chen, 2004, point out), the conditional expectation of the scalar dissipation rate must be modeled.

10.6.1 The Fokker–Planck Equation and pdf Modeling of Turbulent Reactive Flows

In recent years, probability density function methods have been developed for turbulence modeling both with and without chemical reaction. Recommended readings for those wishing to pursue these topics include Pope (1985), Chapter 12 in Pope (2000), and Chapter 6 in Fox (2003). Fox points out that one of the principal advantages of full (or transported) pdf modeling in turbulent reacting flows is that the chemical production term *does not require any closure approximations*. Moreover, transported pdf models provide more information than one obtains from the second-order modeling based upon the Reynolds-averaged Navier–Stokes equations. Consequently, we provide this brief introduction to serve as a gateway to further study of the turbulent transport of scalars in reactive flows.

The Fokker–Planck (FP) equation describes the evolution of a probability density function in space and time. It is convenient for us to think about how FP equations arise in the following way: Assume we were interested in the behavior of a particle immersed in a fluid. It would be subjected to drag, buoyancy, gravity, and so on. Naturally, it would interact with the molecules of the fluid phase—after all this is how momentum is transferred. But suppose the particle size was such that its motion was affected perceptibly by collisions with *individual* molecules; this is, of course, thermal or Brownian motion. Now if we wanted to write an accurate description of the motion of this very small particle, we would need to deal with a *many–many* body problem. That in itself is formidable, but we must also remember that an accurate initial condition would be needed for every single entity. This information is simply not available to us; we must look for alternatives. One possibility is the approach taken in statistical mechanics. While we may not be able to discern what an individual entity is doing, in the aggregate we will have a fairly good idea. This *ensemble* averaging is reasonable for macroscopic systems because even small ones contain ridiculously large numbers of molecules.

For one spatial dimension, the FP equation is

$$\frac{\partial}{\partial t} f(x,t) = -\frac{\partial}{\partial x}[D_1(x,t,f)f(x,t)] + \frac{\partial^2}{\partial x^2}[D_2(x,t,f)f(x,t)], \quad (10.60)$$

where f is the density function and D_1 and D_2 are, respectively, the drift and diffusion coefficients. Note that this partial differential equation has been written in such a way that it *could* be nonlinear. To better understand how this equation might be useful to us, consider a particle (or particles) distributed in a 1D region of fluid. The variable x represents some property, perhaps position or velocity. If it were position, then the probability that the particle of interest would be located in the interval $(a < x < b)$ would be

$$P\{a < x < b\} = \int_a^b f(x)dx. \quad (10.61)$$

Given an initial density function and appropriate functional choices for D_1 and D_2, we can use (10.60) to compute how f will be redistributed in space and time. In other words, the Fokker–Planck equation is an *equation of motion for the probability density function.*

If the FP equation were to be applied to a density function in three space, we would write

$$\frac{\partial f}{\partial t} = -\sum_{i=1}^{3} \frac{\partial}{\partial x_i}[D_{1i}(x_1, x_2, \ldots)f] + \sum_{i=1}^{3}\sum_{j=1}^{3} \frac{\partial^2}{\partial x_i \partial x_j}[D_{2ij}(x_1, x_2, \ldots)f]. \quad (10.62)$$

Note that the drift coefficient is a vector and the diffusion coefficient is a second-order tensor.

Let us look at an example to get a better sense of how the FP equation can be of use to us. Suppose we have an inert scalar (perhaps a tracer) that is initially concentrated in a small subset of the region extending from $x = -5$ to $x = +5$. It is distributed such that

$$\sum_{i=1}^{N} f_i(x,t)\Delta x_i = 1, \quad (10.63)$$

where N is some small number. We take the drift coefficient D_1 to be a linear function of x and assume a constant value for the diffusion coefficient D_2. These choices constitute the Ornstein–Uhlenbeck process (see Risken, 1989) and the analytic solution for (10.60) is known (for this case, the FP equation ultimately produces a Gaussian distribution). However, we anticipate an interest in drift and diffusion coefficients that may be functionally dependent upon x or f in more complicated ways, so we will use a numerical procedure with that in mind. The results of these example computations are shown in Figure 10.8.

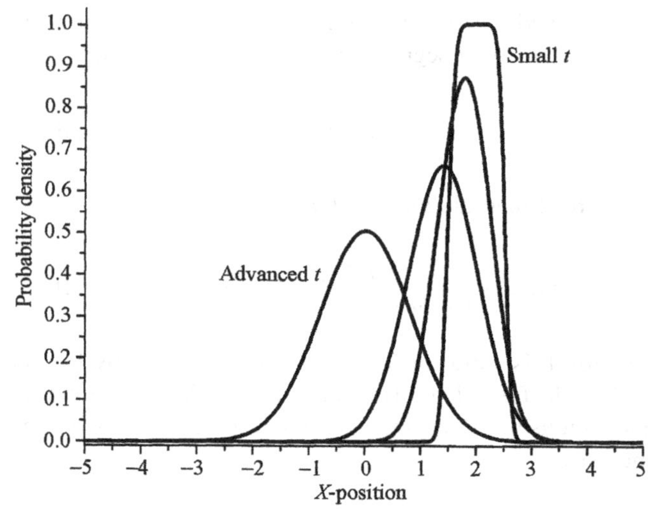

FIGURE 10.8. Computed results from the FP equation with a constant diffusion coefficient and a drift coefficient written as a linear function of x (the Orstein–Uhlenbeck process). The probability is initially clustered at about $x \approx 2$.

By changing the functional form of the drift and diffusion coefficients, one can obtain varied pdf evolutions. For example, suppose that the drift coefficient has a maximum at the center of the interval (corresponding to $x = 0$), decreasing exponentially in both positive and negative x-directions. Let the diffusion coefficient also have a maximum at the center, falling to zero at the limits of the interval; in particular, take $D_2 = A_0 \cos(\pi x/10)$. For this scenario, if we begin with the same initial distribution of probability that we employed above, we find quite different behavior as demonstrated by Figure 10.9.

Now let us consider how the FP equation is to be used in the modeling of scalar transport. Fox (1992) suggested that the FP equation might be employed to model turbulent molecular mixing. Evidence suggests that distribution of a scalar quantity by larger eddies results initially in a layer-like (or lamellar) structure. Thus, a lamella of high concentration would be immediately adjacent to a layer of very low (or zero) concentration and so on. *If* the limiting form for the scalar pdf in turbulent mixing is Gaussian, then the FP equation is a logical framework as indicated qualitatively by the computed examples above. Fox recommended that the FP closure be used in conjunction with the pdf

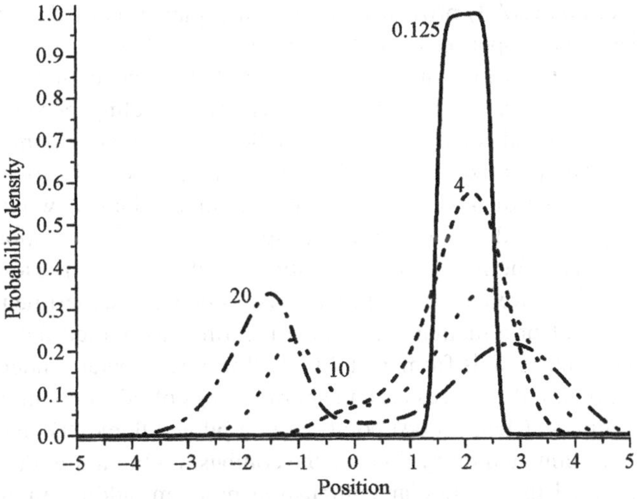

FIGURE 10.9. Computed results from the Fokker–Planck equation assuming the drift coefficient decreases exponentially (from its maximum at the center of the interval). The diffusion coefficient is taken as $A_0 \cos(\pi x/10)$.

balance equations to construct a model for turbulent reactive flows.

In this context, the simulations of turbulent mixing of a passive scalar carried out by Eswaran and Pope (1988) are especially significant. They used DNS (actually the pseudospectral method) to explore the evolution of an initial (scalar) distribution in homogeneous isotropic turbulence. Their computations showed that a scalar pdf beginning with a double delta-function distribution (simulating a nonpremixed condition) would evolve toward a Gaussian distribution. The conditions employed for this simulation effort were idealized and one must exercise caution in extrapolating these results.

10.6.2 Transported pdf Modeling

We previously observed that a complete specification for turbulent flow with chemical reaction would require that we have knowledge of the velocity field, the composition(s), and the temperature, everywhere, and at all time t. Such a level of detail is simply not available to us through any currently practical mechanism. Suppose, on the other hand, that we had a statistical description of the process in the form of a pdf for the velocity vector *and* a pdf for the set of scalar quantities (compositions and temperature) for that process. Pope (1985) notes that a complete one-point statistical description of such a process is contained in the *joint* pdf for velocity and these scalar quantities. When we speak of the joint *velocity–composition* pdf, we are of course implying that both velocity and composition are continuous random variables. We adopt Pope's notation by representing the velocity–composition joint pdf with $f_{U\phi}(V, \psi)$. The fact that a one-point pdf is to be used means that there is no direct information on the velocity field. And although the chemical production term is treated in an exact way, the closure problem is not completely eliminated. As Fox (2003) points out, scalar transport due to velocity fluctuations must be approximated, and a transported pdf micromixing model (such as the FP approach just described) must be developed to represent the decay of the scalar variance. The joint pdf transport equation has the form

$$\frac{\partial f_{U\phi}}{\partial t} + V_i \frac{\partial f_{U\phi}}{\partial x_i} = -\frac{\partial}{\partial V_i}\left[\langle A_i \,|\, V, \psi\rangle f_{U\phi}\right] - \frac{\partial}{\partial \psi_i}\left[\langle \Theta_i \,|\, V, \psi\rangle f_{U\phi}\right], \tag{10.64}$$

where A_i is the substantial time derivative of velocity. The details of the derivation are shown by Pope (1985). This partial differential equation indicates that the evolution of the joint pdf occurs in physical space (x_i) due to the velocity field (V_i), in velocity phase space due to the conditional expectation $\langle A_i | V, \psi\rangle$, and in composition phase space due to the conditional expectation $\langle \Theta_i | V, \psi\rangle$. These conditional expectations must be modeled before eq. (10.64) can be solved. Fox shows that

$$\langle A_i | \, V, \psi\rangle = \left\langle \left(\nu \frac{\partial^2 U_i}{\partial x_j \partial x_j} - \frac{1}{\rho}\frac{\partial p}{\partial x_i}\right) | V, \psi \right\rangle - \frac{1}{\rho}\frac{\partial \langle p\rangle}{\partial x_i} + g_i \tag{10.65}$$

and

$$\langle \theta | \, V, \psi\rangle = \left\langle \Gamma \frac{\partial^2 \phi}{\partial x_j \partial x_j} | V, \psi \right\rangle + S(\psi). \tag{10.66}$$

Γ is the diffusivity for the scalar, ϕ. The viscous dissipation and fluctuating pressure terms on the right-hand side of (10.65) must be closed by model. Similarly, the molecular mixing term on the right-hand side of (10.66) must be closed to complete the model. Fox points out that these closure problems, as usual, are the main challenges confronting transported pdf modeling.

As we noted in (10.64), both velocity and composition are treated as random variables. This is not mandatory. Fox (2003) shows that the transported pdf equation can be written just for the composition pdf:

$$\begin{aligned}\frac{\partial f_\phi}{\partial t} + \langle U_i\rangle \frac{\partial f_\phi}{\partial x_i} + \frac{\partial}{\partial x_i}\left[\langle u_i | \, \psi\rangle \, f_\phi\right] &= -\frac{\partial}{\partial \psi_i}\left[\left\langle \Gamma_i \nabla^2 \phi_i' \middle| \, \psi \right\rangle f_\phi\right] \\ &\quad - \frac{\partial}{\partial \psi_i}[(\Gamma_i \nabla^2 \langle \phi_i\rangle + S_i(\psi)) f_\phi]. \end{aligned} \tag{10.67}$$

This equation describes transport of the composition pdf due to convection by the mean flow $\langle U\rangle$, convective transport by the (conditioned) velocity fluctuations $\langle u \rangle$, and by molecular mixing and chemical reaction.

Let us now consider the actual steps involved in solving a transported pdf problem. In the case of eq. (10.67), one must know the mean velocity field and the turbulence field; in addition, the analyst must have a molecular mixing model and a closure for $\langle u_i | \psi \rangle$. The latter is usually achieved with a gradient transport model:

$$\langle u_i | \psi \rangle = -\frac{\Gamma_T}{f_\phi}\frac{\partial f_\phi}{\partial x_i}. \tag{10.68}$$

The turbulent diffusivity, Γ_T, in (10.68) must be obtained from the spatial distributions of turbulent energy and dissipation, k and ε. The evolution of the composition pdf is normally determined using the Monte Carlo particle method, and it is important that we recognize the differences between an actual fluid system and a *particle representation* of it. We use a large number of particles, each with its own position, velocity, composition, and so on. Pope (2000) notes that such a particle representation can describe a real fluid system only in a limited way. Since each particle represents a mass of fluid, the particle system cannot portray the instantaneous velocity, but only the mean velocity field. Of course, in ideal circumstances, the pdf for *particle* velocity would equal the *fluid* velocity pdf. Similarly, one would hope that the moments of the distributions would also be the same.

Development of an adequate molecular mixing model is one of the principal challenges confronting pdf computations. Numerous alternatives have been explored in the literature, including coalescence–dispersion (CD) models, interaction by exchange with the mean (IEM), the Fokker–Planck (stochastic diffusion) model, and the use of Euclidean minimum spanning trees (EMST). The latter was developed by Subramaniam and Pope (1998) and has been employed by Wang and Chen (2004), among others. One simple idea that is common to several mixing models is that the scalar relaxes toward the mean. Using the format employed by Fedotov et al. (2003),

$$\frac{d\phi}{dt} = -\frac{1}{\tau}(\phi - \langle\phi\rangle), \tag{10.69}$$

where τ is a characteristic time associated with the turbulence. Of course, viewed deterministically, this equation implies an exponential decay of the scalar to its mean value. Subramaniam and Pope observe that this approach violates the "localness" of mixing, that is, the idea that the composition characteristics in proximity to a fluid particle affect the mixing. They elaborate on the criteria that must be satisfied by a mixing model in order to adequately represent the physics of the process. Some of these requirements are obvious. For example, the local mass fraction(s) must be in an allowable region; clearly, they cannot be either negative or greater than one. Fox (2003) provides a thorough explanation of the six desirable properties of molecular mixing models.

The pdf modeling approach introduced above may be of greatest value in flame (combustion) modeling because the chemical source term is handled without approximation. Nonpremixed combustion problems have been the focus of a series of TNF (turbulent nonpremixed flames) workshops carried out under the auspices of the Combustion Research Facility at Sandia National Laboratories. Barlow (2006) showed a series of comparisons between experimental data (for a methane–air flame identified as piloted flame "D") and models from TNF4 that allow one to better understand both the successes and shortcomings of pdf modeling. Wang and Chen (2004) point out that piloted flame "D" has been simulated many times in the combustion literature; they revisited this particular combustion problem, adding more detailed chemistry. They used the parabolized Navier–Stokes equations (neglecting turbulent transport in the mean flow direction), a multiple timescale k–ε model for the turbulent flow closure, and the EMST model of Subramaniam and Pope (1998) for the molecular mixing closure. They presented scatter plot comparisons for temperature, CH_4 mass fraction, CO mass fraction, and NO mass fraction. Their results are generally good, although some problems resulting from the deficiencies of the small-scale mixing model are noted. The student with further interest in pdf modeling is encouraged to read their paper carefully; Wang and Chen point out clearly where the problems and the prospects lie. In particular, they found that the detailed reaction mechanisms were successfully integrated into pdf modeling; at the same time, their work makes it clear that the molecular mixing closure remains as one of the main problem areas for more broadly applied pdf modeling.

10.7 THE LAGRANGIAN VIEW OF TURBULENT TRANSPORT

It is useful, both conceptually and physically, to think a little more about the turbulent transport of scalars from a Lagrangian viewpoint. Consider an entity (perhaps a small particle or marker) placed in a turbulent flow at a particular initial position at $t = 0$. It will "wander" with time depending upon its velocity; we will characterize that velocity in three space as u_i. We expect this velocity to change with time in some fashion as well. Where will our particle be after time t?

$$X_i(x_0, t) = x_0 + \int_0^t u_i(x_0, t)dt. \tag{10.70}$$

Before we go further, we must qualify this statement. Whether a particle faithfully follows the fluid motion depends upon both its size and its density. If a particle is much larger than the Kolmogorov microscale η, recall

$$\eta = \left(\frac{\nu^3}{\varepsilon}\right)^{1/4}, \tag{10.71}$$

then its trajectory will reflect only the influence of the larger eddies. In the type of scalar transport processes we want to consider here, the entities or particles will be very small and we need not worry about this restriction. We will also assume that the turbulence is homogeneous and isotropic, although in real flows this would be unusual to say the least.

If a marker is released from a point source in a quiescent fluid, dispersion will occur due solely to molecular diffusion. Einstein (1905) found that the mean square displacement for this case could be described by

$$\frac{d\overline{X^2}}{dt} = 2D_{AB}. \tag{10.72}$$

Note that this equation indicates that the dispersion of the marker will increase linearly in time. We can compare this with the dispersion of a marker in homogeneous isotropic turbulence. Taylor (1921) found that the mean square displacement could be characterized as

$$\frac{d\overline{X^2}}{dt} = 2\overline{u^2}\int_0^t R_L(t)dt. \tag{10.73}$$

The right-hand side contains the mean square velocity fluctuations $(\overline{u^2})$ and the integral of the Lagrangian correlation coefficient R_L. Two limiting cases can immediately be examined: At small time t, $R_L \approx 1$ and at large times, $R_L \approx 0$. Consequently, the initial *rate* of dispersion is proportional to time and $\overline{X^2}$ itself increases as $\sim t^2$. For large times, the rate of dispersion is a constant. At this point, we need to recognize that the typical data we collect for turbulent flows are Eulerian, that is, they are normally obtained by placing an instrument or probe at a particular spatial position. What we actually need to know is how our small entity or marker is dispersed as it moves with the fluid. Hinze (1975) suggests a similarity between this turbulent dispersion and the Brownian motion created by the random thermal motions of molecules. We must, however, exercise caution here in our use of the word "random." Though nonlinear stochastic processes may superficially appear random, we recognize that for the phenomena of interest, the complete set of governing partial differential equations can in fact be written down. It certainly appears as though the problems of interest to us are fully—if not practically—deterministic. Furthermore, although we assumed homogeneous isotropic conditions for convenience, the real turbulent flows normally have a preferred orientation.

Let us return to eq. (10.73). If we are able to characterize both the mean square velocity fluctuations and the Lagrangian correlation coefficient, we can determine the mean square displacement of the transported entity for any time t. Hanratty (1956) employed Taylor's suggestion by setting

$$R_L(t) = \exp\left(\frac{-t}{\tau_L}\right), \tag{10.74}$$

where the Lagrangian integral timescale is

$$\tau_L = \int_0^\infty R_L(t)dt. \tag{10.75}$$

Note that the exponential form used for the Lagrangian correlation coefficient is merely a convenient approximation—nothing more. Manomaiphiboon and Russell (2003) compared four alternative function forms for R_L, including the exponential equation (10.74). The other forms examined were

$$R_L(t) = \exp\left(\frac{-|t|}{2\tau_L}\right)\cos\left(\frac{t}{2\tau_L}\right), \tag{10.76}$$

$$R_L(t) = \exp\left(\frac{-\pi t^2}{4\tau_L^2}\right), \tag{10.77}$$

and

$$R_L(t) = \exp\left(\frac{-\pi t^2}{8\tau_L^2}\right)\cos\left(\frac{t^2}{2\tau_L^2}\right). \tag{10.78}$$

A proposed functional form for R_L must meet the criteria described by Manomaiphiboon and Russell; the correlation coefficient must

- Be equal to 1 at the origin and rapidly decay to 0 as t increases.
- Have a first derivative equal to zero at the origin.
- Produce a well-defined integral timescale upon integration.
- Yield a spectrum (by Fourier transformation) that is consistent with known functional limits.

The reader is encouraged to compare the shapes of the four forms for R_L and assess the suitability of each. For example, it is obvious that eq. (10.76) fails to satisfy the requirement that the derivative be zero at the origin. However, Manomaiphiboon and Russell note that this may not be a serious limitation with regard to turbulent diffusion.

If we proceed with the exponential form, we obtain

$$\frac{d\overline{X^2}}{dt} = 2\overline{u^2}(e^{-t/\tau}). \tag{10.79}$$

Of course, such an equation would allow us to calculate the mean square displacement as a function of time, given the

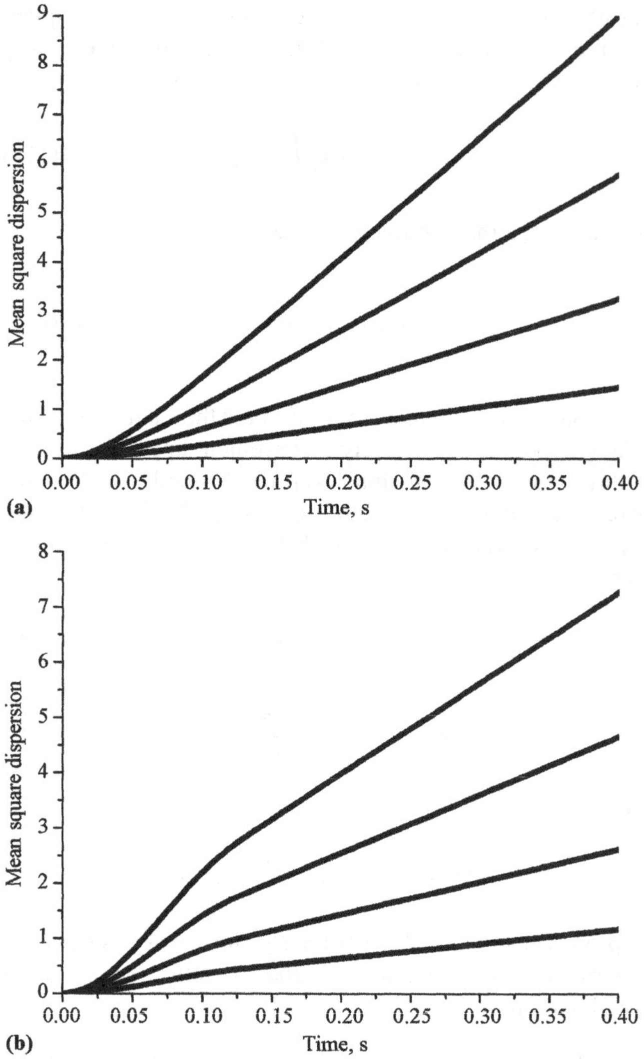

FIGURE 10.10. Behavior of the mean square dispersion with time for an exponential Lagrangian correlation coefficient (a) and a correlation coefficient with a negative tail (b). These results are computed for relative turbulence intensities of 4, 6, 8, and 10%.

characteristics of the turbulent field. However, it is worthwhile for us to question what the result would be using a more realistic form for the Lagrangian correlation coefficient R_L. Some computed data are given in Figure 10.10 that show how the mean square dispersion increases with time for a series of turbulence intensities.

Taylor (1921) speculated that a Lagrangian correlation coefficient with a negative tail might result from a sort of "regularity" in the flow (perhaps periodic vortex shedding). He also noted that some of L. F. Richardson's (1921) time exposure photographs of paraffin vapor plumes revealed a "necking-down" that Taylor attributed to a negative tail in the correlation.

Schlien and Corrsin (1974) reported experimental measurements using thermal markers produced with electrically heated platinum wires in grid-generated turbulence. They were able to calculate the Lagrangian correlation coefficient which was found to have a different shape (especially near the origin) than the Eulerian coefficient. They also found that the Lagrangian microscale is larger than its Eulerian counterpart. Let us make it absolutely clear why this discussion of R_L matters so much to us: If the form of R_L is known, we can determine the mean square displacement for the turbulent transport of a scalar such as temperature or concentration.

Hanratty (1956) attempted a Lagrangian analysis of heat transfer between two parallel walls, one with a thermal energy source present at $t=0$ and the other with a thermal energy sink of equivalent strength. Hanratty's intent was to examine the effects of history in the transport of thermal energy markers (or "particles"). For positive t's, the flux at both walls was set to zero. A considerable simplification was effected by assuming a *uniform velocity profile* and homogeneous isotropic turbulence—neither, of course, possible for flows between parallel walls. These simplifications result in a very appealing governing equation for the process:

$$\frac{\partial T}{\partial t} = \overline{u^2} f(t) \frac{\partial^2 T}{\partial y^2}, \tag{10.80}$$

if the function $f(t)$ can be related to the mean square displacement, then we can readily obtain solutions for this problem. Hanratty found by assuming a probability distribution for the displacement of "particles" that

$$f(t) = u^2 \tau (1 - e^{-t/\tau}). \tag{10.81}$$

Thus, the effects of history upon the rate of dispersion are taken into account, through the Lagrangian correlation coefficient. It is evident from this model that an increase in the mean square velocity fluctuations will result in more effective dispersion of the thermal energy. Conversely, a decrease in the Lagrangian integral timescale will lessen the effectiveness of the turbulent "diffusion" process and constrain the dispersion of thermal energy. In Figure 10.11, the effects of the mean square fluctuations are revealed (all other parameters of the problem held constant).

The problem with the above analysis, of course, was the assumption of uniform flow with homogeneous and isotropic turbulence—not at all realistic for the flow through a channel. Recognizing this, Papavassiliou and Hanratty (1995) updated the original work from 1956; they noted that the determination of trajectories of individual thermal energy markers requires "detailed instantaneous" description of the turbulence. Consequently, they used the pseudo-spectral method described by Orszag and Kells (1980) to obtain a direct numerical solution (DNS) for the turbulent flow. A tracking algorithm developed by Kontomaris et al. (1992) was used to determine the trajectories of the individual markers. The curves shown in Figure 10.12 represent ensemble averages

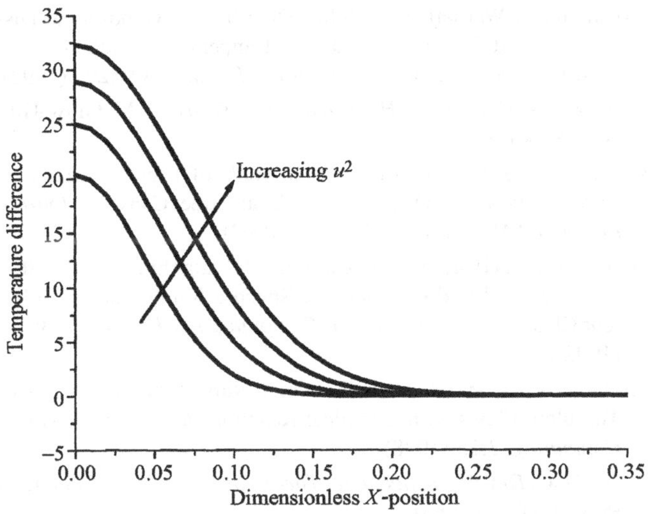

FIGURE 10.11. Comparison of computed results for different values of the mean square velocity fluctuations. The mean fluid velocity (between the parallel walls) is assumed to be uniform and the turbulence is homogeneous and isotropic. Naturally, an increase in turbulence intensity results in increased dispersion.

of the trajectories of 16,129 individual markers released at the wall of a channel with $Re = 2660$.

Individual trajectories for 2 of the more than 16,000 markers are shown in Figure 10.13. Of course, the average transport of heat "particles" away from the wall is determined from the ensemble of individual trials. The second moment of the transverse particle displacement is limited by the opposing channel wall. Conversely, the second moment of the axial displacement will continue to increase without bound.

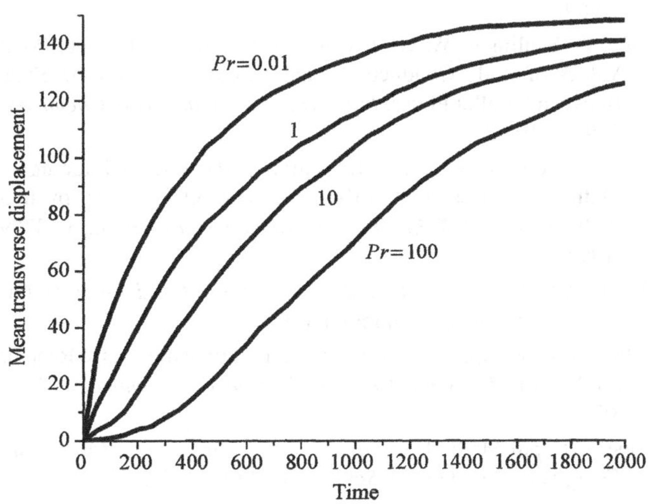

FIGURE 10.12. Mean transverse displacement of thermal markers released at the wall for the Prandtl numbers ranging from 0.1 to 100, as adapted from Papavassiliou and Hanratty (1995). Note how larger the Prandtl numbers inhibit the movement of the markers away from the wall.

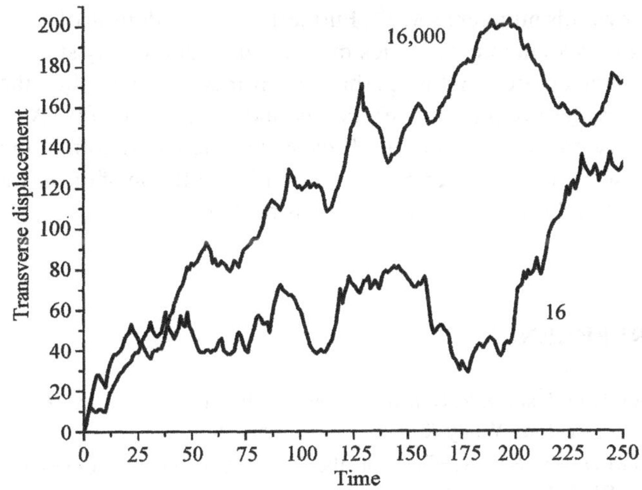

FIGURE 10.13. Individual trajectories (transverse displacement) for markers 16 and 16,000 for $Pr = 0.1$ from the simulation by Papavassiliou and Hanratty. This figure was adapted from their results and the axes have been reversed.

10.8 CONCLUSIONS

Although heat and mass transfer processes occurring in the steady turbulent flows in ducts can be modeled with elementary procedures, the challenges posed by the combination of chemical reactions with complex nonisothermal turbulent flows are immense. Moreover, experimental measurements in such cases are often quite difficult to obtain, making model validation or verification virtually impossible.

The unsatisfactory state of the art for turbulent reacting flows leads one to think about an attack based upon first principles, and the direct numerical simulation comes to mind. DNS has been applied to the homogeneous turbulent flows of fairly small Reynolds numbers. However, the addition of the continuity equation(s) for reacting scalars (concentration) greatly increases the complexity of the calculation. Fox (2003) notes that such efforts have been limited to the small Damköhler numbers; liquid-phase problems with fast chemistry are not feasible. We should point out some interesting observations regarding the direct numerical simulation of turbulent reacting flows made by Leonard and Hill (1988). They estimated that to merely save velocity vectors and three scalars for the construction of a 30 s animation sequence would require about 9×10^9 words (or about 36 GB) of storage. One can, of course, look at snapshots of the computed results but the evolution of the computed field(s) in time can often reveal aspects of flow structure not otherwise apparent.

We may hope for increased computational power, leading to better DNS and eliminating the need for closure approximations; those closure methods known to be based upon questionable physics will not be missed. However, we have previously noted that the number of required numerical operations (for turbulent flow simulations) scales with

Reynolds number as $Re^{9/4}$. Furthermore, the addition of more complex chemical kinetics may require significantly smaller characteristic lengths (perhaps even much smaller than the Kolmogorov microscale), compounding the difficulty. As a consequence, it is not at all clear that increased computing power alone can ever make the complete solution of turbulent heat and mass transport problems routine.

REFERENCES

Baldyga, J. and J. R. Bourne. *Turbulent Mixing and Chemical Reactions*, John Wiley & Sons, Chichester (1999).

Barlow, R. S. Overview of the TNF Workshop, International Workshop on Measurement and Computation of Turbulent Non-Premixed Flames, TNF8 (2006).

Bird, R. B., Stewart, W. E., and E. N. Lightfoot. *Transport Phenomena*, John Wiley & Sons, New York (1960).

Bird, R. B., Stewart, W. E., and E. N. Lightfoot. *Transport Phenomena*, 2nd edition, John Wiley & Sons, New York (2002).

Bourne, J. R. Mixing in Single-Phase Chemical Reactors. In: *Mixing in the Process Industries* (N. Harnby, M.F. Edwards, and A.W. Nienow, editors), 2nd edition, Butterworth-Heinemann, Oxford (1992).

Bourne, J. R. and H. L. Toor. Simple Criteria for Mixing Effects in Complex Reactions. *AIChE Journal*, 23:602 (1977).

Brodkey, R. S. and J. Lewalle. Reactor Selectivity Based on First-Order Closures of the Turbulent Concentration Equations. *AIChE Journal*, 31:111 (1985).

Cebeci, T. and P. Bradshaw. *Physical and Computational Aspects of Convective Heat Transfer*, Springer-Verlag, New York (1984).

Corrsin, S. On the Spectrum of Isotropic Temperature Fluctuations in an Isotropic Turbulence. *Journal of Applied Physics*, 22:469 (1951).

Corrsin, S. The Isotropic Turbulent Mixer: Part II. Arbitrary Schmidt Number. *AIChE Journal*, 10:870 (1964).

Dutta, A. and J. M. Tarbell. Closure Models for Turbulent Reacting Flows. *AIChE Journal*, 35:2013 (1989).

Einstein, A. *Annalen der Physik*, 17:549 (1905).

Eswaran, V. and S. B. Pope. Direct Numerical Simulations of the Turbulent Mixing of a Passive Scalar. *Physics of Fluids*, 31:506 (1988).

Fedotov, S., Ihme, M., and H. Pitsch. Stochastic Mixing Model with Power Law Decay of Variance. *CTR Annual Research Briefs*, 285 (2003).

Fox, R. O. The Fokker–Planck Closure for Turbulent Molecular Mixing: Passive Scalars. *Physics of Fluids*, A4:1230 (1992).

Fox, R. O. *Computational Models for Turbulent Reacting Flows*, Cambridge University Press, Cambridge (2003).

Hanratty, T. J. Heat Transfer Through a Homogeneous Isotropic Turbulent Field. *AIChE Journal*, 2:42 (1956).

J. O. Hinze. *Turbulence*, 2nd edition, McGraw-Hill, New York (1975).

Jayesh and Z. Warhaft. Probability Distribution, Conditional Dissipation, and Transport of Passive Temperature Fluctuations in Grid-Generated Turbulence. *Physics of Fluids*, A4:2292 (1992).

Kays, W. M. *Convective Heat and Mass Transfer*, McGraw-Hill, New York (1966).

Kim, J. J. and J. J. Baik. A Numerical Study of Thermal Effects on Flow and Pollutant Dispersion in Urban Street Canyons. *Journal of Applied Meteorology*, 38:1249 (1999).

Kontomaris, J., Hanratty, T. J., and J. B. McLaughlin. An Algorithm for Tracking Fluid Particles in a Spectral Simulation of Turbulent Channel Flow. *Journal of Computational Physics*, 103:231 (1992).

Leonard, A. D. and J. C. Hill. Direct Numerical Simulation of Turbulent Flows with Chemical Reaction. *Journal of Scientific Computing*, 3:25 (1988).

Leslie, D. C. *Developments in the Theory of Turbulence*, Clarendon Press, Oxford (1983).

Libby, P. A. and F. A. Williams, editors. *Turbulent Reacting Flows*, Academic Press, London (1994).

Lin, H. T. The Analogy Between Fluid Friction and Heat Transfer of Laminar Forced Convection on a Flat Plate. *Warme- und Stoffubertragung*, 29:181 (1994).

Manomaiphiboon, K. and A. G. Russell. Evaluation of some Proposed Forms of Lagrangian Velocity Correlation Coefficient. *International Journal of Heat and Fluid Flow*, 24:709 (2003).

R. C. Martinelli. Heat Transfer to Molten Metals. *Transactions of the ASME*, 69:947 (1947).

O'Brien E. E. The Probability Density Function (pdf) Approach to Reacting Turbulent Flows. In: *Turbulent Reacting Flows* (P.A. Libby, and F.A. Williams, editors). Springer-Verlag, Berlin (1980).

Orszag, S. A. and L. C. Kells. Transition to Turbulence in Poiseuille and Plane Couette Flow. *Journal of Fluid Mechanics*, 96:159 (1980).

Page, F., Schlinger, W. G., Breaux, D. K., and B. H. Sage. Point Values of Eddy Conductivity and Viscosity in Uniform Flow Between Parallel Plates. *Industrial and Engineering Chemistry*, 44:424 (1952).

Papavassiliou, D. V. and T. J. Hanratty. The Use of Lagrangian Methods to Describe Turbulent Transport of Heat from a Wall. *Industrial & Engineering Chemistry Research*, 34:3359 (1995).

Patankar, S. V. *Numerical Heat Transfer and Fluid Flow*, Hemisphere Publishing, Washington (1980).

Patterson, G. K. Application of Turbulence Fundamentals to Reactor Modeling and Scaleup. *Chemical Engineering Communications*, 8:25 (1981).

Pope, S. B. pdf Methods for Turbulent Reacting Flows. *Progress in Energy and Combustion Science*, 11:119 (1985).

Pope, S. B. *Turbulent Flows*, Cambridge University Press, Cambridge (2000).

Prandtl, L. Eine Beziehung zwischen Warmeaustausch und Stromungswiderstand der Flussigkeiten. *Zeitschrift für Physik*, 11:1072 (1910).

Rayleigh, Lord. On the Suggested Analogy Between the Conduction of Heat and Momentum During the Turbulent Motion of a Fluid (with an Appendix by T. E. Stanton). *Technical Report of the British Aeronautical Research Committee,* 497 (1917).

Richardson, L. F. Some Measurements of Atmospheric Turbulence. *Philosophical Transactions of the Royal Society of London A*, 221:1 (1921).

Risken, H. *The Fokker–Planck Equation*, 2nd edition, Springer-Verlag, Berlin (1989).

Schlien, D. J. and S. Corrsin. A Measurement of Lagrangian Velocity Autocorrelation in Approximately Isotropic Turbulence. *Journal of Fluid Mechanics*, 62:255 (1974).

Stanton, T. E. On the Passage of Heat Between Metal Surfaces and Liquids in Contact with Them. *Transactions of the Royal Society*, 190A:67 (1897).

Subramaniam, S. and S. B. Pope. A Mixing Model for Turbulent Reactive Flows Based on Euclidean Minimum Spanning Trees. *Combustion and Flame*, 115:487 (1998).

Taylor, G. I. Diffusion by Continuous Movements. *Proceedings of the Royal Society of London A*, 151:421 (1921).

Toor, H. L. Mass Transfer in Dilute Turbulent and Non-Turbulent Systems with Rapid Irreversible Reactions and Equal Diffusivities. *AIChE Journal*, 8:70 (1962).

Toor, H. L. Turbulent Mixing of Two Species with and without Chemical Reactions. *Industrial & Engineering Chemistry Fundamentals*, 8:655 (1969).

von Karman, T. The Analogy Between Fluid Friction and Heat Transfer. *Transactions of the ASME*, 61:705 (1939).

Wang, H. and Y. Chen. PDF Modeling of Turbulent Non-Premixed Combustion with Detailed Chemistry. *Chemical Engineering Science*, 59:3477 (2004).

11

TOPICS IN MULTIPHASE AND MULTICOMPONENT SYSTEMS

11.1 GAS–LIQUID SYSTEMS

11.1.1 Gas Bubbles in Liquids

Multiphase processes involving gases and liquids are ubiquitous in the chemical process industries, and our intent is to introduce a few important basics. Let us begin with the bubble behavior in liquids, which will be prominently affected by surface tension σ. A bubble surrounded by liquid will have an elevated equilibrium pressure that is described by the Laplace equation:

$$P_i - P = \frac{2\sigma}{R}. \tag{11.1}$$

For the air–water interface, σ is about 72 dyn/cm (0.072 N/m). Small bubbles yield large pressure differences; for an air bubble in water with $R = 0.02$ cm, $\Delta p = 7200$ dyn/cm^2 or about 7 cm of water. As R diminishes, P_i can become very large indeed. To illustrate, Polidori et al. (2009) observe that a CO_2 bubble will begin to rise in champagne when its diameter reaches about 10–50 μm. At 20 μm, (11.1) indicates a pressure difference of about 92,000 dyn/cm^2 (recall that ethanol lowers the surface tension in aqueous systems).

Now consider the pair of photographs illustrating jet aeration in Figure 11.1; air bubbles are being introduced into a water jet issuing into an acrylic plastic tank. In Figure 11.1a, the airflow rate has been increased by a factor of 2.5.

Observe the variety of bubble sizes and shapes apparent in Figure 11.1; many of the smaller bubbles are (nearly) spherical, while the slightly larger bubbles might be better described as ellipsoidal. At the higher gas rate (the bottom image), there are many larger bubbles that have formed by coalescence, some being quite near the edge of the jet. The regimes of bubble shapes (for bubbles rising through liquids) can be characterized with three dimensionless parameters, Reynolds number, Morton number, and Eotvos (approximately pronounced Ert-versh) number:

(a)

(b)

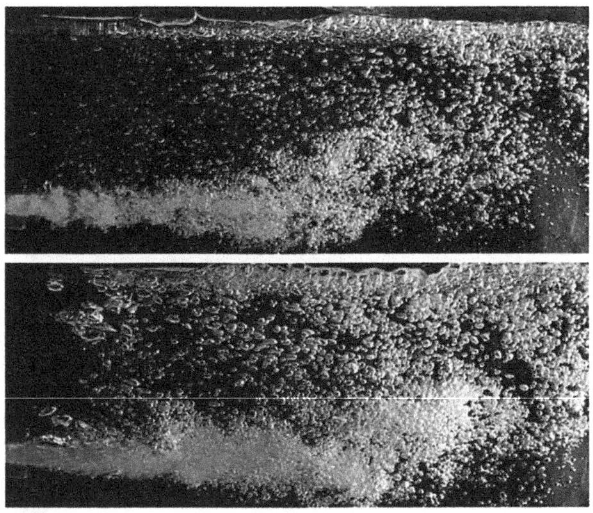

FIGURE 11.1. Air bubbles produced by jet aeration in water. The gas rate in the lower image is 2.5 times larger than in the upper photo; note the appearance of the larger bubbles at the elevated airflow rate (images courtesy of the author).

$$Re = \frac{dV\rho}{\mu}, \quad Mo = \frac{g\mu^4\Delta\rho}{\rho^2\sigma^3}, \quad \text{and} \quad Eo = \frac{gd^2\Delta\rho}{\sigma}. \tag{11.2}$$

Transport Phenomena: An Introduction to Advanced Topics, By Larry A. Glasgow

We are, of course, familiar with *Re*. The Morton number incorporates inertial, gravitational, viscous, and surface tension forces and the Eotvos number (also known as the Bond number *Bo*) compares buoyancy and surface tension. Consider a force balance made upon a *small* spherical bubble rising through a quiescent liquid; we isolate velocity (*V*) and the drag coefficient (*f*) such that

$$V^2 f = \frac{8}{3} Rg \frac{(\rho_f - \rho_b)}{\rho_f}. \tag{11.3}$$

In the case of an air bubble with a diameter of 1 mm rising through water at 25°C,

$$V^2 f = \left(\frac{8}{3}\right)(0.05)(980)\frac{(0.9971 - 0.00118)}{0.9971}$$
$$= 130.5\,\mathrm{cm}^2/\mathrm{s}^2.$$

The reader may wish to verify that the terminal rise velocity of this 1 mm bubble would be about 12 cm/s, yielding a Reynolds number of 120. *However, the reader is also cautioned that as the Reynolds number approaches about 100, the drag coefficient may deviate significantly from that of a rigid sphere.* In fact, at a Reynolds number of 100, Haberman and Morton (1953) found that the drag coefficient ranged over nearly an order of magnitude, depending upon the Morton number (the *Mo* for their data ranged from 1×10^{-2} to 2×10^{-11}). The Morton and Eotvos numbers for our example above are, respectively, $Mo = 2.6 \times 10^{-11}$ and $Eo = 0.136$. These values correspond to the spherical shape regime according to the map provided by Clift et al. (1978) (p. 27). If we were to somehow maintain *Re* but increase *Eo* to about 0.5, we would find ellipsoidal (or wobbling ellipsoidal) bubble shapes. The bubble size and shape profoundly affect terminal rise velocity; extensive experimental data have been obtained by Haberman and Morton and their results have been adapted and presented graphically (Figure 11.2). Note that for the usual range of air bubble sizes seen in water, the rise velocities will be on the order of 10–30 cm/s. We also need to be aware of the fact that the presence of surface-active contaminants can dramatically reduce the rise velocity, in some cases by a factor of 2 or more.

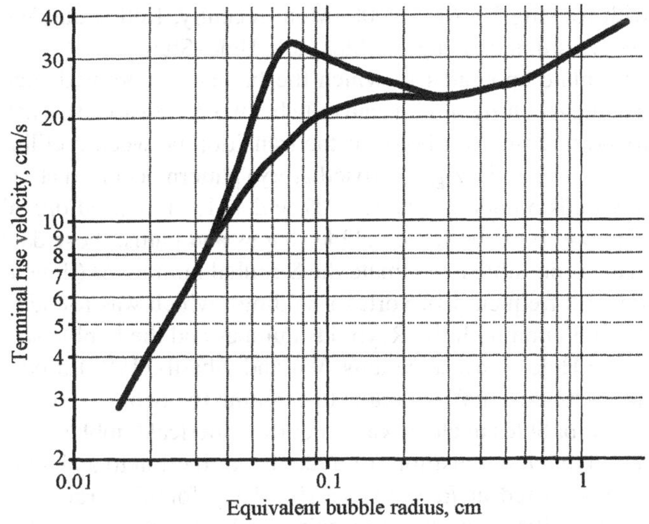

FIGURE 11.2. Approximate envelope for terminal rise velocities of air bubbles in water at 20°C as adapted from Haberman and Morton (1953). The upper bound corresponds to distilled water and the lower bound is for tap (contaminated) water.

The shapes of rising bubbles are categorized (in order of increasing size) as spherical, ellipsoidal, spherical cap, and skirted spherical cap. In addition, rising bubbles can exhibit wobbling or oscillatory behavior depending upon the relative velocity and the nature of the flow in their wake. Fan and Tsuchiya (1990) produced a wonderful monograph that describes the relationships between the rising bubble behavior and the flow about the bubble and in its wake. They note that the increased pressure at the stagnation point at the top of the bubble and the decrease in local pressure as the liquid flows around the object result in changes in curvature, which we can see immediately in a qualitative way by examining the Laplace equation (11.1). Accordingly, we can at least roughly interpret the transition from spherical to ellipsoidal shapes. However, as Fan and Tsuchiya note, the variation in dynamic pressure alone does not explain the appearance of spherical cap bubbles; to grasp how this shape emerges (and changes) for larger bubbles, we must consider the effect of recirculation both in the wake and in the interior of the bubble. We should also note that bubble shape (and behavior) is dynamically influenced by vortex shedding (at larger *Re*).

Let us think about recirculation in the wake in the following way: Suppose a larger nominally spherical bubble begins rising through a viscous liquid. A toroidal vortex forms in the immediate wake and it is fixed (i.e., remains stationary with respect to the gas–liquid interface at the bottom of the bubble). The flow pattern in that vortex will be outward (radially directed) along the bottom of the bubble, downward directed at the outside edge, and upward directed near the center. The result will be a tendency to pull the interface down at the outside edge, and push the interface up near the bottom center. The effects of this liquid flow pattern may be reinforced by recirculation inside the bubble as well. The net result is a spherical cap (or skirted cap shape). The transition from ellipsoidal to spherical cap shape occurs at a Weber number of about 20, as indicated by the extensive data of Haberman and Morton (1953).

Rising bubbles are also influenced by vortex shedding at the sufficiently large Reynolds numbers. Haberman and Morton identified three different types of motion for rising bubbles: a rectilinear path for cases in which $Re < 300$, a spiral motion for $300 < Re < 3000$, and a rectilinear motion with rocking for $Re > 3000$. More recently, Kelley and Wu (1997) studied rising bubbles in a Hele–Shaw cell (a parallel plate apparatus in which the bubble is confined such that the resulting motion can only be two dimensional). They found that the threshold for the transition between rectilinear motion and a zig-zag (oscillatory) pattern occurred at the Reynolds numbers between 137 and 171. They used digital imaging to get both the bubble shapes and paths; these data made it possible to estimate the Strouhal number St (dimensionless frequency of vortex shedding), which was found to depend upon both the Reynolds number and the bubble size in the Hele–Shaw apparatus. Wu and Gharib (1998) used a three-dimensional apparatus to examine the behavior of rising air bubbles in clean water. For the spherical bubbles, they found that the transition from rectilinear motion to a zig-zag path occurred at $Re = 157$ (± 10). They found a transition from rectilinear motion to a spiral pathway that occurred at $Re = 564$ (± 10) for the ellipsoidal bubbles. For the spherical bubbles, they found Strouhal numbers ranging from about 0.08 to 0.12 for Reynolds numbers ranging from 200 to about 600, respectively.

11.1.2 Bubble Formation at Orifices

Bubble formation has been intensively studied because of its practical importance to the process industries. One critical application is in biochemical reactors (or fermentors) where bubbles are sparged into the liquid to provide both oxygen and mixing. Clift et al. (1978) reviewed earlier work that had been carried out for the bubble formation under both the constant flow and constant pressure conditions. They noted that bubble formation at orifices is disconcertingly complex, with bubble volume depending upon perhaps 10 or more parameters. An extremely important effect is *tied to the volume of the chamber or reservoir immediately upstream from the orifice.* If this gas volume is large relative to the bubble volume, then the variation in gas flow does not affect chamber pressure. At the low gas flow rates, bubble volume is independent of gas flow; at intermediate rates, bubble volume increases but the frequency of formation is nearly constant. At the higher gas flow rates (characteristic of many industrial processes), bubble breakage and coalescence events may occur in proximity to the orifice.

Some experimental results obtained for air bubble formation (in distilled water) at a single, 1 mm diameter orifice are shown in Figure 11.3. In this work, hole pressure was measured as a function of time; at very low flow rates, bubble formation was intermittent, with a sequence of four or five bubbles forming over a time span of about 300 ms, followed by a period of inactivity of comparable duration. At slightly larger (but still low) gas rates, bubble formation was purely periodic, occurring at a frequency of about 32 or 33 Hz, as indicated in Figure 11.3b. At modest flow rate, the frequency of the pressure fluctuations is just slightly

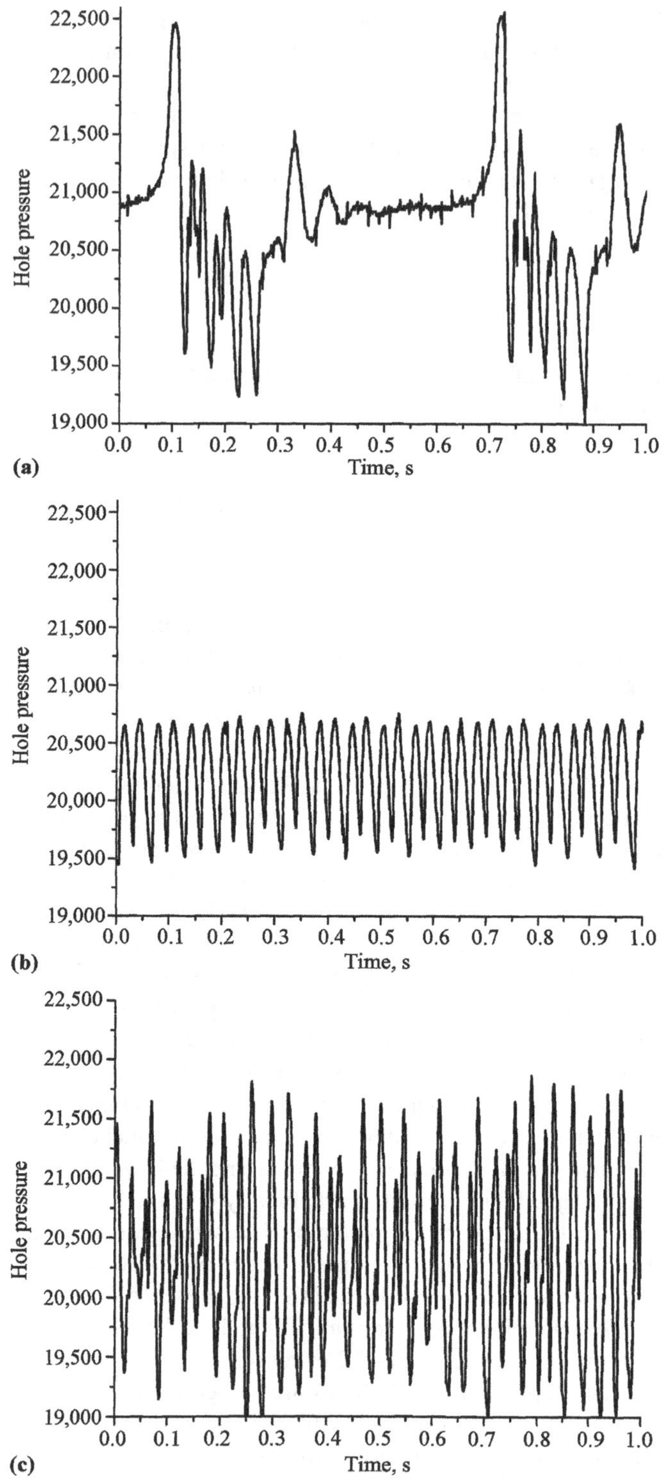

FIGURE 11.3. Hole pressure (dyn/cm^2) measured for the formation of air bubbles at a 1 mm diameter orifice using distilled water for low (a), intermediate (b), and modest (c) gas flow rates. These data underscore the startling complexity of bubble formation at orifices.

higher (about 40 Hz), the mean amplitude of the pressure oscillations is doubled, and the signal is considerably more complicated.

It is useful to consider the information that might be revealed by the phase space portraits of the dynamic behaviors evident in Figure 11.3. In the case of the intermediate gas flow rate (Figure 11.3b), it is clear that a plot of dp/dt against $p(t)$ will exhibit the limit-cycle behavior. At the low flow rates (Figure 11.3a), the phase space portrait will have several distinct lobes, a larger one corresponding to the formation of the initial bubble, with smaller features associated with the subsequent bubble train and recovery. We will return to this general topic (the dynamical behavior of nonlinear systems) in Section 11.1.3.

Numerous efforts have been made to model the bubble formation process. The usual starting point is the Rayleigh–Plesset equation (which we will describe in detail in the next section); for the examples of bubble formation modeling, see Kupferberg and Jameson (1969) and Marmur and Rubin (1976). Unfortunately, completely satisfactory modeling of the bubble formation process has proven elusive for the following reasons: (1) At the higher gas rates, the flow through the orifice is turbulent. (2) The shape of the forming bubble may not be spherical. (3) The flow induced in the liquid phase may be turbulent. (4) Inertial forces in the gas may be important. Note that of the difficulties listed above, (2) is especially problematic. As an initially spherical bubble grows, buoyancy overwhelms surface tension and the base of the bubble necks down (a tail forms). At the instant of detachment from the orifice, the bubble may be quite elongated (vertically). Many modelers have struggled with this aspect of bubble formation, and some have resorted to the use of an empirical detachment criterion as a consequence.

Let us elaborate a little on the difficulties associated with bubble formation modeling. Figure 11.4 (a single frame from a high-speed video recording made at 1000 fps) shows air bubbles immediately above a sparger plate with a single, 0.51 mm diameter orifice. The liquid phase is an aqueous solution of glycerol (50%, with a viscosity of 6 cp and a surface tension of 69.9 dyn/cm). The shapes of the bubbles in this sequence are to be noted and particular attention should be paid to the bubble at the bottom of the image, which is about to detach and leave the sparger plate. The dramatic elongation seen at the top of this bubble is characteristic of bubble formation (at low gas rates) in viscous liquid media when the forming bubble is affected by the departure of an immediately preceding one. The point, of course, is that bubbles rarely form in isolation; the formation of a single spherical bubble in process applications would be quite unusual.

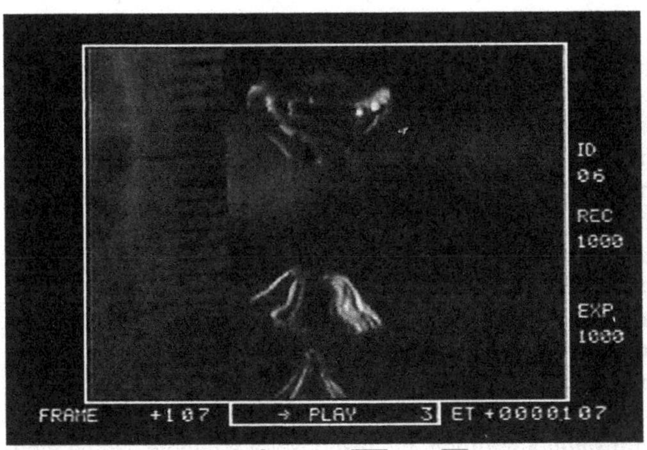

FIGURE 11.4. Single frame from a high-speed (1000 fps) recording of air bubble formation in a 50% solution of glycerol (image courtesy of the author).

11.1.3 Bubble Oscillations and Mass Transfer

We turn our attention to an individual gas bubble, surrounded by a liquid of infinite extent. We envision a process by which the bubble oscillates in response to an applied disturbance. These oscillations take two general forms: pulsation with spherical symmetry (sometimes referred to in the literature as the "breathing" mode), and shape oscillations that include what are known as Faraday waves. The latter result from the application of a driving force with sufficient amplitude; for more details, see Leighton (1994) and Birkin et al. (2001). Birkin et al. provide a remarkable photograph of surface (Faraday) waves on a large (about 4.5 mm) tethered bubble; the 15-point symmetry around the periphery of the bubble is striking. Maksimov and Leighton (2001) observe that the greatest shape distortions occur when the frequency of the driving force (an acoustic field) matches the resonant frequency of the bubble. The frequency of the resulting surface waves then approaches *one-half* of the frequency of simple spherical pulsation. This is confirmed by extensive experimental data, including mass transfer measurements.

Let us now focus upon the "breathing" mode (spherical pulsation). Consider a spherical bubble of mean radius R that is subjected to a disturbance. Lamb (1932) shows that by neglecting viscosity of the liquid and the density of the gas, the Laplace equation can be used to obtain

$$\omega^2 = (n+1)(n-1)(n+2)\frac{\sigma}{\rho R^3}. \tag{11.4}$$

The most important mode of vibration corresponds to $n = 2$, so the frequency of oscillation (in Hz) is given by

$$f = \frac{\sqrt{3}}{\pi}\sqrt{\frac{\sigma}{\rho R^3}}. \tag{11.5}$$

Let us now suppose that we are concerned with an air bubble surrounded by water. In this case, $\sigma = 72$ dyn/cm and

$\rho = 1$ g/cm^3; we find the following:

R (cm)	f (Hz)
0.0125	3347
0.025	1183
0.05	418
0.1	148
0.2	52

Note that these frequencies are all in the acoustic range. Indeed, the topics we are discussing can be characterized as subsets of the field, *acoustic cavitation*. Bubbles certainly are noisy as confirmed by everyday experience, and we can expect them to respond (and perhaps resonate energetically) to sound waves of suitable frequency. Readers interested in bubble-generated noise should consult the original work of Minnaert (1933), and those interested in the history of cavitation problems in marine propulsion should explore the career of Sir Charles Algernon Parsons (the story of the development of the turbine-powered *Turbinia* is fascinating).

In 1917, Rayleigh published his derivation of a model for the pressure developed in a liquid resulting from cavity collapse. We now retrace his analysis (Rayleigh, 1917). Let R be the radius of the spherical cavity and u be the velocity of the fluid *outside* the cavity. The total kinetic energy is then

$$\frac{1}{2}\rho\int_R^\infty 4\pi r^2 u^2 dr. \qquad (11.6)$$

The velocity of the fluid can be related to the velocity of the cavity's boundary (U) since $u/U = R^2/r^2$. Therefore, the kinetic energy integral (11.6) is simply $2\pi\rho U^2 R^3$. This kinetic energy is set equal to the work done by the motion, $(4\pi P/3)(R_0^3 - R^3)$, noting that $U = dR/dt$:

$$\frac{dR}{dt} = \sqrt{\frac{2P}{3\rho}\left(\frac{R_0^3}{R^3} - 1\right)}^{1/2}. \qquad (11.7)$$

We observe from eq. (11.7) that as the radius of the cavity becomes very small, the velocity of the cavity's surface, U, becomes *very* large. Rayleigh noted that this was unphysical, so he subtracted the work of compression (assuming that gas filled the cavity and that the compression was isothermal) such that

$$\frac{dR}{dt} = \sqrt{\frac{2}{\rho}\left[\frac{P}{3}\left(\frac{R_0^3}{R^3} - 1\right) - P_i\frac{R_0^3}{R^3}\ln\left(\frac{R_0}{R}\right)\right]}^{1/2}. \qquad (11.8)$$

If we set $U = 0$ and let $\alpha = R_0/R$, then

$$\frac{P}{P_i} = \frac{3\ln\alpha}{(1 - (1/\alpha^3))}. \qquad (11.9)$$

The ratio of the pressures can then be calculated by assuming values for α , and a few numerical results are given in the table that follows:

$\alpha = R_0/R$	Ratio of Pressures, P/P_i
0.25	0.06601
0.5	0.29706
0.75	0.62979
2	2.3765
4	4.2249
8	6.2505
16	8.3198
32	10.3975

Rayleigh's analysis included three major simplifications; he neglected both the surface tension and the viscosity of the liquid phase and assumed that the pressure at a distance was constant. Plesset (1949) adapted Rayleigh's work to include surface tension; the governing equation (which is the starting point for many investigations of dynamic bubble behavior) is now known as the Rayleigh–Plesset equation:

$$\frac{P_i - P_\infty}{\rho} = R\frac{d^2R}{dt^2} + \frac{3}{2}\left(\frac{dR}{dt}\right)^2 + \frac{4\nu}{R}\frac{dR}{dt} + \frac{2\sigma}{\rho R}. \qquad (11.10)$$

We should make note of some of the more important assumptions used to develop the Rayleigh–Plesset equation:

1. We have a single bubble in an infinite liquid medium.
2. The bubble is spherical for all t.
3. R is small compared to the acoustic wavelength.
4. There are no additional body forces.
5. The density of the liquid is large but its compressibility is small.

This nonlinear second-order ordinary differential equation can be solved to obtain $R(t)$ if the dynamic behavior of the pressure difference is known or specified. We note, however, that the Rayleigh–Plesset equation exhibits some intriguing features; as one might expect with a nonlinear differential equation, there is a rich array of behaviors only partially explored. Such efforts are complicated by the fact that we are unable to use analytic solutions for guidance; the few that are known have dealt with highly simplified cases—see, for example, Brennan (2005).

For the cases in which the external pressure oscillates with *small* amplitude, the response of a bubble can be modeled with the linearized approximation, as described by Prosperetti (1982). The radius of the bubble is taken as $R(t) = R_0(1 + X(t))$ and $X(t)$ can be described with the familiar (see Chapter 1) oscillator equation:

$$\frac{d^2X}{dt^2}+2\beta\frac{dX}{dt}+\omega_0^2X=\frac{P_\infty}{\rho R_0^2}e^{i\omega t}. \qquad (11.11)$$

The damping factor β is a function of frequency and for a gas–vapor bubble in water, $\beta \sim 10^5\ \mathrm{s}^{-1}$. ω_0 is the natural frequency of the bubble. There are several factors that contribute to the damping of the bubble oscillations, including heat and mass transfer and the viscosity of the liquid phase. Thermal effects can be particularly important for *cavitation* bubbles where, as Plesset (1949) observed, the vapor in the bubble comes from a localized phase change. Consequently, the thermal energy requirement for cavitation bubble formation can be estimated:

$$Q_{\mathrm{req}}=\frac{4}{3}\pi R^3\rho_V\Delta H_V, \qquad (11.12)$$

where ρ_V is the density of the vapor. Let us illustrate with a simple calculation. Suppose a cavitation bubble in water grows to $R=2$ mm in about 0.002 s. We will take the vapor density to be about 0.00074 g/cm^3. Therefore, Q_{req} is about 0.0143 cal. The thermal energy required for generation of the bubble must be extracted from a layer of immediately adjacent liquid water. We can obtain a crude estimate for the thickness of this layer: $\delta \approx \sqrt{\alpha t}$, where α is the thermal diffusivity of water, about 0.00145 cm^2/s. Thus, $\delta \approx 0.0017$ cm, and the mean temperature *decrease* for this immediately adjacent water layer is about 16°C. This local disparity in temperature creates opportunity for significant heat transfer *from* the bubble *to* the liquid. See Prosperetti (1977) and Plesset and Prosperetti (1977) for further discussion of thermal effects (and the relationship to the damping factor) and the impact of mass transfer upon bubble behavior.

Since the Rayleigh–Plesset equation must be solved numerically, we should take a moment to discuss the problem this presents. Let us begin by noting the variations in magnitude of the coefficients on the right-hand side of eq. (11.10). It is clear that we can expect the usual difficulties posed by *stiff* differential equations. You may recall that stiffness arises from an incompatibility between the eigenvalues and the time-step size. We can think of this in the following way: A stiff system has a *very broad* distribution of time constants; in order to resolve the behavior of the system at large times, we must use a *very small* step size. This in turn can lead to amplified round-off or truncation errors. Furthermore, whatever integration procedure is used, it must exhibit the required stability. For these reasons, explicit, forward marching techniques (like Runge–Kutta) are generally not very useful. Implicit or semi-implicit methods (including Rosenbrock, implicit Runge–Kutta, and backward difference) must be used. The reader with deeper interest in such problems should consult Hairer and Wanner (1996), Finlayson (1980),

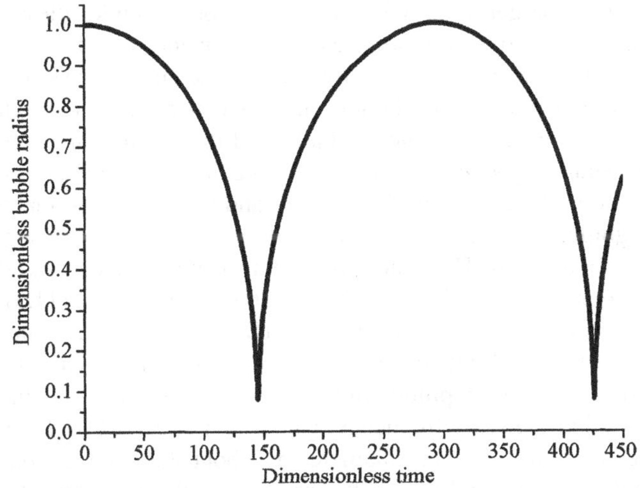

FIGURE 11.5. Computed results for the Borotnikova–Soloukhin example (Figure 11.7) in which a bubble is exposed to an instantaneous jump in pressure to 50 atm. Note that the compression phase bottoms out at about 8% of the initial radius. The dimensionless time is the product of the radian frequency ω and time t.

and Cash (1979). The RADAU5 (Fortran) code, using an implicit Runge–Kutta technique, has been made available for free distribution by Hairer and Wanner and a backward difference method (or BDM) code was provided by Scraton (1987).

Let us now use (11.10) to see how a bubble responds to an applied disturbance. We will numerically explore a case reported by Borotnikova and Soloukhin (1964) in which a bubble, initially at rest, is subjected to an instantaneous increase in external pressure (a step function with a height of 50 atm). We anticipate seeing initial compression, followed by rebound, with periodic repetitions. Following Borotnikova and Soloukhin, we will neglect surface tension and assume that the internal gas compression is adiabatic. The bubble's response, in terms of dimensionless variables, is shown in Figure 11.5. One can gain greater appreciation for the wide range of behaviors produced by the Rayleigh–Plesset equation (for a variety of disturbance types) by examining the other Borotnikova–Soloukhin results reported in Figures 1 through 7 of their paper.

We observed in the introduction to this chapter that many unit operations in chemical engineering practice involve mass transfer between gas bubbles and liquid media. Therefore, it is appropriate for us to think about characteristics of such systems that might be exploited to enhance the interphase transport. These features are, of course, apparent: We should focus upon interfacial area, concentration difference (driving force), and relative velocity. It has occurred to many investigators that pressure (bubble) oscillations might be used to both increase the interfacial area and create the interfacial movement (or relative velocity). See Waghmare (2008) for an overview of the use of vibrations to enhance mass transfer in multiphase systems.

Ellenberger and Krishna (2002), for example, clearly demonstrated the importance of low-frequency oscillation to both bubble size and mass transfer for the air–water system in a bubble column. The gas phase was introduced through a single capillary orifice initially and the oscillations were generated by sinusoidal motion of a flexible membrane at the bottom of the column. Ellenberger and Krishna found that a significant reduction in bubble size occurred at a frequency of about 70–80 Hz (where the mean bubble size decreased from about 3.6 to 2.2 mm). Note that according to eq. (11.5), $f = 61$ Hz if $d = 3.6$ mm and 128 Hz if $d = 2.2$ mm. Naturally, the amplitude of the oscillation also has a critical role. At 100 Hz, an amplitude of 0.001 mm did not affect bubble size, but the same frequency with an amplitude of 0.01 mm reduced mean bubble diameter by about 45%. Of course, both the gas holdup and the product of the mass transfer coefficient and the interfacial area are increased by the oscillations. Perhaps of even greater interest are the local maxima observed as the vibration frequency was increased. This effect was attributed to resonance resulting from reflection of the sinusoidal disturbances at the top of the gas–liquid dispersion.

Sohbi et al. (2007) examined the effect of pressure oscillations upon the absorption–reaction of carbon dioxide in a bubble column containing an aqueous solution of calcium hydroxide. They found, as expected, that the higher frequency pulsations decreased bubble size and increased mass transfer. The lower frequency pulsations did not improve mass transfer, although the authors did not report the amplitude of the oscillations, so it is impossible to generalize their results.

In addition to the enhanced mass transfer in devices such as bubble columns, it has been demonstrated that oscillation can also be used to advantage in electrochemical processes. Birkin et al. (2001) reported a study in which a 25 μm (diameter) Pt electrode could be positioned near a stationary bubble (trapped under a solid surface) in a solution of $Fe(CN)_6$ and $Sr(NO_3)_2$. The bubble was excited acoustically and the effects were detected electrochemically. A significant increase in mass transfer coefficient (to the microelectrode) was detected even at large distances (100×the electrode diameter).

Let us make some closing observations for this section. Though bubble oscillations have demonstrated effectiveness for enhancing interphase transport, there remains a principal difficulty with respect to exploration of the phenomenon: The increases in mass transfer are caused mainly by motions of the bubble surface, *dR/dt*. For small bubbles, these oscillations may be of high frequency and low amplitude, making *direct* observation quite difficult. Holt and Crum (1992) devised an experimental technique that makes use of the Mie scattering allowing them to directly measure even small motions of the bubble surface. They were able to obtain phase space portraits (*dR/dt* against *R*(*t*)) for air bubbles ranging in size (*R*) from about 50 to 90 μm, driven at frequencies of about 24 kHz. Their technique allowed *direct observation* of the transition between radial (spherical) and shape oscillations. Furthermore, they were able to demonstrate a "bursting" behavior (or intermittency) that accompanied larger amplitude driving pressures. Holt and Crum noted that such behavior is commonly observed in driven *nonlinear* systems. Naturally, the linearized model for bubble oscillations, eq. (11.11), cannot provide any insight into such behavior.

11.2 LIQUID–LIQUID SYSTEMS

11.2.1 Droplet Breakage

In this section, we turn our attention to the deformation and breakage of drops of one liquid suspended in another liquid. The two liquids are immiscible and their viscosities may be different; however, we are going to limit our discussion mainly to the case in which the densities of the liquids are similar. In this way we can eliminate the effects of buoyancy upon droplet deformation. This general subject matter is crucial to emulsification and solvent extraction.

Let us begin by contemplating how suspended droplets respond to highly ordered (laminar) flows. Although we do not expect the resulting phenomena to be of great importance to unit operations in the chemical process industries, they may assist us with our interpretation of the physics of more complicated situations. One of the most important investigations carried out in this context was the work of G. I. Taylor (1934); he devised a "four-roller" apparatus consisting of four cylinders (2.39 cm diameter) placed near the inside corners of a box filled with viscous syrup. The cylinders on one diagonal (upper left to lower right) rotated clockwise, and on the other diagonal counter-clockwise. The result was a hyperbolic flow field for which

$$v_x = Cx \quad \text{and} \quad v_y = -Cy. \qquad (11.13)$$

The value of C, of course, was determined by the speed of rotation of the cylinders. Positioned at the center of the apparatus, a deformable body would elongate horizontally and compress vertically (assuming an ellipsoidal shape with length L and height h). The extent of the deformation could be adjusted by changing the speeds of rotation of the cylinders. Any deviation in position of the droplet (the suspended entity) was countered by slight changes in the speeds of rotation of the cylinder(s). Taylor had a camera positioned to record the shapes of the droplets during the course of the experiments.

For a *slightly* deformed drop, the stress condition at the interface results in

$$P_i - P = \sigma\left(\frac{1}{R_1} + \frac{1}{R_2}\right) + c, \qquad (11.14)$$

where R_1 and R_2 are the radii of curvature. In his earlier work, Taylor (1932) found that for the flow in proximity to a suspended drop of viscosity μ_d,

$$P_i - P = \frac{1}{2} C\mu \frac{19\mu_d + 16\mu}{\mu_d + \mu}\left(\frac{x^2 - y^2}{A}\right) + c, \quad (11.15)$$

where A is the radius of the spherical drop. Taylor equated the pressure differences given by (11.14) and (11.15) and then found the shape of a slightly deformed drop for which the variation in $(1/R_1 + 1/R_2)$ is proportional to $(x^2 - y^2)/A^2$.

The resulting criterion was

$$\frac{1}{2} C\mu \frac{19\mu_d + 16\mu}{\mu_d + \mu} = \frac{4\sigma b}{A^2}. \quad (11.16)$$

The photographic record obtained in Taylor's experiments made it easy to measure the horizontal length (L) and the vertical height (h) of the deformed, ellipsoidally shaped drop. Since $(L - h)/(L + h) = b/A$, eq. (11.16) can be written as

$$\frac{L - h}{L + h} = \frac{2C\mu A}{\sigma}\left[\frac{19\mu_d + 16\mu}{16(\mu_d + \mu)}\right]. \quad (11.17)$$

Note that the quotient formed by the combination of viscosities will be nearly 1.0 even in cases where μ_d and μ differ substantially. Therefore, it is reasonable to write

$$\frac{L - h}{L + h} \cong \frac{2C\mu A}{\sigma} = F. \quad (11.18)$$

Taylor found that this relationship accurately represented the experimental results for the case in which $\mu_d/\mu = 0.9$ (and $\sigma \cong 8$ dyn/cm) until F exceeded about 0.3. Remember, the relationship (11.15) was developed for *small* deformations. The droplet (with an initial diameter of 1.44 mm) became highly elongated and burst as $F \cong 0.39$. Taylor's experiments were important because they provided the first quantitative study relating applied stress, deformation, and droplet breakage.

One must recognize that the hyperbolic flow field that Taylor employed, while very useful for droplet positioning, is not very much like the typical flows in which processes requiring droplet breakage are carried out. Naturally, we would like to know how a droplet responds to (more realistic) turbulent flow conditions. In particular, suppose a suspended entity encounters a thin shear layer perhaps associated with the flow ejected by a radial-discharge impeller in a stirred tank. It seems *very unlikely* that the deforming droplet will assume the ellipsoidal (and ultimately lenticular) shapes seen in Taylor's work. To illustrate the differences, let us examine the case in which a neutrally buoyant oil droplet, initially spherical, is allowed to enter a very strong shear layer formed by a turbulent jet issuing from a rectangular slot.

Single-frame, multiple flash photography was used to obtain a record of the entrainment–deformation–breakage process and examples are provided in Figure 11.6. The time interval between flashes for these two examples was 83 ms (0.083 s) and the Reynolds number of flow through the rectangular slot was about 1720, corresponding to an average velocity of 61.4 cm/s. The jet (water) issues from the wall on the left-hand side of the images and is horizontally directed.

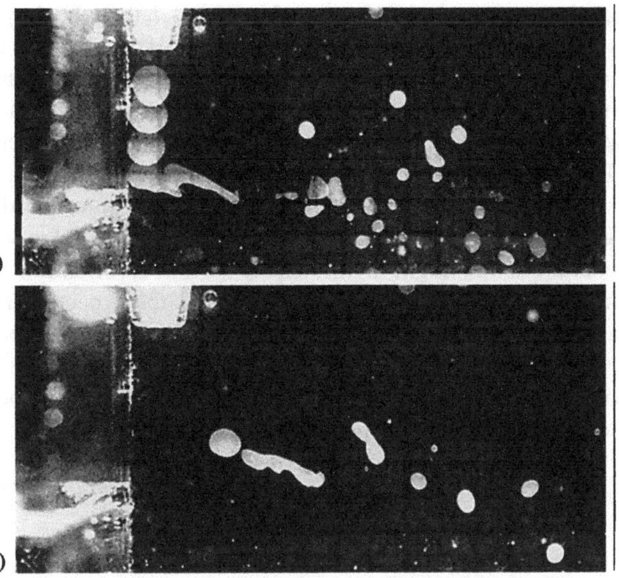

FIGURE 11.6. Examples of neutrally buoyant oil drops experiencing deformation and breakage through interaction with a thin shear layer. The oil viscosity at 25°C was 1.34 cp and the surface tension was 32.5 dyn/cm. The droplets were formed at a pipette tip and subsequently entrained in the horizontal jet (photos courtesy of the author).

In Figure 11.6b, the parent droplet diameter was 3.16 mm and its surface area was about 0.314 cm^2. As you can see, the surface area indicated by the deformed image (prior to breakage) was about 0.95 cm^2. The work performed against surface tension was about 20.6 dyn cm and this occurred in 0.083 s. The "wavy" deformation apparent on the underneath side of the (elongating) droplets as they begin to interact with the upper edge of the turbulent jet should be noted. The photographic evidence presented in Figure 11.6 provides the following picture: When a suspended entity or droplet encounters a strong shear layer (as generated by a turbulent jet), an extensional strain produces elongation of the parent drop. Because this is an inhomogeneous turbulent flow, eddies at the edges of the turbulent jet may act upon the elongating drop and produce additional localized deformations. Under severe conditions, a breakage event may produce many daughter droplets with a wide range of sizes.

Let us continue this discussion by looking at the idealized case for turbulent flows: the liquid droplet suspended in *homogeneous isotropic* turbulence. It is clear in this case that a definite relationship must exist between the entity (droplet) diameter d and the eddy size l if deformation and breakage are to occur. We envision a process in which the suspended droplet encounters a turbulent eddy. If $l \gg d$, then the droplet is merely entrained by the fluid motion. If $l \ll d$, then the droplet is quite unaffected by the encounter. It is reasonable to assume that the critical eddy, with respect to deformation, will be of a scale roughly comparable to the droplet diameter. Hinze (1955) observed that in this case, variations in dynamic pressure occurring at the surface of the droplet would lead to "bulgy" deformation that is, the droplet would develop protuberances that might in turn lead to further deformation and breakage. Naturally, if one could quantify the expected variation in dynamic pressure in a turbulent flow, then it would be possible to develop a breakage criterion. We presume that the critically sized eddies are in the inertial subrange where the Kolmogorov law applies:

$$E(\kappa) = \alpha\varepsilon^{2/3}\kappa^{-5/3}. \tag{11.19}$$

The energy of these eddies can be estimated as we discussed in Chapter 5:

$$[u(\kappa)]^2 \approx \alpha\varepsilon^{2/3}\kappa^{-2/3}, \text{ and since } \kappa \approx \frac{2\pi}{d}, \text{ we find}$$

$$[u(d)]^2 \approx \frac{1}{2}\varepsilon^{2/3}d^{2/3}. \tag{11.20}$$

It is reasonable to assume that breakage will occur when the dynamic pressure fluctuations exceed the restoring force arising from surface tension. Let us emphasize: We are talking about eddies small enough to create a dynamic pressure difference over a length scale corresponding to the drop diameter d. Accordingly, by rough force balance,

$$\frac{c\rho}{2}\varepsilon^{2/3}d^{2/3} \approx \frac{4\sigma}{d} \tag{11.21}$$

and

$$d_c \approx \left[\frac{8\sigma}{c\rho}\varepsilon^{-2/3}\right]^{3/5}. \tag{11.22}$$

Thus, we conclude that if the conditions of this analysis are met, then the stable droplet diameter should depend upon the dissipation rate per unit mass as $d_c \propto \varepsilon^{-2/5}$. A decrease in dissipation rate by a factor of 10 should yield an increase in stable droplet diameter by a factor of 2.5. The form of eq. (11.22) has appeared in equations developed and used by many investigators, for example, Hesketh et al. (1991) cite their result for the breakage of bubbles *and* drops in turbulent pipe flows:

$$d_c \approx \left(\frac{We_c}{2}\right)^{0.6}\frac{\sigma^{0.6}}{(\rho_c^2\rho_d)^{0.2}}\varepsilon^{-2/5}. \tag{11.23}$$

If Taylor's inviscid approximation ($\varepsilon \approx Au^3/l$) is used to replace the dissipation rate per unit mass, then $d_c \sim u^{-1.2}$. There is evidence that this particular power law form is not applicable in low-energy flows and some pipe flows (which are neither isotropic nor homogeneous). Rozentsvaig (1981) pointed out that the contribution of viscous shear may be significant to droplet breakage in pipe flows, and he modified the model in an attempt to reconcile it with the published experimental data.

There is also a *lower* limit to the size of droplets that can be formed in turbulence. Recall that the Kolmogorov microscale is given by $\eta = (\nu^3/\varepsilon)^{1/4}$ and the corresponding velocity scale is $v(\eta) = (\varepsilon\nu)^{1/4}$. If we form a Reynolds number with these quantities, we find $Re_\eta = 1$; the inertial forces associated with the *dissipative* eddies simply are not strong enough to produce droplet breakage. A more reasonable threshold can be established by requiring

$$Re = \frac{d_{\min}v(d)}{\nu} \approx 10, \tag{11.24}$$

which fixes the value of the velocity for a given droplet diameter. The variation of dynamic pressure over the droplet surface is set equal to the restoring force (per unit area) due to surface tension. Levich (1962) found that the resulting lower limit for droplet size is

$$d_{\min} \approx \frac{c\rho\nu^2}{\sigma}, \tag{11.25}$$

where c is on the order of 50–100. As a practical matter, it is difficult to produce droplets in a liquid–liquid comminution process that are much smaller than η.

It is appropriate for us to point out some of the limitations of the preceding analysis of stable droplet size. It has been observed by a number of investigators, including Kostoglou and Karableas (2007) that a "stable" droplet size may not really exist. Such observations are based upon the experimental fact that the drop size distribution may continue to change with time indefinitely. Why should this occur? First, the dissipation rate at particular locations fluctuates, and it is possible that some infrequent fluctuations could be *very* large. Furthermore, in many types of process equipment, the dissipation rate varies with position, for example, in stirred tank reactors it would not be unusual to find ε near the impeller blade tips to be $\sim$100$\times$ greater than the average value determined from the total power input to the tank. Finally, we note that the dynamic pressure fluctuations may (at certain spatial positions and at certain moments in time) greatly exceed our estimated *average* value obtained from eq. (11.20). Hence, the droplet size distributions in dispersion processes may continue to change, though slowly, for a very long time.

The literature of droplet breakage in turbulent flows is vast, and the interested reader is urged to consult the very extensive bodies of work produced by D. Ramkrishna (and coworkers), H. F. Svendsen (see Luo and Svendsen, 1996), N. R. Amundson (and coworkers), and L. L. Tavlarides (and coworkers).

11.3 PARTICLE FLUID SYSTEMS

11.3.1 Introduction to Coagulation

Coagulation is a process by which smaller, fluid-borne particles collide and affiliate to form aggregates. It is widely employed in solid–liquid separations (such as water and wastewater treatment and mineral processing), where colloidal particles are brought together under the influence of Brownian motion (and subsequently as growth occurs, by fluid motions) to produce larger entities that can be removed by sedimentation and/or filtration. Coagulation is also important in atmospheric phenomena, including the dynamic behavior of pollutant aerosols in urban areas, as well as the transport and fate of ash clouds from volcanic eruptions. In the chemical process industries, aerosol behavior figures prominently in spray-applied coatings, cooling tower operation, injection of fuel in burners (combustors), spray drying, and so on.

11.3.2 Collision Mechanisms

The behavior of systems of fluid-borne particles will be affected by the entity–entity collisions and the evolution of the particle size distribution (psd). It is essential, therefore, to understand the mechanisms and rates of coagulation processes occurring for suspended entities in moving fluids.

Following standard practice in the literature, the collision rate between particles of types i and j can be written as

$$N_{ij} = \beta(v_i, v_j)n_i n_j, \tag{11.26}$$

where β is the collision frequency function between particles of the corresponding volumes (v_i and v_j) and n_i is the number density of particles of type i. β has dimensions of cm^3/s. The entity–entity collision can be driven by thermal motion of the fluid molecules (Brownian motion), by fluid motion (both laminar and turbulent), and by differential sedimentation (requiring a difference in size or density).

The collision frequency function for Brownian coagulation was developed by Smoluchowski (1917). For aerosols, if the participating particle size is significantly larger than the mean free path of the gas molecules (≈ 0.06 μm in air at 0°C) and if the Stokes–Einstein diffusion coefficient is employed, then

$$\beta(v_i, v) = \frac{2kT}{3\mu}\left(\frac{1}{v_i^{1/3}} + \frac{1}{v_j^{1/3}}\right)\left(v_i^{1/3} + v_j^{1/3}\right). \tag{11.27}$$

For a monodisperse system (all entities have the same size), $v_i = v_j$, and then $\beta = 8kT/3\mu$. This is valid for the continuum regime where the Knudsen number (Kn) is less than 0.1. In this case, the initial rate of disappearance of particles is given by

$$\frac{dn}{dt} = -\frac{4kT}{3\mu}n^2. \tag{11.28}$$

A collision efficiency factor (λ) can be incorporated into eq. (11.28) to account for the possibility that not all collisions result in aggregate formation; see, for example, Swift and Friedlander (1964). Computed collision efficiencies in hydrosols have been compared by Kusters et al. (1997); for *solid* spherical entities, λ decreases sharply with the increasing particle size.

An attractive feature of (11.28) is that it is easily solved to yield

$$\frac{n}{n_0} = \frac{1}{(4kT/3\mu)n_0 t + 1}. \tag{11.29}$$

Thus, for example, we can estimate the time required for the number concentration of particles in an aerosol to be reduced to $n_0/2$ at 20°C:

Initial Concentration Per cm^3, n_0	$t_{1/2}$ (s)
1×10^8	33.6
1×10^7	336
1×10^6	3357

The *actual* rate of particle disappearance in aerosols will be affected by the breakdown of continuum theory (as very small particles approach each other), deviations from sphericity, and the consequences of electrical charge. Shahub and Williams (1988) reported that van der Waals, viscous, and electrostatic forces interact in a complex way and significantly alter the coagulation rate (from that predicted by classical theory). For electrostatic forces, weakly bipolar atmospheric aerosols yield a net effect that is nearly a wash. However, Friedlander (2000) indicates that a strongly charged (bipolar) aerosol will yield a greatly enhanced coagulation rate. The collision rate correction factor W (sometimes referred to as the Fuchs stability function) is given by

$$W = \frac{1}{y}(e^y - 1), \quad \text{where} \quad y = \frac{z_i z_j e^2}{\varepsilon_0 kT(R_i + R_j)}. \tag{11.30}$$

z is the number of charges on the colliding particles, e is the fundamental electrostatic unit of charge, and ε_0 is the dielectric constant of the medium (air: 1.0006). To illustrate, consider a hypothetical pair of 2 μm particles in air, each carrying 20 charges, but of opposite sign (please note that *small* particles with $d < 0.1$ μm cannot carry more than one charge). For this example, $y = -5.69$ and $W = 0.175$; the collision rate enhancement is $1/W$, which is a factor of 5.7. If, on the other hand, ions of like charge are preferentially adsorbed upon the particle surface, coagulation can be very effectively inhibited. Vemury et al. (1997) performed simulations on systems with (initially) symmetric bipolar charge distributions, as well as upon aerosols with asymmetric bipolar charging. They found that the rate of coagulation was increased in the symmetric case when the particles were highly charged. In the asymmetric case, the initial rate of disappearance of primary particles was greater, but this was attributed to the effects of electrostatic dispersion (in asymmetric charging, positive and negative charges do not balance, resulting in the transport of some particles to the walls of the confining vessel) rather than enhanced coagulation. For a discussion of how ionic additives (such as alkali metals) can be used to affect coagulation rates in aerosols, see Xiong et al. (1992).

We now deviate briefly from our discussion of collision mechanisms to discuss charge effects for particle interactions in *aqueous* systems. Many naturally occurring particulate materials, including clays, silica, and quartz, develop a negative surface charge when immersed in water. For clays, the negative surface charge arises from crystal imperfections. In other cases, a surface charge may be the result of preferential adsorption of specific ions; see van Olphen (1977) for amplification. The presence of the surface charge results in the formation of the double layer, an enveloping atmosphere of ions that can result in a repulsive force as two such particles approach each other. It is, of course, this mechanism that can give a hydrophobic colloid stability; it is possible to prepare a hydrosol that is stable for months, if not years. We should observe that the commonly used terms, hydrophobic and hydrophilic, are not appropriately descriptive. van Olphen notes that hydrophobic particles are in fact wet by water; thus, we should be a little concerned when we employ a term that implies that a particulate material "repels" water (or solvent).

This ionic atmosphere surrounding a charged entity is profoundly affected by both the charge and concentration of ions in solution. To better understand this, consider the Debye length, a measure of the thickness of this "atmosphere."

$$l_D = \left[\frac{4\pi e^2}{\varepsilon_0 kT}\sum n_i z_i^2\right]^{-1/2}. \tag{11.31}$$

In this equation, e is the unit of charge, ε_0 is the dielectric constant of the medium, k is the Boltzmann constant, and n and z are respectively the number concentration and charge of the ions in solution. Let us examine the effect of concentration of symmetric electrolytes upon the Debye length in Figure 11.7. We will note immediately that we can compress the double layer by adding an electrolyte to the solution; furthermore, this effect increases with the valence of the electrolyte. The reader interested in quantifying the effect of counterion valence upon coagulation should investigate the *Schulze–Hardy rule*. Note that compression of the double layer suppresses the repulsive interaction and increases the probability of permanent contact (aggregation) as two charged entities approach.

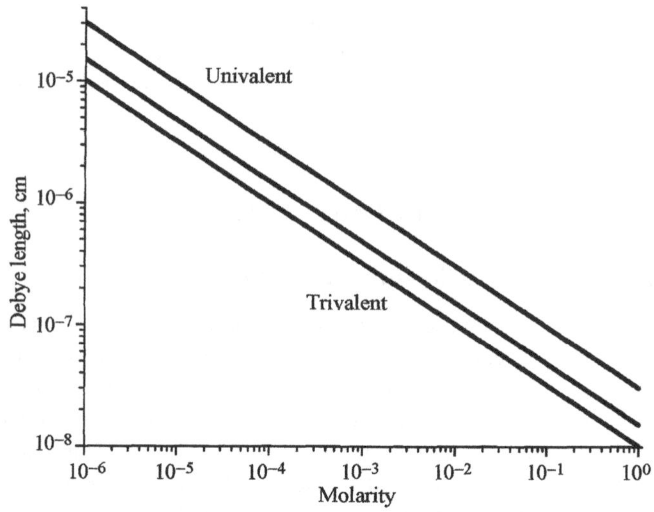

FIGURE 11.7. The Debye length for an aqueous solution of symmetric (uni-, di-, and trivalent) electrolytes as a function of concentration. Note that 10^{-8} cm is 1 Å.

Now consider what happens when the distance between two charged entities is reduced to the point where the double layers begin to interact. Of course, this has the effect of elevating the potential at intermediate points (between the approaching surfaces). For simplicity, we restrict our attention to parallel planar double layers. Please be aware that extensive computations have been performed and tabulated for this type of interaction by Devereux and de Bruyn (1963). The distribution of potential for approaching planar surfaces (separated in the y-direction) is governed by

$$\frac{d^2\psi}{dy^2} = \frac{4\pi e}{\varepsilon}\left[n^- z^- \exp\left(\frac{z^- e\psi}{kT}\right) - n^+ z^+ \exp\left(-\frac{z^+ e\psi}{kT}\right)\right]. \tag{11.32}$$

Let one charged surface be located at $y = 0$ and the other at $y = 2b$. We assume that the surfaces have the same potential ψ_0, although this is certainly not necessary. But selection of these boundary conditions ensures that the minimum potential will be located at $y = b$. Equation (11.32) is readily solved and some computed results are shown in Figure 11.8. We recognize immediately that a large surface potential combined with small separation distance results in a very steep $\psi(y)$; this is crucial, since the derivative of the potential is directly related to the pressure arising from the interaction of the two double layers as indicated by Overbeek (1952).

Let us make perfectly clear the intent of the immediately preceding discussion: We can reduce the barrier to particle–particle contact and aggregation either by compressing the double layer (through electrolyte addition) or by neutralizing the surface charge of the approaching particles.

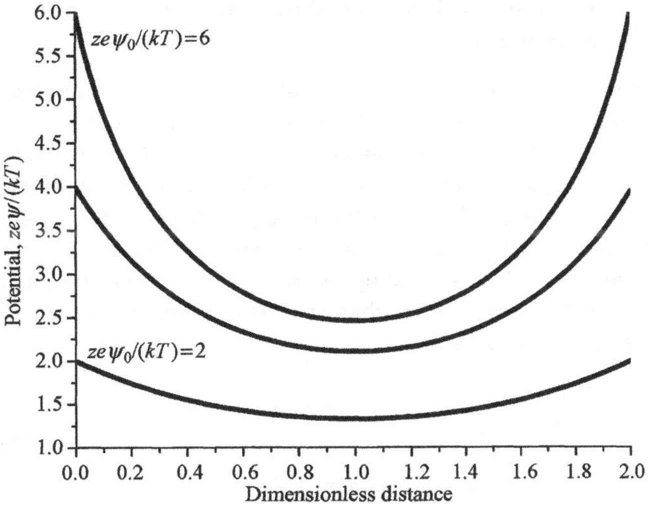

FIGURE 11.8. Distribution of potential between (equally) charged, parallel, planar surfaces, separated by a distance of $2b$. The surface charges $ze\psi_0/(kT)$ for the three curves are 2, 4, and 6.

In many practical applications we do both. The reader should also recognize that when we speak of *rapid* coagulation, we refer to a process in which the potential barrier has been removed, that is, every particle–particle encounter results in a permanent affiliation.

Now we are in a position to resume our discussion of collision mechanisms. Fluid motion can also drive interparticle collisions; in much of the older literature, this process is referred to as "othokinetic" flocculation. The collision frequency function for particles i and j in a laminar shear field with a velocity gradient dU/dz was derived by Smoluchowski (1917):

$$\beta(v_i, v_j) = \frac{4}{3}(R_i + R_j)^3 \frac{dU}{dz}. \tag{11.33}$$

And again, the rate of disappearance of monodisperse particles can be written as a simple ordinary differential equation (assuming that the dispersed-phase volume fraction $\phi = \pi d^3 n/6$ is constant):

$$\frac{dn}{dt} = -\frac{4\phi}{\pi}\frac{dU}{dz}n. \tag{11.34}$$

Note that the introduction of ϕ has rendered (11.34) linear with respect to particle number concentration n. This equation has been tested many times for hydrosols, usually in some type of Couette device with (nearly) uniform velocity gradient. For the concentric cylinder apparatuses, dU/dz can be assigned a single value that can be varied by changing the speed of the (outer) cylinder. Equation (11.34) is also easily integrated, yielding

$$n = n_0 \exp\left(-\frac{4\phi}{\pi}\frac{dU}{dz}t\right). \tag{11.35}$$

This result is, however, not likely to be of utility for many particulate systems for two reasons: Only rarely can the flow field in either aerosols or hydrosols be described as a simple laminar current, and in many cases, the dispersed-phase volume fraction is not constant (as small particles affiliate, fluid becomes trapped in the interstitial spaces of the structure).

Saffman and Turner (1956) developed the collision frequency function for small particles in *isotropic* turbulence:

$$\beta(v_i, v_j) = 1.3\left(\frac{\varepsilon}{\nu}\right)^{1/2}(R_i + R_j)^3. \tag{11.36}$$

ε is the dissipation rate per unit mass and ν is the kinematic viscosity of the fluid. Note the similarity of this equation to (11.33). A few words regarding the dissipation rate are in order. Recall from Chapter 5 that for isotropic turbulence, the dissipation rate is

$$\varepsilon = 2\nu s_{ij}s_{ij}, \tag{11.37}$$

where s_{ij} is the fluctuating strain rate. The strain rate is difficult to determine because it requires measurement of velocities with spatial separation. However, it is a critical parameter of turbulent flows; for a given fluid, it determines the eddy size(s) in the dissipation range of wave numbers. By definition, the wave number that corresponds to the beginning of the dissipation range (in the three-dimensional spectrum of turbulent energy) is $\kappa_d = 1/\eta$, where the Kolmogorov microscale is given by $\eta = (\nu^3/\varepsilon)^{1/4}$. Therefore, in air with $\varepsilon = 100\,\text{cm}^2/\text{s}^3$, $\eta \cong 0.077$ cm and $\kappa_d \cong 13\,\text{cm}^{-1}$; for water with the same dissipation rate, $\eta = 0.01$ cm and $\kappa_d = 100\,\text{cm}^{-1}$. Under normal laboratory conditions, the dissipation rate is often in the range of 10–$10^4\,\text{cm}^2/\text{s}^3$; in geophysical flows, ε can be *much* larger. The dissipation rate can also be estimated with Taylor's inviscid approximation: $\varepsilon \approx Au^3/l$. For pipe flows, Delichatsios and Probstein (1975) used the relation $\varepsilon \approx 4v^{*3}/d_{\text{pipe}}$, where v^* is the shear, or friction, velocity. This relationship for dissipation rate came from the experimental work carried out by Laufer (1954). In atmospheric turbulence, the dissipation rate is inversely proportional to height in neutral air: $\varepsilon = v^{*3}/K_a z$. For the unstable air, ε decreases with height near the surface, becoming constant near the top of the surface layer that is the lowest part of the planetary boundary layer. Panofsky and Dutton (1984) note that in daytime with strong winds, surface layer simplifications are valid to a height of about 100 m.

In cases where dispersed particles differ in size and mass, interparticle collision can also occur by turbulent inertia and by differential sedimentation. The collision frequency functions for these two cases respectively are

$$\beta(v_i, v_j) = 5.7\left(R_i^3 + R_j^3\right)\left|\tau_i - \tau_j\right|\frac{\varepsilon^{3/4}}{\nu^{1/4}}, \tag{11.38}$$

where τ is a characteristic time (mass of particle/$6\pi\mu R$), and

$$\beta(v_i, v_j) = \pi\alpha(R_i + R_j)^2(V_i - V_j), \qquad (11.39)$$

where V_i and V_j are the settling velocities of the particles. Unless (or until) there is a considerable disparity in the sizes of the particles, these collision mechanisms will be minor contributors to the processes of interest. In many particulate processes, we might expect (11.39) to become increasingly important with time, but of little significance initially. In the practical coagulation of hydrosols, (11.38) is not likely to be important; by the time a significant difference in mass develops, the entities have entered a quiescent region (a sedimentation zone) where the dissipation rate is very small. For the cases in which the difference in entity volumes is really large, the collision rate for differential sedimentation may be less than indicated by (11.39). Williams (1988) noted that the presence of large aggregates may distort the velocity field and affect the trajectories of approaching particles.

11.3.3 Self-Preserving Size Distributions

Swift and Friedlander (1964) and Friedlander and Wang (1966) developed a technique for solving certain types of coagulation problems based upon a similarity transformation. They observed that after long times, the solutions to such problems may become independent of the initial particle size distribution. Thus, $n(v, t) = (N^2/\phi)\psi(v/\bar{v})$, where $\bar{v}$ is the average particle volume. ψ is a dimensionless function that is invariant with time. The particle size distribution must also satisfy the following: $N = \int_0^\infty n(v, t)dv$, that is, the total number of particles must be obtained by integrating the distribution over all possible volumes. In addition, the dispersed-phase volume fraction can be determined:

$$\phi = \int_0^\infty n(v, t)v\,dv. \qquad (11.40)$$

Finally, it is usually taken that the distribution function is zero for both $v = 0$ and $v \to \infty$. Friedlander (2000) shows results for the Brownian coagulation case and also provides a comparison with experimental data obtained with a tobacco smoke aerosol. The agreement is reasonable. The principal problem with this technique is that while a transformation may be found for the collision kernel of interest, an appropriate solution may not necessarily exist.

An important question in this context is the length of time required for the size distribution to become self-preserving (T_c). Vemury et al. (1994) report that for the Brownian coagulation in the continuum regime the dimensionless time constant, τ_C was found to be on the order of 12–13; since $T_c = \tau_C/K_C n_0$ and $K_C = 2kT/3\mu$, one can estimate the time required given a specific medium and an initial number concentration of particles. For the air at 20°C with $n_0 = 1 \times 10^7$ particles per cm^3, $T_c \approx 8000$ s.

11.3.4 Dynamic Behavior of the Particle Size Distribution

Processes of the type being discussed here lend themselves to analysis by population balance. In the chemical process industries, population balances were first used for the analysis of crystal nucleation and growth by Hulburt and Katz (1964), among others. For many dispersed-phase processes, we can expect aggregation and aggregate breakage to occur simultaneously; in its simplest form for *aggregation only*, we describe the rate of change of (the number density of) particles of volume v as

$$\frac{dn(v)}{dt} = \frac{1}{2}\int_0^v \beta(v-u, u)n(v-u)n(u)du - n(v)\int_0^\infty \beta(v, u)n(u)du. \qquad (11.41)$$

The first term on the right-hand side corresponds to a birth (generation of particles with volume v) term due to encounters between particles with volumes smaller than v. The prefactor 1/2 is necessary to avoid double counting. The second term is a loss term arising from the growth occurring when particles of volume v affiliate with all (and any) other particles. If the hydrodynamic environment is such that the breakage of aggregates may occur, then two additional terms are necessary: one generation term due to the breakage of larger volume ($v \to \infty$) particles, and one loss term due to the breakage of particles of volume v. Even for the "apparently" simple problems, obtaining agreement between model and experimental data can be daunting. To illustrate, Ding et al. (2006) tested 16 different models (different size dependencies for aggregation and breakage) in their work on flocculation of activated sludge.

For aerosols, additional problems arise. In cases with charged particles, we can also expect electrostatic deposition (a process that is extremely important in painting and coating operations). Furthermore, small airborne particles will be carried about by eddies of all sizes (from integral to dissipative scales). In decaying and/or inhomogeneous turbulent flows, the general problem is quite intractable. Some alternative approaches will be discussed later. Friedlander (2000) notes that if the Reynolds decomposition and time averaging are employed with the general population balance for turbulent flows, the result is

$$\begin{aligned}\frac{\partial \bar{n}}{\partial t} + V\nabla\bar{n} + \frac{\partial}{\partial v}(\bar{n}\,\bar{q}) + \frac{\partial}{\partial v}\left(\overline{n'q'}\right) &= -\nabla\overline{n'V'} + D\nabla^2\bar{n} \\ &+\tfrac{1}{2}\int_0^v \beta(v^*, v-v^*)n(v^*)n(v-v^*)dv^* \\ &- \int_0^\infty \beta(v, v^*)n(v)n(v^*)dv^* \\ &+\tfrac{1}{2}\int_0^v \beta(v^*, v-v^*)\overline{n'(v^*)n'(v-v^*)}dv^* \\ &- \int_0^\infty \beta(v, v^*)\overline{n'(v)n'(v^*)}dv^* - V_s\frac{\partial\bar{n}}{\partial z}.\end{aligned} \qquad (11.42)$$

The familiar problem of closure rears its head again. The turbulent fluxes are often represented as though they were mean field, gradient transport processes; for example, for the turbulent diffusion term,

$$\overline{n'V_i'} \approx -D_{\mathrm{T}}\frac{\partial \bar{n}}{\partial x_i}, \qquad (11.43)$$

where D_{T} is an eddy diffusivity. However, we should remember that such analogies have little physical basis; coupling between the turbulence and the mean field variables is usually weak.

A dynamic equation that includes aggregation and sedimentation for a system that is spatially homogeneous (well mixed) can be written as

$$\frac{dn(v)}{dt} = \frac{1}{2}\int_0^v \beta(v, v-\bar{v})n(v)n(v-\bar{v})d\bar{v} - n(v) \times \int_0^\infty \beta(v, \bar{v})n(\bar{v})d\bar{v} - \frac{V_s(v)}{h}n(v), \qquad (11.44)$$

where $n(v)$ is the particle size distribution (number concentration as a function of volume), β is the collision frequency function, v_s is the settling velocity, and h is the vertical "depth" of the system. Note that (11.44) does not include diffusion or convective transport. If the settling particles follow Stokes law and if buoyancy is neglected, then

$$\frac{4}{3}\pi R^3\rho_p g = 6\pi\mu R V_s. \qquad (11.45)$$

However, the right-hand side of (11.45) might need to be modified for *smaller* particles in aerosols to account for the noncontinuum effects. If the particle diameter is comparable to the mean free path in the gas, then the drag obtained from the Stokes law is too large. This is usually corrected in the following way: $F = 6\pi\mu RV/C$, where C is the Cunningham correction factor. Seinfeld (1986) provided a table of values for the Cunningham correction factor for air at 1 atm pressure and 20°C; for a particle with a diameter of 0.1 μm, the Stokes drag should be divided by 2.85. Thus, V_s would be increased by 285%.

Farley and Morel (1986) recast eq. (11.44) in discrete form for application to a limited number of logarithmically spaced particle classes:

$$\frac{dn_k}{dt} = \frac{1}{2}\sum_{i+j=k} \alpha(i,j)\beta(i,j)n_i n_j - n_k\sum_{i=1}^{m}\alpha(i,k)\beta(i,k)n_i - \frac{V_s(k)}{h}n_k, \qquad (11.46)$$

where $\alpha = 1$ if $i \neq j$ and 2 if $i = j$. With a discrete model of this type, a collision does not necessarily produce a particle in the next larger class; consequently, particle volume may not be conserved with eq. (11.46) even if the disappearance by sedimentation is removed. One method of compensation is to use weighting fractions so that only a portion of $i-j$ collisions yields production in higher classes. Additional collision frequencies can be added to (11.46) to account for the turbulence-induced coagulation or other phenomena. However, Williams (1988) notes that there is no *a priori* reason to assume that the resultant coagulation kernel should merely be the sum of the individual mechanisms. The most attractive aspect of the modeling approach described above is that influences of the initial particle size distribution, settling velocities, and collision efficiencies could be very rapidly compared, at least qualitatively. A simulation program was developed to illustrate this; the algorithm considers Brownian motion and uses eight particle classes with mean diameters corresponding to 0.375, 0.75, 1.5, 3, 6, 12, 24, and 48 μm. This is a logarithmic spacing as recommended by Gelbard and Seinfeld (1978). The graphs provided in Figure 11.9 give some indication of the wide variations possible in the evolution of the particle size distribution.

A comparison of these preliminary results with those computed by Lindauer and Castleman (1971) indicates that the simple simulation performs surprisingly well. However, a number of modifications would clearly be appropriate, including allocating the classes or bins according to $v_{n+1} = 2v_n$. For spherical particles or entities, this corresponds to $d_{n+1} = 1.26d_n$. Therefore, covering particle diameters ranging from 0.4 to 10 μm would require 15 classes and extension to 40 μm would require 21 classes. This alteration should make it easier to achieve conservation of volume, where appropriate.

11.3.5 Other Aspects of Particle Size Distribution Modeling

Gelbard et al. (1980) observed that numerical solutions for dynamic aerosol balances require approximation of the continuous size distribution by some finite set of classes or sections. They addressed the question as to whether a "sectional representation" can in fact produce an accurate solution for a dynamic aerosol problem. They were able to show that for the limiting case in which the section size (or class interval) decreases, the finite representation reduced to the classic coagulation equation. By comparison with experimental (power plant plume) data, they demonstrated that the discrete approximation yielded satisfactory results.

Direct numerical simulation has become (at least somewhat) feasible due to the recent increases in computing power. Reade and Collins (2000), for example, devised a simulation for a "periodic" volume (a particle whose trajectory causes it to leave through a bounding surface immediately reenters the domain on the opposite side) using 262,144 initial particles. They considered isotropic turbulence with a Reynolds number (based upon the Taylor microscale) of 54. Their results

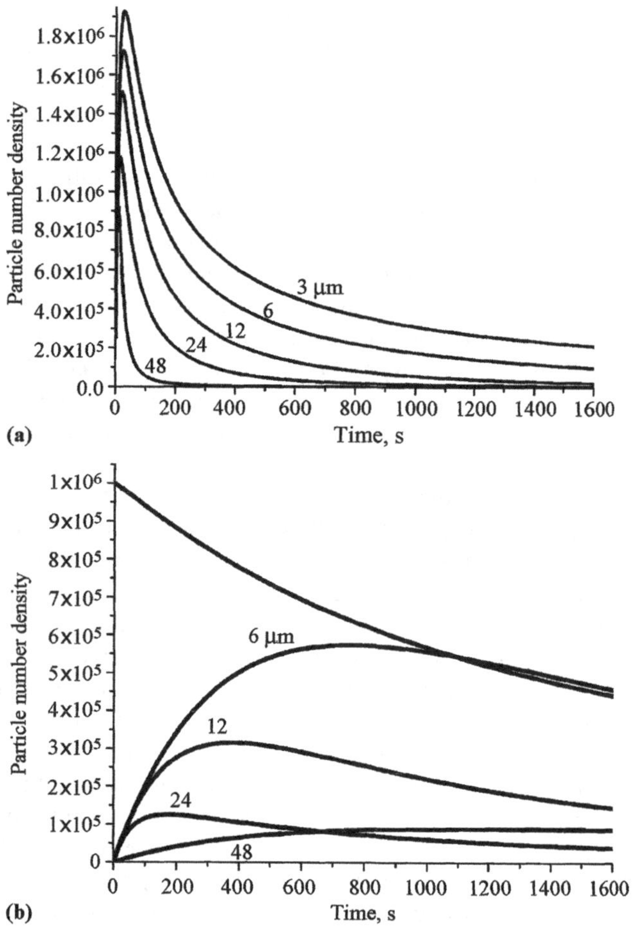

FIGURE 11.9. Comparison of simulation results showing the changes in population of the five largest particle classes. In (a) the particles are initially placed in the 0.75 μm class and the loss of larger particles by settling is enhanced. In (b) the initial particles are fewer in number and spread among the first four classes, but loss by sedimentation is suppressed. The reduced particle number density and the inhibited settling used in (b) result in sluggish dynamics.

show that a finite Stokes number St^1 results in a much broader particle size distribution than does either limiting case ($St = 0$ or $St \to \infty$). Furthermore, they found that the standard deviation of the psd decreased with the increasing St. Reade and Collins used their results to test collision kernels written in power law form (collision diameter raised to a power p). They found that dynamic psd behavior could not be adequately represented with a constant value of p; the conclusion is that the dynamic behavior of real particles may not correspond closely to the idealized collision mechanisms.

Sandu (2002) employed a discretization of the coagulation equation in which the integral terms were approximated by Newton–Cotes sums. A polynomial of order n was used to interpolate the function at the nodes (collocation). This resulted in a system of coupled ordinary differential equations that was solved with a semi-implicit Gauss–Seidel iteration. The technique was said to offer improved accuracy over earlier approaches.

Fernandez-Diaz et al. (2000) improved the semi-implicit technique developed by Jacobson et al. (1994) that produced unwanted numerical diffusion (unphysical broadening of the particle size distribution). Fernandez-Diaz et al. attacked this problem by devising different partition coefficients for the bins; they noted that the coagulation of i- and j-type particles might not necessarily result in a new entity of volume $v_i + v_j$. In fact, the new entity could have a volume corresponding to $(v_i + v_j)$ where the volumes were both from the bottom (minimum) of the original bins and from the top (maximum) of each. Therefore, they assumed that each bin could be characterized by the geometric mean of its limits, that is, $v_k = \sqrt{v_{k-} v_{k+}}$. This results in each bin having a width of 1 in the new size space. In addition, particles were uniformly distributed throughout the bin and the volumes of the bins varied as $v_x = v_1[1 + b(x-1)]^a$, where a and b were appropriately chosen. It appeared that this technique better approximated populations in the larger entity sizes than that achieved with geometrical spacing of bins.

11.3.6 A Highly Simplified Example

Let us briefly contemplate a situation in which a cloud of particles is introduced impulsively into an enclosure. We will formulate a highly simplified model that provides partial connection between the particle number density and the fluid mechanics (dissipation rate). We expect the results to be more qualitative than quantitative, but we note that differential sedimentation could be added and the model could be compartmentalized (with exchange between the subunits) to handle highly inhomogeneous turbulence. If we presume that the collision kernels are additive (which is suspect, as noted previously) and neglect particle size variation, then

$$\frac{dn}{dt} = -\left(\frac{4}{3}\frac{kT}{\mu} + 5.2\left(\frac{\varepsilon}{\nu}\right)^{1/2} R^3\right) n^2, \qquad (11.47)$$

with

$$\frac{d}{dt}\left(\frac{3}{2}u^2\right) = -\varepsilon \approx -A\frac{u^3}{l}. \qquad (11.48)$$

In eq. (11.48), the dissipation rate is represented with Taylor's inviscid approximation; u is a characteristic velocity and l is

[1] St is the ratio of the stop distance and a characteristic dimension of the system; it is important in inertial deposition. For example, for particle impact upon a cylindrical fiber, $St = \rho_p d_p^2 V / 18\mu d$.

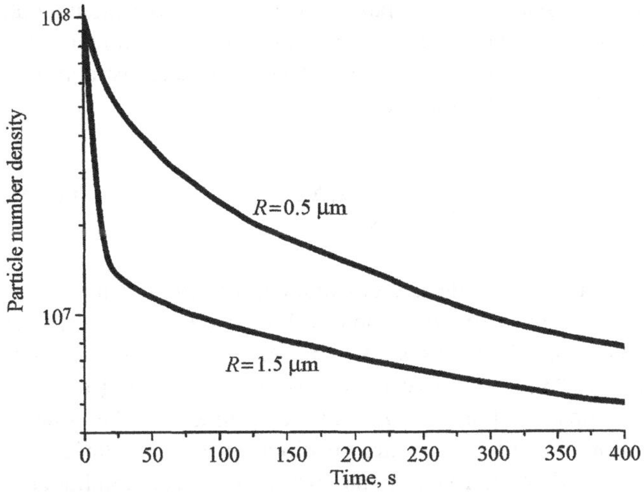

FIGURE 11.10. Illustration of the effects of particle size upon the (simultaneous) solution of eqs. (11.47) and (11.48). Clearly, turbulence is very effective in the initial rate of reduction of *larger* particles (with $R = 1.5\ \mu m$); the times required for an order of magnitude reduction can be compared: $t_{10\%}(1.5)/t_{10\%}(0.5) \approx 85/290 = 0.29$.

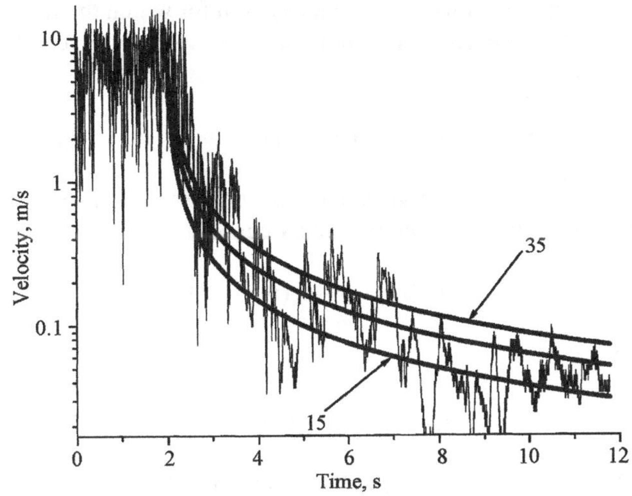

FIGURE 11.11. Results from a simplified model for decaying turbulence in an enclosure (a box) using Taylor's inviscid approximation for the dissipation rate. The three curves are for integral length scales (l's) of 15, 25, and 35 cm. Actual experimental data obtained with hot wire anemometry for decaying turbulence in a box are shown for comparison.

the integral length scale. Such a model would be valid only initially and only for the initial period of decay (of turbulence in a box); for advanced times, the dissipation rate estimate would need to be replaced with an equation of the type

$$\varepsilon \approx C\nu u^2/l^2. \qquad (11.49)$$

Tennekes and Lumley (1972) recommend making the transition to the final period of decay at

$$Re = \frac{ul}{\nu} = 10. \qquad (11.50)$$

This modeling approach might be useful for qualitative purposes such as assessment of the initial effects of dissipation rate, particle number density, and particle size. It would also be possible to include a loss term in (11.47) to account for the deposition onto surfaces, should that be necessary. Some computed results appear in Figure 11.10.

An important question in this context is whether eq. (11.48) can adequately represent the decay of turbulent energy in enclosures. We simply note that there are experimental data to suggest that (11.48) is at least semiquantitative. In eq. (11.48), the constant A has been set to 1.5 as indicated by experiment. The integral length scale l is generally taken to correspond to the size of the largest eddies present in the flow. In enclosures, the smallest of the principal dimensions, length, width, and height (L, W, h), would be a rough approximation. For the apparatus used to test the simplified model, the minimum dimension (size) was about 36 cm. Equation (11.48) was solved for integral lengths of 15, 25, and 35 cm and the results are shown in Figure 11.11. It was discovered that the curve for 20 cm corresponded reasonably well with the experimental (CTA) data (i.e., at $t = 4$ s, $u \approx 0.2$ m/s; at $t = 6$ s, $u \approx 0.1$ m/s; and at $t = 10$ s, $u \approx 0.05$ m/s) obtained for the decaying turbulent flow in this particular small box.

The available data suggest that eq. (11.48) is an appropriate approximation for turbulent energy decay, at least for systems of small scale. We should also observe that the Reynolds number, as given by eq. (11.50), would still have a value of about 500 at $t = 12$ s; the final period of decay would begin when the velocity u was about 0.08 cm/s. Based upon the results shown, $u \approx 0.08$ cm/s would not be attained until $t \approx 500$ s. At that point, Taylor's approximation for ε would have to be replaced by eq. (11.49).

11.4 MULTICOMPONENT DIFFUSION IN GASES

11.4.1 The Stefan–Maxwell Equations

Recall that in Chapter 8 we restricted our attention to binary systems for which the diffusional fluxes were assumed to be Fickian. The limitation of this approach is apparent in multicomponent diffusion problems where the concentration gradient for species "1" must be written in terms of the fluxes of all species. Our recourse for such problems can be found in the Stefan–Maxwell (SM) equations, which can be developed from the kinetic theory of gases (the interested reader may consult Taylor and Krishna, 1993). We will set the background for the SM equations with an approach outlined by B. G. Higgins (2008).

We initially consider a binary system for which the molar flux of "1" relative to the molar average velocity v^* can be written as

$$J_1 = c_1(v_1 - v^*) = -cD_{12}\nabla x_1. \tag{11.51}$$

Of course, c is the total molar concentration, so $x_1 = c_1/c$. Therefore, we can write for components "1" and "2",

$$D_{12}\nabla x_1 = -x_1(v_1 - v^*)$$

and

$$D_{21}\nabla x_2 = -x_2(v_2 - v^*). \tag{11.52}$$

Since $x_2 = 1 - x_1$ and $D_{12} = D_{21}$, we write

$$D_{12}\nabla x_1 = x_2(v_2 - v^*). \tag{11.53}$$

The molar average velocity v^* can be isolated:

$$v^* = v_2 - \frac{D_{12}}{x_2}\nabla x_1, \tag{11.54}$$

and therefore

$$D_{12}\nabla x_1\left(1 + \frac{x_1}{x_2}\right) = -x_1(v_1 - v_2). \tag{11.55}$$

We now multiply by x_2 and divide by the diffusivity:

$$\nabla x_1 = -\frac{x_1 x_2(v_1 - v_2)}{D_{12}}. \tag{11.56}$$

It is to be noted that the gradient of x_1 depends upon the difference in *species velocities*. If there were no differences between the species velocities, there would be of course no diffusive flux. For a gaseous mixture of n species, the Stefan–Maxwell equations can be written in a manner analogous to eq. (11.56):

$$\nabla x_i = \sum_{j=1}^{n} \frac{x_i x_j(v_j - v_i)}{D_{ij}}, \quad \text{where} \quad j \neq i. \tag{11.57}$$

The principal difficulty is clear: the Stefan–Maxwell equations give the concentration (or mole fraction) gradient in terms of the fluxes of all other species. In our work, we usually want the inverse, that is, we would like to obtain the flux in terms of the concentration gradient! The computational burden in multicomponent diffusion problems posed by the SM equations is significant. Consequently, much effort has been spent developing Fickian approximations for the SM equations. For example, one approach that has appeared in the literature utilizes the Fickian model with effective diffusivities (D_{eff}) that depend upon the concentrations of all other species. In many cases (particularly where experimental data are limited), the validity of such methods is unknown.

It is common practice to replace the species velocities in eq. (11.57) with molar fluxes:

$$\nabla x_i = \sum_{j=1}^{n} \frac{1}{cD_{ij}}(x_i N_j - x_j N_i). \tag{11.58}$$

Let us now illustrate an elementary approach to a simple multicomponent diffusion problem. Suppose we have a ternary system in which gases "1" and "2" are diffusing through species "3." This diffusional process is occurring between positions $z=0$ and $z=L$, and we assume that the concentrations for all species are specified at the boundaries. We use the SM equation template to write the three simultaneous differential equations, using molar concentrations:

$$\frac{dc_1}{dz} = \frac{x_1 N_2 - x_2 N_1}{D_{12}} + \frac{x_1 N_3 - x_3 N_1}{D_{13}}, \tag{11.59a}$$

$$\frac{dc_2}{dz} = \frac{x_2 N_1 - x_1 N_2}{D_{12}} + \frac{x_2 N_3 - x_3 N_2}{D_{23}}, \tag{11.59b}$$

$$\frac{dc_3}{dz} = \frac{x_3 N_1 - x_1 N_3}{D_{13}} + \frac{x_3 N_2 - x_2 N_3}{D_{23}}. \tag{11.59c}$$

Now suppose "3" is stagnant such that $N_3 = 0$. We obtain an initial estimate for the molar flux of "2" assuming the diffusion process is Fickian. Using this value for N_2, we solve the differential equations (11.59a–c), searching for the "best" value for N_1. Then, we fix that value of N_1 and solve the equations seeking an improved N_2. This process is repeated until a satisfactory solution is obtained. Note that what is required is a two-dimensional search (employing a univariant method) that involves repeated solution of the ODEs. We will illustrate this process with a modification of an example originally presented by Geankoplis (1972). A significant difference is that we want to explore the effects of changing diffusivities upon the solution. Our initial parametric choices are summarized in the following table; the temperature is 375K and the total pressure is 0.65 atm.

	$x_i(z=0)$ Position	$x_i(z=L)$ Position	Diffusivities	
Species 1	0.08	0.00	1–3	2.00
Species 2	0.00	0.35	2–3	2.00
Species 3	0.92	0.65	1–2	2.00

For the specified conditions, the total molar concentration is about 2.11×10^{-5} gmol per cm^3. The molar flow

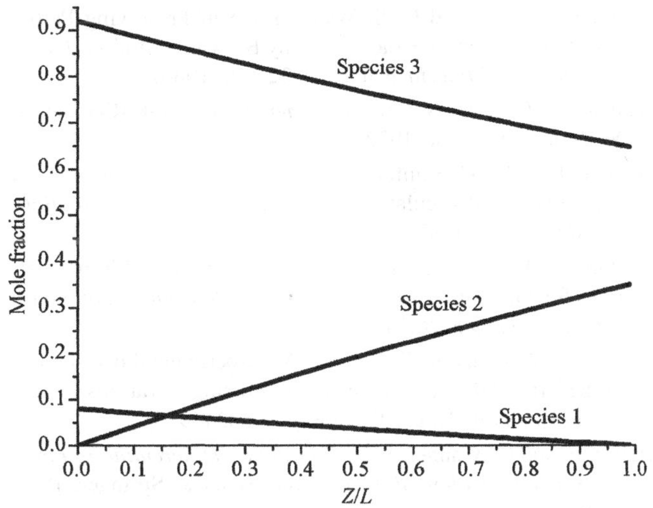

FIGURE 11.12. A Stefan–Maxwell example with equal diffusivities.

rate for component "1" assuming a Fickian process is about 3.379×10^{-6} gmol/(cm^2 s); however, the *correct* flux is only 84% of that value. The computed concentration profiles are illustrated in Figure 11.12.

Now, suppose the preceding example is repeated but with quite different diffusivities.

	$x_i(z=0)$ Position	$x_i(z=L)$ Position	Diffusivities	
Species 1	0.08	0.00	1–3	2.00
Species 2	0.00	0.35	2–3	1.00
Species 3	0.92	0.65	1–2	0.50

An initial estimate of the molar flux using Fick's law for the binary case (1–3) is exactly the same as before, but this time the *correct* flux is just 41.7% of the approximate value. These examples illustrate the importance of accounting for the resistance offered by the presence of multiple chemical species; these additional constituents, to quote Taylor and Krishna, "get in the way" of the transport process. Use of the Stefan–Maxwell equations permits us to correct the diffusional fluxes.

Finally, we look at a specific numerical example using data collected by Carty and Schrodt (1975) for a system consisting of acetone (1), methanol (2), and air (3). They used a Stefan tube operated at 328.5K and a pressure of 0.9805 atm. They cited diffusivity values D_{13}, D_{23}, and D_{12} of 0.1372, 0.1991, and 0.0848 cm^2/s, respectively. Repeating their calculations, we found slightly different values for the fluxes of species "1" and "2": 1.790×10^{-7} and 3.138×10^{-7} gmol/(cm^2 s), respectively. The results of the computations, however, agreed very nicely with their experimental data, as shown in Figure 11.13.

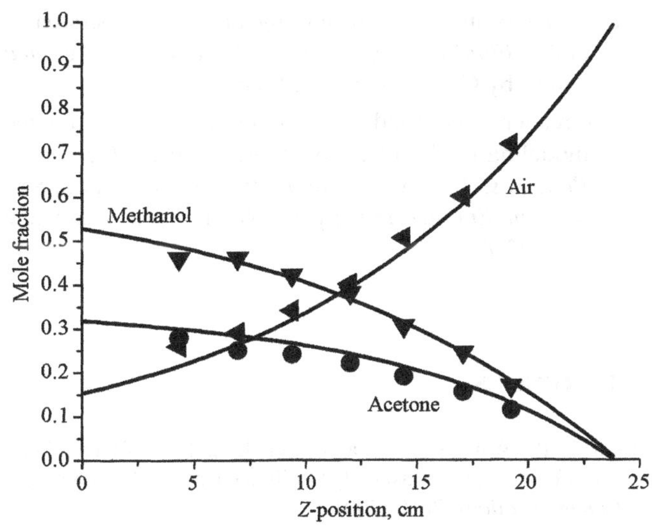

FIGURE 11.13. Solution of the SM equations for the acetone–methanol–air system, compared with experimental data adapted from Carty and Schrodt (1975). It is to be noted that Carty and Schrodt also provided a comparison of their data with the approximate solution obtained using Toor's (1964) method. The SM equations provide *much* better agreement with the experimental data.

11.5 CONCLUSION

This chapter is merely the barest of introductions to a few selected multiphase and multicomponent problems in transport phenomena. The objective is to stimulate the interest of students in these areas, which are important to many facets of contemporary chemical engineering research and practice. Because this book represents the *actual* two-semester advanced transport phenomena course sequence that I teach every year, the content reflects what we try to accomplish in about 90 lectures. Naturally, there are many fascinating topics that must be omitted and I am troubled by the realization that an advanced student—looking for some specific assistance—might not find what he/she needs here. Therefore, I would like to draw the reader's attention to some resources that might be useful for some additional exploration of multiphase phenomena.

For readers interested in gas–solid flows and fluidization: *Principles of Gas–Solid Flows,* by L. S. Fan and C. Zhu, Cambridge University Press (1998).

For readers interested in the breakup of drops and bubbles, capillarity, electrolytic systems, and behavior of dispersions: *Physicochemical Hydrodynamics,* by V. G. Levich, Prentice-Hall (1962).

For readers interested in cavitation: *Cavitation and Bubble Dynamics,* by C. E. Brennen, Oxford University Press (1995).

For readers interested in mixing and gas dispersion in tanks: *Fluid Mixing and Gas Dispersion in Agitated Tanks,* by G. B. Tatterson, McGraw-Hill (1991).

For readers interested in population balances and the modeling of discrete (countable) entities: *Population Balances: Theory and Applications to Particulate Systems in Engineering,* by D. Ramkrishna, Academic Press (2000).

REFERENCES

Birkin, P. R., Watson, Y. E., and T. G. Leighton. Efficient Mass Transfer from an Acoustically Oscillated Gas Bubble. *Chemical Communications*, 2650 (2001).

Borotnikova, M. I. and R. I. Soloukhin. A Calculation of the Pulsations of Gas Bubbles in an Incompressible Liquid Subject to a Periodically Varying Pressure. *Soviet Physics—Acoustics*, 10:28 (1964).

Brennan, C. E. *Fundamentals of Multiphase Flow*, Cambridge University Press, Cambridge (2005).

Carty, R. and T. Schrodt. Concentration Profiles in Ternary Gaseous Diffusion. *Industrial & Engineering Chemistry Fundamentals*, 14:276 (1975).

Cash, J. R. *Stable Recursions*, Academic Press, London (1979).

Clift, R., Grace, J. R., and M. E. Weber. *Bubbles, Drops, and Particles*, Academic Press, Boston (1978).

Delichatsios, M. A. and R. F. Probstein. Coagulation in Turbulent Flow: Theory and Experiment. *Journal of Colloid and Interface Science*, 51:394 (1975).

Devereux, O. F. and P. L. de Bruyn. *Interaction of Plane Parallel Double Layers*, MIT Press (1963).

Ding, A., Hounslow, M., and C. Biggs. Population Balance Modeling of Activated Sludge Flocculation. *Chemical Engineering Science*, 61:63 (2006).

Ellenberger, J. and R. Krishna. Improving Mass Transfer in Gas–Liquid Dispersions by Vibration Excitement. *Chemical Engineering Science*, 57:4809 (2002).

Fan, L. S. and K. Tsuchiya. *Bubble Wake Dynamics in Liquids and Liquid–Solid Suspensions*, Butterworth-Heinemann, Stoneham, MA (1990).

Farley, K. J. and F. M. M. Morel. Role of Coagulation in the Kinetics of Sedimentation. *Environmental Science and Technology*, 20:187 (1986).

Fernandez-Diaz, J. M., Gonzalez-Pola Muniz, C., Rodriguez Brana, M. A., Arganza Garcia, B., and P. J. Garcia Nieto. A Modified Semi-Implicit Method to Obtain the Evolution of an Aerosol by Coagulation. *Atmospheric Environment*, 34:4301 (2000).

Finlayson, B. A. *Nonlinear Analysis in Chemical Engineering*, McGraw-Hill, New York (1980).

Friedlander, S. K. *Smoke, Dust, and Haze*, 2nd edition, Oxford University Press (2000).

Friedlander, S. K. and C. S. Wang. The Self-Preserving Particle Size Distribution for Coagulation by Brownian Motion. *Journal of Colloid and Interface Science*, 22:126 (1966).

Geankoplis, C. J. *Mass Transport Phenomena*, Holt, Rinehart and Winston, New York (1972).

Gelbard, F. and J. H. Seinfeld. Numerical Solution of the Dynamic Equation for Particulate Systems. *Journal of Computational Physics*, 28:357 (1978).

Gelbard, F., Tambour, Y., and J. H. Seinfeld. Sectional Representations for Simulating Aerosol Dynamics. *Journal of Colloid and Interface Science*, 76:541 (1980).

Haberman, W. L. and R. K. Morton. An Experimental Investigation of the Drag and Shape of Air Bubbles Rising in Various Liquids. *David W. Taylor Model Basin Report* 802 (1953).

Hairer, E. and G. Wanner. *Solving Ordinary Differential Equations II: Stiff and Differential-Algebraic Systems*, Springer, Berlin (1996).

Hesketh, R. P., Etchells, A. W., and T. W. Fraser Russell. Experimental Observations of Bubble Breakage in Turbulent Flow. *Industrial and Engineering Chemistry Research*, 30:835 (1991).

Higgins, B. G. ECH 256 Course Notes, UC Davis (2008).

Hinze, J. O. Fundamentals of the Hydrodynamic Mechanism of Splitting in Dispersion Processes. *AIChE Journal*, 1:289 (1955).

Holt, R. G. and L. A. Crum. Acoustically Forced Oscillations of Air Bubbles in Water: Experimental Results. *Journal of the Acoustical Society of America*, 91:1924 (1992).

Hulburt, H. M. and Katz, S. Some Problems in Particle Technology: A Statistical Mechanical Formulation. *Chemical Engineering Science*, 19:555 (1964).

Jacobson, M. Z., Turco, R. P., Jensen, E. J. and O. B. Toon. Modeling Coagulation Among Particles of Different Composition and Size. *Atmospheric Environment*, 28:1327 (1994).

Kelley, E. and M. Wu. Path Instabilities of Rising Air Bubbles in a Hele–Shaw Cell. *Physical Review Letters*, 79:1265 (1997).

Kostoglou, M. and A. J. Karableas. On the Breakage of Liquid–Liquid Dispersion in Turbulent Pipe Flow: Spatial Patterns of Breakage Intensity. *Industrial & Engineering Chemistry Research*, 46:8220 (2007).

Kupferberg, A. and G. J. Jameson. Bubble Formation at a Submerged Orifice above a Gas Chamber of Finite Volume. *Transactions of the Institution of Chemical Engineers*, 47:T241 (1969).

Kusters, K. A., Wijers, J. G., and D. Thoenes. Aggregation Kinetics of Small Particles in Agitated Vessels. *Chemical Engineering Science*, 52:107 (1997).

Lamb, H. *Hydrodynamics*, 6th edition, Dover Publications, New York (1932).

Laufer, J. The Structure of Turbulence in Fully Developed Pipe Flow. *NACA Report 1174* (1954).

Leighton, T. G. *The Acoustic Bubble*, Academic Press, London (1994).

Lindauer, G. C. and A. W. Castleman Jr. Behavior of Aerosols Undergoing Brownian Coagulation and Gravitational Settling in Closed Systems. *Aerosol Science*, 2:85 (1971).

Luo, H. and H. F. Svendsen. Theoretical Model for Drop and Bubble Breakup in Turbulent Dispersions. *AIChE Journal*, 42:1225 (1996).

Maksimov, A. O. and T. G. Leighton. Transient Processes Near the Acoustic Threshold of Parametrically-Driven Bubble Shape Oscillations. *Acta Acoustica*, 87:322 (2001).

Marmur, A. and E. Rubin. A Theoretical Model for Bubble Formation at an Orifice Submerged in an Inviscid Liquid. *Chemical Engineering Science*, 31:453 (1976).

Minnaert, M. On Musical Air Bubbles and the Sounds of Running Water. *Philosophical Magazine*, 16:235 (1933).

Overbeek, J. Th. G. The Interaction Between Colloidal Particles. In: *Colloid Science* (H. R. Kruyt, editor), Elsevier, Amsterdam (1952).

Panofsky, H. A. and J. A. Dutton. *Atmospheric Turbulence*, Wiley-Interscience (1984).

Plesset, M. S. The Dynamics of Cavitation Bubbles. *Journal of Applied Mechanics*, 16:277 (1949).

Plesset, M. S. and A. Prosperetti. Bubble Dynamics and Cavitation. *Annual Review in Fluid Mechanics*, 9:145 (1977).

Polidori, G. Jeandet, P. and G. Liger-Belair. Bubbles and Flow Patterns in Champagne. *American Scientist*, 97:294 (2009).

Prosperetti, A. Thermal Effects and Damping Mechanisms in the Forced Radial Oscillations of Gas Bubbles in Liquids. *Journal of the Acoustical Society of America*, 61:17 (1977).

Prosperetti, A. Bubble Dynamics: A Review and Some Recent Results. *Applied Scientific Research*, 38:145 (1982).

Rayleigh, Lord. On the Pressure Developed in a Liquid During the Collapse of a Spherical Cavity. *Philosophical Magazine*, 34:94 (1917).

Reade, W. C. and L. R. Collins. A Numerical Study of the Particle Size Distribution of an Aerosol Undergoing Turbulent Coagulation. *Journal of Fluid Mechanics*, 415:45 (2000).

Rozentsvaig, A. K. Breakup of Droplets in Turbulent Shear Flow of Dilute Liquid–Liquid Dispersions. *Journal of Applied Mechanics and Technical Physics*, 22:797 (1981).

Saffman, P. G. and J. S. Turner. On the Collision of Drops in Turbulent Clouds. *Journal of Fluid Mechanics*, 1:16 (1956).

Sandu, A. A Newton–Cotes Quadrature Approach for Solving the Aerosol Coagulation Equation. *Atmospheric Environment*, 36:583 (2002).

Scraton, R. E. *Further Numerical Methods in Basic*, Edward Arnold, London (1987).

Seinfeld, J. H. *Atmospheric Chemistry and Physics of Air Pollution*, Wiley-Interscience (1986).

Shahub, A. M. and M. M. R. Williams. Brownian Collision Efficiency. *Journal of Physics D*, 21:231 (1988).

Smoluchowski, M. V. Versuch einer Mathematischen Theorie der Koagulationskinetik. *Zeitschrift fuer Physikalische Chemie*, 92:129 (1917).

Sohbi, B., Emtir, M., and M. Elgarni. The Effect of Vibration on the Absorption with Chemical Reaction in an Aqueous Solution of Calcium Hydroxide. *Proceedings of the World Academy of Science, Engineering and Technology*, 23:311 (2007).

Swift, D. L. and S. K. Friedlander. The Coagulation of Hydrosols by Brownian Motion and Laminar Shear Flow. *Journal of Colloid Science*, 19:621 (1964).

Taylor, G. I. The Viscosity of a Fluid Containing Small Drops of Another Fluid. *Proceedings of the Royal Society of London A*, 138:41 (1932).

Taylor, G. I. The Formation of Emulsions in Definable Fields of Flow. *Proceedings of the Royal Society of London A*, 146:501 (1934).

Taylor, R. and R. Krishna. *Multicomponent Mass Transfer*, John Wiley & Sons, New York (1993).

Tennekes, H. and J. L. Lumley. *A First Course in Turbulence*, MIT Press (1972).

Toor, H. L. Solution of the Linearized Equations of Multicomponent Mass Transfer: 1. *AIChE Journal*, 10:448 (1964).

van Olphen, H. *An Introduction to Clay Colloid Chemistry*, 2nd edition, Wiley-Interscience, New York (1977).

Vemury, S., Kusters, K. A., and S. E. Pratsinis. Time-Lag for Attainment of the Self-Preserving Particle Size Distribution by Coagulation. *Journal of Colloid and Interface Science*, 165:53 (1994).

Vemury, S., Janzen, C., and S. E. Pratsinis. Coagulation of Symmetric and Asymmetric Bipolar Aerosols. *Journal of Aerosol Science*, 28:599 (1997).

Waghmare, Y. G. Vibrations for Improving Multiphase Contact. PhD Dissertation, LSU (2008).

Williams, M. M. R. A Unified Theory of Aerosol Coagulation. *Journal of Physics D* 21:875 (1988).

Wu, M., and M. Gharib. Path Instabilities of Air Bubbles Rising in Clean Water. *Repository: arXivUSA* (1998).

Xiong, Y., Pratsinis, S. E., and S. V. R. Mastrangelo. The Effect of Ionic Additives on Aerosol Coagulation. *Journal of Colloid and Interface Science*, 153:106 (1992).

PROBLEMS TO ACCOMPANY TRANSPORT PHENOMENA: AN INTRODUCTION TO ADVANCED TOPICS

Problem 1A. Partial Differential Equations and the Conservation of Mass

Identify each of the following partial differential equations by type and determine (as completely as possible) what phenomenon is being described for each case.

$$\frac{1}{\mu}\frac{dp}{dz} = \frac{\partial^2 v_z}{\partial x^2} + \frac{\partial^2 v_z}{\partial y^2}$$

$$\rho C_p \frac{\partial T}{\partial t} = k\left[\frac{\partial^2 T}{\partial y^2} + \frac{\partial^2 T}{\partial z^2}\right] \qquad \nabla^2 C_A = 0.$$

The variables are assumed to have their usual meaning.

Then, starting with an appropriate volume element (shell) in *cylindrical coordinates*, perform a mass balance and derive the continuity equation for a *compressible* fluid. Simplify your result for the following scenario: The laminar Couette flow between concentric cylinders in which the fluid motion is driven solely by the rotation of the *inner* cylinder.

Problem 1B. Practice with the Product Method or Separation of Variables

Consider the elliptic partial differential equation:

$$\frac{\partial^2 \beta}{\partial x^2} + \frac{\partial^2 \beta}{\partial y^2} = 0.$$

Use the product method and show that $4\exp(-3x)\cos(3y)$ is *a solution* given that $\beta(x, \pi/2) = 0$ and $\beta(x, 0) = 4\exp(-3x)$.

Problem 1C. Vorticity Vector in Cylindrical Coordinates

In cylindrical coordinates, $\nabla x V$ is

$$\frac{1}{r}\frac{\partial v_z}{\partial \theta} - \frac{\partial v_\theta}{\partial z} \qquad \frac{\partial v_r}{\partial z} - \frac{\partial v_z}{\partial r} \qquad \frac{1}{r}\frac{\partial}{\partial r}(r v_\theta) - \frac{1}{r}\frac{\partial v_r}{\partial \theta}.$$

Find expressions for the vorticity for the Hagen–Poiseuille flow, for the Poiseuille flow through an annulus, and for the Couette flow between concentric cylinders in which the inner cylinder is rotating and the outer cylinder is at rest.

Problem 1D. Solution of Parabolic Partial Differential Equation

Find the solution for the following partial differential equation:

$$\frac{\partial \psi}{\partial t} = 2\frac{\partial^2 \psi}{\partial y^2},$$

where y ranges from 0 to 5 with the boundary conditions $\psi(0, t) = 0$, $\psi(5, t) = 0$. The initial condition is $\psi(y, 0) = my + b$, where m and b are constants.

Problem 1E. Some Vector and Tensor Review Questions

What do we mean when we say that a velocity field is solenoidal?

The stress tensor is symmetric. Is that the same as saying we have conservation of angular momentum?

Transport Phenomena: An Introduction to Advanced Topics, By Larry A. Glasgow

What is the relationship between dilatation and the divergence of a velocity field?

A fluid motion in the $x-y$ plane is irrotational. If $v_x = a + by + cy^2$, what is v_y?

Problem 1F. Nonlinear Relationships Between Stress and Strain

The Ostwald-de Waele (or power law) model relates stress to strain in a nonlinear manner:

$$\tau_{yx} = -m\left|\frac{dv_x}{dy}\right|^{n-1}\frac{dv_x}{dy}.$$

If $n < 1$, the fluid is a pseudo-plastic (shear-thinning); if $n > 1$, the fluid is dilatant. Now, suppose we have a steady pressure-driven flow in the x-direction between parallel plates (with $y=0$ located at the center and the planar surfaces at $y=\pm h$). The governing equation is

$$\frac{dp}{dx} = -\frac{\partial \tau_{yx}}{\partial y}.$$

Therefore, since dp/dx is a constant,

$$\frac{1}{m}\frac{dp}{dx} = \frac{d}{dy}\left\{\left|\frac{dv_x}{dy}\right|^{n-1}\frac{dv_x}{dy}\right\}.$$

Solve this nonlinear differential equation for two cases, $n=4$ and $n=1/2$, and sketch the velocity distributions $v_x(y)$ from $y=0$ to $y=h$. Note that for this range of y's, the velocity is decreasing, that is, dv_x/dy is negative. The applicable boundary conditions are

$$\text{at } y=0,\ v_x = V_{\max} \quad \text{and} \quad \text{at } y=h,\ v_x = 0.$$

Problem 1G. Properly Posed Boundary Value Problems

If we say that a boundary value problem, consisting of a partial differential equation with appropriate boundary and initial conditions, is *properly posed*, what exactly do we mean? You may refer to a source like Weinberger, *A First Course in Partial Differential Equations* (Wiley, 1965).

Problem 1H. The Product Method Applied to Unbounded Regions

Situations in mathematical physics that are described by the elliptic partial differential equation

$$\frac{\partial^2 \psi}{\partial x^2} + \frac{\partial^2 \psi}{\partial y^2} = 0$$

are often referred to as "potential" problems. Can the product method (separation of variables) be used to solve such problems in unbounded regions? Provide an illustration, if possible.

Problem 1I. Different Forms of the Navier–Stokes Equation

The Navier–Stokes equation(s) can be written in three different forms: nonconservation, conservation, and control volume–surface integral. Describe the essential differences and provide an example of an appropriate application for each.

Problem 1J. Half-Range Fourier Series

Consider the linear function $f(x)=2x$, for $0<x<3$. Expand the function in a half-range Fourier sine series and prepare a graph that illustrates the quality of the representation as the number of terms is increased. Recall that

$$a_n = \frac{2}{L}\int_0^L f(x)\sin\frac{n\pi x}{L}dx.$$

Could the same function be represented with a half-range Fourier cosine series? What would the essential differences be?

Problem 1K. The Method of Characteristics

What is the "method of characteristics" and to what type of flow problem has it been generally applied? Is this technique widely used today? Why not?

Problem 1L. Uniqueness and the Equations Governing Fluid Motion

When we speak of uniqueness in the context of a partial differential equation, we mean that there is at most one function $\varphi(x,y,z,t)$, satisfying the PDE. In recent years, there has been much interest in the connection between nonuniqueness (for the Navier–Stokes equation) and the transition from laminar to turbulent flow. Search the recent literature and prepare a brief report of an investigation of nonuniqueness in fluid flow.

Problem 1M. Approximate Solution of Boundary Value Problem by Collocation

Consider the boundary value problem

$$\frac{d^2y}{dx^2} - y(x) = 1,$$

with $y(0) = y(1) = 0$. Find the analytic solution for this differential equation. Then, let $y(x)$ be approximated by

$$y(x) \cong a_1\phi_1(x) + a_2\phi_2(x) + a_3\phi_3(x) + \cdots.$$

Let $\phi_1 = x(x-1), \phi_2 = x^2(x-1)$, and so on. Note that these trial functions satisfy the boundary conditions. Truncate the expansion given above and use the collocation method (see Appendix H) to find the coefficients a_1 and a_2. Do this first by placing the collocation points at the ends of the interval $x = 0$ and $x = 1$. Then, repeat the process, using $x = 1/3$ and $x = 2/3$. Which of the approximations gives the better results?

Problem 1N. A "Simple" Example from Mechanics

Consider the second-order ordinary differential equation (the equation of motion for a frictionless pendulum):

$$\frac{d^2\theta}{dt^2} + \frac{g}{L}\sin\theta = 0, \tag{1}$$

where g is the acceleration of gravity and L is the pendulum length. At rest, $\theta = 0$, so motion can be initiated by moving the pendulum to a new angular position, say $\pi/4$ rad. Two points are immediately clear: The pendulum will oscillate between angular positions $+\pi/4$ and $-\pi/4$, and a characteristic time for the system is $\sqrt{L/g}$. Suppose, however, we wished to solve (1). We *might* observe that the equation can be integrated once to yield

$$\frac{1}{2}\left(\frac{d\theta}{dt}\right)^2 - \frac{g}{L}\cos\theta = C. \tag{2}$$

At the pendulum's position of maximum displacement, $d\theta/dt = 0$, so we can determine the constant of integration: $C = -(g/L)\cos\theta_{\max}$. Consequently, we can rearrange (2) to obtain

$$\frac{d\theta}{dt} = \sqrt{\frac{2g}{L}[\cos\theta - \cos\theta_{\max}]}. \tag{3}$$

This equation can be rewritten for our purposes:

$$dt = \sqrt{\frac{L}{2g}}\frac{d\theta}{[\cos\theta - \cos\theta_{\max}]^{1/2}}. \tag{4}$$

We define $k = \sin(\theta_{\max}/2)$ and use trigonometric identities to rewrite eq. (4) as

$$dt = \sqrt{\frac{L}{g}}\frac{d\phi}{\sqrt{1-k^2\sin^2\phi}}. \tag{5}$$

Note that $\cos\theta - \cos\theta_{\max} = 2k^2\cos^2\phi$. If we wanted to determine the time required for the pendulum to swing from the equilibrium position ($\theta = 0$) to some new angular position ϕ_1, we can do so by integration:

$$t_{\text{req}} = \sqrt{\frac{L}{g}}\int_0^{\phi_1}\frac{d\phi}{\sqrt{1-k^2\sin^2\phi}}. \tag{6}$$

The integral on the right-hand side of eq. (6) is an *elliptic* integral. The time required for a *complete oscillation* is the period P:

$$P = 4\sqrt{\frac{L}{g}}\int_0^{\pi/2}\frac{d\phi}{\sqrt{1-k^2\sin^2\phi}}. \tag{7}$$

The definite integral in eq. (7) is a *complete elliptic integral of the first kind* of modulus k. Values for this definite integral can be found in the literature, for example, for the specific modulus value $k = 1/\sqrt{2}$, this integral (from 0 to $\pi/2$) is 1.8541. The reader with further interest in elliptic integrals may wish to see page 786 *et seq.* in the *Handbook of Tables for Mathematics*, revised 4th edition, CRC Press, 1975.

We now revise our pendulum model; we would like to include dissipative effects (damping) and some kind of periodic forcing function (so we have a *driven* pendulum). We also employ a dimensionless time by incorporating $\tau = \sqrt{L/g}$. The three governing equations are as follows:

$$\frac{d\theta}{dt} = \omega, \tag{8a}$$

$$\frac{d\omega}{dt} = -\frac{\omega}{C} - \sin\theta + A\cos\phi, \tag{8b}$$

and

$$\frac{d\phi}{dt} = \omega_D. \tag{8c}$$

Note that C is the damping coefficient, A is the forcing function amplitude, and ω_D is the frequency at which the pendulum is being driven. Alternatively, we could of course write

$$\frac{d^2\theta}{dt^2} = -\frac{1}{C}\frac{d\theta}{dt} - \sin\theta + A\cos\omega_D t. \tag{9}$$

Although the model does not appear to be especially complex, there are *three* parameters to be specified: the damping coefficient, the forcing function amplitude, and the drive frequency. Thus, an exhaustive parametric exploration would be challenging. Fortunately, Baker and Gollub (*Chaotic Dynamics: An Introduction*, Cambridge University Press, 1990) have provided us with detailed guide to this problem that will significantly simplify our task. We set $C = 2$, $A = 0.9$, and $\omega_D = 2/3$ and solve the system (8a–c) numerically—it is to be noted that the behavior we wish to explore may not develop quickly! (Figure 1N).

Confirm the computation carried out above, and then repeat the process for both $A = 1.07$ and $A = 1.15$ and prepare plots illustrating dynamic system behavior. How does the system evolve as A increases? You may also wish to consult Gwinn and Westervelt, *Physical Review Letters*, 54:1613 (1985).

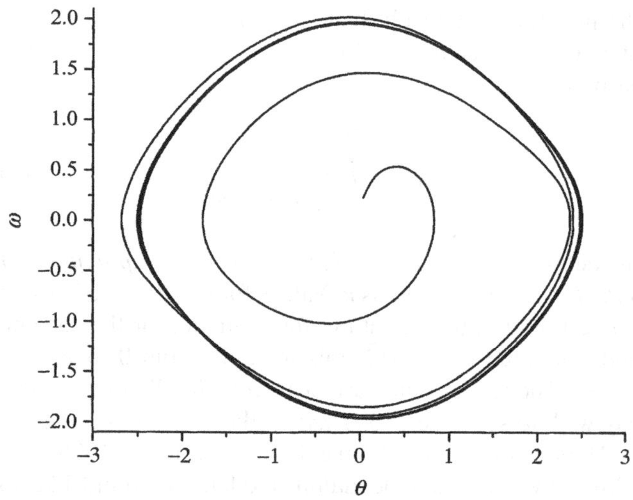

FIGURE 1N. Driven pendulum example, with $A = 0.9$.

Problem 2A. Inviscid Irrotational Flow in Two Dimensions

Consider the complex potential given by

$$w(z) = az^{2/3}, \quad \text{where} \quad z = x + iy.$$

Construct the streamlines for this flow on an appropriate figure, indicating flow direction, and describe the flow field. It may be useful to recall that

$$r = \sqrt{x^2 + y^2} \text{ and that } y = r\sin\theta.$$

Next, write out the appropriate Navier–Stokes equations for viscous flow in this situation. If one were to solve these equations at the modest Reynolds number, what would the essential differences be? Prepare a sketch illustrating this anticipated (viscous) flow field, emphasizing the expected differences between it and the potential flow.

Problem 2B. Potential Flow Past a Wedge

A wedge in the shape of a right triangle is placed in a wind tunnel as illustrated in Figure 2B. Compute the two-dimensional potential flow about this object and obtain an estimate of the lift generated by the body (if any). Finally, comment on the desirability of this shape for vehicle profiles, that is, is this a good shape for trucks and/or cars? The horizontal side of the wedge is 25 cm long and the vertical side is 7 cm. The moving air approaches the wedge at a velocity of 5 m/s.

Note that the governing equation for ψ is of the Laplace type. You will probably want to seek a numerical solution using an iterative technique.

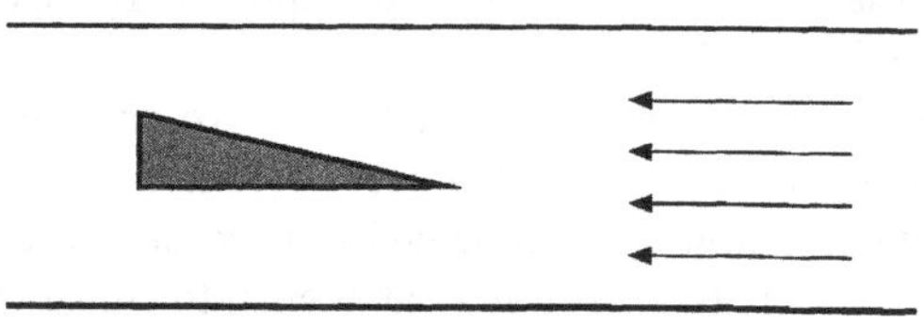

FIGURE 2B. Potential flow past a wedge.

Problem 2C. Potential Flow Past a Vertical Plate

Milne-Thompson (*Theoretical Hydrodynamics,* 1960) provided the complex potential for flow past a vertical plate of height $2c$:

$$w(z) = U(z^2 + c^2)^{1/2},$$

where $w = \phi + i\psi$ and $z = x + iy$. Construct the streamlines for this flow on an appropriate figure and indicate flow direction. Next, write out the appropriate Navier–Stokes equations for this flow (at the modest Reynolds number). If one were to solve these equations, what essential differences would be noted? Sketch the anticipated viscous flow and draw attention to the differences between the potential and viscous flow fields.

Problem 2D. Potential Flow Past an Inverted "L"

Consider a two-dimensional potential flow past an inverted "L" as shown in Figure 2D. The inverted "L" extends half-way across the height of the channel. Assume a uniform velocity of approach of 20 cm/s and a channel height of 20 cm. Compute the flow field for this case and prepare a suitable plot, clearly showing the expected streamlines. Recall that we demonstrated that the stream function ψ is governed by the Laplace equation:

$$\frac{\partial^2\psi}{\partial x^2} + \frac{\partial^2\psi}{\partial y^2} = 0.$$

This equation is very easily solved iteratively using the Gauss–Seidel method; you simply apply the algorithm we developed in Appendix C.

After you have found your solution and prepared the requested figure, write down the appropriate components of the Navier–Stokes equation for this problem and prepare an additional sketch that underscores the expected differences between the potential and viscous flow solutions.

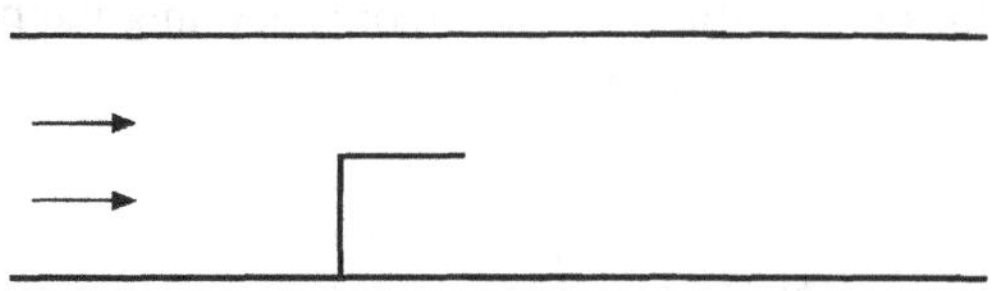

FIGURE 3D. Flow past an inverted "L."

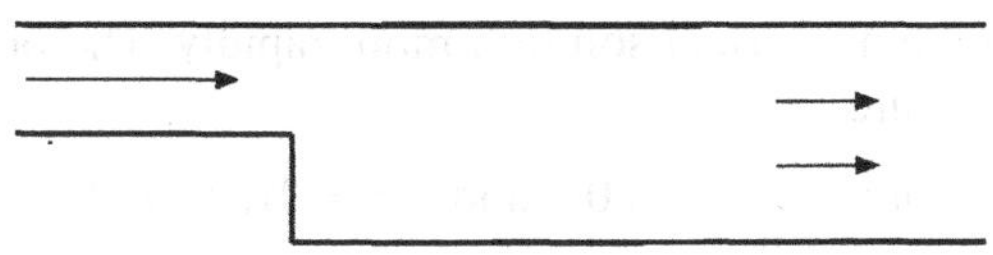

FIGURE 2E. Potential flow off of a step.

Problem 2E. Potential Flow Over a Rearward-Facing Step

Consider a two-dimensional potential flow over a rearward-facing step. The channel has a 10 cm height before the step and a 20 cm height after (i.e., the flow area doubles). The approach velocity is 20 cm/s. Solve the Laplace equation for the stream function

$$\frac{\partial^2 \psi}{\partial x^2} + \frac{\partial^2 \psi}{\partial y^2} = 0,$$

using the method of your choice and plot the resulting streamlines. The flow arrangement is depicted in Figure 2E.

Next, consider a horizontal line constructed 8 cm *below* the upper wall. Determine the pressure along that line using the Bernoulli equation and prepare a plot illustrating the result. If the flow occurring in this apparatus had viscous character, how might the pressure differ from that revealed by your calculations? Be very specific with your answer.

Problem 2F. The Edmund Fitzgerald Disaster

Additional background and detail for this problem can be obtained from the *NTSB-MAR-78-3 (Report)*, *Shipwrecks of Lake Superior* by James R. Marshall, and from Julius Wolff's *Lake Superior Shipwrecks.*

The ore carrier *Edmund Fitzgerald* left Superior, Wisconsin on November 9, 1975, beginning a voyage that would result in a multimillion dollar loss to the Northwestern Mutual Life Insurance Company and the deaths of 29 men. For non-mariners, it is hard to believe that an inland lake could produce such a tragedy. The *Fitzgerald* was a large ore carrier, built specifically for the transport of taconite mined in northern Minnesota. She was 729 ft long, 75 ft wide, and 39 ft in depth. Fully laden, she drew 27 ft of water. This means, of course, that any wave bigger than about 12 ft would put "green" water on deck. Although it was known that a strong weather system was approaching Lake Superior, the projection indicated only snow squalls, a northeast wind, and 15 ft waves. What actually occurred on November 10 was a brutal gale that ultimately led to 90 mph winds and 35 ft waves; this was a combination of forces that somehow ripped the *Fitz* into two >300 ft pieces and deposited her on the bottom of Lake Superior in 530 ft of water.

A (Great Lakes) bulk carrier is essentially an undivided (no solid, only screen bulkheads) rectangular box with numerous large hatches on top and ballast tanks running along the bottom outside corners. The ballast tanks can be filled with water for necessary trim and to provide bite for the propeller and effective turning with the rudder. The *Fitz* had six pumps for removal of water from ballast tanks, four rated at 7000 gpm and two auxiliary pumps rated at 2000 gpm. Sometime around 3 p.m. on November 10, she sustained an injury that turned out to be mortal. It seems likely that one of the four following scenarios must have played out:

1. A massive piece of flotsam came on deck, damaged the hull, and destroyed vents for one or more ballast tanks (most mariners dismiss this theory).
2. Hatch covers were improperly secured, allowing water to enter the cargo area.
3. The hull sustained a major stress fracture.
4. The *Fitz* ran onto a shoal near Caribou Island and "hogged" puncturing the hull and one or more tanks (NOAA chart 14960 of 1991 shows a region with depth of only 30 ft that extends about 8600 yard north of Caribou Island).

At about 3:20 p.m., Captain McSorley (master of the *Fitzgerald* since 1972) radioed the *SS Arthur Anderson* and reported that the *Fitz* had vent damage and a starboard list. Furthermore, McSorley reported that he was running two pumps, trying to remove water from ballast tanks (presumably at 14,000 gpm). As afternoon turned to evening, the *Fitz* began to settle by the bow, but because of the enormous seas, this may not have been detected by her crew. Sometime just after 7:10 p.m., a phenomenon known to Lake Superior sailors as the "three sisters" occurred; this involved the formation of three large rogue waves of unbelievable size, perhaps 40 ft trough to crest. These waves put something on the order of approximately 8000–15,000 ton of water on board the forward deck of the doomed ship (her rated gross tonnage was 13,612). The weight drove her nose down into the base of another wave and she headed for the bottom like a submarine in a crash dive (at 46°59.9′ N, 85°06.6′ W). Her initial surface speed was roughly 10 mph when the catastrophic plunge began. *She did not break into two on the surface*; the *NTSB-MAR-78-3 Report* is quite clear on this point. She was just 17 miles from the safety of Whitefish Bay when the end came for the ship and crew. Based upon information provided here, answer the following questions to the best of your ability:

(a) What size was the hole (or breach) in the hull of the *Fitzgerald*?

(b) At what speed was the hull traveling when the bow struck bottom (at 530 ft)?

(c) What would be the estimated speed of propagation of a large (40 ft) wave on Lake Superior (see Chapter IX in Lamb's *Hydrodynamics*, 6th edition, 1945)?

(d) Based upon your answer to (c), if the *Fitzgerald* had not checked down (slowed to allow the *Anderson* to keep track of her), would she have been able to reduce the weight of water on deck?

Problem 3A. Laminar Flow in a Triangular Duct

Steady laminar flows in noncircular ducts (flow in the x-direction) are governed by the equation

$$\frac{1}{\mu}\frac{dp}{dx} = \left[\frac{\partial^2 v_x}{\partial y^2} + \frac{\partial^2 v_x}{\partial z^2}\right]. \tag{1}$$

Since this constitutes a classic Dirichlet problem, a significant number of solutions are known; in fact, many of them appear in Berker (*Encyclopedia of Physics*, Vol. 8, 1963). For an equilateral triangle (length of each side, a), the velocity distribution is

$$v_x(y, z) = \frac{-dp/dx}{2\sqrt{3}a\mu}\left(z - \frac{\sqrt{3}}{2}a\right)(3y^2 - z^2), \tag{2}$$

where the origin is placed at the upper vertex; the y-axis is horizontal and the z-axis extends vertically toward the base of the equilateral triangle. We would like to consider a laminar flow in an isosceles triangle (triangular duct) where the base has a length of 15 cm and the two equal sides are 10.61 cm in length. Find the velocity distribution, the average velocity, and the Reynolds number for the flow (of water) that results from a pressure gradient corresponding to

$$\frac{p_0 - p_L}{L} = 0.0159\,\text{dyn/cm}^2 \text{ per cm}.$$

Equation (1) is a Poisson-type partial differential equation and it is well suited to the Gauss–Seidel iterative solution method.

Problem 3B. Laminar Flow in an Open Rectangular Channel

We would like to examine a relatively simple laminar open-channel flow of water; this should serve as a good review of some elementary concepts in fluid mechanics. Consider a rectangular channel, open at the top, that is inclined with respect to horizontal (at 0.2°) such that a steady flow occurs under the influence of gravity. Find the velocity distribution in the channel, and use appropriate software to plot the velocity contours. The square channel is 12 in. wide but the liquid depth is just 8 in. It is to be ensured that the provided notation (flow in the x-direction, with $y = 0$ corresponding to the channel floor) is used and the governing equation is put into dimensionless form.

Note that the governing equation is of the Poisson type. One might seek an analytic solution, but you can probably find the numerical solution more rapidly. The boundary conditions are

$$\text{at } y = 0,\ z = 0 \quad \text{and} \quad z = W,\ v_x = 0$$

$$\text{at } y = h,\ \frac{\partial v_x}{\partial y} = 0\,\text{(almost)}.$$

Problem 3C. Flow in the Bottom Half of a Cylindrical Duct

Let us consider steady flow in a half-filled cylindrical duct (with $d = 10$ cm); in rectangular coordinates, the governing equation can be written as

$$\frac{\partial^2 v_z}{\partial x^2} + \frac{\partial^2 v_z}{\partial y^2} = -\frac{\rho g_z \sin\phi}{\mu}.$$

Take the specific gravity of the liquid to be 1, the viscosity (μ) to be 4 cp, $g_z = 980\,\text{cm/s}^2$, and $\sin(\phi) = 0.001$. Find both the average and maximum velocities in the duct and plot the velocity distribution. Note that an approximate boundary condition at the free surface is

$$\frac{\partial v_z}{\partial y} \cong 0.$$

Obviously, the problem could also be written in cylindrical coordinates:

$$\frac{\partial^2 v_z}{\partial r^2} + \frac{1}{r}\frac{\partial v_z}{\partial r} + \frac{1}{r^2}\frac{\partial^2 v_z}{\partial \theta^2} = -\frac{\rho g_z \sin\phi}{\mu}.$$

Of course, each approach has advantages (and liabilities).

Problem 3D. Steady Laminar Flow in a Rectangular Duct

Consider laminar flow of water through a rectangular duct with a width measuring 18 in. and a depth of 6 in. Let the imposed pressure drop be 7.0×10^{-4} dyn/cm² per cm. If the temperature is 70°F, find

1. The velocity distribution.
2. The average velocity $\langle v_z \rangle$.
3. The Reynolds number *Re*.
4. The shear stress distribution across the *bottom* boundary (the duct floor).

The governing partial differential equation for this flow problem is of the Poisson type:

$$\frac{1}{\mu}\frac{dp}{dz} = \frac{\partial^2 v_z}{\partial x^2} + \frac{\partial^2 v_z}{\partial y^2}.$$

You should recognize immediately that either the Gauss–Seidel or extrapolated Liebmann (SOR) methods will work for this problem.

Problem 3E. Start-Up Flow in a Cylindrical Tube

Consider a viscous fluid initially at rest in a cylindrical tube. At $t=0$, a pressure difference is imposed and the fluid begins to move in the positive z-direction. The governing equation for this case is

$$\rho\frac{\partial v_z}{\partial t} = -\frac{\partial p}{\partial z} + \mu\left[\frac{1}{r}\frac{\partial}{\partial r}\left(r\frac{\partial v_z}{\partial r}\right)\right]. \tag{1}$$

We would like to solve this equation explicitly to gain experience with the technique. Let $R=3$ cm, $\nu=0.01$ cm^2/s, and $\rho=1$ g/cm^3. Find the velocity distributions at $\nu t/R^2=0.075$, 0.15, and 0.75. The constant pressure drop is 0.04074 dyn/cm^2 per cm. Present your results graphically by plotting v_z/V_{max} as a function of r/R. Find the Reynolds number corresponding to each value of t. This problem has been solved analytically by Szymanski (*Journal de Mathematiques Pures et Appliquies*, Series 9, 11:67, 1932), you can check your results by consulting the corresponding figure (3.2) in Chapter 3. Note that when (1) is put into finite difference form, the dimensionless grouping

$$\frac{\nu\Delta t}{(\Delta r)^2} \tag{2}$$

will arise. You must make sure that it has a small value, less than 0.5 is required for numerical stability (you might use something less than 0.1 to provide better resolution).

Problem 3F. Transient (Start-Up) Flow Between Parallel Planes: Part 1

A viscous fluid is initially at rest between two semi-infinite parallel planes (separated by a distance b). At $t=0$, a pressure gradient is imposed upon the fluid and motion ensues in the x-direction. The governing equation is

$$\frac{\partial v_x}{\partial t} = -\frac{1}{\rho}\frac{\partial p}{\partial x} + \nu\frac{\partial^2 v_x}{\partial y^2}. \tag{1}$$

Show that the steady-state solution has the form

$$v_{xss} = \frac{1}{2\mu}\frac{dp}{dx}(y^2 - by). \tag{2}$$

Let $v_x = v_1 + v_{ss}$; use the steady-state solution to eliminate the inhomogeneity in (1). Then, propose that

$$v_1 = f(y)g(t)$$

and solve the problem with separation of variables. Finally, use your analytic solution to obtain velocity profiles at $t=0.001, 0.01, 0.1$, and 1 s. Prepare a suitable figure showing the results. The liquid properties are as follows:

$$\mu = 4.75\text{ cp} \qquad \rho = 1.15\text{ g/cm}^3.$$

Take $b=1$ cm and $dp/dx=-75$ dyn/cm^2 per cm. What is the average velocity as $t\to\infty$?

Problem 3G. Transient Viscous Flow Between Parallel Planes: Part 2

A viscous fluid is initially at rest between two semi-infinite parallel planes (separated by a distance b). At $t=0$, the upper plate begins to slide in the positive x-direction with a constant velocity V_0 (15 cm/s). The governing equation is

$$\frac{\partial v_x}{\partial t} = \nu\frac{\partial^2 v_x}{\partial y^2}. \tag{1}$$

Find the steady-state solution and then let $v_x = v_1 + v_{xss}$. Next, propose that $v_1 = f(y)g(t)$ and solve the problem with separation of variables. Finally, use your analytic solution to obtain velocity profiles at $t=0.05$, 0.5, and 5 s. How many terms are required in the infinite series for convergence? Prepare a suitable figure showing the results. The liquid properties are as follows: $\mu=4.75$ cp, $\rho=1.15$ g/cm^3. Take $b=1$ cm.

Problem 3H. Unsteady (Start-Up) Flow in an Annulus

First, consult Problem 4D.4 in Bird et al. (2002) and examine Problem 3E. We would like to look at the start-up flow in a concentric annulus; the specific gravity of the liquid is 1.15 and the viscosity is 5 cp. At $t=0$, a pressure gradient of (−) 0.02 dyn per square cm per cm is imposed upon the resting fluid contained in the annulus; the two radii (inner and outer) are 1.05 and 3 in., respectively. Determine the time(s) required for the fluid to achieve 25, 65, and 99% of its ultimate velocity. Prepare a plot illustrating the velocity distribution once the 99% level is attained. Use the method of your choice for solution.

Problem 3I. Transient Couette Flow Between Concentric Cylinders

Consider the case in which a viscous fluid is contained between concentric cylinders; the outer cylinder is rotating at a constant 100 rpm and the inner cylinder is fixed and stationary. The radii are 9 and 10 cm for the inner and outer cylinders, respectively. Thus, the annular gap is exactly 1 cm. The fluid contained within has a viscosity of 2 cp and a density of 1 g/cm^3.

At $t=0$, the rotation of the outer cylinder is stopped completely. Prepare a plot that shows the evolution of the velocity profile as the fluid comes to rest. About four profiles will be necessary to adequately illustrate the process. Do you see anything (connected with the shape of those profiles) that might be cause for concern with regard to the stability of such a flow?

Problem 3J. Flow Inside a Rectangular Enclosure: A Variation of Problem 3B.6 on Page 107 in Bird et al. (2002)

Consider a rectangular enclosure filled with a viscous oil. The lower surface moves with constant velocity V_0 in the x-direction; the upper surface (at $y=\delta$) is fixed and stationary. We will examine the flow under steady-state conditions, *far from the ends* of the apparatus. Because the ends are sealed, there must be an extensive region in which the velocity is negative, that is, the net flow inside the enclosure must be zero. Under these conditions, the governing equation is

$$\frac{1}{\mu}\frac{\partial p}{\partial x} = \frac{\partial^2 v_x}{\partial y^2}.$$

Find an expression for the velocity distribution in the enclosure. Then, if $V_0 = 750$ cm/s, what must *dp/dx* be to yield no net flow? Prepare a figure illustrating the velocity distribution from $y=0$ to $y=\delta$ for this case. The viscosity and specific gravity of the oil are 89 cp and 0.9, respectively. The gap between the planar surfaces (δ) is 4 mm.

Problem 3K. Viscous Flow Near a Wall Suddenly Set in Motion

Examine the parabolic partial differential equation that describes viscous flow in Stokes' first problem:

$$\frac{\partial v_x}{\partial t} = \nu\frac{\partial^2 v_x}{\partial y^2}.$$

As we noted previously, this problem can be solved readily through use of the substitution

$$\eta = \frac{y}{\sqrt{4\nu t}},$$

resulting in $(v_x/V_0) = \text{erfc}(\eta)$.

However, we would now like to explore an explicit numerical procedure that can later be adapted to other types of problems. Let the i index refer to y-position and let j refer to time. *One* finite difference representation for the governing equation can be written as

$$\frac{v_{i,j+1} - v_{i,j}}{\Delta t} \cong \nu\frac{v_{i+1,j} - 2v_{i,j} + v_{i-1,j}}{(\Delta y)^2}.$$

Clearly, this can be rearranged to solve for the velocity on the new time step; a solution can be achieved by simply forward marching in time. However, this elementary explicit procedure does have an important limitation as we observed previously; the parameter appearing on the right-hand side is restricted such that

$$\frac{(\Delta t)(\nu)}{(\Delta y)^2} \leq \frac{1}{2}.$$

Find the numerical solution for this problem for $t=24$ s, given that $V_0 = 10$ cm/s and that $\nu = 0.15$ cm^2/s. Use the following value for nodal spacing: $\Delta y = 0.1$ cm and choose three time steps: 0.033, 0.02, and 0.01 s. Compare the three solutions graphically with the known analytic solution. Are your computational results adequate?

Problem 3L. Unsteady Poiseuille Flow Between Parallel Planes

A viscous fluid initially at rest is contained between stationary planar surfaces. The lower surface corresponds to $y=0$ and the upper plane is located at $y=b$. The flow is initiated at $t=0$ by the imposition of a pressure gradient *dp/dx*. The appropriately simplified equation of motion is

$$\frac{\partial v_x}{\partial t} = -\frac{1}{\rho}\frac{dp}{dx} + \nu\frac{\partial^2 v_x}{\partial y^2}.$$

It proves to be convenient to begin by finding the steady-state solution, which is

$$v_x = \frac{A}{2\nu}(y^2 - by), \quad \text{where} \quad A = \frac{1}{\rho}\frac{dp}{dx}.$$

Now let $v_x(y,t) = v_{x1} + (A/2\nu)(y^2 - by)$, that is, allow the velocity of the fluid be represented by both transient and steady-state parts. When this form is introduced into the original equation, the pressure term is eliminated, leaving us with

$$\frac{\partial v_{x1}}{\partial t} = \nu\frac{\partial^2 v_{x1}}{\partial y^2}.$$

We now apply separation of variables in the usual fashion, obtaining

$$v_{x1} = C_1\exp(-\nu\lambda^2 t)[\alpha\cos\lambda y + \beta\sin\lambda y].$$

The Newtonian no-slip condition requires that the velocity disappear at $y=0$, consequently, we must set $\alpha=0$. Obviously, the same must be true at $y=b$ as well. This means that either the leading constant must be zero or $\sin(\lambda b)=0$. The latter is the only logical choice and of course there are an infinite number of possibilities:

$$\lambda_n = \frac{n\pi}{b}, \quad \text{where } n = 1, 2, 3 \ldots.$$

Of course, no single value will produce a solution. We use superposition to write

$$v_x(y,t) = \frac{A}{2\nu}(y^2 - by) + \sum_{n=1}^{\infty} C_n \exp\left(-\frac{\nu n^2 \pi^2 t}{b^2}\right) \sin\frac{n\pi y}{b}.$$

The initial condition requires that fluid be at rest for $t=0$. This means that the transient and steady-state parts must combine such that

$$0 = \frac{A}{2\nu}(y^2 - by) + \sum_{n=1}^{\infty} C_n \sin\frac{n\pi y}{b}.$$

The final step is the selection of coefficients (C's) that cause the series to converge to the desired solution. Recognizing that we have a half-range Fourier sine series, we note

$$C_n = -\frac{2}{b}\int_0^b \frac{A}{2\nu}(y^2 - by)\sin\frac{n\pi y}{b}dy.$$

It is fairly easy to show that

$$C_n = \frac{Ab^2}{\nu n^3 \pi^3}[-2(-1)^n + 2].$$

Notice that the even coefficients are zero. We tend to think of our work as finished at this point, but one should always consider the question of convergence. How many terms must be retained in order to reach sufficient accuracy? Let $b=2$ cm and the kinematic viscosity have a value of 1 cm^2/s. Suppose that the imposed pressure gradient $(1/\rho)(dp/dx) = -55$ cm/s^2. How long will it take the centerline velocity to reach 25, 50, 75, and 90% of its ultimate value? Plot the entire velocity distribution for the 50% case. How many terms were required for convergence? Would the numerical solution give exactly the same results?

Problem 3M. Transient Viscous Flow with Immiscible Fluids

Two immiscible fluids are initially at rest in a rectangular duct (for which $W \gg h$). The light fluid (which is on top) has a density of 0.88 g/cm^3 and a viscosity of 2.5 cp. The heavy fluid has the corresponding property values of 1.47 g/cm^3 and 8 cp. At $t=0$, a pressure gradient is imposed upon the fluid such that $dp/dz = -4.8356$ dyn/cm^2 per cm. We would like to compute the velocity distributions in the duct at $t=0.5$, 3, and 6 s. The duct extends in the y-direction from $y=0$ to $y=b$ where $b=3$ cm. Each fluid occupies exactly one-half of the duct, so the interface is located at $y=b/2$. The governing equation has the form

$$\rho\frac{\partial v_z}{\partial t} = -\frac{\partial p}{\partial z} + \mu\left[\frac{\partial^2 v_z}{\partial y^2}\right].$$

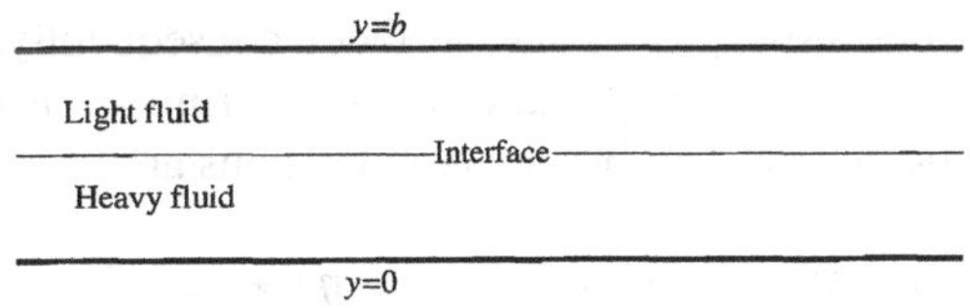

FIGURE 3M. Viscous flow of immiscible fluids.

What is the velocity at the interface between the two fluids after 10 s? Note that the momentum flux at the interface must be continuous, so

$$\tau_1 = -\mu_1 \left.\frac{\partial v_z}{\partial y}\right|_{y=y_i} = \tau_2 = -\mu_2 \left.\frac{\partial v_z}{\partial y}\right|_{y=y_i}.$$

A sketch of the initial setup is shown in Figure 3M. Particular attention needs to be paid to the shape of the velocity distribution near the interface. This will be important to us later.

Problem 3N. Flow in a Microchannel with Slip at the Wall

Consider a pressure-driven flow through a square microchannel, 18 μm on each side. The fluid is an aqueous medium and the pressure drop is 5300 dyn/cm^2 per cm of duct length. The governing equation for the flow is of the Poisson type:

$$0 = -\frac{\partial p}{\partial z} + \mu\left[\frac{\partial^2 v_z}{\partial x^2} + \frac{\partial^2 v_z}{\partial y^2}\right].$$

Find the velocity distribution, the average velocity, and the Reynolds number for this flow using the conventional no-slip boundary condition at the walls. Then, suppose that slip occurs due either to the presence of a gas layer at the boundary or an atomically smooth surface. The boundary condition at the walls must be changed to something like

$$V_0 = L_s \left.\frac{\partial v_z}{\partial y}\right|_{y=0},$$

where L_s is referred to as the slip or extrapolation length. Rework the duct flow problem from above assuming the slip length is 1.25 μm. Find the new velocity distribution, the average velocity, and the Reynolds number (Δp remains the same of course).

Problem 4A. Approximate Solutions for the Blasius Equation

The Blasius equation for the laminar boundary layer on a flat plate is a third-order nonlinear ordinary differential equation,

$$f''' + \frac{1}{2}ff'' = 0.$$

It describes flow past a flat surface; consequently, it has numerous practical applications for determination of drag force. The appropriate boundary conditions are

$$\text{at } \eta = 0,\ f = f' = 0, \text{ and as } \eta \to \infty,\ f' = 1.$$

η is given by $\eta = y\sqrt{(V_\infty/\nu x)}$. As usual, the similarity variable and the stream function are defined in such a way as to produce

$$f' = \frac{V_x}{V_\infty}.$$

Use regular perturbation to find an approximate analytic solution for the Blasius equation and compare your result graphically with the available numerical solution. Provide a plot of both f and f' for $0 \le \eta \le 6$. Is perturbation an appropriate technique for this problem?

Problem 4B. Solution of the Blasius Equation for the Boundary Layer on a Flat Plate

One of the more significant developments in fluid mechanics in the twentieth century was successful treatment of the laminar boundary layer on a flat plate. Blasius accomplished this using the similarity transform in 1908. The transform (scaling) variable is

$$\eta = y\sqrt{\frac{V_\infty}{\nu x}}$$

and the stream function, expressed in terms of f, is

$$\psi = \sqrt{\nu x V_\infty} f(\eta).$$

The transformation (applied to Prandtl's equation) results in the *ordinary* differential equation,

$$\frac{d^3 f}{d\eta^3} + \frac{1}{2} f \frac{d^2 f}{d\eta^2} = 0,$$

with the boundary conditions: at $\eta = 0, f = f' = 0$, and for $\eta \to \infty, f' = 1$.

Find the correct numerical solution for the Blasius equation and then present your results graphically for the entire range of η (both f and f'). Now, consider the flat surface of a race car traveling at 125 mph: Find the thickness of the boundary layer and the drag force at distances from the leading edge ranging from 10 to 100 cm. If the surface of the vehicle was porous and if fluid was drawn through it (pulled from the boundary layer into the interior of the vehicle), how would drag be affected?

The similarity transformation itself should be of interest to you (historically, they were very valuable because the transformation results in a reduction in the number of independent variables). Many significant problems in fluid mechanics were successfully handled by this technique in the first third of the twentieth century. A number of methods have been employed in efforts to identify similarity variables; these include separation of variables, transformation groups, the free parameter method, and dimensional analysis. The second of these, for example, generally involved the following process:

1. Selection of a transformation group.
2. Determination of the general form of the group invariants.
3. Application of the group to the differential equation(s) to identify the specific form of the invariants.
4. Test by trial (Can auxiliary conditions be written in terms of the similarity variables?).

If you have further interest in similarity transformations, you may refer to *Similarity Analyses of Boundary Value Problems in Engineering* by Hansen (1964).

Problem 4C. Additional Solutions of the Falkner–Skan Equation

A fascinating extension of laminar boundary-layer theory was the work of Falkner and Skan (*Aeronautical Research Council, R&M* 1314, 1930) on the family of wedge flows. Recall that the included angle for the wedge was $\pi\beta$ radians. The Falkner–Skan equation has the form

$$f''' + ff'' + \beta(1 - f'^2) = 0,$$

with the boundary conditions:

$$f(0) = f'(0) = 0 \text{ and as } \eta \to \infty,\ f'(\eta) = 1.$$

The potential flow on the wedge is given by $U(x) = u_1 x^m$ and m and β are related by

$$\beta = \frac{2m}{m+1}.$$

The similarity variable and the stream function are

$$\eta = y\sqrt{\frac{m+1}{2}\frac{u_1}{\nu}x^{(m-1)/2}} \quad \text{and}$$

$$\psi = \sqrt{\frac{2}{m+1}}\sqrt{\nu u_1 x^{(m+1)/2}}\, f(\eta).$$

The nonlinear ordinary differential equation given above has caught the attention of numerous applied mathematicians since Hartree published his solutions in 1937. Nearly 20 years later, Stewartson (1954) described additional reverse flow solutions for certain negative included angles. As Stewartson noted, this condition is somewhat artificial; the governing equation is not really capable of fully describing reverse flow.

Indeed, this behavior is not expected given that the usual "Prandtl assumptions" were employed, but it does underscore the nonunique character of the Navier–Stokes equation. Find two solutions for the case in which $\beta = -0.0925$; provide a graphical comparison and identify the value of η corresponding to the largest *negative* velocity. Also, identify the location of maximum strain rate for *both* solutions.

Problem 4D. Simple Problem with Separation

We recognize the limitations of laminar boundary-layer theory; flow in regions near both the stagnation and separation points clearly violates Prandtl's underlying assumptions. Consequently, it is instructive to look at a case where separation can be fully dealt with at reasonable cost with regard to computational time and effort. Consider steady laminar flow in a two-dimensional channel over a (forward-facing) step. The appropriate components of the Navier–Stokes equation can be written as

$$\rho\left(v_x\frac{\partial v_x}{\partial x}+v_y\frac{\partial v_x}{\partial y}\right)=-\frac{\partial p}{\partial x}+\mu\left[\frac{\partial^2 v_x}{\partial x^2}+\frac{\partial^2 v_x}{\partial y^2}\right]$$

and

$$\rho\left(v_x\frac{\partial v_y}{\partial x}+v_y\frac{\partial v_y}{\partial y^2}\right)=-\frac{\partial p}{\partial y}+\mu\left[\frac{\partial^2 v_y}{\partial x^2}+\frac{\partial^2 v_y}{\partial y^2}\right].$$

1. Rewrite the problem in terms of the stream function, vorticity, and the velocity vector components.
2. Solve your equation(s) numerically using the method of your choice and present your results by preparing a plot of the streamlines in the channel. Note that you must use a spatial resolution adequate for the flow features that you wish to examine, namely, possible regions of separation.
3. One of the major problems confronting an analyst in problems of this type is specification of the outflow boundary condition. Explain (clearly).
4. In this flow field, where does the vorticity vector component have the largest value?

Assume a channel height of 6 cm and a one-third cut step 2 cm high. The fluid is water and the mean velocity of approach is 4 cm/s.

Problem 4E. The Poiseuille Flow in the Entrance Section of Parallel Plates

Entrance flows are particularly important in heat and mass transfer applications, and while it might not seem appropriate, boundary-layer methods have been used successfully in such cases. One example is the developing flow between parallel plates. Schlichting used a modified boundary-layer approach to treat this problem in 1934 (*ZAMM*, 14:368, 1934). His technique is also described in *Boundary-Layer Theory* on pp. 176–177 in the 6th edition and pp. 185–186 in the 7th edition. Much later, Wang and Longwell (*AIChE Journal*, 10:323, 1964) revisited this problem, finding numerical solutions that did not rely upon the boundary-layer assumptions. We would like to compare the two approaches.

1. Prepare a brief written description of the essential features of the two approaches, emphasizing how the governing equations differ.
2. At first glance, it might appear that the boundary layer on a flat plate could be used directly (in the case of Schlichting's method) for solution. However, there is a complication related to the core that must be taken into account. Explain.
3. Wang and Longwell show results for two cases. The early profiles for case 1 display a concavity in the middle of the distributions, whereas case 2 does not. What accounts for the difference?
4. Wang and Longwell used a modified independent variable in their analysis. Why? How would one choose a numerical value for the constant c?

Problem 4F. The Biharmonic Equation in Plane Flow and Stokes' Paradox

Recall that for creeping fluid motion in two dimensions, the stream function is governed by the biharmonic equation

$$\nabla^4\psi = 0. \tag{1}$$

In cylindrical coordinates, this is

$$\left(\frac{\partial^2}{\partial r^2}+\frac{1}{r}\frac{\partial}{\partial r}+\frac{1}{r^2}\frac{\partial^2}{\partial \theta^2}\right)^2\psi = 0. \tag{2}$$

We imagine a flow of uniform velocity (at very large distance) approaching a cylinder from left to right. In order to provide this uniform upstream flow, it is necessary that

$$\psi \propto r\sin\theta \text{ as } r\to\infty. \tag{3}$$

Van Dyke (1964) observes that the form of (3) leads us to seek a solution using the product

$$\psi = \sin\theta\cdot f(r). \tag{4}$$

We must impose the no-slip condition at the surface of the cylinder since this is a viscous flow; therefore,

$$\psi(r=R,\theta)=0 \quad\text{and}\quad \left.\frac{\partial\psi}{\partial r}\right|_{r=R}=0. \tag{5}$$

When Stokes investigated this problem in 1851, he discovered that no *global* solution could be found (satisfying all the necessary conditions). He attributed the difficulty to the notion that behind a moving body, the influence of momentum transfer would be felt at continually increasing distance, that is, the problem would always be transient.

- Show that $\psi = C\sin\theta(r\ln r - (r/2) + (1/2r))$ is a solution for the biharmonic equation.
- Try to find a suitable value for C.
- Explain Stokes' paradox and describe why Stokes' conclusion regarding the difficulty appears to be wrong.
- What is it about this particular situation that—no matter how small the Reynolds number—makes the inertial terms in the Navier–Stokes equation important?
- Shaw (2007) found a "patch" for Stokes' paradox and verified it by comparing the analytic C_D with the experimental data. Describe Shaw's approach.

Here are some useful references for this problem:

Langlois, W. *Slow Viscous Flow,* Macmillan (1964).

Shaw, W. T. A Simple Resolution of Stokes Paradox, Working Paper, Department of Mathematics, King's College, London (2007).

Stokes, G. G. On the Effect of the Internal Friction of Fluids on the Motion of Pendulums, *Transactions of the Cambridge Philosophical Society*, 9:8 (1851).

Van Dyke, M. *Perturbation Methods in Fluid Mechanics*, Academic Press (1964).

White, F. M. *Viscous Fluid Flow*, 2nd edition, McGraw-Hill (1991).

Problem 4G. Stokes' Law and Oseen's Correction

Stokes' law for the drag force acting upon a sphere (with creeping fluid motion) is

$$F_D = 6\pi\mu R V_\infty. \tag{1}$$

Extensive data support the validity of this relationship as long as the Reynolds number Re is less than about 0.1. Langlois (1964) showed that for slow viscous flow around a sphere, the importance of the inertial forces could be assessed by examining the ratio in eq. (2). See below.

It is to be noted that *this Reynolds number* (Re) is based upon the sphere's radius R instead of diameter. Suppose we now focus our attention upon regions close to the sphere's surface where $r \to R$. The ratio in these circumstances is negligible, that is, inertial forces are unimportant near the surface. But if we turn our attention to *large* values of r, then

$$\frac{\zeta}{\nu} \approx Re\frac{r}{R}\sqrt{\frac{\sin^2\theta + \cos^2\theta}{1 + (4\cos^2\theta/\sin^2\theta)}} = Re\frac{r}{R}\sqrt{\frac{1}{1 + 4\cot^2\theta}}. \tag{3}$$

Suddenly Stokes' assumptions regarding inertial forces look suspect. Oseen (*Arkiv foer Matematik, Astronomi, och Fysik*, 6:154, 1910) recognized this problem and sought a correction by including a *linearized* inertial term. Thus, in plane flow, the Navier–Stokes equation

$$v_x\frac{\partial v_x}{\partial x} + v_y\frac{\partial v_x}{\partial y} = -\frac{1}{\rho}\frac{\partial p}{\partial x} + \nu\left[\frac{\partial^2 v_x}{\partial x^2} + \frac{\partial v_x}{\partial y^2}\right] \tag{4}$$

would have the left-hand side approximated by

$$V_\infty\frac{\partial v_x}{\partial x}. \tag{5}$$

Obtain Oseen's solution for the stream function for slow viscous flow around a sphere from the literature and plot $\psi(r,\theta)$. What is the essential difference between Oseen's solution and Stokes' result for the flow field around a sphere? What is the approximate Reynolds number *limit* for applicability of Oseen's correction?

Problem 4H. Investigation of the Development of a Vortex Street

Consider a stationary rectangular object (block) centered in the gap between two parallel plates. At $t = 0$, the plates begin to move with a constant velocity V_0. As the Reynolds number increases, a pair of fixed vortices will appear on the downstream side of the block. If the velocity increases further, the vortices will be alternately shed from the block. We would like to explore this scenario, using the paper of Fromm and Harlow (Numerical Solution of the Problem of Vortex Street Development, *Physics of Fluids*, 6:975, 1963) as a guide. We will let the distance between the parallel plates be H and the vertical height of the block be b; we will set $H/b = 6$ for our computations. Initially, we will focus upon $Re = 40$, where $Re = V_0 b/\nu$. Note that we *would* have a plane of symmetry at the centerline *if we restrict our attention to smaller Reynolds numbers.* However, our intent is to look at transient behavior when the wake (initially with fixed vortices) is no longer

$$\frac{\zeta}{\nu} = Re\left(\frac{r}{R} - 1\right)\sin\theta\sqrt{\frac{\left(1 + (R/4r) + (R^2/4r^2)\right)^2\sin^2\theta + \left(1 - (R/2r) - (R^2/2r^2)\right)^2\cos^2\theta}{\sin^2\theta + 4\cos^2\theta}}. \tag{2}$$

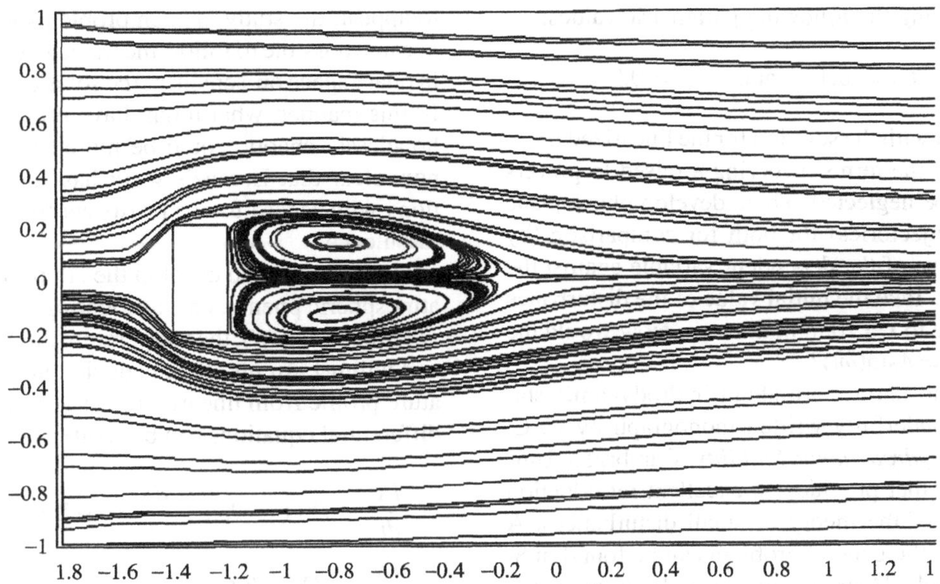

FIGURE 4H. Illustration of computed results for $Re = 40$ with $H/b = 5$. These streamlines were computed with COMSOL™. Note that the fixed vortices extend (in the downstream direction) a distance greater than $2b$.

stable. This is a two-dimensional problem that is best worked with the vorticity transport equation:

$$\frac{\partial \omega}{\partial t} + v_x \frac{\partial \omega}{\partial x} + v_y \frac{\partial \omega}{\partial y} = \nu \left[\frac{\partial^2 \omega}{\partial x^2} + \frac{\partial^2 \omega}{\partial y^2} \right]. \qquad (1)$$

By utilizing the stream function, the definition of vorticity can be written as a Poisson-type partial differential equation:

$$\frac{\partial^2 \psi}{\partial x^2} + \frac{\partial^2 \psi}{\partial y^2} = -\omega. \qquad (2)$$

Of course, the velocity vector components are obtained from the stream function:

$$v_x = \frac{\partial \psi}{\partial y} \text{ and } v_y = -\frac{\partial \psi}{\partial x}. \qquad (3)$$

Flow can be initiated by impulsively moving the walls and, of course, this will create vorticity at the upper and lower boundaries. A simple solution procedure is now apparent: Obtain explicitly a new vorticity distribution from (1). Use the new vorticity distribution to determine the stream function by solving the Poisson equation (2) iteratively. Use the stream function to obtain the velocity vector components everywhere in the flow field by (3). Increment time, and repeat. Fromm and Harlow found that they could stimulate the vortex shedding process by introducing a small perturbation; they did this by artificially increasing the value of ω at three mesh points immediately upstream of the block. Once we are confident that our solution procedure yields the correct results for small Re (a pair of fixed, symmetric vortices), we would like to experiment with such a disturbance (this will be good experience for us, leading to Chapter 5). Keep in mind that the convective transport of vorticity in (1) must be handled appropriately. An example of the expected flow field (plotted streamlines) is shown in Figure 4H for the *steady* case with $H/b = 5$ and a Reynolds number of 40.

Problem 5A. Linearized Stability Theory Applied to Simple Mechanical Systems

Much effort was expended to develop linearized hydrodynamic stability theory at the beginning of the twentieth century. The objective, of course, was to predict the onset of turbulence (i.e., transition from laminar to turbulent flow). In this regard, the theory of small disturbances has been only partially successful. While it has been applied to a number of boundary-layer flows (including the Blasius and Falkner–Skan flows), it has failed completely for the Hagen–Poiseuille flow (finding no instability at any Reynolds number). It is now thought that finite disturbances at the tube inlet may drive the instability in this case. We can examine a simplified problem to familiarize ourselves with the basic concepts. Consider the case of a frictionless cart attached to a wall with a nonlinear spring. If we include viscous damping, the governing equation might appear as

$$\frac{d^2 X}{dt^2} + k_1 \frac{dX}{dt} + C_1 X + C_3 X^3 = F(t). \qquad (1)$$

Let $X = X_0 + \varepsilon$, where ε is a "small" disturbance. Substitute this into the equation above, and subtract out the terms that satisfy the base equation (1). What is left is the disturbance equation. Solve it using the following parametric values:

$$k_1 = 0.1, \quad C_1 = 0.1, \quad \text{and} \quad C_3 = 0.9.$$

Compare your result with the solution for the linearized problem; assume that all the terms involving ε raised to powers greater than 1 can be neglected. Then, develop phase-plane portraits (system trajectories) for both for comparison (by plotting the derivative of the dependent variable against the dependent variable). Take the initial value of the disturbance to be 1 and integrate to $t = 20$ in both cases. How would the results differ if $F(t) = A\sin(\omega t)$?

Should you like to learn more about hydrodynamic stability, there is a wonderfully written monograph by C. C. Lin, *The Theory of Hydrodynamic Stability* (Cambridge University Press, 1945) that provides an excellent introduction to the development of the theory of small disturbances. A broader treatment of the general problem can be found in S. Chandrasekhar's book, *Hydrodynamic and Hydromagnetic Stability*, which was published in 1961 by Dover Publications.

Problem 5B. Practice with Construction of Phase-Plane Portraits

Suppose we construct a function from the product of periodic functions like sine and cosine. In particular, we let $y(t) = \sin(w_1 t)\cos(w_2 t)$; the system trajectory can be developed by cross-plotting $y(t)$ and dy/dt. Construct a system trajectory yourself for the following function:

$$y(t) = 2\sin(4t)\cos(0.75t) + 1.4\sin(0.2t)\cos(8.3t).$$

What are the essential features of the phase-plane portrait?

Problem 5C. Deterministic Chaos: The Lorenz Problem

The sequence—instability, amplification of disturbances, and transition to turbulence—is incompletely understood. In fact, it is possible (but not likely) that the Navier–Stokes equations breakdown at higher *Re*, meaning that the classical hydrodynamical theory may be incomplete. Nevertheless, a picture that many accept has been put forward by O. E. Lanford:

> The mathematical object which accounts for turbulence is an attractor or a few attractors, of reasonably small dimension, imbedded in the very-large-dimensional state space of the fluid system. Motion on the attractor depends sensitively on initial conditions, and this sensitive dependence accounts for the apparently stochastic time dependence of the fluid.

The publication of Edward Lorenz's paper "Deterministic Nonperiodic Flow" in *Journal of the Atmospheric Sciences* (20:130, 1963) did not initially stimulate great interest. However, in the 1970s and 1980s, when graphics terminals began to appear, the study of such problems was revolutionized. It became possible to follow the trajectory of a nonlinear system in phase space on-screen, as the solution was being computed. In this manner, what might have previously appeared to be hopelessly chaotic could be more readily appreciated. It is now clear that Lorenz's work has some profound implications with regard to our prospects for adequately modeling turbulence.

Lorenz set out to develop the simplest possible model for atmospheric phenomena, accounting for the intensity of convective motion (X), the temperature difference between rising and falling currents (Y), and deviation of the vertical temperature profile from linearity (Z). The resulting set of ordinary differential equations can be written as

$$\frac{dX}{dt} = Pr(Y - X), \quad \frac{dY}{dt} = -XZ + rX - Y, \quad \text{and}$$

$$\frac{dZ}{dt} = XY - bZ.$$

We will take $Pr = 10$ and $b = 8/3$. For initial conditions (X,Y,Z), select (0,1,0) and then obtain the projected (on the Y–Z and X–Y planes) system trajectory by numerical solution of the differential equations (setting $r = 28$). The result is a "portrait" of a strange attractor. What are the most important conclusions that one might draw from this study? What is the effect of setting $r = 24$ and then 27? The type of behavior that we are seeing here has sometimes been explained in the popular press as the "butterfly effect." Explain precisely what the implications are with regard to the full and complete modeling of turbulent phenomena.

Note: For a simple mechanical system that oscillates with decaying amplitude, the phase space trajectory (2D) will be an inward spiral—this is characteristic of dissipative systems. The point in phase space to which the trajectory is drawn is called an "attractor." If a frictionless system oscillates with constant amplitude, the phase space portrait will be an ellipse (limit-cycle); such systems are said to be conservative because the phase "volume" remains constant.

Problem 5D. Stability Investigation Using the Rayleigh Equation

We begin by observing that the Rayleigh equation

$$\phi'' - \left[\frac{V''_x}{V_x - c} + \alpha^2\right]\phi = 0$$

will have particular value *if* the solution corresponds to the limiting case for the Orr–Sommerfeld equation when *Re* is very large (μ very small). To give shape to this discussion, we examine the shear layer between two fluids moving in opposite directions; following Betchov and Criminale (*Stability of Parallel Flows,* Academic Press, 1967), the velocity distribution is assumed to have the form

$$V_x = V_0 \tanh\left(\frac{y}{\delta}\right).$$

Refer to Figure 5.4 to see this shear layer at the interface between two fluids moving in opposite directions. For this case, we have

$$\frac{dV_x}{dy} = \frac{V_0}{\delta}\frac{1}{\cosh^2 \frac{y}{\delta}} \quad \text{and} \quad \frac{d^2 V_x}{dy^2} = -\frac{8V_0}{\delta^2}\frac{e^X - e^{-X}}{(e^X + e^{-X})^3},$$

where $X = y/\delta$. We can spend a little time profitably here by carrying out some numerical investigations of this problem. We arbitrarily set $\delta = 1$, $\alpha = 0.8$, and $V_0 = 1$; we start the integration at $y = -4$ and carry it out to $y = +4$. We know that the amplitude function must approach zero at large distances from the interface. If we can find a value of c that results in meeting these conditions, we would identify an eigenvalue. We will start with $c = 0$ and let $\phi(-4) = 0$; the latter is an approximation since the amplitude function is certainly small but not really zero at $y = -4$.

Begin by computing $\phi(y)$ for $\alpha = 0.8$ and $c = 0$; note that we *cannot obtain a solution* for this eigenvalue problem with these values. This is clear, because we cannot obtain the expected symmetry between negative/positive values of y. In fact, Betchov and Criminale show that the eigenvalue for this α is $c_r = 0$ and $c_i = 0.1345$. Continue this exercise by increasing the value of α and repeating the process. Search for a solution using values of α ranging from 0.98 to 1.02. Identify the correct eigenvalue (if you can) in this range. Construct a figure that illustrates how the amplitude function behaves for this range of α 's.

Problem 5E. Closure and the Reynolds Momentum Equation

It will clearly be necessary for many flows of engineering interest to use the Reynolds momentum equation to obtain some type of result. The development of the logarithmic velocity distribution using mixing length theory is an example. Any effort to model the Reynolds "stresses" with mean flow parameters must be viewed with suspicion, and any result thus obtained will still require empirical determination of parameters. It is worthwhile, therefore, to investigate existing closure schemes simply to become familiar with the options that are available. Prepare a brief historical sketch of methods that have been developed to achieve closure in turbulence modeling (using the RANS); your work should not exceed three typewritten pages, but should include sufficient detail so that a neophyte could gain an appreciation for the scope of the closure problem in turbulence.

Problem 5F. Turbulent Pipe Flow at Re = 500,000

John Laufer's experimental study "The Structure of Turbulence in Fully Developed Pipe Flow" is available as *NACA Report 1174*. He made extensive measurements in a 9.72 in. diameter brass tube using hot wire (90% platinum–10% rhodium) anemometry. The following data were obtained at a Reynolds number of 500,000 (based upon the centerline velocity 100 ft/s).

s/R, Dimensionless	V/V_{max} ($V_{max} \approx$ 100 ft/s)
0.0005	0.118
0.001	0.171
0.0015	0.22
0.002	0.269
0.0025	0.327
0.003	0.39
0.0035	0.44
0.004	0.492
0.006	0.529
0.010	0.6
0.0164	0.636
0.025	0.664
0.05	0.72
0.1	0.781
0.2	0.83
0.3	0.878
0.4	0.917
0.5	0.93
0.6	0.964
0.7	0.977
0.8	0.989
0.9	0.994

Use the available data to find V^*, where $V^* = \sqrt{(\tau_0/\rho)}$. Prepare a semilogarithmic plot of the measurements above in the form of $V^+(s^+)$, where $V^+ = \overline{V}/V^*$ and $s^+ = sV^*/\nu$. Use data in the turbulent core to fit Prandtl's logarithmic equation. What is the "best" value of the "universal" constant κ? Can you identify a "laminar sublayer" where $V^+ = s^+$? If so, how far does it extend? Next, plot the data using Schlichting's empirical curve fit:

$$\frac{\overline{V}}{V_{max}} = \left(1 - \frac{r}{R}\right)^{1/n}.$$

Based upon Laufer's data for $Re = 500{,}000$, what is the "best" value for n? Finally, the rule of thumb in turbulent pipe flow is that the average velocity is about 80% of the maximum. What is that ratio for these data?

Laufer also measured the pressure along the pipe axis, obtaining the following:

z/D	4	8	12	16
$(P-P_e)/q$	0.04	0.08	0.1198	0.1596

Please note that q is the dynamic pressure at the pipe centerline and P_e is the mean pressure at the pipe exit.

Problem 5G. Second-Order Closure Models

Search the recent literature and find an application of second-order closure ($k-\varepsilon$) modeling. Write a brief two-paragraph summary of the work. Then, answer the following questions:

1. How were the parametric choices made?
2. Is the performance of the model realistic based on what you would expect in this flow?
3. How was the adequacy of the modeling assessed?
4. Did the authors use their own code or a commercially available package?
5. Can this particular model be extended to other, perhaps related, flows? If so, which?
6. What do the authors characterize as the principal contribution of their work?

Problem 5H. Decaying Turbulence in a Box

Consider the data shown below (for decaying turbulence in a box); a hot wire anemometry has been used to measure the velocity of air circulating in a box. At $t \approx 2.39$ s, the energy supply (a centrifugal blower) was shut off and the flow begins to decay. Note that the approximate mean velocity prior to shutdown was about 6 m/s. Within just 6 or 7 s, the mean velocity has fallen to about 0.06 m/s.

Assuming the integral length scale l is about 25.5 cm, the initial value of the Reynolds number is

$$Re_l = \frac{ul}{\nu} = \frac{(600)(25.5)}{(0.151)} = 1 \times 10^5.$$

The decay process shown in Figure 5H is initiated at about 2.39 s. Note that the mean velocity at the end of each time segment was about 25, 6, and 2 cm/s, respectively. That is, at $t = 12.29$ s, the average velocity has fallen to about 2 cm/s. Of course, this point is about 12.29–2.39 = 9.9 s into the decay period. The sample interval was 0.002 s such that the Nyquist frequency is 250 Hz.

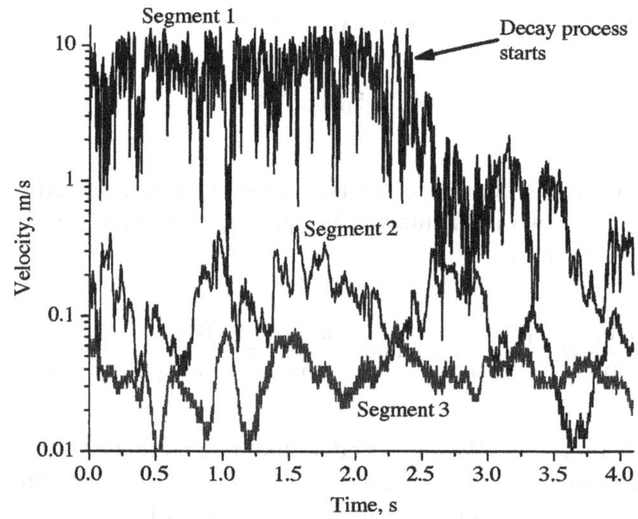

FIGURE 5H. Experimental data shown in three segments, each corresponding to 4.096 s.

1. Find the autocorrelation coefficient and the power spectrum for the initial data (from $t = 0$ to $t = 2.39$ s).
2. Estimate the initial value for the Kolmogorov microscale η.
3. Model the decay process using Taylor's inviscid approximation for the dissipation rate per unit mass: $\varepsilon \approx A(u^3/l)$. When will the kinetic energy of the turbulence ($(3/2)u^2$) fall to 0.1% of its initial value?
4. According to your model, when will this process enter the final period of decay (which is approximately earmarked by $Re_l = ul/\nu = 10$)?
5. During the final period of decay, the estimate for the dissipation rate per unit mass must be replaced by $\varepsilon \approx c\nu u^2/l^2$. What is the approximate value of c?

Problem 5I. Time-Series Data and the Fourier Transform

Consider the time-series data provided to you separately. These data were obtained from impact tube (ID = 0.95 mm) measurements made on the centerline of a free turbulent (air) jet. In one case, the flow was unobstructed and in the other, an aeroelastic oscillator was positioned in the flow field. We would like to use the Fourier transform to identify important periodicities present in the data (in the case of the oscillator, this should not be too difficult). This is an extremely valuable technique in the study of turbulence and nonlinear phenomena in general. Recall that the autocorrelation for a time-varying signal, $u(t)$, is given by

$$\rho(\tau) = \frac{\overline{u(t)u(t+\tau)}}{\overline{u^2}},$$

and the power spectrum (one-sided) is defined as

$$S(\omega) = \frac{1}{\pi}\int_0^\infty \rho(\tau)\cos(\omega\tau)d\tau.$$

S can be thought of as the distribution of signal energy in frequency space.

Prepare figures that will allow easy comparison of the computed frequency spectra. You are free to use the Fourier transform (FFT) package or software of your choice.

Can you identify any particularly important frequencies for the unobstructed case?

Problem 5J. Time-Series Data and the Fourier Transform

It is natural to think of the power spectrum in connection with measurements of velocity (or dynamic pressure) in turbulent flows. As we have seen, the Fourier transform can be used to identify important periodicities in time-series data. Consider the use of an impact tube in conjunction with a pressure transducer; such an arrangement has been employed to make measurements in a turbulent free jet (air) where the mean velocity was approximately 13 m/s. Two cases were examined, one in which the impact tube was aligned with the center of the jet and the flow was unobstructed, and in the second, an elastically supported rectangular slat was placed in between the jet orifice and the impact tube. In this latter case, aeroelastic oscillations occurred, as anticipated (you may recall the history of the Tacoma Narrows suspension bridge's failure).

If the inside diameter of the impact tube is 0.91 mm ($T = 22.5°C$), what could the dissipation rate be at the point of measurement if the equipment is to be capable of resolving the full spectrum of eddy sizes (scales)?

Use a Fourier transform program of your choice to calculate the power spectra for the two data sets that are being supplied to you in separate files. Provide a graphical comparison of the results. What are the effective frequency ranges for the two data sets? In the case of the aeroelastic oscillator, virtually all the signal energy will be concentrated around a single frequency. What is it?

Problem 5K. Time-Series Data for Aerated Jets

Two-phase turbulent jets are common throughout the process industries. For the air–water system (jet aeration),

(a)
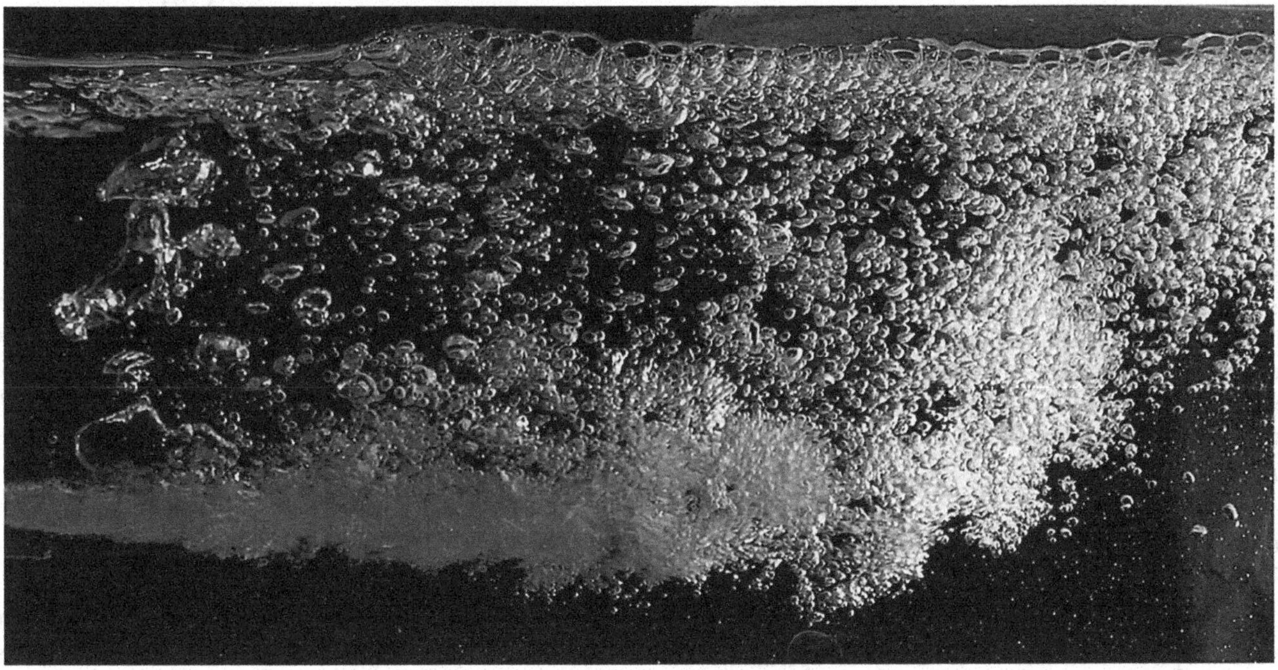

(b)
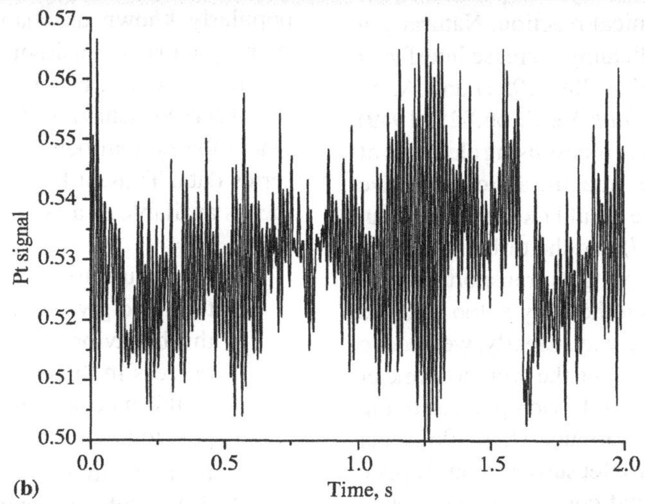

FIGURE 5K. Illustration of jet aeration (a) and typical pressure measurements (b).

typical operation produces a flow similar to that shown in Figure 11.1a and b.

A small-diameter impact tube has been used in conjunction with a pressure transducer to obtain data for this type of flow (but with larger bubbles). Excerpts from these data are shown in Figure 5K(b) and the table that follows.

0.001	0.534	0.5135	0.5175
0.002	0.535	0.518	0.5205
0.003	0.5345	0.5215	0.5255
0.004	0.534	0.525	0.5325
0.005	0.533	0.526	0.54
0.006	0.533	0.526	0.5455
0.007	0.5315	0.5245	0.5495
0.008	0.5315	0.522	0.5505
0.009	0.531	0.518	0.5475
0.01	0.531	0.5135	0.5425
0.011	0.5305	0.5085	0.535
0.012	0.5305	0.503	0.5275
0.013	0.529	0.498	0.521
0.014	0.528	0.4955	0.516
0.015	0.5275	0.497	0.513

The first column is time, followed by three columns of data (each with 2048 entries). The time interval (Δt) for sampling was 0.001 s, therefore, the Nyquist cut-off frequency is $f_c = 1/(2\Delta t) = 500$ Hz. Furthermore, since only $3 \times 2048 = 6144$ points have been recorded, we will not be able to detect periodic phenomena occurring slower (less frequently) than about 6 Hz. Use the Fourier transform to find the power spectrum for these data and plot the autocorrelation coefficient. Estimate the integral timescale from your graph of $\rho(\tau)$.

Problem 5L. Breakage of Fluid-Borne Entities in Turbulence

In the chemical process industries, the breakage of suspended droplets is critical to a variety of operations that involve mass transfer and/or chemical reaction. Naturally, a reduction in droplet size can significantly increase interfacial area. J. O. Hinze (*AIChE Journal*, 1:289, 1955) and A. N. Kolmogorov (*Doklady Akademi Nauk SSSR*, 66:825, 1949) were among the first to examine this process using elements of the statistical theory of turbulence. Imagine a droplet of size d suspended in a turbulent flow; we would like to think about interactions between the droplet (d) and the turbulent eddies (L). If $L \gg d$, then the droplet simply gets transported without any deformation. If $L \ll d$, then the eddy is too small to affect the droplet in any substantive way. Clearly, we need to focus upon cases where the eddy size and the droplet diameter are comparable, that is, where $L \approx d$. Levich pointed out in *Physicochemical Hydrodynamics* (Prentice-Hall, 1962) that the variation in velocity near the droplet surface would create differences in dynamic pressure that could be large enough to produce droplet deformation. The key equations describe the dynamic pressure variation and the pressure difference across the interface (the Laplace relation):

$$Q = k_f \frac{\rho(u_1^2 - u_2^2)}{2} \quad \text{and} \quad p_i - p_0 = \frac{2\sigma}{R}.$$

Now, suppose that the critically sized eddies lie in the inertial subrange of the three-dimensional spectrum of turbulent energy, where $E(\kappa) = \alpha\varepsilon^{2/3}\kappa^{-5/3}$. A characteristic velocity for these eddies can be determined:

$$u(\kappa) \approx [\kappa E(\kappa)]^{1/2} = \alpha^{1/2}\varepsilon^{1/3}\kappa^{-1/3}.$$

Since the critical wave number is related to the droplet size by $\kappa = 2\pi/d$,

$$[u(d)]^2 = \frac{\alpha\varepsilon^{2/3}d^{2/3}}{(2\pi)^{2/3}}.$$

Therefore, a simple force balance can be used to determine a threshold droplet size:

$$d = A\left(\frac{\sigma}{\rho}\right)^{3/5}\varepsilon^{-2/5}.$$

Show that this relationship is correct, and use it to determine the droplet size(s) expected for the agitation of a lean dispersion of benzene in water, where $\sigma \cong 35$ dyn/cm. Obtain reasonable values for the expected range of dissipation rates from the extensive STR (stirred tank reactor) literature. Is there a lower limit for benzene droplet size? Explain.

Problem 5M. Turbulence, Determinism, and Nonlinear Systems

Many nonlinear systems display evolution in time that is irregular and/or unpredictable. This behavior has become popularly known as chaos. One of the characteristics of such systems is sensitivity to initial conditions, referred to as SIC. However, it is not always readily apparent whether the observed behavior is truly chaotic, particularly in cases where the system behavior is obtained in the form of time-series data. Thus, it has become very important to have the means available to address this question.

1. One route to chaos is period doubling. Define this term and give some examples of systems that exhibit this behavior. Recall we concluded that the transition process in the Hagen–Poiseuille flow does not occur by this mechanism. Explain and offer support for your position.
2. In the study of the transition to turbulence, systems that exhibit an evolutionary (or spectral) transition

process are of great interest to theoreticians. Why? What are examples of such systems? Describe some of the tools that might be used in a study of such a process.

3. In the January 1998 issue of *Atlantic Monthly,* William H. Calvin describes how gradual warming of the planet could lead to drastic and abrupt cooling, with catastrophic effect upon civilization in Europe. In particular, failure of the northernmost loop of the North Atlantic current could produce (in about a decade) a severe drop in temperature that might result in food shortages for about 650 million people. That such events have occurred in the past seems clear, based upon data obtained from ice corings from Greenland. Suppose appropriate measurements produced time-series data (for annual temperature and atmospheric composition); what tests could you perform that might help identify characteristics of appropriate climatic models? That is, *How will you determine whether the global climate should be regarded as chaotic?*
4. The Lyapunov exponent has been used to estimate the divergence of system trajectories on (or about) an attractor. Is there any realistic way that it could be used in the context of the global climate? Explain carefully.

Problem 5N. Statistical Theory of Turbulence, Correlations, and Spectra

A (1–1) spatial correlation (with separation in the "2" or y-direction) has been measured (A. J. Reynolds, *Turbulent Flows in Engineering,* Wiley-Interscience, 1974) for grid-generated turbulence (mesh size, 3 in. × 3 in.) in a wind tunnel and the data for a mean velocity of 15 ft/s are provided in the following table.

Spatial Separation r (in.)	Correlation Coefficient $R_{11}(r)$
0.035	0.981
0.05	0.962
0.07	0.928
0.1085	0.851
0.197	0.716
0.284	0.565
0.512	0.370
1.00	0.180
2.00	0.036
3.00	– 0.022
4.00	– 0.026
6.00	– 0.015
8.00	0.00

Use these data to find the integral length scale l and the Taylor microscale λ. Is there any way to estimate the Kolmogorov microscale (η) from the available information? If so, do so.

Then, use the data above to find the one-dimensional wave number spectrum,

$$\phi_{11}(\kappa_2) = \frac{1}{2\pi}\int_{-\infty}^{+\infty} \exp(-i\kappa_1 r)dr.$$

Assume that the correlation coefficient is an even function. Does the spectrum exhibit an inertial subrange (where $\phi_{11} \propto \kappa_1{}^{-5/3}$)? If so, how extensive is it? Can you identify the wave number range that corresponds to the energy-containing eddies? If so, what is it? Finally, can you tell where the dissipation range begins in your spectrum? If you can, does that wave number correspond (inversely) to your estimate of η?

Problem 5O. Velocity Measurements for the Turbulent Flow in a Pipe

John Laufer carried out a very meticulous study of turbulent flow of air through a 9.72 in. diameter tube (The Structure of Turbulence in Fully Developed Pipe Flow, *NACA Report 1174,* 1954). He studied two Reynolds numbers 50,000 and 500,000, both based upon the centerline (maximum) velocity. He used the hot wire anemometry to measure point velocity; a reconstruction of his data for flow *close to the pipe wall* is given in the following table.

Dimensionless Position s/R	V/V_{max} ($Re = 500{,}000$)	V/V_{max} ($Re = 50{,}000$)
0	0	0
0.000275	0.098	0.0115
0.00055	0.17	0.024
0.000825	0.22	0.0363
0.0011	0.265	0.0483
0.001375	0.329	0.0608
0.0018	0.385	0.0792
0.0028	0.44	0.118
0.004	0.491	0.176
0.006	0.537	0.261
0.008	0.570	0.322
0.010	0.59	0.384
0.012	0.612	0.4198
0.014	0.63	0.46
0.016	0.638	0.486
0.018	0.645	0.51
0.02	0.65	0.52
0.024	0.665	0.55
0.028		0.575
0.032		0.591
0.036		0.605

Use these data to

1. Estimate the shear (or friction) velocities.
2. Prepare *appropriate* plots of $v^+(s^+)$.

3. Determine if logarithmic equations can be fit to any portion(s) of the data.
4. Fit the "corrected" ($\tanh^{-1}$) equation (5.56) to appropriate portions of the data and determine (best values for) the constants of integration.
5. Estimate the friction factor ($F = AKf$) using these data and compare with values from the Moody chart (was Laufer's pipe hydraulically smooth?).

Problem 5P. The Burgers Model of Turbulence

J. M. Burgers proposed a simplified model of turbulence (Mathematical Examples Illustrating Relations Occurring in the Theory of Turbulent Fluid Motion, *Akad. Amsterdam*, 17:1, 1939) with the hope that such a system (since it shared some of the characteristics of the Navier–Stokes equations) might provide new insight into turbulence. Burgers' model consists of

$$\frac{dU}{dt} = P - u^2 - \nu U \qquad (1)$$

and

$$\frac{du}{dt} = Uu - \nu u. \qquad (2)$$

Note that P is a source term, or driving force, analogous to pressure. Time t is the only independent variable, but the system is nonlinear through u^2 and Uu. If these equations are multiplied by U and u, respectively, and added together, one obtains an "energy" equation:

$$\frac{1}{2}\frac{d}{dt}\left[U^2 + u^2\right] = PU - \nu(U^2 + u^2). \qquad (3)$$

If the disturbance quantity u is zero, then it can be shown that a "laminar" solution exists if $P < \nu^2$ (the reader may refer to Chapter VII in A. Sommerfeld's book *Mechanics of Deformable Bodies*, Academic Press, 1950).

Bec and Khanin (Burgers Turbulence, submitted to *Physics Reports*, 2007) note that recent years have seen renewed interest in Burgers' model; they report applications in statistical mechanics, cosmology, and hydrodynamics. Of particular interest are recent efforts to explore "kicked" Burgers turbulence, where the model is subjected to impulsive forcing functions applied either periodically or randomly. Our intent is to study eqs. (1) and (2) numerically, beginning with the case in which the fluctuation u is initially perturbed with a constant. Assume initial values of U and u corresponding to 0 and 0.02, respectively. Let $P/\nu \approx 3$; solve the equations to obtain Figure 5P:

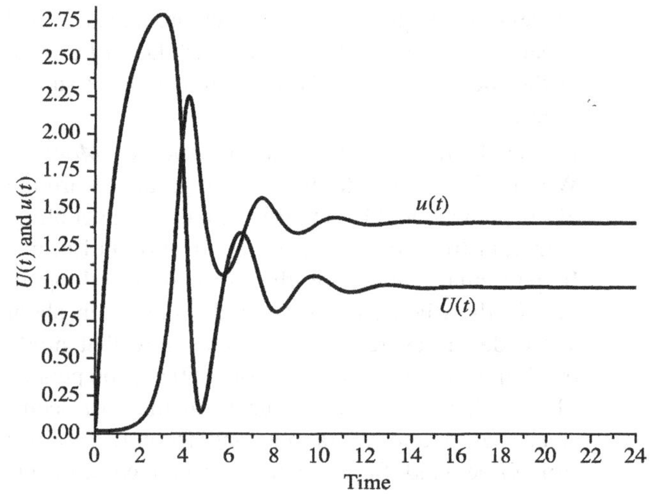

FIGURE 5P. Illustration of the numerical solution of the Burgers model with u initially perturbed.

Next, introduce a periodic disturbance (or kick) to the model by assigning u a random value between 0 and 1 at a set interval. How is the response of the model changed? Does it make any difference if the disturbance is applied periodically or at a random interval? What is the effect of changing the interval between disturbances upon the solution? Consult the literature to determine whether chaotic behavior can ever emerge from the Burgers model.

Problem 6A. Transient Conduction in a Mild Steel Bar

Consider a steel bar of length L at an initial temperature of 300°C. At $t = 0$, two large thermal reservoirs are applied to the ends of the bar, instantaneously imposing a temperature of 0°C at both $y = 0$ and $y = L$. The temperature in the interior of the bar is governed by the parabolic partial differential equation:

$$\frac{\partial T}{\partial t} = \alpha \frac{\partial^2 T}{\partial y^2}.$$

Clearly, this is a candidate for separation of variables; letting $T = f(y)g(t)$ leads to

$$T = C_1 \exp(-\alpha\lambda^2 t)[A \sin \lambda y + B \cos \lambda y].$$

However, for all positive t's, we have $T = 0$ at both $y = 0$ and $y = L$; therefore, $B = 0$ and $\sin(\lambda L) = 0$. Consequently, $\lambda = n\pi/L$ with $n = 1,2,3,\ldots$. The solution for this problem then takes the form

$$T = \sum_{n=1}^{\infty} A_n \exp(-\alpha\lambda_n^2 t) \sin \lambda_n y.$$

We apply the initial condition: at $t = 0$, $T = 300$ for all y,

$$300 = \sum_{n=1}^{\infty} A_n \sin \lambda_n y.$$

This is a half-range Fourier sine series; by the Fourier theorem, we have

$$A_n = \frac{2}{L}\int_0^L 300 \sin\frac{n\pi y}{L}dy.$$

Find and plot the temperature distributions in the steel bar for $t = 15$, 30, 60, and 120 s. When will the center temperature in the bar become 7.5°C? Let $L = 15$ cm.

Problem 6B. Conduction in a Type 347 Stainless Steel Slab

The thermal conductivity of type 347 stainless steel varies significantly with temperature as shown by the four data points (adapted from Kreith, *Principles of Heat Transfer*, 2nd edition, 1965) given below:

T (°F)	32	212	572	932
k (Btu/(h ft °F))	8.0	9.31	11.0	12.8

Naturally, the question of how this variation affects transient conduction is of pressing interest in heat transfer. We begin by assuming that we have a two-dimensional slab of 347 that measures 20 cm × 20 cm. The stainless steel is initially at a uniform temperature of 60°F, but at $t = 0$, the front face is suddenly heated to 900°F. The left and top faces are insulated such that $q = 0$. The right face loses thermal energy to the surroundings and the process is adequately described by Newton's law of cooling: $q = h(T_s - T_\infty)$. By experiment we know that $h = 1.95$ Btu/(h ft^2 °F). If the thermal conductivity were constant, then the appropriate equation would be simply

$$\frac{\partial T}{\partial t} = \alpha\left[\frac{\partial^2 T}{\partial x^2} + \frac{\partial^2 T}{\partial y^2}\right].$$

We would like to determine how $k(T)$ will affect heat flow into the slab. Find the evolution of the temperature distribution for both cases (constant and variable k) and prepare contour plots for easy comparison.

Problem 6C. Global Warming and Kelvin's Estimate of the Age of the Earth

A great debate between physicists and geologists was initiated in 1864 by Lord Kelvin when he estimated the age of the earth using the known geothermal gradient. His conclusion, an age less than 100 million years, was in conflict with the geologic evidence of stratification. We now know that the increase in melting temperature with pressure and the production of thermal energy by radioactive decay account for Kelvin's underestimate.

More recently, Lachenbruch and Marshall (*Science*, 234:689, November 1986) have obtained extensive temperature data from oil wells drilled in the Arctic. These temperature logs indicate recent warming of the permafrost at the surface. Such data may prove to be an irrefutable indicator of global climate change brought about by the activities of man. Indeed, there is no assurance that such changes will not lead to extinctions (of polar bears, for one example). See Jarvis (Trouble in the Tundra, *Chemical & Engineering News*, Vol. 87, No. 33, pp. 39, 2009) for an updated view of warming in the Arctic; the recent proliferation of "thermokarsts" is a troubling development.

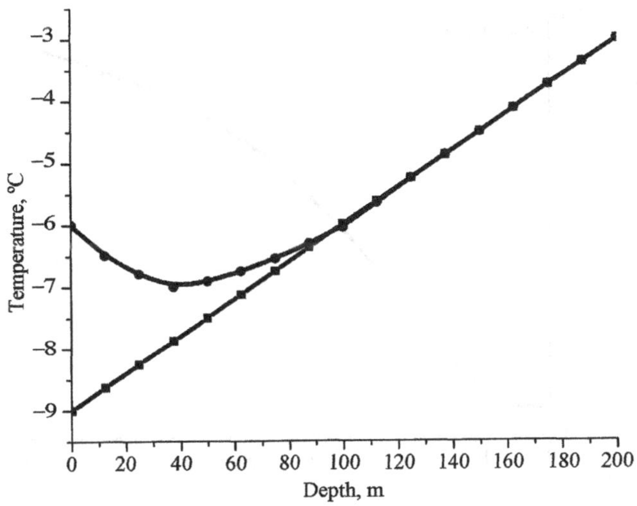

FIGURE 6C. Data adapted from Lachenbruch and Marshall, *Science*, 234:689 (1986).

Develop your own transient model of the surface temperature perturbation that will reproduce the essential characteristics of the Awuna (1984) temperature profile cited on page 691 of the Lachenbruch and Marshall report (Figure 6C). Then, extrapolate your model 50 years (from the publication date). What will the temperature profile near the surface look like in 2036?

Problem 6D. Transient Conduction in a Cylindrical Billet

Consider an experiment in which we can examine transient conduction in a solid cylindrical billet. In the laboratory, a cylindrical specimen ($L = 6$ in. and $d = 1$ in.) is removed from an ice water bath (3 or 4°C) and plunged into a heated constant temperature bath maintained at 72°C. The center temperature of each sample is recorded as a function of time, resulting in temperature histories as illustrated in Figure 6D for Plexiglas® and stainless steel. For the former, use the appropriate figure in Chapter 6 (6.11) to estimate the thermal diffusivity of acrylic plastic; do so at 50 s intervals for t's ranging from 50 to 400 s. Do you have reason to believe that any of your estimates are more reliable than others? Note that the center of the Plexiglas® sample attains only 49°C in

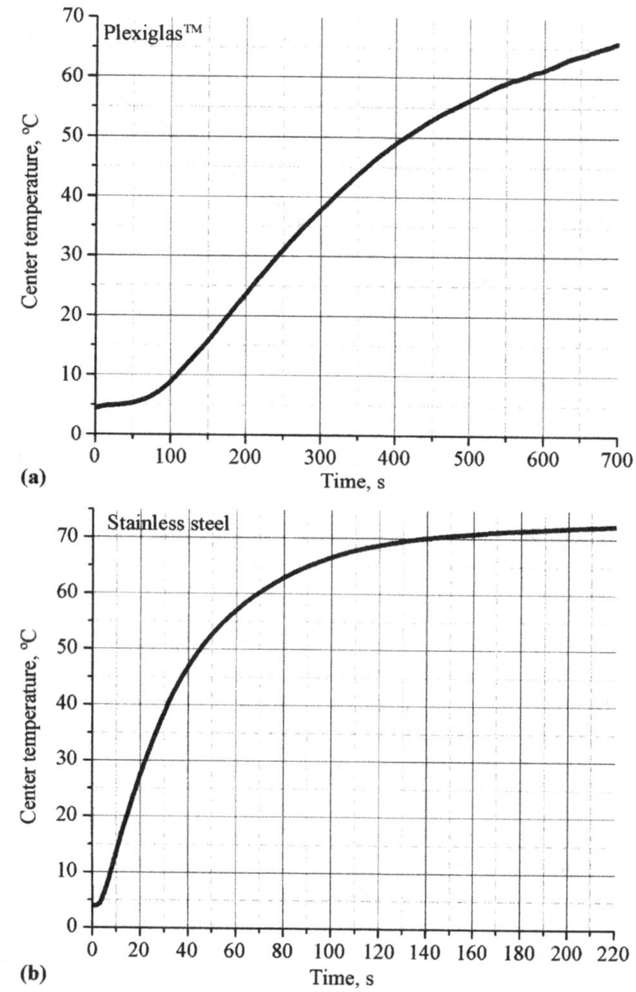

FIGURE 6D. Temperature histories for two cylindrical specimens.

400 s! The data for the stainless steel sample must be treated differently since the main resistance to heat transfer is now located outside the sample. We will define the dimensionless temperature as

$$\theta = \frac{T - T_b}{T_i - T_b},$$

where T_b is the temperature of the heated bath and T_i is the initial temperature of the specimen.

In both cases, the governing partial differential equation can be written as

$$\rho C_p \frac{\partial \theta}{\partial t} = k\left[\frac{1}{r}\frac{\partial}{\partial r}\left(r\frac{\partial \theta}{\partial r}\right)\right]$$

(if we neglect axial conduction). Although the solutions have the same functional form

$$\theta = \sum_{n=1}^{\infty} A_n \exp(-\alpha\lambda^2 t) J_0(\lambda_n r),$$

the boundary condition (at $r = R$) for the stainless steel cylinder must be written as

$$-k\left.\frac{\partial T}{\partial r}\right|_{r=R} = -h(T_{r=R} - T_\infty).$$

This leads to the transcendental equation $\lambda_n R J_1(\lambda_n R) = (hR/k) J_0(\lambda_n R)$, where hR/k is the Biot modulus.

Assume that the thermal conductivity of (type 304) stainless steel is known: $\alpha = 0.156\ \text{ft}^2/\text{h}$. Find the value of the heat transfer coefficient h that gives best agreement with the experimental data.

Problem 6E. Temperature Distribution in an Aluminum Rod Heated at One End

Consider a horizontal aluminum rod with one end inserted into a brass cylinder that can be rapidly filled with low-pressure saturated steam. At $t = 0$, saturated steam is admitted to the brass drum and the end of the metal rod is instantaneously heated to about 120°C. The 1 in. diameter aluminum rod has copper-constantan thermocouples embedded at z-positions of 1.5, 4.5, 11, 17, 24.5, 32, 47, 62.5, 77.5, and 93 cm. In this way, we can monitor the temperature $T(z, t)$. One model for this scenario can be written as

$$\alpha\frac{\partial^2 T}{\partial z^2} - \frac{2h}{\rho C_p R}(T - T_\infty) = \frac{\partial T}{\partial t},$$

where the heat loss from the surface of the rod is being accounted for in an approximate way (the ambient temperature is about 25°C). It is convenient to define a dimensionless temperature θ:

$\theta = (T - T_\infty)/(T_0 - T_\infty)$, where T_0 is the temperature at the hot end of the rod for all $t > 0$.

Therefore, the model may be rewritten as $\alpha(\partial^2\theta/\partial z^2) - (2h/\rho C_p R)\theta = (\partial\theta/\partial t)$.

We would like to compare this model to experimental data and find the "best" possible value for the heat transfer coefficient h. *It is to be noted that this analysis can be performed in several different ways* (Figure 6E)!

We do have, among the alternatives, an *approximate* analytic solution (assuming constant h) available:

$$\theta = \frac{1}{2}\left[\exp^{\sqrt{(2h/\alpha\rho C_p R)}z}\operatorname{erfc}\left(\frac{z}{\sqrt{4\alpha t}} + \sqrt{\frac{2h}{\rho C_p R}t}\right) + \exp^{-\sqrt{(2h/\alpha\rho C_p R)}z}\operatorname{erfc}\left(\frac{z}{\sqrt{4\alpha t}} - \sqrt{\frac{2h}{\rho C_p R}t}\right)\right].$$

Find the "best" possible value for h and prepare a graphical comparison with the experimental data shown in Figure 6E. Should the heat transfer coefficient be constant or vary with position (temperature)? Explain your reasoning.

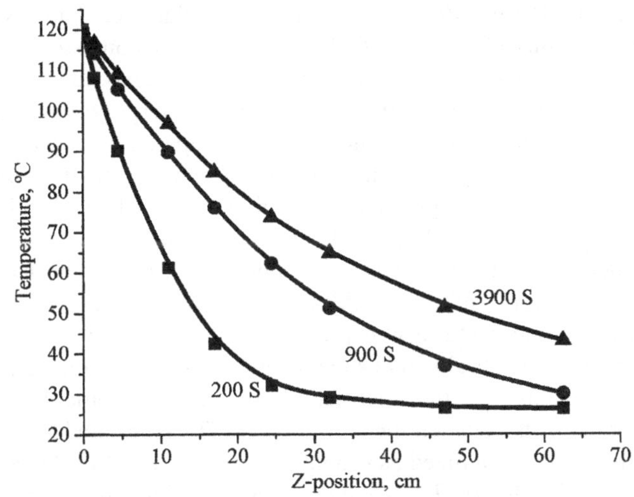

FIGURE 6E. Characteristic experimental results for the 1 in. aluminum rod; the data are temperature profiles at 200, 900, and 3900 s.

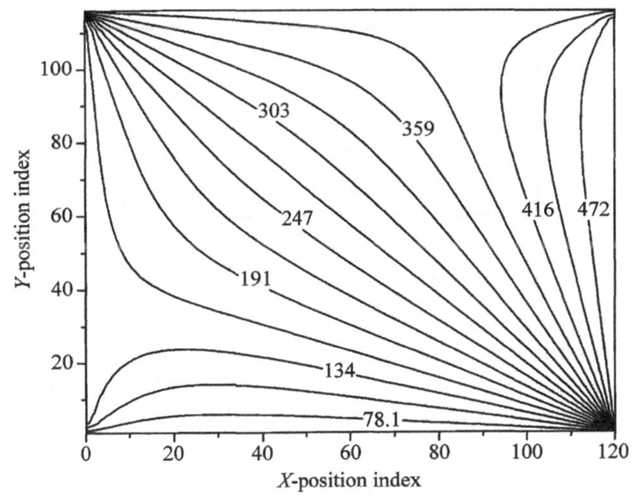

FIGURE 6F. Typical computed temperature field for a two-dimensional slab.

Problem 6F. An Introduction to Steady Two-Dimensional Conduction

The governing equation for this case is $(\partial^2 T/\partial x^2) + (\partial^2 T/\partial y^2) = 0$.

Now we let the i index represent x and j represent y. One finite difference representation for this Laplace equation is

$$\frac{T_{i+1,j} - 2T_{i,j} + T_{i-1,j}}{(\Delta x)^2} + \frac{T_{i,j+1} - 2T_{i,j} + T_{i,j-1}}{(\Delta y)^2} = 0.$$

If we use a square mesh for the discretization, then $\Delta x = \Delta y$ and we have

$$T_{i,j} = \frac{1}{4}(T_{i+1,j} + T_{i-1,j} + T_{i,j+1} + T_{i,j-1}).$$

Accordingly, we have a simple iterative means of solution (Gauss–Seidel or better, SOR). A program was written for a square domain, 40 cm on each side. The edge temperatures are maintained as follows: top, 400°C; bottom, 60°C; left side, 150°C; and right side, 500°C. The resulting temperature field is shown in Figure 6F.

Some extremely interesting changes can be made to the program very easily. For example, suppose we would like one boundary (say, the bottom) to be insulated. Thus, across the x-axis we need $dT/dy = 0$. A second-order forward difference for the first derivative can be written as $(dT/dy)_{i,j} = (1/2\Delta y)(-3T_{i,j} + 4T_{i,j+1} - T_{i,j+2})$. Since this is zero, we can immediately solve for the temperature on the bottom row (x-axis): $T_{i,j} = (1/3)(4T_{i,j+1} - T_{i,j+2})$. What changes would you expect to see in the figure above as a result? Note that this technique could be applied to a three-dimensional solid just as easily. We could also incorporate a source term or Neumann or Robin's-type boundary conditions, if desired.

Perform your own analysis of 2D conduction for a square slab of material with edge temperatures (T,B,L,R) of 600, 175, 75, and 690°C. Prepare an appropriate contour plot as shown in the example above. Then, repeat the analysis but with the bottom of the slab insulated. Compare the results.

Problem 6G. Transient Conduction in an Iron Slab

We would like to investigate the evolution of the temperature distribution in a semi-infinite slab of iron (>99.99%) when one face is instantaneously elevated from 90 to 900K. Prepare two solutions, one assuming constant k and the other taking the temperature dependence of k into account. The data given in the following table are provided for your reference. Assume we are particularly interested in the temperature profiles at $t = 5$ min and $t = 50$ min.

Temperature (K)	Thermal Conductivity (W/cm K)
90	1.46
150	1.04
200	0.94
300	0.803
400	0.694
600	0.547
800	0.433
900	0.380

Note: To obtain k in cgs, divide above values by 4.184.

Problem 6H. Steady-State Conduction in a Rectangular Slab

Consider a rectangular slab of aluminum measuring 40 cm × 20 cm. Three of the edges are maintained at constant

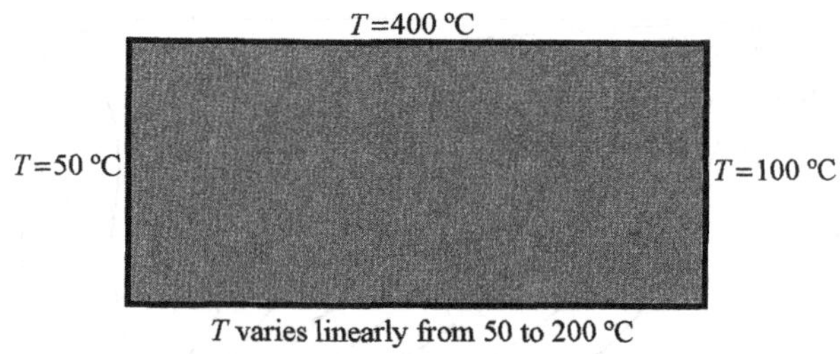

FIGURE 6H. Conduction in an aluminum slab.

temperatures, as shown in Figure 6H. Along the fourth edge, the temperature varies in the manner prescribed. Naturally, the governing equation in this case is

$$\nabla^2 T = 0 \quad \text{or} \quad \frac{\partial^2 T}{\partial x^2} + \frac{\partial^2 T}{\partial y^2} = 0.$$

Find the temperature distribution in the interior of the slab by suitable means (clearly, Gauss–Seidel and SOR are among the possibilities) and prepare a contour plot showing the behavior of the isotherms. Then, investigate the thermal conductivity of aluminum. Is it temperature dependent? How much variation is there? What are the consequences if it becomes necessary to write $k = k(T)$? Explain.

Problem 6I. Destruction of the Shuttle Challenger: The Mission 51-L Disaster

On January 28, 1986 the space shuttle Challenger exploded just 73 s after liftoff, killing the seven crew members and delaying crucial future flights by years, in some cases. The disaster occurred in part because of technological hubris and in part because political concerns took precedence over sound engineering judgment. The shuttle, stacked for launch, consists of the orbiter vehicle, a large external fuel tank, and two SRBs (solid rocket boosters). The culprit in the 1986 disaster was a tang/clevis field joint sealed against combustion gas blowby by zinc chromate putty and two DuPont Viton fluoroelastomer O-rings. It is now clear that the dynamic loads associated with fuel ignition and vehicle motion may have caused the gap at the primary O-ring to widen by as much as 0.029 in. (about one-tenth of the ring's normal thickness). The photographic record shows that smoke issued from the aft field joint in the right-hand SRB just 0.678 s after SRB ignition; this evidence suggests that burn-through of the putty, insulation, O-rings, and accompanying grease began even before the vehicle left the launch pad. Indeed, at 59 s into the flight, a jet of flame appeared from this very same area and directed in such a way as to impinge upon the external fuel tank. About 5s later the tank was breached and hydrogen began to escape. At 73 s, the fuel tank exploded, destroying the orbiter and resulting in the two SRBs moving erratically outward in opposite directions. To understand how this came about, it is necessary to examine the construction of the SRBs.

Each SRB is 149 ft long and 146 in. in diameter. The casing contains about 450,000 kg of propellant consisting of aluminum powder, ammonium perchlorate, iron oxide powder, polybutadiene acrylic acid acrylonitrile terpolymer, and an epoxy curing agent. The fuel was prepared and cast in 600 lb batches by Morton Thiokol in Utah. Then, the four main cylindrical segments were shipped by rail to Florida for field assembly. The inside surface of the motor case is coated with a nitrile-butadiene rubber insulation (to protect it for recovery and reuse). Although the system had experienced 24 successful flights previously, it later came out that some previous flights had shown signs of thermal damage at the field joints, with either actual erosion or in some cases soot deposits between the two O-rings. The tang-clevis field joints were recognized as problem areas and *NASA had been warned by Morton Thiokol engineers not to launch the shuttle in cold ambient temperatures* because the O-rings lost their resiliency in the cold, and could not rapidly conform to the gap in response to the combustion pressure. Later tests revealed that rapid dynamic sealing was not achieved at 25°F and was marginal even at temperatures 20°F higher! Therein lies the fatal problem. The night prior to launch was exceptionally cold, with the temperature approaching 20°F. In fact, at launch time, 11:38 a.m., the air temperature was only 36°F. Thus, a key question concerns the temperature profile $T(r, t)$ in the vicinity of the aft field joint.

This is rather difficult to model accurately because the tang and clevis joints were actually secured by 180 steel pins each 1 in. diameter and 2 in. long. The outside end of each pin was flush with the external casing surface, and the inside end corresponded approximately to the location of the two O-rings. Moving inward, a layer of zinc chromate putty filled the gap in insulation between field-assembled segments and extended to actual contact with the solid propellant. We can take this distance to be about 2–3 in. The thermal conductivity of the putty is about 0.000496 cal/(cm s °C) and the thermal conductivity of the propellant is approximately 0.000162 cal/(cm s °C). The propellant is in the form of an annular solid within the casing; the central void is of course required for combustion gases. We will take the radii corresponding to the inner and outer surfaces of the propellant to be 1 and 5.92 ft, respectively. The putty (and insulation) extends to 6.15 ft and the outer surface of the casing (at the joint) corresponds to $R = 6.317$ ft. We will assume that the (air) temperature history is initiated at noon the previous day; at that time the entire assembly had a uniform temperature of about 55°F. The ambient temperature then varied as shown in Figure 6I.

Naturally, the key question concerns the radial temperature profile in the SRB; find $T(r,t)$ at the moment of launch. It seems pretty obvious that at the location of the O-rings, the temperature could not have been significantly different from 36°F. After ignition, the contrast in temperatures was (and is) really extreme since the solid fuel bums at 3200°C. For

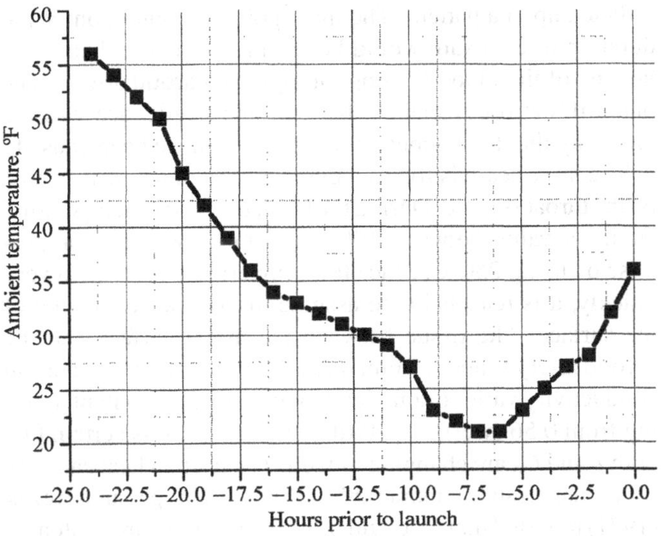

FIGURE 6I. Approximate ambient temperature history for *Challenger* prior to launch.

simplicity, assume that the flux of thermal energy was nearly zero at the inside surface of the solid annular propellant prior to launch.

Problem 6J. Heat Transfer and the Columbia Disaster

Note: For a definitive account of the tragedy, refer to *Columbia Accident Investigation Board Report*, Vol. 1, August 2003.

On February 1, 2003, the space shuttle *Columbia* broke apart above Texas showering debris over an area of about 2000 square miles. The catastrophe resulted in the deaths of the crew members: Husband, McCool, Anderson, Brown, Chawla, Clark, and Ramon, and it raised the specter of the *Challenger* disaster of 1986. Everyone realized that space flight was inherently dangerous, but NASA had sold the shuttle concept as a means of providing quick, cheap, and frequent access to earth orbit. The reality, of course, is that budget restrictions led to a compromise vehicle, for example, one that used solid rocket boosters to generate about 85% of the required thrust. A pair of SRBs is capable of providing the needed 6×10^6 lb of thrust, but the SRBs are uncontrollable (in the sense that once ignited, they burn until the fuel is exhausted). They also vary; the batch production of the aluminum powder/ammonium perchlorate fuel oxidizer and the segmented assembly never results in two SRBs having identical performance. Despite the deficiencies of the shuttle stack system, the program has yielded just two horrific accidents in more than 20 years of operation. NASA images of the crew and the launch of *Columbia*, STS 107 are available online.

The STS 107 dedicated science mission was launched on January 16, 2003 at 10:39 a.m. About 81.7 s after launch, a piece of foam insulation detached from the external tank and struck the Orbiter on the left wing, somewhere between panels 5 and 9. The insulation fragment was about 24 in. long, 15 in. wide, and weighed about 1.67 lb. It was tumbling at 18 revolutions per second, and when it struck the Columbia's wing, it did so with a relative velocity of over 500 mph. The insulation fragment came from the bipod attachment (between the shuttle and the fuel tank); this area was monitored by video camera during the launch of *Discovery*, July 26, 2005.

On January 23, Mission Control sent an image and a video clip of the debris impact upon the left wing to Husband and McCool. According to the *Columbia Accident Investigation Report*, Vol. 1, Mission Control also relayed the message that there was "absolutely no concern for entry." This mindset doomed *Columbia*; though the possible significance of the impact upon the wing was understood by NASA, no actual evaluation of the results of such impacts had been undertaken.

A critical consequence of the debris strike became apparent on January 17, although the event itself remained undetected until the postaccident review. During the morning hours of January 17, a small object drifted slowly away from the shuttle and re-entered the earth's atmosphere about 2 days later. Later testing revealed that the only plausible object with an equivalent radar cross section was a piece of reinforced carbon–carbon (RCC) composite from the leading edge of *Columbia's* left wing. It was determined that the fragment must have had a surface area of about 140 in^2. The Thermal Protection System on the left wing had been breached and the vehicle and the crew were destined for destruction. Impact resistance had not been part of the specifications for the RCC (leading edge) panels.

At 8:15 a.m. on February 1, Husband and McCool fired the maneuvering engines for 2.5 min to slow the Orbiter and begin re-entry. At 8:44, Entry Interface (EI) was attained (an altitude of 400,000 ft). In about 4 min, a sensor on a left-wing spar began showing an abnormally high strain. At about 8:53, signs of debris shedding from the vehicle were noted over California and about 1 min later four hydraulic sensors in the left wing went off-scale low (ceased to function). At about 8:59, outputs from the tire pressure sensors (left wing landing gear) were lost and 17 s later, the last (fragmentary) communication from *Columbia* was received. Visual observation at 9 a.m. indicated that the Orbiter was coming apart. The Modular Auxiliary Data System (MADS) recorder continued to function during 9:00:19.44; these data were not transmitted to the ground but the recorder itself was recovered near Hemphill, Texas. This finding was critical to the investigation because the MADS data showed that 169 of 171 sensor wires in the left wing had burned through by the time MADS quit working.

Other data also confirmed that significant damage to the left wing had occurred. At about 500 s after EI, the roll and yaw forces began to diverge from nominal operation. Even more telling, images recorded by scientists at Kirtland Air

Force Base (near Albuquerque, NM) clearly show an unusual disturbance on the left wing. As the drag increased on the left side, *Columbia's* flight control system compensated by firing all four right yaw jets, but at 970 s after EI control was lost and the vehicle began to tumble at a speed of about 12,000 mph—with the predictable result.

The catastrophic end of STS 107 was a sobering reminder that the space shuttle system was (and is) really more about development and flight test than it was about routine operations. The only positive result may be that aspects of the NASA culture that contributed to the accident may be changed for the better.

For students of transport phenomena, the disaster poses several intriguing questions:

1. When the foam separated from the external tank, the shuttle stack velocity was 1586 mph; when it struck the left wing of the Orbiter 0.161 s later, it was moving at a velocity of only ~1022 mph (creating a relative velocity of 560 mph). Explain how this could occur.
2. What is meant by "ballistic coefficient?" The volume of the foam piece was thought to have been about 1200 in^3. What would its ballistic coefficient have been?
3. Estimate how much energy was transmitted to the *Columbia's* wing by the foam piece.
4. The RCC panels on the wings were designed to accommodate a leading edge temperature of about 3000°F. *If* heat transfer behind the RCC occurred only by conduction and only through the aluminum structural members, how far could the heat penetrate in ~500 s (disregarding the fact that aluminum melts at 1220°F)? Would this be sufficient to explain the observed sensor cable burn-through?
5. The accident investigation concluded that there must have been some "sneak" flow entering the wing through the breach in the leading edge. This means that gas flow at about 2300–3000°F was occurring inside the wing. Given an Orbiter altitude of 210,000 ft, what characteristics of the hot gas were critical to heat transfer between the gas and the structural members? Be quantitative.

Problem 6K. Heat Transfer Resulting from Laser Burn in the Human Throat

Surgeons often use lasers as excisional tools to perform laryngectomies; cancers of the larynx and pharynx have been treated—generally with few complications—since the mid-1990s. However, complications have arisen in a few cases when the localized heating has affected blood flowing in the carotid artery. You have been retained as an expert witness in a malpractice case in which permanent brain damage was inflicted upon a patient. The main point of contention: What duration of exposure would be required to cause dangerous heating of the blood flowing through the carotid artery? The plaintiff's attorney claims that extreme negligence was the only way that the patient's injuries could have been caused.

The laser beam is focused upon an area of about 1–2 mm^2 on the throat surface. During the burn, the surface temperature attains a value between 100 and 400°C; we can compromise by using 250°C. Since this temperature is attained very quickly, it is reasonable to assume instantaneous heating of the surface. The tissue between the throat surface and the carotid artery is about 7 mm thick. Unfortunately, the thermal conductivity varies dramatically with moisture content, ranging from 0.56 (wet) to 0.20 (dry) J/(s m °C); it is certain that both ρ and C_p are changing as well. The normal heat capacity for human tissue is about 0.85 cal/(g °C). Cooper and Trezek (1971) reported that C_p could be related to moisture content in human tissue by $C_p = [M + 0.4(100 - M)]x41.9\,\mathrm{J/(kg\,K)}$, where M is percent water. Therefore, if $M = 35\%$, then $C_p = 2556$ J/(kg K), or about 0.61 cal/(g °C).

The blood flowing in the artery is a *Casson fluid*, that is, it is shear thinning like a pseudoplastic, but has a definite yield stress value. The viscosity of human blood approaches a constant value of about 3 cp for shear rates above about 100 s^{-1}. The usual temperature of blood is 37°C and the flow velocity in the carotid artery for an adult is about 28 cm/s with a typical cross-sectional area of 33 mm^2 (cardiac output is normally about 6 L/min).

It seems likely that the simplest possible model that can be used for this problem will be written as

$$\frac{\partial}{\partial t}(\rho C_p T) = \frac{\partial}{\partial y}\left[k\frac{\partial T}{\partial y}\right].$$

Estimate the duration of exposure that would be required to heat the interior surface of the artery to a dangerous level, say 50°C. That is, how long must the laser be fixed upon a specific spot to cause serious permanent injury? As a first approximation, we might relate k to moisture content and moisture content to local temperature (e.g., one might assume that the moisture content is zero for local temperatures exceeding 100°C).

Problem 6L. Transient Cooling of a Smoothbore Projectile

In the era of wooden warships, it was common practice to heat cannonballs prior to firing at the enemy. This would result in the diversion of some sailors from gunnery to firefighting as the consequence of hits from "hot shot." Suppose a solid iron sphere ($d = 4$ in.) is heated to 1400°F and then fired at a muzzle velocity of 500 ft/s through air at a uniform temperature of 70°F. Find the temperature distribution *inside* the cannonball after 1, 3, 6, and 10 s of flight, assuming constant velocity.

About how far must the projectile travel before it loses its ability to ignite wood?

A number of assumptions are necessary in order to work this problem. You may like to begin by looking at the Ranz and Marshall (1952) correlation for spheres:

$$Nu_m = \frac{h_m d}{k} = 2 + 0.6 Re^{1/2}\, Pr^{1/3}.$$

The guns that fired such projectiles were smoothbores (no rifling inside the barrel). This means that the cannonball might leave the muzzle with some small (modest) rate of rotation. Is this a sufficient reason to neglect angular variations in T?

Carefully list the assumptions you employ and provide an explanation (reasoning) for each.

Problem 6M. Heat Losses from a Wire with Source Term (Electrical Dissipation)

We would like to consider heat losses from an 8 AWG copper wire suspended between two large supports each maintained at 90°F. The wire has a diameter of 128 mil (0.128 in.), and according to the National Board of Fire Underwriters, it can safely carry a current of 40 A. However, we are going to allow it to carry a current large enough to produce a maximum temperature (at the center) of 1000°F. Our purpose is to explore modeling options with a view toward identifying one with really good performance. We do have the following data for copper:

$$k = 220\,\text{Btu h}^{-1}\,\text{ft}^{-1}\,{}^{\circ}\text{F}^{-1} \quad \text{and}$$

$$k_e = 510{,}000\,\text{ohm}^{-1}\,\text{cm}^{-1},$$

but the possible variation $k(T)$ has not been assessed. Suppose we make a balance on a segment of wire length Δz:

$$\pi R^2 q_z|_z - \pi R^2 q_z|_{z+\Delta z} - 2\pi R \Delta z q_s + \pi R^2 \Delta z S_e = 0,$$

where the terms represent axial conduction (in and out), loss at the surface by means unspecified, and production by electrical dissipation, respectively. Note that we have neglected the possibility of radial variation of temperature. This is a point that we will come back to later. The steady-state balance, with the loss attributed to radiation, might result in the ordinary differential equation:

$$\frac{d^2T}{dz^2} - \frac{2\sigma}{kR}(T^4 - T_0^4) + \frac{S_e}{k} = 0,$$

where the production term $S_e = I^2/k_e$ and I is the current density, A/cm^2. Find the temperature distribution in the wire for this case and the maximum allowable current; then address the following questions:

1. Should the production term be written as a function of temperature for copper?
2. Is radiation really the dominant loss mechanism?
3. If free convection is important, how would you modify the model to account for it? And would your temperature profile change significantly as a result?
4. What circumstances might lead to significant $T(r)$? And how would the differential equation be modified to account for radially directed conduction?
5. Finally, would you expect to see any important differences if you actually solved the model for $T(r,z)$?

Problem 6N. Heat Transfer in Jet Impingement Baking

One strategy used in the food processing industry to reduce baking time and save energy is jet impingement baking. In this method, a jet of heated air is directed downward upon the top of the "biscuit." Typical air temperatures range from about 100–250°C, and the jet velocities are often on the order of 20–30 m/s. Naturally, this results in a much larger heat transfer coefficient, particularly near the stagnation point on the top of the "biscuit." However, as the axisymmetric stagnation flow approaches the corner (top edge), h is much lower. The flow off of the "step" results in separation and produces another region of low h. We would like to model the temperature distribution in the interior of the biscuit as a function of time. The biscuit diameter is 15 cm and its height s is 4 cm. The bottom boundary is isothermal at 202°C and the problem is axisymmetric such that $\left.\frac{\partial T}{\partial r}\right|_{r=0} = 0$. The heat transfer coefficient varies linearly from the top center, where $h = 185$ W/(m^2 K), to a lower value at the top, outside corner where $h = 42$ W/(m^2 K). On the vertical surface (edge), h decreases from 42 W/(m^2 K) to 26 at the bottom. The temperature of the hot air jet is 202°C and the initial biscuit temperature is 6°C. Find the temperature distribution inside the biscuit at $t = 5$, 10, and 15 min. Assume that the thermal conductivity of the biscuit is constant at 0.00055 cal/(cm s °C), the specific gravity is 1.22, and the heat capacity is 0.48 cal/(g °C). Of course, these values would change as moisture is lost (and the product texture changes) during the baking process, but these changes will be neglected for our analysis.

Problem 6O. Temperature Distribution in a Circular Fin

We would like to determine the temperature distribution in an aluminum fin (a circular fin of width w) mounted upon a hot cylinder. The radius of the cylinder R is 0.32808 ft and the outer edge of the fin (at βR) corresponds to 0.4429 ft. Thus, $\beta = 1.35$. The purpose of the fin, of course, is to discard thermal energy to the surrounding air. The governing differential equation for this problem is

$$\frac{d^2\theta}{dr^2}+\frac{1}{r}\frac{d\theta}{dr}-\frac{2h}{wk}\theta=0.$$

The surface temperature of the heated cylinder is 437°F and the ambient temperature is 77°F. The fin is made of aluminum with $k=121$ Btu/(h ft °F). Assume that air is moved past the fin with such a velocity that the average heat transfer coefficient is $h=19$ Btu/(h ft^2 °F); this value applies both on the flat surfaces and at the (curved) edge. Find the temperature distribution $T(r)$, and the total heat cast off by the fin per hour. We would like to make sure that we use a Robin's-type boundary condition (by equating the fluxes) at $r=\beta R$. Finally, is there an easy way to determine whether $T=T(r,z)$, that is, because h is large, might there be a significant temperature difference *across* the fin?

How would you assess that concern?

Problem 7A. Heat Transfer for the Fully Developed Flow in an Annulus

Consider an annular region formed by two concentric cylinders with radii R_1 and R_2. Water enters the annulus at a uniform temperature of 70°F and with an average velocity of 1.75 cm/s. At $z=0$, the fluid encounters a heated *inner* surface (maintained at a constant 150°F). This heated surface extends for a distance of 3 ft; beyond that point, the inner surface is insulated such that $q_r(r=R_1)=0$. Find the temperature distributions and the Nusselt number at z-positions of 0.5, 1, 2, and 3 ft. The annular gap is 1.25 cm with $R_1=3.75$ cm. The outer surface is maintained at 70°F for all z-positions. The governing equation is

$$\rho C_p v_z\frac{\partial T}{\partial z}=k\left[\frac{1}{r}\frac{\partial}{\partial r}\left(r\frac{\partial T}{\partial r}\right)+\frac{\partial^2 T}{\partial z^2}\right].$$

Is it acceptable to omit axial conduction?

Problem 7B. Heat Transfer from Pipe Wall to Gas Mixture

We are interested in heat transfer from a pipe wall to a mixture of helium and carbon dioxide. The gas has a mean velocity of 0.4 cm/s in 10 cm (diameter) pipe, 1.4 m long; it enters the heated section at a uniform temperature of 22°C and the walls of the pipe are maintained at a constant 84°C. Determine the value of the Nusselt number at the following z-positions: 10, 20, 50, and 125 cm. The thermal diffusivity of the gas mixture can be taken as a constant, 0.065 cm/s, and the thermal conductivity is about 0.045 Btu/h ft °F. The equation you must solve is

$$\rho C_p v_z\frac{\partial T}{\partial z}=k\left[\frac{1}{r}\frac{\partial}{\partial r}\left(r\frac{\partial T}{\partial r}\right)\right].$$

Problem 7C. Revisiting the Classical Graetz Problem

The governing equation for the Graetz problem is

$$2\langle v_z\rangle\left(1-\frac{r^2}{R^2}\right)\frac{\partial T}{\partial z}=\alpha\left[\frac{1}{r}\frac{\partial}{\partial r}\left(r\frac{\partial T}{\partial r}\right)\right].$$

It is useful to recast the equation in dimensionless form yielding

$$[1-r^{*2}]\frac{\partial\theta}{\partial z^*}=\frac{1}{Re\,Pr}\frac{1}{r^*}\frac{\partial}{\partial r^*}\left(r^*\frac{\partial\theta}{\partial r^*}\right).$$

We would like to consider the laminar flow of water through a 1 cm diameter tube at $Re=150$. The inlet water temperature is 60°F and the tube wall is maintained at 140°F. Find the bulk fluid temperatures and Nusselt numbers at axial positions corresponding to $10R$, $20R$, $50R$, and $100R$.

Hausen (*Verfahrenstechnik Beih. Z. Ver. Deut. Ing.*, 4:91, 1943) suggested that the mean Nusselt number (over a length z) for the Graetz problem was adequately represented by

$$Nu=3.66+\frac{0.0668(Pe/(z/d))}{1+0.04((z/d)/Pe)^{-2/3}}.$$

Does Hausen's correlation seems to agree with your results?

Problem 7D. Free Convection from a Vertical Heated Plate

Free convection on a vertical heated plate was considered in 1881 by Lorenz, but it was not until Ostrach's work in 1953 that accurate numerical solutions were obtained. This is a particularly interesting heat transfer problem because it is evident that the velocity profile must contain a point of inflection. Accordingly, one must be concerned about the transition to turbulence. Eckert and Jackson conducted an experimental study of this situation in 1951 and concluded that transition occurs when the product $GrPr$ is between 10^8 and 10^{10}. At the same time, it is also necessary that $GrPr$ be greater than 10^4 so that the boundary-layer approximation will be valid. Pohlhausen (1921) found a similarity transformation for this problem by defining a new independent variable as $\eta=(y/x)(Gr_x/4)^{1/4}$ and a dimensionless temperature as $\theta=(T-T_\infty)/(T_s-T_\infty)$.

By introducing the stream function $\psi=4\nu(Gr_x/4)^{1/4}f(\eta)$, he was able to obtain the two coupled nonlinear ordinary differential equations:

$$f'''+3ff''-2f'^2+\theta=0$$

and

$$\theta''+3Pr\,f\theta'=0.$$

We would like to solve these equations, assuming the fluid of interest is water with $T = 25°C$. Prepare a graph illustrating both the temperature and velocity profiles. If the wall temperature is 40°C, estimate the position at which transition is likely to occur and evaluate the local Nusselt number at that value of x. LeFevre (Laminar Free Convection from a Vertical Plane Surface, *MERL Heat 113*, 1956) has proposed an empirical interpolation formula that applies for *any Pr*:

$$Nu_x = \frac{hx}{k} = \left(\frac{Gr_x}{4}\right)^{1/4} \frac{0.75\,Pr^{1/2}}{(0.609 + 1.221\,Pr^{1/2} + 1.238\,Pr)^{1/4}}.$$

Are the results of your computations in agreement with this equation?

Problem 7E. Heat Transfer to a Falling Film of Water

Consider heat transfer between a vertical heated wall and a flowing liquid film of water; the fluid flows in the z-direction under the influence of gravity; the film extends from the wall ($y = 0$) to the free surface at $y = \delta$.

For this situation,

$$v_z = V_{\max}\left(\frac{2y}{\delta} - \frac{y^2}{\delta^2}\right)$$

and the energy equation can be reduced to

$$\rho C_p v_z \frac{\partial T}{\partial z} \cong k \frac{\partial^2 T}{\partial y^2}.$$

Starting with the correct equation (and using the correct velocity distribution), introduce the appropriate dimensionless variables and determine a numerical solution with the method of your choice. Compare your results graphically with those calculated from eq. (12B.4–8) in Bird et al. (2002). Assume that the heated wall is maintained at a constant temperature (T_s) of 135°F and that the uniform initial liquid temperature is 55°F. The falling film thickness is approximately constant at 0.9 mm. Note that the maximum (free surface) velocity is given by

$$V_{\max} = \frac{\delta^2 \rho g}{2\mu}.$$

Problem 7F. The Rayleigh–Benard Convection in a Two-Dimensional Enclosure

We would like to solve a Rayleigh–Benard problem so that we can better understand the evolution of the convection rolls in enclosures. Find and plot the stream function at dimensionless times of 0.03, 0.15, 0.375, and 0.8 for a rectangular enclosure in which the width-to-height ratio is 2.375. The equations (which are developed in Chapter 7) are summarized here for your convenience:

Energy:

$$\frac{\partial \theta}{\partial t^*} + \frac{\partial (v_x^* \theta)}{\partial x^*} + \frac{\partial (v_y^* \theta)}{\partial y^*} = \frac{1}{Pr}\left(\frac{\partial^2 \theta}{\partial x^{*2}} + \frac{\partial^2 \theta}{\partial y^{*2}}\right)$$

Vorticity:

$$\frac{\partial \Omega}{\partial t^*} + \frac{\partial (v_x^* \Omega)}{\partial x^*} + \frac{\partial (v_y^* \Omega)}{\partial y^*} = Gr \frac{\partial \theta}{\partial x^*} + \frac{\partial^2 \Omega}{\partial x^{*2}} + \frac{\partial^2 \Omega}{\partial y^{*2}}.$$

Note the similarities between the two equations; of course, the implication is that we can use the same procedure to solve both. We must use a stable differencing scheme for the convective terms, and the method developed by Torrance (1968) is known to work well for both natural convection and rotating flow problems. You may like to start with an array size of 38 × 16, which corresponds to 608 nodal points. Obviously, better resolution is desirable, but if you bump up to 57 × 24, the total number of required storage locations is 9576 (you must have both vorticity and temperature on old and new time-step rows). The generalized solution procedure follows:

1. Calculate stream function from the vorticity distribution using SOR.
2. Find the velocity vector components from the stream function.
3. Compute vorticity on the new time-step row explicitly.
4. Calculate temperature on the new time-step row explicitly.

If you stay with an array size of (38,16), the optimal relaxation parameter value is 1.74 by direct calculation. If you change the number of nodal points, then you must recalculate this factor. The other parametric values we wish to employ are

$$Pr = 6.75 \qquad Gr = 1000$$

$$\Delta x^* = 0.0667 \qquad \Delta t^* = 0.0005.$$

Note that the time-step size has not been optimized. You may be able to use a slightly larger value. Finally, remember that this solution procedure can be used for a variety of two-dimensional problems in transport phenomena if the right-hand boundary is handled properly (in our case, it is a line of symmetry).

Problem 7G. Heat Transfer in the Thermal Entrance Region

Recall the analysis of heat transfer for fully developed laminar flows in circular tubes; we found for constant heat flux, $Nu = 4.3636$ and for constant wall temperature, $Nu = 3.658$. It stands to reason that the Nusselt number in the thermal entrance region should be larger. We would like to analyze the case of constant wall temperature using the Leveque approach for laminar flow in the entrance region of a circular tube. Find an expression for the Nusselt number and evaluate it numerically, making use of the following information:

Values for integral: $\int_0^w \exp(-w^3)dw$:

w	$\int_0^w \exp(-w^3)dw$
0.1	0.100
0.2	0.200
0.3	0.298
0.5	0.485
0.7	0.645
0.9	0.765
1.2	0.861
1.5	0.889
1.9	0.893
3.0	0.893

Problem 7H. Natural Convection from Horizontal Cylinders

The long horizontal cylinder is an extremely important geometry in heat transfer because of common use in process engineering applications. When such a cylinder is hot, it will lose thermal energy by free convection (among other mechanisms). The first successful treatment of this problem was carried out by R. Hermann (1936), Free Convection and Flow Near a Horizontal Cylinder in Diatomic Gases, *VDI Forschungsheft*, 379 (see also NACA Technical Memorandum 1366). Hermann used a boundary-layer approach (in fact, he extended Pohlhausen's treatment of the vertical heated plate) despite the fact that no similarity solution is possible in this case. The equations he employed (excluding continuity) follow:

$$v_x \frac{\partial v_x}{\partial x} + v_y \frac{\partial v_x}{\partial y} = \nu \frac{\partial^2 v_x}{\partial y^2} + g\beta(T - T_\infty)\sin\left(\frac{x}{R}\right)$$

and

$$v_x \frac{\partial T}{\partial x} + v_y \frac{\partial T}{\partial y} = \alpha \frac{\partial^2 T}{\partial y^2},$$

where, in usual boundary-layer fashion, the x-coordinate represents distance along the surface of the cylinder and y is the normal coordinate measured from the surface into the fluid.

1. Consider Hermann's analysis. What are the main limitations? What is the consequence of a very small Grashof number? Very large Gr?
2. Formulate this problem in cylindrical coordinates, noting the (dis)advantages.
3. Describe how you might solve this problem in cylindrical coordinates (if you have the time, try it).

Problem 7I. Effects of $\mu(T)$ Upon Heat Transfer in a Tube

The viscosity of olive oil changes significantly with temperature; data from Lange's *Handbook of Chemistry*, revised 10th edition (McGraw-Hill, 1961) are reproduced here:

Temperature (°C)	Viscosity (cp)
15.6	100.8
37.9	37.7
65.7	15.4
100.0	7.0

Suppose we have a fully developed laminar flow of olive oil through a 2 cm diameter cylindrical tube where the oil has a uniform temperature of 15°C. The Reynolds number is 117.5 At $z=0$, the oil enters a heated section in which the wall temperature is maintained at 100°C. Obviously, the reduction in viscosity near the wall will affect the shape of the velocity profile; the energy and momentum equations are coupled. We would like to determine the evolution of the velocity and temperature profiles by computation. We would also like to calculate the change in Nusselt number; recall that for a fully developed laminar flow in a tube with constant wall temperature, $Nu=3.658$. Find $v_z(r,z)$ and $T(r,z)$ at $z=60$, 180, and 300 cm. The governing equations can be written as

$$\frac{dp}{dz} = -\frac{1}{r}\frac{d}{dr}(r\tau_{rz})$$

and

$$\rho C_p v_z \frac{\partial T}{\partial z} = k\left[\frac{1}{r}\frac{\partial}{\partial r}\left(r\frac{\partial T}{\partial r}\right)\right].$$

We will assume that ρ, C_p, and k are all constant. The density of olive oil is 0.915 g/cm^3 at 15°C, the thermal conductivity is 0.000452 cal/(cm s °C), and the heat capacity is approximately 0.471 cal/(g °C).

Problem 7J. Variation of the Olive Oil Problem

Olive oil flows under the influence of pressure between two parallel planar surfaces. The oil enters with a uniform temperature of 15°C; the average velocity at the entrance is 2.25 cm/s. Both walls (located at $y=0$ and $y=b$) are maintained at 85°C. Find the *pressure* at z-positions corresponding to $z/b=20$, 100, and 500. Let $b=0.55$ cm; use the property data given in Problem 7I.

Problem 7K. Modified Graetz Problem in Microchannel with Production

Begin this problem by reading Jeong and Jeong (Extended Graetz Problem Including Streamwise Conduction and

Viscous Dissipation in Microchannel, *International Journal of Heat and Mass Transfer*, 49:2151, 2006). We will assume fully developed laminar flow (in the x-direction) through the rectangular microchannel. The origin is placed at the center of the channel and the parallel walls are located at $y = +H$ and $y = -H$. We assume that the viscosity and the flow rate are such that production of thermal energy by viscous dissipation is a real possibility. Therefore, the governing equation is written as

$$v_x \frac{\partial T}{\partial x} = \alpha \left[\frac{\partial^2 T}{\partial x^2} + \frac{\partial^2 T}{\partial y^2} \right] + \frac{\mu}{\rho C_p} \left(\frac{\partial v_x}{\partial y} \right)^2. \quad (1)$$

Note that the axial conduction term has been retained in this equation. Whether this is necessary will depend upon the product *RePr*. The reader is referred to Jeong and Jeong (2006) for a discussion as to when this inclusion might be required in microchannel flows. The velocity distribution in the duct (since $W \gg 2H$) is given by

$$v_x = \frac{1}{2\mu} \frac{dp}{dx} (y^2 - H^2). \quad (2)$$

We will incorporate eq. (2) into the governing equation, initially neglecting axial conduction. By computing the bulk mean fluid temperature as a function of x-position, we can equate the fluxes and determine the Nusselt number as a function of (dimensionless) x-position. You might consider initially omitting production to more easily verify your computational scheme. Use the following parametric values (all cgs units):

$$H = 0.1\,\text{cm} \quad C_p = 0.56$$

$$\rho = 0.802 \quad k = 0.00034 \quad \mu = 0.04,$$

and take $dp/dx = -2000\,\text{dyn/cm}^2$ per cm. This pressure drop will yield a centerline velocity of 250 cm/s. Assume the fluid enters at a uniform temperature of 15°C with the walls maintained at 45°C. Compute the evolution of the Nusselt number and the temperature distribution in the x-direction. Some typical results for $T(x,y)$ with $Re = 1336.6$ (consistent with Jeong and Jeong who define the Reynolds number: $Re = (4H\langle v_x \rangle \rho)/\mu$) are given in Figure 7K to allow you to check your work.

Next (once you have verified your computational scheme), we would like to examine the results shown in Figure 5 in Jeong and Jeong. Adjust the parameters of this problem to obtain $RePr = 1 \times 10^6$ and $Br = 0.2$. At what value of x does the Nusselt number begin to increase? Can the Brinkman number be this large in a practical microchannel problem? What conditions would be necessary to make $Br = 0.2$?

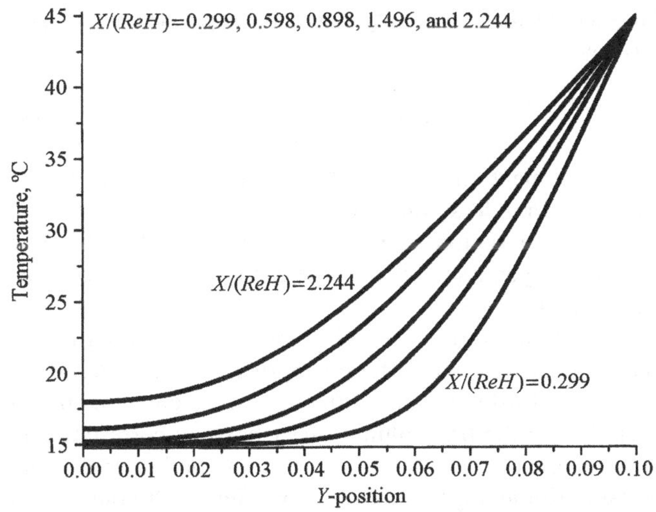

FIGURE 7K. Typical results for $Re = 1336.6$. Over the range of x-positions covered in this Figure, the Nusselt number decreases from 12.3 to 7.83.

Problem 7L. Heat Transfer in the Entrance Region of a Rectangular Duct

Consider a rectangular duct where the centerline corresponds to the x-axis. The planar walls are located at $y = +b$ and $y = -b$ and it may be assumed that the channel width is much greater than its height: $W >> 2b$. Both the velocity and the temperature of the entering fluid are uniform ($v_x = V_0$ and $T = T_0$) at the entrance. The walls of the duct are maintained at an elevated temperature T_w. We would like to explore a numerical approach to this combined entrance region problem with the objective of finding the Nusselt number as a function of x-position. Our plan is to re-examine the procedure employed by Hwang and Fan (Finite Difference Analysis of Forced Convection Heat Transfer in Entrance Region of a Flat Rectangular Duct, *Applied Scientific Research,* A-13:401, 1963). Their calculations were carried out on an IBM 1620, so we should be able to refine the mesh considerably (the IBM 1620 used 6-bit data representation and it could perform 200 multiplications in 1 s).

Hwang and Fan employed the following equations:

$$v_x \frac{\partial v_x}{\partial x} + v_y \frac{\partial v_x}{\partial y} = -\frac{1}{\rho} \frac{dp}{dx} + \nu \frac{\partial^2 v_x}{\partial y^2}, \quad (1)$$

$$\frac{\partial v_x}{\partial x} + \frac{\partial v_y}{\partial y} = 0, \quad (2)$$

$$v_x \frac{\partial T}{\partial x} + v_y \frac{\partial T}{\partial y} = \alpha \frac{\partial^2 T}{\partial y^2}. \quad (3)$$

They noted that continuity could be expressed in integral form as

$$2bV_0 = 2\int_0^b v_x dy. \tag{4}$$

Take a moment and contemplate the proposed model. It is clear that Prandtl's equations are being employed, that is, this entrance region problem is being treated with some boundary-layer assumptions. This cannot be completely correct. Explain.

The solution procedure to be employed was described in detail by Bodoia and Osterle (*Applied Scientific Research*, A-10:265, 1961). A finite-difference representation for (1) is applied to the first column (the x-position corresponding to the entrance); it is used to determine both velocity and pressure (implicitly) on the $x + \Delta x$ column. Of course, it is assumed that pressure is a function of x only. Continuity is used to compute the y-component of the velocity vector on the $x + \Delta x$ column, and then the process is repeated. The technique, therefore, is a semi-implicit, forward marching method. Note that for the convective transport terms, a first-order forward difference is used for $\partial v_x/\partial x$ and pressure, and a second-order central difference is used for $\partial v_x/\partial y$. The viscous transport term is centrally differenced, but on the $x + \Delta x$ column. Of course, this technique would not work if the areas of recirculation were present in the flow; fortunately that is not a problem in this case. Results presented by Bodoia and Osterle show that the hydrodynamic development is virtually complete when $X = 0.05$, where the dimensionless x-position is defined by

$$X = \frac{\nu x}{2b\langle v_x\rangle}.$$

Does this result agree with other available data?

Problem 8A. Unsteady Evaporation of a Volatile Organic Liquid

Consider an enclosure in which 2,2-dimethylpentane is spilled upon the floor; the temperature in this process area is 40°C. Find the (vertical) concentration profiles at $t = 10$ min, 30 min, and 2 h. Use two different analyses: First, assume that this situation is governed by

$$\frac{\partial C_A}{\partial t} = D_{AB}\frac{\partial^2 C_A}{\partial y^2},$$

with the solution

$$\frac{C_A}{C_{A0}} = 1 - \text{erf}\left(\frac{y}{\sqrt{4D_{AB}t}}\right).$$

Compare this result with that obtained from Arnold's analysis (Studies in Diffusion III: Unsteady-State Vaporization and Absorption, *Transactions AIChE*, 40:361, 1944). In this case, use the *correct* governing equation:

$$\frac{\partial x_A}{\partial t} = D_{AB}\left\{\frac{\partial^2 x_A}{\partial y^2} + \frac{\partial x_A}{\partial y}\left[\frac{1}{1 - x_{A0}}\left.\frac{\partial x_A}{\partial y}\right|_{y=0}\right]\right\},$$

with the solution

$$\frac{x_A}{x_{A0}} = \frac{1 - \text{erf}(\eta - \phi_0)}{1 + \text{erf}(\phi_0)}.$$

The prevailing pressure is 1 atm and the enclosure can be taken to be very tall (y-direction). The value of ϕ_0 depends upon the volatility of species "A" and we can use the initial condition to show

$$\frac{x_{A0}}{1 - x_{A0}} = \sqrt{\pi}\cdot\phi_0\exp(\phi_0^2)(1 + \text{erf}(\phi_0)).$$

Therefore,

x_{A0}	0	0.1	0.2	0.4	0.6	0.8	0.9
ϕ_0	0	0.0586	0.1222	0.2697	0.4608	0.7506	1.0063

Problem 8B. Transient Diffusion in a Porous Slab

A rectangular slab of a porous solid material 1 cm thick is saturated with pure ethanol. At $t = 0$, the slab is immersed in a very large reservoir of water (thoroughly agitated). The void volume of the slab corresponds to about 50%; the effective diffusivity is thought to be 22% of the value in the free liquid. How long will it take for the mole fraction of ethanol at the center of the slab to fall to 0.022? Because of the energetic stirring, it may be assumed that resistance to mass transfer in the water phase at the surface is nearly zero. The following data are available at 25°C:

D_{AB} for Ethanol (A) and Water (B):

x_A	D_{AB} (cm²/s)
0.05	1.13×10^{-5}
0.10	0.90
0.275	0.41
0.50	0.90
0.70	1.41
0.95	2.20

Find two answers for this problem: one assuming that the diffusivity can be taken as a constant, and the other in which the concentration dependence is taken into account in your calculations. Note that for the first case, the problem can be handled using the product method. If it were absolutely essential that this process (the centerline reduction of ethanol) be accelerated, what steps would you consider? For the case in which $D = f(x_A)$, the governing equation must be written as

$$\frac{\partial x_A}{\partial t} = \frac{\partial}{\partial y}\left(D\frac{\partial x_A}{\partial y}\right).$$

Obviously, the exact nature of the resulting equation will depend upon your functional choice for *D*.

Problem 8C. Gas Absorption into a Falling Liquid Film

The manufacture of cellulosic fibers and films was initiated in 1891 and continues to the present day. A persistent problem in this industry has been the liberation of hydrogen sulfide (due to the use of sulfuric acid in the spinning bath). Obviously, absorption would be one possibility for dealing with this problem. Consider a wetted wall device in which water flows down a flat vertical surface; hydrogen sulfide is to be removed from the gas phase by contact with the liquid film. The entire apparatus is to be maintained at 25°C. The diffusivity of hydrogen sulfide in water is 1.61×10^{-5} cm²/s and the solubility is approximately 0.3375 g per 100 g water.

If the contact time is slight, then the H_2S penetration should be small. Consequently, an approximate model for this process can be written as

$$V_{max}\frac{\partial C_A}{\partial z} = D_{AB}\frac{\partial^2 C_A}{\partial y^2}.$$

This model is attractive because

$$\frac{C_A}{C_A^*} = \mathrm{erfc}\left(\frac{y}{\sqrt{4D_{AB}z/V_{max}}}\right).$$

However, for the apparatus being contemplated here, the exposure time is not necessarily short and the penetration of hydrogen sulfide into the liquid film may be significant. Suppose that the water film is 0.02 cm thick such that the maximum (free surface) velocity is just less than 20 cm/s. Use a more suitable model to determine whether the simplified solution is appropriate if the absorber apparatus employs a vertical wall 1.75 m high (long). Compare concentration distributions at *z*-positions (origin at top of absorber wall) of 10, 80, and 150 cm. Also, look at the total absorption over 1.75 m. Can the simple model be used in this case?

Problem 8D. Transient Diffusion with Impermeable Regions Inserted

Consider transient two-dimensional mass transfer (contamination) in a square region measuring 49 × 49 cm. The governing equation (using dimensionless concentration) is

$$\frac{\partial C}{\partial t} = D_{AB}\left(\frac{\partial^2 C}{\partial x^2} + \frac{\partial^2 C}{\partial y^2}\right).$$

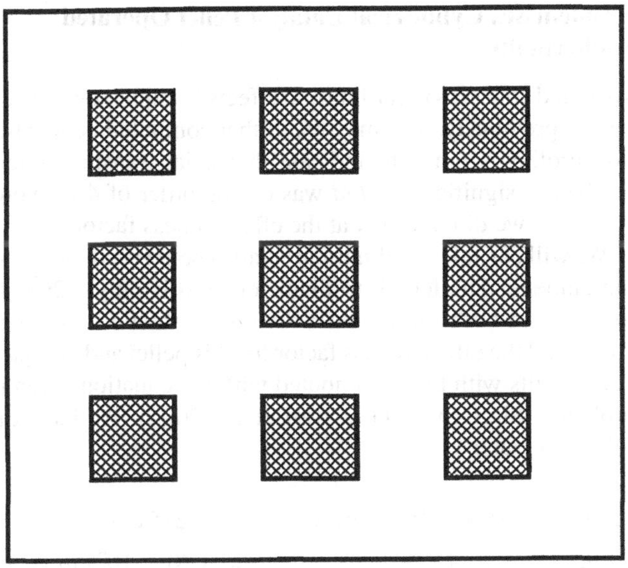

FIGURE 8D. Diffusion region with nine impermeable blocks inserted.

Initially, the field contains *no* contaminant. For all $t > 0$, the concentration on the left-hand boundary will be $C(x=0,y) = 1$. The bottom boundary is completely impermeable such that

$$\left.\frac{\partial C}{\partial y}\right|_{y=0} = 0.$$

The contaminant will be lost through the right-hand and top boundaries; for example on the

$$\text{right-hand side}: \quad \frac{\partial C}{\partial x} = \beta C|_{x=L},$$

where $\beta = -0.25$. A similar relationship applies to the top except that the derivative is written with respect to the *y*-direction. Assume that the diffusion coefficient has an effective value of 6.0×10^{-5} cm²/s. Find the concentration distributions at $t = 3{,}500{,}000$ s and 5,184,000 s (about 40 and 60 days, respectively. *Now*, place nine impermeable blocks in the domain as shown in Figure 8D. These are regions in which $D_{AB} = 0$. This technique has been used previously to simulate transport through a porous medium. Note that for this case, 18% of the original field has been occluded. Repeat the previous analysis and determine the effects of the blockages upon the development of the concentration distributions. Provide a graphical comparison of your results. Comment upon the suitability of (and problems encountered with) this technique for examining the spread of contaminants through porous media.

Problem 8E. Cylindrical Catalyst Pellet Operated Isothermally

We noted concerns regarding end effects in (squat) cylindrical pellets previously. You may recall that computed concentration profiles seemed to indicate that axial transport might not be too significant if *L/d* was on the order of 4 or more. However, we did not look at the effectiveness factor.

We will consider a cylindrical catalyst pellet used for cracking cumene. The length-to-diameter ratio is only 2.267, so transport in both the *r*- and *z*-directions should be considered. Find the effectiveness factor for this pellet and compare your results with those computed with the equation given in problem 18D.1 (p. 581) in Bird et al. (2002). We have the following data:

$$K_1 = 2.094 \times 10^{-7}\,\text{cm/s} \qquad R = 0.215\,\text{cm}$$
$$L = 0.975\,\text{cm} \qquad a = 3.90 \times 10^6\,\text{cm}^{-1}$$
$$C_{as} = 3.0 \times 10^{-4}\,\text{g mol/cm}^3$$

The effective diffusivity should range from 1×10^{-5} to 5×10^{-3} cm^2/s (an interesting problem could be formulated by allowing different values for D_{eff} in the *r*- and *z*-directions—how might that come about?).

Problem 8F. The Loschmidt or Shear-Type Diffusion Cell

Consider an apparatus consisting of two cylinders that can be aligned vertically to provide a continuous pathway with length 2*L*. Initially, one-half of the apparatus is filled with carbon dioxide and the other half is filled with methane. Both are at $p=1$ atm and 25°C. At $t=0$, the two halves are brought into alignment and diffusion commences. The governing equation is

$$\frac{\partial x_A}{\partial t} = D_{AB}\frac{\partial^2 x_A}{\partial z^2}.$$

Since the ends of the apparatus are impermeable to "A", we have the boundary conditions:

$$\text{For } z = +L \text{ and } -L,\ \frac{\partial x_A}{\partial z} = 0.$$

By applying the product method, we find that

$$x_A = C_1 \exp(-D_{AB}\lambda^2 t)[A \sin\lambda z + B\cos\lambda z].$$

The boundary conditions allow us to show that $\cos\lambda L = 0$. Consequently, the constant of separation must assume the values: $\pi/2L$, $3\pi/2L$, $5\pi/2L$, and so on. Thus, the solution can be written more usefully as

$$x_A = \sum_{n=1}^{\infty} A_n \sin\frac{(2n-1)\pi z}{2L}\exp\left(-\frac{(2n-1)^2\pi^2}{4L^2}D_{AB}t\right).$$

Naturally, as *t* becomes very large, $x_A \to 1/2$ over the entire apparatus. Since the leading coefficients must be evaluated using the initial condition

$$x_A = 1 \text{ for } -L<z<0 \text{ and } x_A = 0 \text{ for } 0<z<+L,$$

it makes sense for us to write the solution in the following form:

$$x_A = \frac{1}{2} + \sum_{n=1}^{\infty} A_n \sin\frac{(2n-1)\pi z}{2L} \times \exp\left(-\frac{(2n-1)^2\pi^2}{4L^2}D_{AB}t\right).$$

For the apparatus in question, $L=12.5$ in. Find and plot the concentration profiles for the following *t*'s: 200, 800, and 1600 s. *When will the average mole fraction of methane (in the methane half, of course) fall to 0.705?*

Problem 8G. Mass Transfer Studies with the Laminar Jet Apparatus

Scriven and Pigford (*AIChE Journal*, 4:439, 1958) measured the absorption of carbon dioxide into water using a laminar jet apparatus in which the exposure time of the fresh liquid could be tightly controlled. It is clear that such experiments could be used in a variety of ways. For example, it should be possible to test the usual assumption of equilibrium at the gas–liquid interface. In addition, such experiments should facilitate accurate determination of diffusivities, should the assumption of interfacial equilibrium prove to be valid. Be aware this problem has normally been treated as a semi-infinite slab and the familiar erfc solution has been used for analysis. However, it is clear that the column of liquid is not really rod-like since the no-slip condition must hold up to the instant the fluid leaves the nozzle assembly. We would like to address the question: Does the obvious variation in velocity affect the absorption process or the depth of penetration of the solute? Scriven and Pigford note that their results differ no more than just a few percent from the ideal jet case. Let us examine a laminar jet for which the nozzle diameter is 1.5 mm and the mean velocity of the jet is 100 cm/s. In the cited work, the authors used a brass nozzle with a diameter of 1.535 mm and a glass receiver with an ID of 1.941 mm. This means that some swelling is certain to occur. Since we cannot rigorously treat the absorption process without knowing the velocity distribution, it seems prudent to tackle it first. An elementary approach might involve the parabolic PDE:

$$\frac{\partial v_z}{\partial t} = \nu \left[\frac{\partial^2 v_z}{\partial r^2} + \frac{1}{r}\frac{\partial v_z}{\partial r} \right]. \tag{1}$$

Thus, we would forward march in time, computing the approximate evolution of the jet. However, this is definitely a steady-state situation and as an alternative, one might contemplate a more appropriate model:

$$v_r \frac{\partial v_z}{\partial r} + v_z \frac{\partial v_z}{\partial z} = \nu \left[\frac{\partial^2 v_z}{\partial r^2} + \frac{1}{r}\frac{\partial v_z}{\partial r} + \frac{\partial^2 v_z}{\partial z^2} \right]. \tag{2}$$

First, find the approximate velocity "distribution" using eq. (1), assuming that the jet must travel a distance of 4.75 cm (nozzle to receiver). Then, consider the following:

1. How could eq. (2) be solved?
2. What boundary conditions would you employ?
3. Would there be any advantage to assuming that the penetration depth was slight such that the problem could be worked in rectangular coordinates?
4. What other equations—in addition to (2)—would have to be utilized to solve the complete problem?

Problem 8H. Diffusivity of Carbon Dioxide in Air

Reid and Sherwood (1966) give the diffusivity of carbon dioxide in air as 0.142 cm^2/s at 276.2K and 0.177 cm^2/s at 317.2K. In 1955, S. P. S. Andrew published a description of a simple method for the determination of gaseous diffusion coefficients (*Chemical Engineering Science*, 4:269–272, 1955); one of the systems he tested was carbon dioxide in air. We would like to use his experimental data to determine D_{AB} for this pair of gases.

Andrew's apparatus consisted of two 2 L spherical flasks connected by a diffusion tube, less than 24 cm long and about 0.7 cm in diameter. This entire assembly was placed in a water bath to equilibrate and maintain temperature. Air was placed in one flask, and a mixture of carbon dioxide and air in the other. A common absorber was used to equalize the pressures of the two flasks. At $t = 0$, a stopcock located at the center of the diffusion tube was opened and equimolar counterdiffusion was initiated.

Andrew reported his results in terms of initial and final concentration differences, where concentration was expressed on a volumetric ratio basis. Here are excerpts from his data:

t (h)	39.66	66	111.5
p (mmHg)	755	765	765
T (K)	293	291	291
ΔC_o	0.1132	0.1136	0.1035
ΔC_t	0.0788	0.0638	0.0384
Q (cm^3)	39.4	56.9	74.5

Note that Q is the quantity of carbon dioxide transferred in time t, expressed as the volume of pure gas at the prevailing total pressure and temperature.

Andrew notes that the length of the diffusion tube must be corrected because of the resistance offered to the diffusing species as it spreads from the end of the tube throughout the flask volume. He estimates that the effective length of the diffusion tube is about 2% greater than the measured value and gives an approximate (corrected) value of 23.4 cm. One solution for this difficulty would be to agitate (stir) the two flasks, but this would significantly complicate the apparatus. The measured cross-sectional area of the diffusion tube was 0.41 cm^2, and the precise volumes of the two flasks were 2.3 and 2.278 L.

Find an analytic solution for this problem, expressing the mole fraction of carbon dioxide (at the lean end of the tube) as a function of time.

Find a numerical solution (using trial and error for selection of the diffusivity) that leads to agreement with Andrew's data.

What are appropriate values for the diffusivity of carbon dioxide in air for the three experimental cases described above?

Problem 8I. Diffusivity of Carbon Dioxide in Seawater

Reid and Sherwood (1966) have provided the following value for the diffusivity of carbon dioxide in water at 25°C:

$$D_{AB} = 2.0 \times 10^{-5}\,cm^2/s.$$

This diffusivity may be one of the more important transport properties from an environmental perspective; it must be a key factor in the absorption of CO_2 by seawater. The reason this is critical has been made clear by a number of recent review articles. For an example, see the piece written by Bette Hileman in *Chemical and Engineering News*, November 27, 1995, pp. 18–23. This writer concluded that ocean levels may rise by 15–95 cm by the year 2100 due to the activities of man that are elevating the mean global temperature. Of course, we did not set out to do this; it is an inadvertent result of industrialization. Nevertheless, it may be a bad time to buy beach property.

Hileman cites NASA data indicating that the mean global temperature has increased by about 0.6 or 0.7°C over the last century. If one were to extrapolate these data linearly (always a risky proposition), he/she might conclude that we could expect another 0.2 or 0.3°C rise by 2030. One thousand years ago, the carbon dioxide concentration in the atmosphere was a little less than 280 ppm. We are now rapidly approaching 400 ppm. We need a very accurate diffusivity in order to estimate how rapidly CO_2 is absorbed into seawater.

Suppose we explore use of the liquid laminar jet apparatus; see Scriven and Pigford, *AIChE Journal*, 4:439 (1958) and 5:397 (1959). Assume the jet nozzle diameter is 1.54 mm and the mean velocity of the seawater jet is 95 cm/s. Using the highly simplified analysis where the governing equation is taken as

$$\frac{\partial C_B}{\partial t} = D_{AB}\frac{\partial^2 C_B}{\partial x^2},$$

estimate the total expected absorption when the jet is exposed to pure CO_2 at $p = 1$ atm and 25°C. The nozzle and receiver are 9 cm apart. Would you expect the values of the diffusivities in pure water and seawater to vary significantly? Explain.

Problem 8J. Nonisothermal Effectiveness Factors for First-Order Reactions

Consider the spherical catalyst pellet with an exothermic chemical reaction (and operating at steady state). The governing equations are

$$\frac{d^2c_A}{dr^2} + \frac{2}{r}\frac{dc_A}{dr} - \frac{k_1 a}{D_{\text{eff}}}c_A = 0$$

and

$$\frac{d^2T}{dr^2} + \frac{2}{r}\frac{dT}{dr} - \frac{k_1 a \Delta H}{k_{\text{eff}}}c_A = 0.$$

Note the obvious similarities between the equations (you might want to review the Damköhler relationship between temperature and concentration). D_{eff} and k_{eff} are the effective diffusivity and thermal conductivity, respectively. It is convenient to characterize the behavior of this system with three dimensionless groups:

Thiele modulus : $\phi = R\sqrt{\frac{k_1 a}{D_{\text{eff}}}}$

Arrhenius number : $\gamma = \frac{E}{RT_s}$

Heat generation parameter : $\beta = -\frac{\Delta H D_{\text{eff}} c_{As}}{k_{\text{eff}} T_s}$.

Among the interesting possibilities for this system are effectiveness factors (η_A's) greater than one and steady-state multiplicity. Using the parametric values $\phi_s = 0.3$, $\gamma = 20$, and $\beta = 0.7$, find and prepare a figure illustrating the three possible concentration distributions in the interior of the spherical pellet. What are the corresponding values for η_A? Are the three concentration profiles equally likely? That is, can we draw any conclusions regarding the relative stability for the three cases?

Problem 8K. Uptake of Sorbate by a Sphere in a Solution of Limited Volume

Consider a porous sorbent sphere placed in a well-agitated solution of limited volume. The solute species ("A") is taken up by the sphere and the concentration of "A" in the liquid phase is depleted. The governing equation for transport in the sphere's interior is

$$\frac{\partial C_A}{\partial t} = D_{AB}\left[\frac{\partial^2 C_A}{\partial r^2} + \frac{2}{r}\frac{\partial C_A}{\partial r}\right]. \tag{1}$$

Note that this equation can be transformed into an equivalent problem in a "slab" by setting $\phi = C_A r$. The total amount of "A" in solution initially is VC_{A0} and the rate at which "A" is removed from the solution can be described by

$$4\pi R^2 D_{AB}\left.\frac{\partial C_A}{\partial r}\right|_{r=R}, \tag{2}$$

therefore, the total amount removed over a time t can be obtained by integration of (2). We would like to try to confirm part of the graphical results presented in Figure 6.4 in J. Crank's *The Mathematics of Diffusion* (Clarendon Press, Oxford, 1975). Use the following parametric values:

$$D_{AB} = 1\times10^{-5}, \quad R = 1, \quad V = 6, \quad C_{A0} = 1.$$

What is the ultimate fraction of solute taken up by the sphere in this case? Does your plot of $M(t)/M_\infty$ against $\sqrt{(D_{AB}t/R^2)}$ correspond to the results provided in Crank's Figure 6.4?

Problem 8L. Edge Effects in Transport Through Membranes

Consider one-dimensional transport of a constituent "A" through a membrane; the process is approximately described by

$$\frac{\partial c_A}{\partial t} = D\frac{\partial^2 c_A}{\partial z^2}.$$

The membrane extends from $z = 0$ to $z = h$. The concentration at $z = 0$ is maintained at c_{A0} for all $t > 0$ and the initial concentration of "A" within the membrane is zero. The fluxes are equated at $z = h$ by setting

$$-D\left.\frac{\partial c_A}{\partial z}\right|_{z=h} = K(c_A(z=h) - c_{A\infty}).$$

Use the product method to find an analytic solution for the case, where K is *very large*.

Now, let us assume that the membrane is supported at the edges by an impermeable barrier (clamping bracket). If the effective diameter of the membrane is only a small multiple of its thickness, then the governing equation must be rewritten as

$$\frac{\partial c_A}{\partial t} = D\left[\frac{\partial^2 c_A}{\partial r^2} + \frac{1}{r}\frac{\partial c_A}{\partial r} + \frac{\partial^2 c_A}{\partial z^2}\right].$$

Obviously, the flux of the permeate will be reduced near the edges where the supporting hardware obstructs transport in the z-direction. We will confine our attention to the case for which $h/R = 1/3$. Let $h = 4$ mm, $D = 2 \times 10^{-5}$ cm^2/s, and $c_A(z=0) = 1$. We assume that transport into the fluid phase at $z = h$ occurs so rapidly that the concentration is effectively zero (there is no resistance to mass transfer in the fluid phase at $z = h$). Under these conditions, the interesting dynamics occur mainly in the first 1000 s or so. Solve this problem by the method of your choice and prepare a figure that shows the flux of permeate at $t = 900$ s as a function of r (setting $z = h$). A rule of thumb for transport through membranes is that edge effects are probably negligible if $h/R \leq 0.2$.

Problem 8M. Modification of Shrinking Core Models for Regeneration of Catalyst Particles

When catalyst pellets become fouled by carbon deposition, they lose their effectiveness. One remedy is regeneration in which the pellet is exposed to elevated temperatures in an oxygen-rich environment. In the resulting combustion process, the carbon is converted to CO_2. This is convenient because for every O_2 diffusing in, a CO_2 diffuses out. If the reaction occurs rapidly, then movement of the carbon "front" in the interior is strictly the result of mass transfer of oxygen. For a spherical particle, the governing equation is simply

$$\frac{\partial C}{\partial t} = D_{\text{eff}} \left[\frac{1}{r^2} \frac{\partial}{\partial r} \left(r^2 \frac{\partial C}{\partial r} \right) \right]. \tag{1}$$

If we assume the process is pseudo-steady state, then (1) can be directly integrated and the flux at the carbon front can be estimated from

$$N|_{r=R_c} = \frac{-D_{\text{eff}} C_s}{(R_C - (R_C^2/R))}, \tag{2}$$

where R is the radius of the catalyst pellet and R_C corresponds to the position of the carbon interface. An unsteady carbon balance can now be written since the rate at which oxygen arrives at the interface is virtually equal to the rate at which carbon disappears:

$$-4\pi R_C^2 \cdot \left| N_{r=R_c} \right| = \frac{d}{dt} \left[\frac{4}{3} \pi R_C^3 \rho_C \phi \right], \tag{3}$$

where ρ_C and ϕ are the molar density and volume fraction of carbon, respectively. Equation (3) can be solved to yield an estimate for the time required to consume all the carbon in the pellet interior:

$$t_{\text{req}} = \frac{\rho_C \phi R^2}{6 D_{\text{eff}} C_S}. \tag{4}$$

Therefore, if $\phi = 0.22$, $\rho_C = 0.0387$ g mol/cm^3, $R = 0.6$ cm, $D_{\text{eff}} = 2 \times 10^{-3}$ cm^2/s, and $C_S = 2.433 \times 10^{-5}$ g mol/cm^3, the required time for regeneration is 175 min.

We would like to modify this elementary analysis for spheres by solving the transient diffusion equation (1) using a variable diffusivity to account for the inability of the oxygen to penetrate the carbon-blocked pores. Let us assume that $D_{\text{eff}} = mC + b$, with $m = 0.82203$ and $b = 2 \times 10^{-19}$ (effectively zero). This means $D_{\text{eff}} = 2 \times 10^{-5}$ cm^2/s when C corresponds to the surface value. Prepare a figure that shows the radial distribution(s) of oxygen as a function of time and find the time required for regeneration. Is this mass transfer model capable of representing the regeneration process? Could a reaction term be added to the balance to improve model performance? Propose a formulation for this term.

Then, repeat your analysis for the case of a cylindrical catalyst pellet for which $L = 2d$. This ratio is clearly not large enough to discount the axial (z-direction) transport of oxygen, so take the governing equation for oxygen transport in the interior to be (if D_{eff} were constant):

$$\frac{\partial C}{\partial t} = D_{\text{eff}} \left[\frac{\partial^2 C}{\partial r^2} + \frac{1}{r} \frac{\partial C}{\partial r} + \frac{\partial^2 C}{\partial z^2} \right]. \tag{5}$$

Assume that the parametric values are the same as above with $R = 0.6$ cm. Will the regeneration time be significantly different in this case (versus the sphere)? Note that the actual equation to be solved in this case is

$$\frac{\partial C}{\partial t} = -\frac{1}{r} \frac{\partial}{\partial r}(r N_{Ar}) - \frac{\partial}{\partial z}(N_{Az}). \tag{6}$$

Problem 8N. Absorption of CO_2 at Elevated Pressures

Carbon dioxide is to be absorbed into an aqueous solution in a 10 L cylinder. The cylinder is charged with 9 L of the aqueous solution, and the 1 L gas space is pressurized with CO_2 to 800 psi. Mass transfer into the liquid phase will occur solely by diffusion and the temperature of the surroundings is maintained at 18°C. The cylinder is to be positioned vertically such that the interfacial area is 410 cm^2. Since the solubility of CO_2 is pressure dependent, the interfacial equilibrium mole fraction will diminish as the absorption proceeds. Some interpolated data for $T = 18$°C are provided in the following table.

CO_2 Pressure (atm)	Mole Fraction x_{A0}
10	0.006
20	0.011
30	0.0151
50	0.0217
75	0.0248

Determine the evolution of the concentration profile in the liquid phase over the first 10 h of the process. You must take the changing pressure in the gas space into account because of its effect upon x_{A0}. Experiments reveal that the cylinder pressure diminishes more rapidly than indicated by the diffusional model. Offer a plausible explanation.

Problem 9A. Mass Transfer in the Laminar Boundary Layer

We would like to examine the combined problem of momentum and mass transfer in the laminar boundary layer on a flat plate. In particular, imagine a large spill of a volatile liquid like methyl ethyl ketone (*mek*) upon a flat impermeable surface. The liquid is exposed to the atmosphere while the airflow approaching the spill is steady at 1 m/s. The governing equations appear to be

$$v_x \frac{\partial v_x}{\partial x} + v_y \frac{\partial v_x}{\partial y} = \nu \frac{\partial^2 v_x}{\partial y^2}$$

and

$$v_x \frac{\partial c_A}{\partial x} + v_y \frac{\partial c_A}{\partial y} = D_{AB} \frac{\partial^2 c_A}{\partial y^2}.$$

Suppose the air temperature is 26.5°C and the prevailing pressure is 1 atm. Estimate the flux of mek to the atmosphere and plot the results from the leading edge of the pool to a position 1 m downstream. If the spill is roughly 1 m × 1 m in size, estimate the total rate of transfer of mek to the gas phase. Neglect any possible deformation of the liquid surface (rippling). Finally, prepare a plot of the concentration distribution at a point 40 cm downstream from the leading edge. One analysis of this problem is presented in Section 20.2 in Bird et al. (2002); you may also want to see Hartnett and Eckert, *Transactions of the ASME*, 79:247 (1957). The following vapor pressure data are available for mek:

Temperature (°C)	Vapor Pressure (mmHg)
14	60
25	100
41.6	200
60	400

Pay particular attention to the shape of your concentration distribution. See anything interesting with broader implications? Are there any important limitations of your analysis?

Problem 9B. Polychlorinated Biphenyl Deposition in Riverine Sediments

In 1865, a chemical similar to PCB was discovered in coal tar; in 1929, Monsanto began to manufacture PCBs. Although the PCB-related health problems had appeared by 1936 (these include chloracne, reproductive disorders, liver disease, and cancer), GE began to use PCBs (as a dielectric fluid) in the manufacture of electrical capacitors at its Ft. Edward plant on the Hudson River in 1947. By 1974, the EPA had discovered that fish from the Hudson River were loaded with PCBs. Finally, in 1976, GE stopped dumping PCBs into the Hudson River and 1 year later, Monsanto stopped production completely. By this time, the environmental damage was both pervasive and ongoing. In 1993, tests of the groundwater and sediments near the GE plant at Hudson Falls revealed 2000–50,000 ppm PCBs; in fact, an "oily liquid" found seeping into a structure near the plant was tested in July of 1993—it turned out to be 72% PCBs! This environmental disaster is the basis for this problem.

Consider a small clay particle (loaded with adsorbed pollutant) released near the river surface. We would like to know where this particle might be deposited (on the river bottom) downstream. Assume that the surface water velocity is 3.25 mph (4.7667 ft/s). The velocity distribution is assumed to vary parabolically from zero at the channel bottom to 4.7667 ft/s at the free surface. The small particle has a diameter of 15 μm and a density of 1.9 g/cm^3. Assume the river channel has a mean depth of 4 ft. The particle settles under the influence of gravity, but its progress is hindered by drag (as given by the Stokes law). Consequently, the force acting in the *y*-direction will be *approximated* by

$$F_y = mg - 6\pi\mu RV,$$

where m is the mass of the particle and V is the velocity in the *y*-direction, *dy/dt*. Assume that the particle is completely entrained in the downstream flow. Where is the particle likely to reach bottom? Then search the literature and report on the extent of partitioning of PCBs between water and suspended clays and humus materials. We are particularly interested in the likelihood that adsorbed PCBs might be released from the sediments (which would constitute an ongoing source, especially if the channel bottom was disturbed).

Problem 9C. Point Source Pollution of a Stream in Near-Laminar Motion

A stream with rectangular cross section (10 ft wide and 1 ft deep) is contaminated at one side very near the free surface. The pollutant enters the stream at a rate of 2.5 g mol/min continuously until a virtual steady-state condition is attained. Find the concentration profile at both 2000 and 8000 yard downstream (from the point of injection). We presume that the governing equation can be written as

$$v_z \frac{\partial C_A}{\partial z} = D \left[\frac{\partial^2 C_A}{\partial x^2} + \frac{\partial^2 C_A}{\partial y^2} \right].$$

The (effective) diffusivity of the contaminant in water is 1.6×10^{-1} cm^2/s. The maximum velocity of the water (center of the channel at the free surface) is about 1.65 ft/s. It should be presumed that the flow is *nearly* laminar. The concentration at the point of injection is about 0.058 g mol/cm^3 (very rough); this should extend over 1.5% over the flow area near the upper corner. Assume that there is no loss from the free surface; then, solve the problem for two cases: (1) No loss at the top or the bottom of the channel, and (2) allow loss by setting the bottom surface concentration to zero. The velocity distribution will be governed by the equation

$$\frac{\partial^2 v_z}{\partial x^2} + \frac{\partial^2 v_z}{\partial y^2} + \frac{\beta}{\mu} = 0.$$

Problem 9D. SO_2 Release from Coal-Fired Power Plant

The mean residence time for sulfur compounds in the atmosphere has been estimated to be between 25 and 400 h. Sulfur dioxide is particularly worrisome, since it has been shown to cause a variety of cardiovascular and cardiorespiratory problems. In fact, prolonged exposure to as little as 0.10 ppm has been known to cause death in humans and animals. Increased hospital admissions have been observed for chronic exposure to concentrations as low as 0.02 ppm. 10 ppm can lead to death in as little as 20 min. As you might imagine, SO_2 emissions have been studied all over the country. Some of the "leading" states for emissions include Ohio, Indiana, Illinois, Missouri, and Tennessee. Consequently, acidification (resulting from acid rain) has been noted in Ontario, Quebec, Nova Scotia, Newfoundland, and the northeastern United States. There have been areas where the summer precipitation routinely had a pH of about 4.

Consider a coal-fired power plant that produces 650 MWe at an overall efficiency of about 28.5%. The plant burns a subbituminous coal from Wyoming with an approximate heating value of 9740 Btu/lb$_m$. This coal has a sulfur content of about 1% by weight and of that, it can be assumed that about 15% of that total sulfur ends up as SO_2 (leaving the plant with the flue gas). The boiler operates with about 6% excess air. The flue gas leaves the plant through a stack 700 ft high at an average temperature of about 260°F. The experimental diffusivity of SO_2 in air at 263K is 0.104 cm^2/s, but it can be assumed that in the atmosphere, the diffusivity has an effective value (corrected to the right temperature) about 50% larger than $D_{AB}(T)$. The ambient temperature is constant at 80°F. At an elevation 700 ft above the ground surface, the wind velocity can be taken to be constant (W to E) at 4.5 mph. *Assume* that the governing equation is

$$V_0 \frac{\partial C_A}{\partial z} = D_{AB}\left[\frac{1}{r}\frac{\partial}{\partial r}\left(r\frac{\partial C_A}{\partial r}\right) + \frac{\partial^2 C_A}{\partial z^2}\right].$$

The solution is provided in Bird et al. (2002) on p. 580. Use it to determine the steady-state distribution of SO_2 downstream from the power plant. Prepare a graphic illustrating the concentration profile on the z-axis. At what value of z do you expect to find interaction of the plume with the ground? Once the plume begins to interact with the ground, the cited solution is no longer valid. Describe the expected complications in detail.

Problem 9E. The Use of Axial Dispersion Models in Biochemical Reactors

Consider an unsteady-state model for the flow of a reactant species in a loop-type (airlift) reactor. The impetus for flow is provided by the introduction of bubbles on one side of the column divider. The flow field on the upflow side of such a reactor is quite complex; the rising bubbles and their accompanying wakes result in chaotic three-dimensional fluid motions. The downflow region, in contrast, tends to be very highly ordered (virtually laminar) at low gas rates. In the particular reactor under study, the flow path for one complete circulation is about 91 or 92 cm (about 46 cm on each side of the column divider). *One* model (balance) for the reactant employing three parameters can be written as

$$\frac{\partial c_A}{\partial t} + v_z \frac{\partial c_A}{\partial z} = D_L \frac{\partial^2 c_A}{\partial z^2} - k_1 c_A \tag{1}$$

A series of experiments was conducted in which an inert tracer was introduced as a pulse at the top of the column divider. Reactant (tracer) concentration was then determined photometrically at a fixed spatial position (near the bottom of the downflow side). The resulting photomultiplier output was recorded and a sample appears in Figure 9E. In this case,

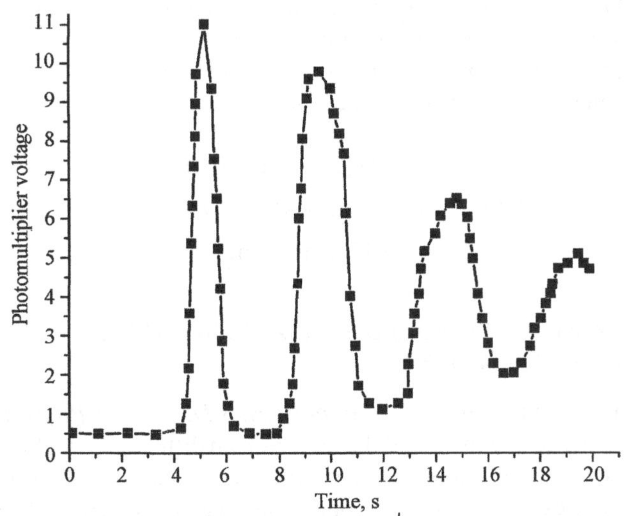

FIGURE 9E. Tracer data obtained with a bench-scale airlift reactor.

the superficial gas velocity on the upflow side was 1.26 cm/s. However, you should remember that typical rise velocities of air bubbles in aqueous media are on the order of 15–30 cm/s. As you can see from the data, the mean circulation velocity is on the order of 16 or 17 cm/s. Fit a model(s) of the type of eq. (1) to the data, determining suitable numerical values for the parameters (it is clearly advantageous to put the equation in dimensionless form). Demonstrate the suitability of your model by graphical comparison with the data. Should two models (one for upflow and one for downflow) be employed?

Problem 9F. Dissolution of Cast Benzoic Acid into a Falling Water Film

Consider the case where a film of water, 1.5 mm thick, flows down a flat vertical surface. Once the velocity profile is fully developed, the water encounters a section of wall consisting of cast benzoic acid. The governing equation for this situation is

$$v_z \frac{\partial c_A}{\partial z} = D_{AB} \frac{\partial^2 c_A}{\partial y^2},$$

where z is the direction of flow and y is the transverse (across the film) direction. If the penetration of the benzoic acid into the liquid film is slight, then one might replace the velocity distribution with a simple linear function of y. However, we would like to test that simplification with a more nearly correct expression for the variation of velocity. Indeed, let us assume that

$$v_z = \frac{\rho g}{2\mu} y^2.$$

Find and graph the concentration profiles at $z = 50$, 100, and 200 cm. Does the change in the functional form of the velocity distribution (from the straight line approximation) lead to a significant difference? The following data are available for the benzoic acid–water system at $T = 14°C$:

$$Sc = 1850 \qquad D_{AB} = 5.41 \times 10^{-6} \text{cm}^2/\text{s}$$

Solubility of benzoic acid : $2.39\ \text{kg/m}^3$

$$c_{A0} = 1.96 \times 10^{-5} \text{g mol/cm}^3.$$

Problem 9G. Pressure-Driven Duct Flow with a Soluble Wall and $D(C_A)$

Kuntz and Lavallee (*Journal of Physics D: Applied Physics*, 37:L5, 2004) considered the non-Fickian diffusion of $CuSO_4$ in aqueous solutions. They characterized cases in which D decreases with increasing concentration as *subdiffusive*; they also observed that moisture transport through certain porous materials occurs *more rapidly* than indicated by Fick's law. They referred to such processes as *superdiffusion* and noted that this occurs in a number of building materials and polymers.

Suppose we have a fully developed pressure-driven flow of water (in the z-direction) between parallel planar surfaces. The upper wall, located at $y = h$, is impermeable. The lower wall (at $y = 0$) consists of a slab of $CuSO_4$, with a solubility of 39 g per 100 g water (about 2.4 mol/L) at the prevailing water temperature. We will assume the diffusivity of $CuSO_4$ in water is adequately represented by

$$D \cong 0.31 + \frac{0.42}{(1 + C)^{2.87}}$$

(where C is mol/L and D is cm^2/s); of course, this means that D decreases by nearly 60% over the concentration range of interest. Let $h = 1$ cm and take the average velocity of the water to be 1.25 cm/s. Find the concentration distributions for z/h of 20, 200, and 2000, taking variable D into account and determine the Sherwood number at each location. If the diffusional process is Fickian with a constant diffusivity of 0.65 cm^2/s, how would the concentration profiles differ? What will the approximate viscosity of the aqueous solution be for this process?

Problem 9H. Mass Transfer with an Oscillating Upper Wall

An effort to increase the mass transfer rate using an oscillating wall is to be investigated. A fluid, initially at rest, begins to move through the space between two parallel planes at $t = 0$. The flow is pressure driven, but is influenced by an oscillating upper wall that moves as prescribed here: $V = V_0 + b\sin(\omega t)$. The flow field between the planar surfaces is governed by

$$\frac{\partial v_z}{\partial t} = -\frac{1}{\rho}\frac{\partial p}{\partial z} + \nu \frac{\partial^2 v_z}{\partial y^2},$$

with $v_z = 0$ at $y = 0$ and $v_z = V_0 + b\sin(\omega t)$ at $y = h$. *Assume* the concentration field is governed by

$$\frac{\partial C}{\partial t} = D\frac{\partial^2 C}{\partial y^2} - v_z \frac{\partial C}{\partial z},$$

and the concentration is initially zero everywhere between the plates. At $t = 0$, a soluble patch on the lower wall is exposed, *dp/dz* is applied, and the upper wall begins to oscillate.

We would like to determine what frequency of oscillation and what intensity of motion (of the upper surface) will be required to positively affect the mass transfer rate. Assume the apparatus consists of planar surfaces, 10 cm long and 1 cm apart. The fluid filling the apparatus has the properties of water. The soluble patch extends from $z = 0.333$ to

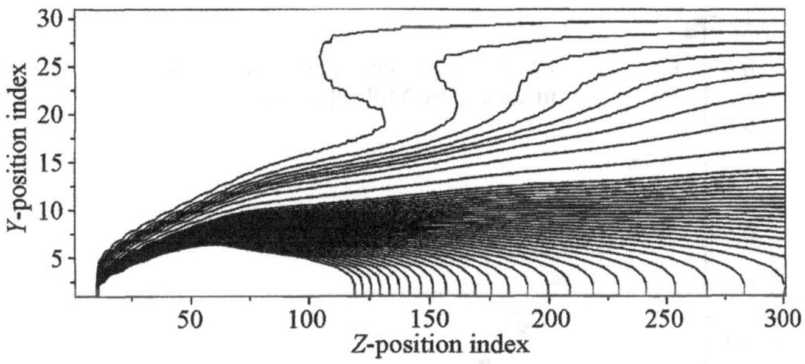

FIGURE 9H. An example of results of a computation illustrating the effects of an oscillating wall upon mass transfer.

1.0 cm along the bottom wall. The wall is impermeable at all other locations. Begin your investigation with the following property values: Let $Sc = \nu/D = 111$, $dp/dz = -0.01$ dyn/cm^2 per cm (for all positive t), $V_0 = -0.145$ cm/s, $b = 0.375$, and $\omega = 0.10$ rad/s. How long does it take for the flow field to acquire its ultimate oscillatory behavior? How should one assess any mass transfer enhancement? Identify the parameters of the problem that are most likely to positively affect performance of the apparatus. Comment on the significance of the oscillation frequency ω. It is to be noted that larger frequencies will not be effective. Would you expect to find an optimal value?

It is to be noted that the dimensionless time $t^* = Dt/h^2$ will have to achieve a value of about 0.3 (or more) in order for the effects of the oscillating wall to become apparent. An illustration of results obtained from a trial computation is given in Figure 9H.

Problem 10A. Heat Transfer for Turbulent Flow in a Pipe

Water enters a straight section of nominal 2 in., schedule 40 steel pipe with an initial (uniform) temperature of 60°F. The Reynolds number for the flow is 45,000 based upon the inlet temperature. The pipe wall is maintained at a constant 200°F and the heated section is 20 ft long. What is the water temperature at exit? Make a series of estimates using the following:

1. Reynolds analogy

$$\ln(T_w - T_{b1})/(T_w - T_{b2}) = 2fL/d.$$

2. Dittus and Boelter correlation

$$Nu_m = 0.023 Re^{0.8}\, Pr^{0.4}.$$

3. Prandtl's analogy (which takes into account the thickness of the "laminar" sublayer)

$$Nu = \frac{(f/2) Re\, Pr}{1 + 5\sqrt{f/2}(Pr - 1)}.$$

4. The result developed by Sandall et al. (*Canadian Journal of Chemical Engineering*, 58:443, 1980) for *constant heat flux*. See pp. 411–414 in Bird et al. (2002). Although this result (13.4–20) was developed for constant heat flux at the wall, we will apply it to our case by dividing the pipe length into increments.

Problem 10B. "Prediction" of Eddy Diffusivity for the Fully Developed Duct Flow

Elementary closure schemes often require a functional representation for the eddy diffusivity ε_M and for the heat transfer problems, a relationship between ε_M and ε_H. One popular approach is to use Nikuradse's mixing length expression,

$$l = R\left[0.14 - 0.08\left(1 - \frac{y}{R}\right)^2 - 0.06\left(1 - \frac{y}{R}\right)^4\right]$$

in conjunction with Van Driest's damping factor:

$$\varepsilon_M = l^2(1 - e^{-y/A})^2 \frac{dV}{dy}.$$

Use these expressions, and an appropriate functional form for $V(y)$, to find ε_M. Does the shape of the eddy diffusivity correspond to available experimental data?

Problem 10C. Martinelli's Analogy

Refer to Martinelli's paper (*Transactions of the ASME*, 69:947, 1947) and prepare a brief description of how the function F_1 was determined. Note that this function depends upon both *Re* and *Pr*. Does F_1 have an apparent physical interpretation? If so, what is it?

Problem 10D. Exploring Analogies Between Heat and Momentum Transfer

Reynolds proposed that heat and momentum transfer mechanisms were the same in turbulent flow in tubes. What lends this idea credence is that

$$\frac{1}{\rho}\frac{\partial P}{\partial z} = \frac{1}{r}\frac{\partial}{\partial r}\left[r(\nu + \varepsilon_M)\frac{\partial V_z}{\partial r}\right]$$

and

$$V_z\frac{\partial T}{\partial z} = \frac{1}{r}\frac{\partial}{\partial r}\left[r(\alpha + \varepsilon_H)\frac{\partial T}{\partial r}\right].$$

Note that the upper case letters represent time-averaged quantities. Remember that these equations imply first-order closure, which means gradient transport models will be used to represent turbulent fluxes that *are not* gradient transport processes! Reynolds' analogy resulted in

$$Nu = \frac{f}{2} Re\, Pr, \text{ or alternatively, } \ln\left[\frac{T_w - T_{b1}}{T_w - T_{b2}}\right] = \frac{fL}{R}.$$

Prandtl improved Reynolds' development by taking the velocity distribution in the "laminar sublayer" into account:

$$Nu = \frac{(f/2) Re\, Pr}{1 + 5\sqrt{(f/2)}(Pr - 1)}.$$

Von Karman took this one more step by making use of the "universal" velocity distribution to obtain:

$$Nu = \frac{(f/2)\mathrm{RePr}}{1 + 5\sqrt{(f/2)}\left[\mathrm{Pr} - 1 + \ln(1 + 5/6(\mathrm{Pr} - 1))\right]}.$$

Compare the Nusselt numbers obtained from these analogies with available experimental data/correlations. Assume Reynolds numbers ranging from 10^4 to 10^6 with water as the fluid.

Problem 10E. Resistance to Heat Transfer

For large Pr, the resistance to heat transfer in turbulent flows is concentrated in the "wall layer." But for small Pr, the situation can be quite different as the resistance is more evenly distributed. What types of fluids have small Pr? Prepare a brief report on the effects of Pr upon the temperature distribution in heat transfer in a turbulent duct flow.

Problem 10F. Temperature Fluctuations in Grid-Generated Turbulence

Mills et al. (Turbulence and Temperature Fluctuations Behind a Heated Grid, *NACA TN 4288*, 1958) carried out a study of temperature fluctuations behind a heated grid in a wind tunnel. They measured both velocity and temperature at dimensionless positions ranging from $x/M = 17$–65 (x is the downstream distance and M is the mesh size for the grid, 1 in.). They employed a mean velocity of 14 ft/s and their data yielded both velocity and temperature correlations, and example of the latter is given in Figure 10F.

The temperature correlation coefficient is defined by

$$\theta(r) = \frac{\overline{T'(x)T'(x+r)}}{\overline{T'^2}}.$$

Note that the distance of separation is rendered dimensionless with the temperature microscale λ_θ. Consequently, if a parabola of osculation was fit to $\theta(r)$, it would intercept the x-axis at 1. The authors noted that the temperature microscale could be estimated from the isotropic decay equation:

$$\frac{d\overline{T'^2}}{dx} = -12\frac{\alpha}{U}\frac{\overline{T'^2}}{\lambda_\theta^2},$$

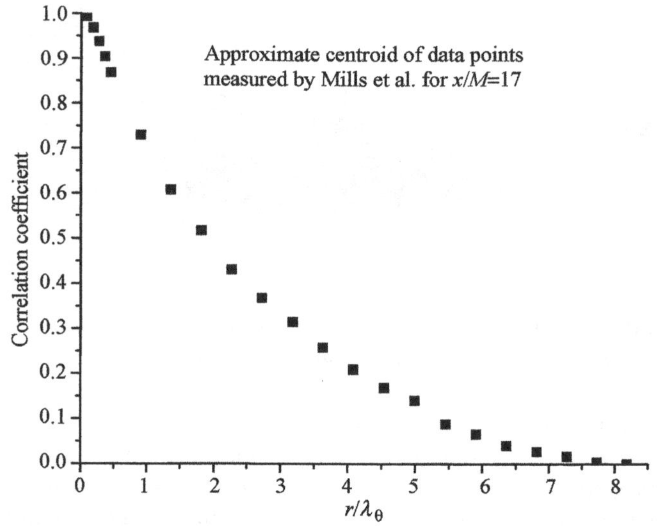

FIGURE 10F. Correlation coefficient data adapted from Mills et al. (1958).

where α is the thermal diffusivity and U is the mean air velocity in the test section. Use the data available in *NACA TN* 4288 to obtain an estimate of the temperature microscale, and then find the spectrum for temperature fluctuations by transforming $\theta(r)$.

Problem 11A. Solutions for the Rayleigh–Plesset Equation

The Rayleigh–Plesset equation is a second-order, nonlinear, ordinary differential equation that describes the motion of the gas–liquid interface of a spherical bubble undergoing collapse and (possibly) rebound. Rayleigh's original development was adapted by Plesset (The Dynamics of Cavitation Bubbles, *Journal of Applied Mechanics*, 16:277, 1949) to include surface tension; the form that we now find throughout the literature is

$$\frac{P_i - P_\infty}{\rho} = R\frac{d^2R}{dt^2} + \frac{3}{2}\left(\frac{dR}{dt}\right)^2 + \frac{4\nu}{R}\frac{dR}{dt} + \frac{2\sigma}{\rho R}.$$

Of course, R corresponds to the radius of the spherical bubble or cavity. The effect of the viscous term is usually small, so it is frequently neglected. We can use this equation to predict how a bubble will respond to changes in the pressure difference (the driving force on the left-hand side). The principal problem with this equation is that it is *stiff* (there is an incompatibility between the eigenvalues and the time-step size). Because of this characteristic, the familiar numerical procedures will not work well for this type of problem. Solve the Rayleigh–Plesset equation for the case in which the ambient pressure undergoes an instantaneous step increase. Use the form of the equation employed by Borotnikova and

Soloukhin (1964)—this will provide an easy means for you to verify your results:

$$\frac{d^2R}{d\tau^2} = \frac{\mu}{R}\left[R^{-3\gamma} + (A_0 + A_1 \cos\tau\right] - \frac{3}{2}\frac{1}{R}\left(\frac{dR}{d\tau}\right)^2. \quad (1)$$

This equation has been put into dimensionless form, so the dependent variable R is now defined as R/R_0. Note that

$$\tau = \omega t, \quad \gamma = 1.4, \quad \mu = 0.837 \times 10^{-6},$$

$$A_0 = 50, \quad \text{and} \quad A_1 = 0.$$

Problem 11.B. Solving the Stefan–Maxwell Equations for a Ternary System

A gaseous system contains components A, B, and C. The diffusivities (cm^2/s) for the system are

$$D_{AC} = 0.135 \qquad D_{BC} = 0.199 \qquad D_{AB} = 0.086.$$

The diffusion path is 22 cm long and the mole fractions at the boundaries are as follows:

Component	Position 1	Position 2
A	0.305	0.001
B	0.585	0.002
C	0.110	0.997

Use the Stefan–Maxwell equations to find the concentration distributions of the three constituents for $0 < z < 22$ cm. Do the computed fluxes differ from your initial (Fickian) estimate?

Problem 11.C. Estimating the Initial Dynamic Behavior of the Particle Number Densities in an Aerosol

We would like to examine the relative effectiveness of Brownian motion and turbulence with regard to the initial rate of disappearance of particles (of different initial size) in an aerosol with decaying turbulence. We will compare three cases using the particle diameters of 0.75, 1.5, and 3.0 μm. Use the following parametric values for all three cases:

$$n_0 = 3 \times 10^7 \text{ particles/cm}^3, \quad l = 40\,\text{cm},$$

$$v = 0.151\,\text{cm/s}, \quad \text{and } T = 25°\text{C}.$$

Assume that the initial dissipation rate (per unit mass) ε is 1×10^5 cm^2/s^3. We will assume that the decay of turbulent energy is adequately represented by

$$\frac{d}{dt}\left(\frac{3}{2}u^2\right) = -\varepsilon = -A\frac{u^3}{l},$$

but remember to check the Reynolds number to make sure that Taylor's inviscid estimate for the dissipation rate is appropriate.

APPENDIX A

FINITE DIFFERENCE APPROXIMATIONS FOR DERIVATIVES

Finite difference approximations allow us to develop algebraic representations for partial differential equations. Consider the following Taylor series expansions:

$$y(x+h) = y(x) + hy'(x) + \frac{h^2}{2}y''(x) + \frac{h^3}{6}y'''(x) + \cdots \tag{A.1}$$

and

$$y(x-h) = y(x) - hy'(x) + \frac{h^2}{2}y''(x) - \frac{h^3}{6}y'''(x) + \cdots. \tag{A.2}$$

If we add the two equations together,

$$y(x+h) + y(x-h) = 2y(x) + h^2y''(x) + f(h^4) + \cdots,$$

and then discard all the terms involving h^4 (and up), we get

$$y''(x) \cong \frac{y(x+h) - 2y(x) + y(x-h)}{h^2}. \tag{A.3}$$

This second-order central difference approximation for the second derivative has a leading error on the order of h^2. If h is small, this approximation should be good. For example, let

$$y = x\sin x, \text{ thus, } \frac{dy}{dx} = \sin x + x\cos x, \text{ and}$$

$$\frac{d^2y}{dx^2} = 2\cos x - x\sin x.$$

Now assuming $x = 0.3$,

$$y = 0.088656, \quad \frac{dy}{dx} = 0.582121, \quad \text{and} \quad \frac{d^2y}{dx^2} = 1.822017;$$

then choose $h = 0.01$:

$$\frac{d^2y}{dx^2} \cong \frac{0.094568 - 2(0.088656) + 0.082926}{(0.01)^2} = 1.820.$$

This is about 0.11% less than the analytic value for the second derivative. By simply combining Taylor series expansions, we can build any number of approximations for derivatives of any order. Furthermore, these approximations can be forward, backward, centered, or skewed. Some of the more useful are compiled below. Note that F stands for forward, C for central, B for backward, and h is convenient shorthand for Δx.

First Order

F $$y_i' = \frac{1}{h}(y_{i+1} - y_i). \tag{A.4}$$

B $$y_i' = \frac{1}{h}(y_i - y_{i-1}). \tag{A.5}$$

Second Order

F $$y_i' = \frac{1}{2h}(-3y_i + 4y_{i+1} - y_{i+2}). \tag{A.6}$$

$$y_i'' = \frac{1}{h^2}(y_i - 2y_{i+1} + y_{i+2}). \tag{A.7}$$

C $$y_i' = \frac{1}{2h}(y_{i+1} - y_{i-1}). \tag{A.8}$$

Transport Phenomena: An Introduction to Advanced Topics, By Larry A. Glasgow

$$y_i'' = \frac{1}{h^2}(y_{i+1} - 2y_i + y_{i-1}). \quad \text{(A.9)}$$

B $$y_i' = \frac{1}{2h}(3y_i - 4y_{i-1} + y_{i-2}). \quad \text{(A.10)}$$

$$y_i'' = \frac{1}{h^2}(y_i - 2y_{i-1} + y_{i-2}). \quad \text{(A.11)}$$

Third Order

F $$y_i' = \frac{1}{6h}(2y_{i+3} - 9y_{i+2} + 18y_{i+1} - 11y_i). \quad \text{(A.12)}$$

$$y_i'' = \frac{1}{h^2}(-y_{i+3} + 4y_{i+2} - 5y_{i+1} + 2y_i). \quad \text{(A.13)}$$

$$y_i''' = \frac{1}{h^3}(y_{i+3} - 3y_{i+2} + 3y_{i+1} - y_i). \quad \text{(A.14)}$$

B $$y_i' = \frac{1}{6h}(11y_i - 18y_{i-1} + 9y_{i-2} - 2y_{i-3}). \quad \text{(A.15)}$$

$$y_i'' = \frac{1}{h^2}(2y_i - 5y_{i-1} + 4y_{i-2} - y_{i-3}). \quad \text{(A.16)}$$

$$y_i''' = \frac{1}{h^3}(y_i - 3y_{i-1} + 3y_{i-2} - y_{i-3}). \quad \text{(A.17)}$$

Fourth Order

F $$y_i' = \frac{1}{12h}(-3y_{i+4} + 16y_{i+3} - 36y_{i+2} + 48y_{i+1} - 25y_i). \quad \text{(A.18)}$$

$$y_i'' = \frac{1}{12h^2}(11y_{i+4} - 56y_{i+3} + 114y_{i+2} - 104y_{i+1} + 35y_i). \quad \text{(A.19)}$$

$$y_i''' = \frac{1}{2h^3}(-3y_{i+4} + 14y_{i+3} - 24y_{i+2} + 18y_{i+1} - 5y_i). \quad \text{(A.20)}$$

$$y_i'''' = \frac{1}{h^4}(y_{i+4} - 4y_{i+3} + 6y_{i+2} - 4y_{i+1} + y_i). \quad \text{(A.21)}$$

C $$y_i' = \frac{1}{12h}(-y_{i+2} + 8y_{i+1} - 8y_{i-1} + y_{i-2}). \quad \text{(A.22)}$$

$$y_i'' = \frac{1}{12h^2}(-y_{i+2} + 16y_{i+1} - 30y_i + 16y_{i-1} - y_{i-2}). \quad \text{(A.23)}$$

$$y_i''' = \frac{1}{2h^3}(y_{i+2} - 2y_{i+1} + 2y_{i-1} - y_{i-2}). \quad \text{(A.24)}$$

$$y_i'''' = \frac{1}{h^4}(y_{i+2} - 4y_{i+1} + 6y_i - 4y_{i-1} + y_{i-2}). \quad \text{(A.25)}$$

B $$y_i' = \frac{1}{12h}(25y_i - 48y_{i-1} + 36y_{i-2} - 16y_{i-3} + 3y_{i-4}). \quad \text{(A.26)}$$

$$y_i'' = \frac{1}{12h^2}(35y_i - 104y_{i-1} + 114y_{i-2} - 56y_{i-3} + 11y_{i-4}). \quad \text{(A.27)}$$

$$y_i''' = \frac{1}{2h^3}(5y_i - 18y_{i-1} + 24y_{i-2} - 14y_{i-3} + 3y_{i-4}). \quad \text{(A.28)}$$

$$y_i'''' = \frac{1}{h^4}(y_i - 4y_{i-1} + 6y_{i-2} - 4y_{i-3} + y_{i-4}). \quad \text{(A.29)}$$

A.1 SOME ILLUSTRATIVE APPLICATIONS

Suppose we have a transient viscous flow in a rectangular duct in which the duct width is much greater than its height. The governing equation can be written as

$$\frac{\partial v_x}{\partial t} = -\frac{1}{\rho}\frac{\partial p}{\partial x} + \nu\frac{\partial^2 v_x}{\partial y^2}. \quad \text{(A.30)}$$

We assume that a pressure gradient is applied at $t=0$ and the fluid begins to move in the x-direction. We let v_x be represented by V for clarity. One possible finite difference representation (letting the indices i and j correspond to y-position and time, respectively) is

$$\frac{V_{i,j+1} - V_{i,j}}{\Delta t} \cong -\frac{1}{\rho}\frac{dp}{dx} + \nu\frac{V_{i+1,j} - 2V_{i,j} + V_{i-1,j}}{(\Delta y)^2}. \quad \text{(A.31)}$$

Next, suppose we have a transient conduction in a two-dimensional slab. The governing equation is

$$\frac{\partial T}{\partial t} = \alpha\left[\frac{\partial^2 T}{\partial x^2} + \frac{\partial^2 T}{\partial y^2}\right]. \quad \text{(A.32)}$$

In this case we will have three subscripts (indices): i, j, and k corresponding to the x- and y-directions and time, respectively. A finite difference representation for this equation might appear as

$$\frac{T_{i,j,k+1} - T_{i,j,k}}{\Delta t} \cong \alpha\left[\frac{T_{i+1,j,k} - 2T_{i,j,k} + T_{i-1,j,k}}{(\Delta x)^2} + \frac{T_{i,j+1,k} - 2T_{i,j,k} + T_{i,j-1,k}}{(\Delta y)^2}\right]. \quad \text{(A.33)}$$

Generally, we would select the same nodal spacing in the x- and y-directions such that $\Delta x = \Delta y$.

Finally, we examine an equation written in cylindrical coordinates; this example is appropriate for conductive heat transfer in the radial direction:

$$\frac{\partial T}{\partial t} = \alpha \left[\frac{\partial^2 T}{\partial r^2} + \frac{1}{r}\frac{\partial T}{\partial r} \right]. \tag{A.34}$$

If the center of the cylinder corresponds to an i-index value of 1 (rather than 0), then we might write:

$$\frac{T_{i,j+1} - T_{i,j}}{\Delta t} \cong \alpha \left[\frac{T_{i+1,j} - 2T_{i,j} + T_{i-1,j}}{(\Delta t)^2} + \frac{1}{(i-1)\Delta r} \frac{T_{i+1,j} - T_{i-1,j}}{2\Delta r} \right]. \tag{A.35}$$

Note that in this case the first derivative of T (with respect to r) has been replaced with a second-order central difference approximation. Finally, observe that the time derivatives that appeared in the preceding examples were replaced by the first-order forward differences. Since the spatial derivatives on the right only involve the *current time index*, we should be aware that an *explicit* algorithm is contemplated. This simply means that we can forward march in time, directly computing all spatial positions on each successive time-step row.

A.2 BOUNDARIES WITH SPECIFIED FLUX

Consider a conduction problem for which the right-hand boundary is insulated, thus $q_x = 0$. Let the nodal point on the boundary be represented by the index n and let the temperatures for $n-2$ and $n-1$ be 50°C and 45°C, respectively. We can determine the temperature at the boundary by setting the derivative equal to zero. However, if we use a first-order backward difference in this situation:

$n-2$	$n-1$	n
50°C	45°C	?°C

then $T_n = 45$°C, a result that is clearly unphysical because the temperature "profile" on this row has a discontinuity in slope. One alternative is to employ eq. (A.10):

$$T_n = \frac{1}{3}(-50 + 4(45)) = 43.333. \tag{A.36}$$

Of course, a third- or fourth-order backward difference could be used as well.

Now suppose we had to use a Robin's-type boundary condition for a solid–fluid interface:

$$-k_s \left.\frac{\partial T}{\partial x}\right|_{x=x_n} = h_f(T_n - T_\infty). \tag{A.37}$$

Assuming $Bi = \Delta\, x h_f / k_s$, one possible expression for T_n is

$$T_n = \frac{2BiT_\infty + 4T_{n-1} - T_{n-2}}{3 + 2Bi}. \tag{A.38}$$

If we select $Bi = 1$ and $T_\infty = 20$°C and use the temperatures given above for the $n-1$ and $n-2$ positions, then

$$T_n = \frac{2(20) + 4(45) - 50}{5} = 34°\text{C}. \tag{A.39}$$

APPENDIX B

ADDITIONAL NOTES ON BESSEL'S EQUATION AND BESSEL FUNCTIONS

Whenever we encounter a radially directed flux in cylindrical coordinates, the operator $(1/r)(\partial/\partial r)\left(r\frac{\partial\phi}{\partial r}\right)$ will arise. Depending upon the exact nature of the problem, this can result in some form of Bessel's differential equation, which for the generalized case can be written as shown in Mickley et al. (1957):

$$r^2\frac{d^2T}{dr^2}+r(a+2br^v)\frac{dT}{dr}+[c+dr^{2s}-b(1-a-v)r^v+b^2r^{2v}]T=0. \tag{B.1}$$

For many applications in transport phenomena, we find that $a=1$, $b=0$, and $c=0$. The nature of the solution is then determined by the sign of d. If $\sqrt{d}$ is real, then the solution is written in terms of J_n or J_n+Y_n. If $\sqrt{d}$ is imaginary, then the solution will be either I_n or I_n+K_n. The order n is determined by $n=(1/s)\sqrt{((1-a)/2)^2-c}$.

As an illustration, consider steady conduction in an infinitely long cylinder with a production term that is linear with respect to temperature. The governing differential equation has the form

$$r^2\frac{d^2T}{dr^2}+r\frac{dT}{dr}+r^2\frac{\gamma T}{k}=0, \tag{B.2}$$

where γ is a positive constant. Note that $a=1$, $b=0$, $c=0$, $s=1$, and $d=\gamma/k$. In this case the solution is

$$T=AJ_0\left(\sqrt{\frac{\gamma}{k}}r\right)+BY_0\left(\sqrt{\frac{\gamma}{k}}r\right). \tag{B.3}$$

For a solid cylindrical domain, $T(r=0)$ would have to be finite and therefore $B=0$. But, of course, for an annular region, no boundary condition could be written for $r=0$ and both terms (A and B) would remain in the solution. Note that if the production term in (B.2) were replaced by a sink (disappearance) term, then γ/k would have been negative and the solution would have been written in terms of the modified Bessel functions I_0 and K_0. To illustrate this, consider a catalytic reaction in a long, cylindrical pellet; the reactant species "A" is being consumed by a first-order reaction. A homogeneous model results in the differential equation:

$$r^2\frac{d^2C_A}{dr^2}+r\frac{dC_A}{dr}-r^2\frac{k_1a}{D_{eff}}C_A=0, \tag{B.4}$$

with the solution

$$C_A=AI_0\left(\sqrt{\frac{k_1a}{D_{eff}}}r\right)+BK_0\left(\sqrt{\frac{k_1a}{D_{eff}}}r\right). \tag{B.5}$$

We need to know something about the behavior of these Bessel functions if we are to apply them correctly. Therefore, Table B.1 of numerical values is being provided for the zero-order Bessel functions of the first and second kinds, as well as the modified Bessel functions I_0 and K_0; more extensive tables are provided in Carslaw and Jaeger (1959). Note that neither Y_0 nor K_0 can be part of the solution for a problem

Transport Phenomena: An Introduction to Advanced Topics, By Larry A. Glasgow

TABLE B.1. An Abbreviated Table of Zero-Order Bessel Functions

r	$J_0(r)$	$Y_0(r)$	$I_0(r)$	$K_0(r)$
0.0	1	$-\infty$	1	∞
0.2	0.99	−1.0811	1.01	1.7527
0.4	0.9604	−0.606	1.0404	1.1145
0.6	0.912	−0.3085	1.092	0.7775
0.8	0.8463	−0.0868	1.1665	0.5653
1.0	0.7652	0.0883	1.2661	0.421
1.2	0.6711	0.2281	1.3937	0.3185
1.4	0.5669	0.3379	1.5534	0.2437
1.6	0.4554	0.4204	1.7500	0.188
1.8	0.34	0.4774	1.9896	0.1459
2.0	0.2239	0.5104	2.2796	0.1139
2.2	0.1104	0.5208	2.6291	0.0893
2.4	0.0025	0.5104	3.0493	0.0702
2.6	−0.0968	0.4813	3.5533	0.0554
2.8	−0.185	0.4359	4.1573	0.0438
3.0	−0.2601	0.3769	4.8808	0.0347
3.2	−0.3202	0.3071	5.7472	0.0276
3.4	−0.3643	0.2296	6.7848	0.022
3.6	−0.3918	0.1477	8.0277	0.0175
3.8	−0.4026	0.0645	9.5169	0.0139
4.0	−0.3971	−0.0169	11.302	0.0112
4.2	−0.3766	−0.0938	13.443	0.0089
4.4	−0.3423	−0.1633	16.010	0.0071
4.6	−0.2961	−0.2235	19.093	0.0057
4.8	−0.2404	−0.2723	22.794	0.0046
5.0	−0.1776	−0.3085	27.239	0.0037
5.2	−0.11029	−0.33125	32.584	0.00297
5.4	−0.04121	−0.34017	39.009	0.002385
5.6	0.02697	−0.33544	46.738	0.00192
5.8	0.0917	−0.317746	56.038	0.00154
6.0	0.15065	−0.28819	67.234	0.00124
6.2	0.20174	−0.24831	80.718	0.001
6.4	0.24331	−0.19995	96.962	0.00081
6.6	0.27404	−0.14523	116.54	0.00065
6.8	0.2931	−0.08643	140.14	0.00053
7.0	0.3001	−0.02595	168.59	0.00042
7.2	0.29507	0.03385	202.92	0.000343
7.4	0.2786	0.09068	244.34	0.000277
7.6	0.2516	0.1424	294.33	0.0002
7.8	0.2154	0.1872	354.69	0.000181
8.0	0.1717	0.2235	427.56	0.000146
8.2	0.1222	0.25012	515.59	0.000118
8.4	0.06916	0.26622	621.94	0.000096
8.6	0.01462	0.27146	750.5	0.000077
8.8	−0.0392	0.26587	905.8	0.000063
9.0	−0.0903	0.2498	1094	0.000051
9.2	−0.13675	0.22449	1321	0.000041
9.4	−0.17677	0.19074	1595	0.000033
9.6	−0.20898	0.15018	1927	0.0000271
9.8	−0.23277	0.10453	2329	0.0000219
10.0	−0.2459	0.05567	2816	0.0000178

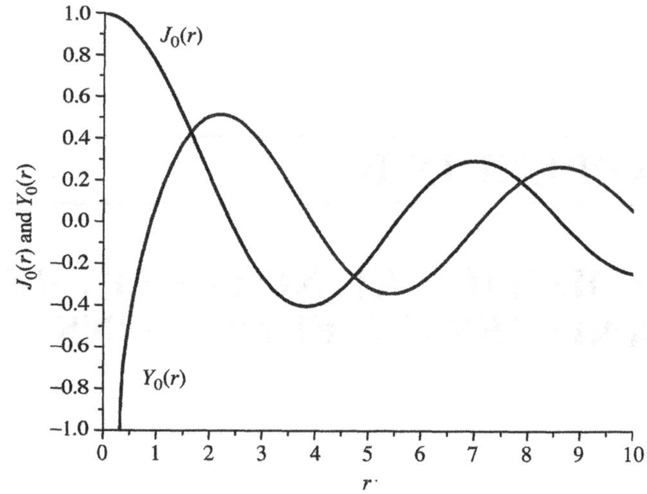

FIGURE B.1. Bessel functions $J_0(r)$ and $Y_0(r)$ for r from 0 to 10.

in cylindrical coordinates if the field variable (V, T, or C_A) is finite at the center ($r=0$).

$J_0(r)$ and $Y_0(r)$ are also shown graphically in Figure B.1.

The need to differentiate the Bessel functions arises frequently, particularly when a boundary condition involves a specified flux (Neumann or Robin's type). For J, Y, and K, we have

$$\frac{d}{dr}\left[Z_p(\alpha r)\right] = -\alpha Z_{p+1}(\alpha r) + \frac{p}{r} Z_p(\alpha r). \tag{B.6}$$

Accordingly, we note that

$$\frac{d}{dr}[J_0(\beta r)] = -\beta J_1(\beta r) \quad \text{since } p = 0. \tag{B.7}$$

For I_p, we have

$$\frac{d}{dr}\left[I_p(\alpha r)\right] = \alpha I_{p+1}(\alpha r) + \frac{p}{r} I_p(\alpha r). \tag{B.8}$$

Application of the initial condition in the analytic solution of parabolic partial differential equations may require that we make use of orthogonality. For example, in cylindrical coordinates where the solution domain is from $r=0$ to $r=R$, we note that

$$\int_0^R r J_n(\lambda_m r) J_n(\lambda_p r) dr = 0 \quad \text{as long as } m \neq p. \tag{B.9}$$

The integral that will remain to be of interest (for order zero, $n=0$) is

$$\int_0^R r J_0(\lambda_n r) J_0(\lambda_n r) dr. \tag{B.10}$$

The solution depends upon the nature of λ. If the λ_n's are from the roots of $J_0(\lambda_n R) = 0$, then the integral above simply has the value:

$$\frac{R^2}{2} J_1^2(\lambda_n R). \tag{B.11}$$

To see how this could come about, consider transient conduction in a cylindrical solid. The governing equation

$$\frac{\partial T}{\partial t} = \alpha \left[\frac{\partial^2 T}{\partial r^2} + \frac{1}{r}\frac{\partial T}{\partial r} \right], \tag{B.12}$$

is solved in the usual fashion by separation of variables: $T = f(r)g(t)$ results in

$$g = C_1 \exp(-\alpha\lambda^2 t) \quad \text{and} \quad f = AJ_0(\lambda r) + BY_0(\lambda r). \tag{B.13}$$

Since T is finite at the center (at $r=0$), $B=0$. We write

$$T = T_\infty + A\exp(-\alpha\lambda^2 t)J_0(\lambda r). \tag{B.14}$$

If the surface of the cylinder is maintained at T_∞ for all t, then it is necessary that $J_0(\lambda R) = 0$. This condition is encountered regularly in applied mathematics. Since J_0 is oscillatory, there are infinitely many zeroes. The first 30 are compiled in Table B.2 along with the values for $J_1(\lambda R)$ and the coefficients (A_n's) from eq. (B.17) with the temperature difference set equal to 1.

Turning our attention back to the problem at hand, we apply the initial condition whereby

$$T_i - T_\infty = A_n J_0(\lambda_n r). \tag{B.15}$$

The initial temperature T_i could be constant or a function of r. We make use of orthogonality to find the A_n's:

$$\int_0^R (T_i - T_\infty) r J_0(\lambda_m r)dr = A_n \int_0^R r J_0(\lambda_n r) J_0(\lambda_m r)dr. \tag{B.16}$$

If T_i and T_∞ are constants, we obtain

$$A_n = \frac{2(T_i - T_\infty)}{\lambda_n R J_1(\lambda_n R)}. \tag{B.17}$$

However, it is essential that we remember that this result is valid only for the simple Dirichlet boundary condition. For a Neumann condition, such as an insulated boundary, we could have λ_n as a root of

$$J'_0(\lambda_n R) = 0. \tag{B.18}$$

In this case, the integral shown as (B.10) has the solution

$$\frac{R^2}{2}\left(1 - \frac{n^2}{\lambda^2 R^2}\right)\{J_0(\lambda_n R)\}^2. \tag{B.19}$$

TABLE B.2. Zeroes for $J_0(\lambda R)$ Along with the Values for $J_1(\lambda R)$ and the Coefficients from (B.17)

n	$\lambda_n R$	$J_1(\lambda_n R)$	A_n from (B.17)
1	2.40483	0.51915	1.60198
2	5.52008	−0.34026	−1.06481
3	8.65373	0.27145	0.85141
4	11.79153	−0.23246	−0.72965
5	14.93092	0.20655	0.64852
6	18.07106	−0.18773	−0.58954
7	21.21164	0.17327	0.54418
8	24.35247	−0.16171	−0.50788
9	27.49348	0.15218	0.47802
10	30.63461	−0.14417	−0.45284
11	33.77582	0.13730	0.43128
12	36.91710	−0.13132	−0.41254
13	40.05843	0.12607	0.39603
14	43.19979	−0.12140	−0.38135
15	46.34119	0.11721	0.36821
16	49.48261	−0.11343	−0.35633
17	52.62405	0.10999	0.34554
18	55.76551	−0.10685	−0.33566
19	58.90698	0.10396	0.32659
20	62.04847	−0.10129	−0.31822
21	65.18996	0.09882	0.31046
22	68.33147	−0.09652	−0.30324
23	71.47298	0.09438	0.29649
24	74.61450	−0.09237	−0.29018
25	77.75603	0.09049	0.28426
26	80.89756	−0.08871	−0.27869
27	84.03909	0.08704	0.27343
28	87.18063	−0.08545	−0.26847
29	90.32217	0.08395	0.26376
30	93.46372	−0.08253	−0.25928

If Newton's "law of cooling" must be equated with Fourier's law at a solid–fluid interface (Robin's-type boundary condition), then the λ_n's will come from the transcendental equation:

$$\lambda_n R J_1(\lambda_n R) = \frac{hR}{k} J_0(\lambda_n R). \tag{B.20}$$

It is to be borne in mind that the dimensionless quotient hR/k *is not the Nusselt number: It is the Biot modulus Bi.* For this third case, the application of orthogonality still results in the integral (B.10), but the solution is now

$$\frac{1}{2\lambda_n^2}\left(\mathrm{Bi}^2 + \lambda_n^2 R^2\right) J_0^2(\lambda_n R). \tag{B.21}$$

Before (B.21) can actually be used, the roots of (B.20) must be available. In many situations unfortunately, the heat transfer coefficient h will not be known with any precision. We should look at an example for illustration: Consider a cylindrical rod of phosphor bronze ($d = 1$ in.) placed in

TABLE B.3. Roots of the Transcendental Equation (B.20) for Selected Bi

Bi	$(\lambda_1 R)$	$(\lambda_2 R)$	$(\lambda_3 R)$	$(\lambda_4 R)$	$(\lambda_5 R)$
0.08	0.396	3.8525	7.0270	10.1813	13.3297
0.10	0.4417	3.8577	7.0298	10.1833	13.3312
0.15	0.5376	3.8706	7.0369	10.1882	13.3349
0.20	0.6170	3.8835	7.0440	10.1931	13.3387
0.30	0.7465	3.9091	7.0582	10.2029	13.3462
0.40	0.8516	3.9344	7.0723	10.2127	13.3537
0.50	0.9408	3.9594	7.0864	10.2225	13.3611
1.00	1.2558	4.0795	7.1558	10.2710	13.3984
2.00	1.5994	4.2910	7.2884	10.3658	13.4719
5.00	1.9898	4.7131	7.6177	10.6223	13.6786

circulating hot water with $h \approx 150\,\text{Btu/(h ft}^2\,^\circ\text{F)}$. The Biot modulus will have a value of about 0.156. Extracting values from the table provided by Carslaw and Jaeger (1959), we find the values given in Table B.3.

For the example above, the first five roots are approximately 0.54, 3.87, 7.04, 10.19, and 13.3. So, *if values for h, k, and R are known,* the needed roots for the transcendental equation (B.20) can be obtained and the problem can be solved.

There are many useful sources of information for Bessel's equation and Bessel functions. A few of them are provided below:

1. Abramowitz, M. and I. A. Stegun. *Handbook of Mathematical Functions*, Dover (1972).
2. Carslaw, H. S. and J. C. Jaeger. *Conduction of Heat in Solids*, 2nd edition, Oxford (1959).
3. Dwight, H. B. *Tables of Integrals and Other Mathematical Data*, 3rd edition, Macmillan (1957).
4. Gray, A., Mathews, G. B., and T. M. MacRobert. *A Treatise on Bessel Functions and Their Applications to Physics*, 2nd edition, Macmillan (1931) and reprinted by Dover (1966).
5. Kreyszig, E. *Advanced Engineering Mathematics*, 3rd edition, Wiley (1972).
6. Mickley, H. S., Sherwood, T. K., and C. E. Reed. *Applied Mathematics in Chemical Engineering*, 2nd edition, McGraw-Hill (1957).
7. Selby, S. M., editor. *Handbook of Tables for Mathematics*, CRC Press (1975).

APPENDIX C

SOLVING LAPLACE AND POISSON (ELLIPTIC) PARTIAL DIFFERENTIAL EQUATIONS

Many equilibrium problems in transport phenomena are governed by elliptic partial differential equations. For the case of steady-state conduction in two dimensions, we have the Laplace equation:

$$\frac{\partial^2 T}{\partial x^2} + \frac{\partial^2 T}{\partial y^2} = 0. \tag{C.1}$$

For steady Poiseuille flow in ducts with constant cross section, we obtain a Poisson equation:

$$\frac{\partial p}{\partial z} = \mu \left[\frac{\partial^2 V_z}{\partial x^2} + \frac{\partial^2 V_z}{\partial y^2} \right]. \tag{C.2}$$

C.1 NUMERICAL PROCEDURE

There are a number of solution techniques that can be applied in such cases; we shall consider laminar flow in a rectangular duct as an example. By using the second-order central difference approximations for the second derivatives (where the i- and j-indices represent the x- and y-directions, respectively), eq. (C.2) can be written as

$$\frac{1}{\mu}\frac{dp}{dz} \cong \frac{V_{i+1,j} - 2V_{i,j} + V_{i-1,j}}{(\Delta x)^2} + \frac{V_{i,j+1} - 2V_{i,j} + V_{i,j-1}}{(\Delta y)^2}. \tag{C.3}$$

If the discretization employs a square mesh ($\Delta x = \Delta y$), then eq. (C.3) can be conveniently written as

$$V_{i,j} \approx \frac{1}{4}\left[V_{i+1,j} + V_{i-1,j} + V_{i,j+1} + V_{i,j-1} - \frac{(\Delta x)^2}{\mu}\frac{dp}{dz} \right]. \tag{C.4}$$

Please note that the term with the *largest* coefficient has been isolated on the left-hand side. This approximation is the basis for a simple Gauss–Seidel iterative computational scheme for the solution of such problems. In this case, of course, the velocity is zero on the boundaries, so we merely apply the algorithm to all the interior points row-by-row. The newly computed values are employed as soon as they become available (which distinguishes the Gauss–Seidel method from the Jacobi iterative method). As an example, consider the case of laminar flow in a rectangular duct 8 cm wide and 4 cm high, the pressure gradient is $-3\ \text{dyn/cm}^2$ per cm and the viscosity is 0.04 g/(cm s). All the nodal velocities will be initialized to zero to start the computation. For the specified pressure gradient, the centerline (maximum) velocity will be about 139 cm/s. The computed velocity distribution is shown in Figure C.1 as a contour plot.

In a computation of this type, a key issue is the number of iterations required to attain convergence. For the example shown here, we can monitor the centerline velocity during the calculations (Figure C.2).

Note that a reasonably accurate value is obtained with about 1000 iterations and after 3000 iterations, the third decimal place is essentially fixed. We can set down the

Transport Phenomena: An Introduction to Advanced Topics, By Larry A. Glasgow

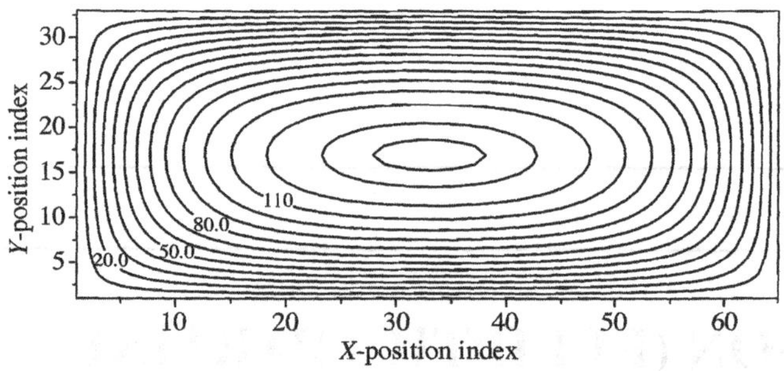

FIGURE C.1. Velocity distribution in a rectangular duct computed with the Gauss–Seidel iterative method.

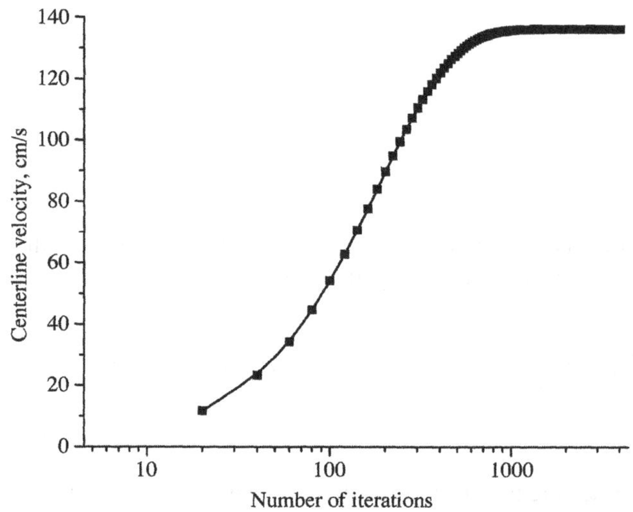

FIGURE C.2. Centerline velocity as a function of the number of iterations for the solution of the Poisson equation (C.2).

programming logic concisely:

```
DIMENSION ARRAY
INITIALIZE FIELD VARIABLE
SET ITERATION COUNTER TO ZERO

J=1 TO N
I=1 TO M

COMPUTE V(I,J)

NEXT I
NEXT J                                   NO

INCREMENT ITERATION COUNTER

TEST CONVERGENCE CRITERION
                                         YES
WRITE V(I,J) TO FILE

END
```

The rate of convergence of iterative solutions can be accelerated significantly through use of the extrapolated Liebmann method (also known as successive overrelaxation, SOR). In this technique, the change that would be produced by a single Gauss–Seidel iteration is increased through use of an accelerating factor that is usually denoted by ω. SOR can be implemented easily in the previous example by a slight modification of (C.4):

$$V_{i,j}^{(\text{new})} \approx V_{i,j} + \frac{1}{4}\omega \left[V_{i+1,j} + V_{i-1,j} + V_{i,j+1} + V_{i,j-1} - 4V_{i,j} - \frac{(\Delta x)^2}{\mu}\frac{dp}{dz} \right]. \tag{C.5}$$

The $V_{i,j}$'s appearing on the right-hand side of C.5 are from the previous iterate. You can see immediately that if $\omega = 1$, this is identically the Gauss–Seidel algorithm. For overrelaxation, ω will have a value between 1 and 2; the rate of convergence is very sensitive to the value of the acceleration parameter. Refer to Smith (1965) for additional discussion. Frankel (1950) has shown that for large rectangular domains such as that used in our example,

$$\omega_{\text{opt}} \approx 2 - \sqrt{2}\pi \left(\frac{1}{p^2} + \frac{1}{q^2} \right)^{1/2}, \tag{C.6}$$

where p and q are the number of nodal points used in the x- and y-directions, respectively. For our case, $p = 65$ and $q = 33$, so $\omega_{\text{opt}} \approx 1.85$. The consequences of a poor choice are shown clearly in Figure C.3, where the number of iterations required to achieve a desired degree of convergence is reported.

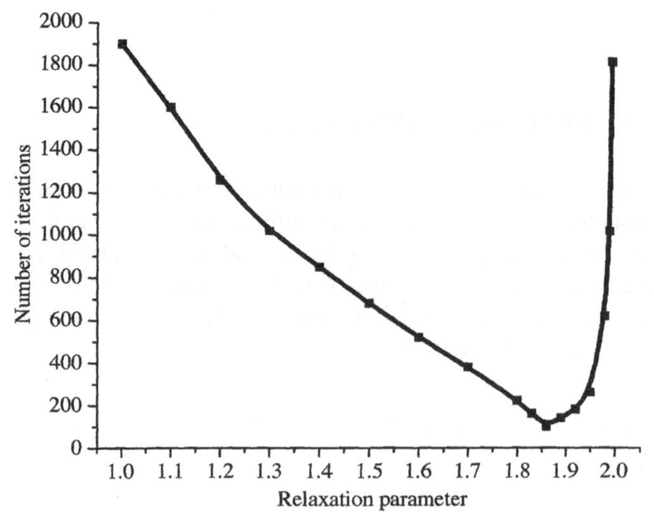

FIGURE C.3. Number of iterations required to achieve $\varepsilon = 2 \times 10^{-7}$ as a function of ω. A Poisson-type equation for the laminar flow in a rectangular duct is being solved and the minimum is located at about $\omega = 1.86$.

It is clear that SOR can significantly reduce the computational effort required to solve the elliptic partial differential equations. However, ω must be chosen carefully to obtain the greatest possible benefit.

C.2 SEPARATION OF VARIABLES (PRODUCT METHOD)

Some problems governed by elliptic equations can be solved analytically. For example, consider a square steel slab, 15 in. on a side (L). We pose a two-dimensional Dirichlet problem with three sides maintained at 50°F and one at 300°F. We want to find the temperature distribution in the interior of the slab. We render the problem dimensionless by setting

$$\theta = \frac{T-50}{300-50}, \quad X = x/L, \quad \text{and} \quad Y = y/L.$$

This results in the two-dimensional Laplace equation:

$$\frac{\partial^2\theta}{\partial X^2} + \frac{\partial^2\theta}{\partial Y^2} = 0. \tag{C.7}$$

By letting $\theta = f(X)g(Y)$, we find

$$\frac{f''}{f} = -\frac{g''}{g} = -\lambda^2. \tag{C.8}$$

The resulting two ordinary differential equations are easily solved, *producing a solution:*

$$\theta = (A\cos\lambda X + B\sin\lambda X)(C\cosh\lambda Y + D\sinh\lambda Y). \tag{C.9}$$

We must have $\theta(X,0) = 0$ and $\theta(0,Y) = 0$, so both C and A must be zero. We must also have $\theta(1,Y) = 0$, which means that

$$\sin(\lambda) = 0 \quad \text{and} \quad \lambda = \pi, 2\pi, 3\pi, \ldots.$$

Thus, (C.9) can be written as the infinite series:

$$\theta = \sum_{n=1}^{\infty} B_n \sin\lambda_n X \sinh \lambda_n Y. \tag{C.10}$$

Finally, we note that at $Y=1$, $\theta=1$ for all X, so that

$$1 = \sum_{n=1}^{\infty} B_n \sin n\pi X \sinh n\pi. \tag{C.11}$$

Equation (C.11) is a half-range Fourier sine series and this allows us to determine B_n by integration:

$$B_n = \frac{2(1-\cos n\pi)}{n\pi \sinh n\pi}. \tag{C.12}$$

The analytic solution is complete but the work required to produce useful results is not. We must now compute the temperature distribution, making sure that we use sufficient terms for convergence of the series. A contour plot of the results is presented in Figure C.4; the upper (hot) surface of the steel slab presents a small problem that is apparent by inspection of these computed data.

The infinite series solution converges rapidly near the center of the slab and slowly near the edges. This is illustrated by the following table that shows n (1,3,5,. . .) in the first column and the computed results for θ in subsequent columns. The second column corresponds to the (X,Y) position, 0.02, 0.98, the third 0.05, 0.95, and so on. The last column is at the center of the slab. Note that $n=25$ is *not sufficient* for ($X=0.02$, $Y=0.98$). In contrast, at ($X=0.5$, $Y=0.5$), we have six correct decimal digits for only $n=7$.

n	0.02, 0.98	0.05, 0.95	0.10, 0.90	0.20, 0.80	0.30, 0.70	0.40, 0.60	0.50, 0.50
1	7.51E-02	0.170107543	0.286908448	0.397379637	0.397183239	0.337323397	0.253714979
3	0.140924543	0.290383458	0.420701116	0.458666295	0.404942483	0.331572294	0.249902710
5	0.198400274	0.372480929	0.473636866	0.458666205	0.402654946	0.331572294	0.250001550
7	0.248286843	0.426451832	0.489956170	0.456538618	0.402731627	0.331588477	0.249998495
9	0.291349798	0.460439056	0.492542595	0.456247538	0.402755320	0.331586838	0.249998599
11	0.328314066	0.480749846	0.491413593	0.456315309	0.402752370	0.331586957	0.249998599
13	0.359859496	0.492073804	0.490079373	0.456341714	0.402752221	0.331586957	0.249998599
15	0.386618018	0.497762531	0.489316851	0.456341714	0.402752280	0.331586957	0.249998599
17	0.409172297	0.500116408	0.489026457	0.456340075	0.402752280	0.331586957	0.249998599
19	0.428055316	0.500646472	0.488973528	0.456339806	0.402752280	0.331586957	0.249998599
21	0.443751246	0.500296175	0.488999099	0.456339866	0.402752280	0.331586957	0.249998599
23	0.456697077	0.499618232	0.489031702	0.456339896	0.402752280	0.331586957	0.249998599
25	0.467284292	0.498908669	0.489051461	0.456339896	0.402752280	0.331586957	0.249998599

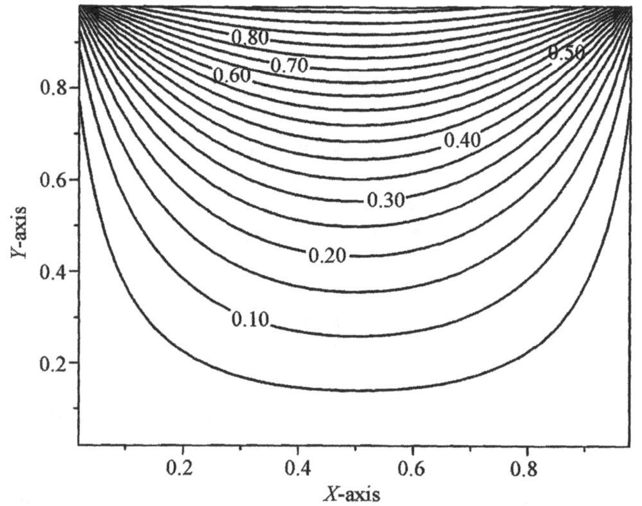

FIGURE C.4. Temperature distribution in a steel slab with the upper surface maintained at $\theta = 1$; the other surfaces are uniformly $\theta = 0$.

There are numerous references for the solution of Laplace and Poisson (elliptic) partial differential equations, including

1. Frankel, S. P. Convergence Rates of Iterative Treatments of Partial Differential Equations, *Mathematical Tables and Other Aids to Computation*, 4:65 (1950).
2. James, M., Smith, G. M., and J. C. Wolford. *Applied Numerical Methods for Digital Computation*, 2nd edition, Harper and Row (1977).
3. Smith, G. D. *Numerical Solution of Partial Differential Equations*, Oxford University Press (1965).
4. Spiegel, M. R. *Fourier Analysis with Applications to Boundary Value Problems*, McGraw-Hill (1974).

Also, several common commercial software packages (an example is Mathcad) have capabilities for simple problems involving elliptic PDEs. Far greater capability is available through ELLPACK, a FORTRAN system for the solution and exploration of elliptic partial differential equations. The ELLPACK project was coordinated by John Rice of Purdue University and it was initiated in 1976. The software contains modules that allow the analyst to choose between different solution procedures; among the included routines are collocation, Hermite collocation, spline Galerkin, and several multipoint iterative techniques. One of the purposes of ELLPACK is the evaluation and comparison of different solution procedures for specific elliptic PDE problems. The interested reader should refer to *Solving Elliptic Problems Using ELLPACK* by J. R. Rice and R. F. Boisvert (Springer-Verlag, New York, 1985). For recent developments in the software, consult the ELLPACK Home Page. One of the really attractive features of ELLPACK is its capability for *nonrectangular* domains—a situation encountered frequently in the engineering applications involving the Laplace and Poisson partial differential equations.

APPENDIX D

SOLVING ELEMENTARY PARABOLIC PARTIAL DIFFERENTIAL EQUATIONS

The simplest equations of this type are often referred to as "conduction" or "diffusion" equations and examples include

$$\text{(a)}\quad \text{momentum}\quad \frac{\partial V_x}{\partial t} = \nu \frac{\partial^2 V_x}{\partial y^2}, \tag{D.1a}$$

$$\text{(b)}\quad \text{heat}\quad \frac{\partial T}{\partial t} = \alpha \frac{\partial^2 T}{\partial y^2}, \tag{D.1b}$$

$$\text{(c)}\quad \text{mass}\quad \frac{\partial C_A}{\partial t} = D_{AB} \frac{\partial^2 C_A}{\partial y^2}. \tag{D.1c}$$

We have numerous options in such cases, including scaling or variable transformation, separation of variables, and a plethora of numerical methods. First, we consider the transformation of eq. (D.1b); we define $\eta = y/\sqrt{4\alpha t}$ and write the left-hand side of (D.1b) as

$$\frac{\partial T}{\partial \eta}\frac{\partial \eta}{\partial t} = T'\left(-\frac{1}{2}\right)\frac{y}{\sqrt{4\alpha}}t^{-3/2}. \tag{D.2}$$

Differentiating the right-hand side of (D.1b) the first time,

$$\frac{\partial T}{\partial \eta}\frac{\partial \eta}{\partial y} = T'\frac{1}{\sqrt{4\alpha}}, \text{ and then again, we obtain } T''\frac{1}{4\alpha t}. \tag{D.3}$$

Substitution into (D.1b) results in

$$-2\eta\frac{dT}{d\eta} = \frac{d^2T}{d\eta^2}, \tag{D.4}$$

an ordinary differential equation. Whether (D.4) can produce a useful solution depends upon the nature of the problem. For the transient conduction in an infinte slab or the viscous flow near a wall suddenly set in motion, it results in the familiar error function solution, for example,

$$\theta = 1 - \text{erf}\left(\frac{y}{\sqrt{4\alpha t}}\right). \tag{D.5}$$

For contrast, we now examine conduction in a *finite* slab of material; let this object extend from $y=0$ to $y=1$. We can have either a uniform initial temperature or a temperature distribution that can be written as a function of y. At $t=0$, both faces are instantaneously heated to some new temperature T_s. Define a dimensionless temperature,

$$\theta = \frac{T - T_s}{T_i - T_s}, \text{ and let } \theta = f(y)g(t). \tag{D.6}$$

The product method yields

$$g' = -\alpha\lambda^2 g \quad \text{and} \quad f'' + \lambda^2 f = 0. \tag{D.7}$$

As expected, we get

$$g = C_1 \exp(-\alpha\lambda^2 t) \quad \text{and} \quad f = A\sin\lambda y + B\cos\lambda y. \tag{D.8}$$

Since B must be zero and $\sin(\lambda) = 0$, we find

$$\theta = \sum_{n=1}^{\infty} A_n \exp(-\alpha\lambda_n^2 t)\sin\lambda_n y. \tag{D.9}$$

Transport Phenomena: An Introduction to Advanced Topics, by Larry A. Glasgow

TABLE D.1. Illustration of Infinite Series Convergence for Small t's

Term No.	$t = 0.001$	$t = 0.005$	$t = 0.025$	$t = 0.125$	$t = 0.625$
1	1.271981	1.266969	1.242205	1.12546	0.6870893
3	0.851322	0.8609938	0.9023096	0.9856378	0.6854422
5	1.099763	1.086086	1.039727	0.9972914	0.6854423
7	0.926459	0.9432634	0.9854355	0.9968604	0.6854423
9	1.05706	1.038121	1.004608	0.9968669	0.6854423
11	0.954341	0.9744126	0.9987616	0.9968669	0.6854423
13	1.037236	1.01695	1.000275	0.9968669	0.6854423
15	0.969256	0.9889856	0.9999457	0.9968669	0.6854423
17	1.025566	1.006978	1.000006	0.9968669	0.6854423
19	0.97864	0.9956936	0.9999966	0.9968669	0.6854423
21	1.017874	1.002573	0.9999977	0.9968669	0.6854423
23	0.985031	0.9985044	0.9999976	0.9968669	0.6854423
25	1.012515	1.000835	0.9999976	0.9968669	0.6854423
27	0.98955	0.9995433	0.9999976	0.9968669	0.6854423
29	1.008694	1.000235	0.9999976	0.9968669	0.6854423
31	0.992785	0.9998772	0.9999976	0.9968669	0.6854423
33	1.005956	1.000056	0.9999976	0.9968669	0.6854423
35	0.995097	0.9999698	0.9999976	0.9968669	0.6854423
37	1.004008	1.00001	0.9999976	0.9968669	0.6854423
39	0.996732	0.9999919	0.9999976	0.9968669	0.6854423
41	1.002642	0.9999996	0.9999976	0.9968669	0.6854423
43	0.997868	0.9999964	0.9999976	0.9968669	0.6854423

If we have a uniform initial temperature T_i, then application of the initial condition results in

$$1 = \sum_{n=1}^{\infty} A_n \sin \lambda_n y, \tag{D.10}$$

a half-range Fourier sine series. By theorem,

$$A_n = \frac{2}{L} \int_0^L f(y) \sin \frac{n\pi y}{L} dy, \tag{D.11}$$

but for our case $L = 1$ and the function $f(y)$ is also 1. The integral (D.10) is zero for even n and equal to $4/(n\pi)$ for $n = 1,3,5,\ldots$. With this example, we have a good opportunity to examine the convergence of the infinite series solution. Let $y = 1/2$, $\alpha = 0.1$, and t range from 0.001 to 0.625 by repeated factors of 5. We shall examine the series for n's from 1 to 43 (Table D.1). Note that for small t's, the series does not converge quickly. However, for $t = 0.125$, we need only five terms and at $t = 0.625$, only three. The results should not be surprising. For very small t's, the temperature profile is virtually half a cycle of a square wave.

D.1 AN ELEMENTARY EXPLICIT NUMERICAL PROCEDURE

Suppose we have a viscous flow near a plane wall set in motion with velocity V_0 at $t = 0$. Letting $V = v_x/V_0$,

$$\frac{\partial V}{\partial t} = \nu \frac{\partial^2 V}{\partial y^2}. \tag{D.12}$$

An explicit algorithm is easily developed for (D.11):

$$V_{i,j+1} = \frac{\Delta t \nu}{(\Delta y)^2}\left[V_{i+1,j} - 2V_{i,j} + V_{i-1,j}\right] + V_{i,j}. \tag{D.13}$$

Equation (D.13) is attractive because of its simplicity; it is easy to understand and program, but it poses a potential problem. To ensure stability, it is necessary that

$$\frac{\Delta t \nu}{(\Delta y)^2} \le \frac{1}{2}.$$

We will illustrate this using (D.13). Choose $\nu = 0.05\ \text{cm}^2/\text{s}$, $\Delta y = 0.1$ cm, and $\Delta t = 0.12$ s; of course, this guarantees that we are over the limit of 1/2. We can put the calculation into a table and monitor the evolution of the nodal velocities, which will reveal the consequence of our choices (Table D.2).

The problem we see here is easy to resolve. We change our parametric choices to yield $\Delta t \nu/(\Delta y)^2 = 0.4$ and repeat the calculation (Table D.3).

This is an important lesson. If we need good spatial resolution, Δy will be small and Δt will need to be *very small*, perhaps prohibitively small. Fortunately, we do have options that will work well for this type of problem.

TABLE D.2. Explicit Computation with Unstable Parametric Choice(s)

t	$i=1$	$i=2$	$i=3$	$i=4$	$i=5$	$i=6$	$i=7$
0	1	0	0	0	0	0	0
Δt	1	0.6	0	0	0	0	0
$2\Delta t$	1	0.48	0.36	0	0	0	0
$3\Delta t$	1	0.72	0.216	0.216	0	0	0
$3\Delta t$	1	0.5856	0.5184	0.0864	0.1296	0	0
$4\Delta t$	1	0.7939	0.2995	0.3715	0.0259	0.0777	0
$5\Delta t$	1	0.6209	0.6394	0.1210	0.2644	0	0.0467
$6\Delta t$	1	0.8594	0.3173	0.5181	0.0197	0.1866	−0.0093
$7\Delta t$	1	0.6185	0.7630	0.0986	0.4189	−0.0311	0.1306

D.2 AN IMPLICIT NUMERICAL PROCEDURE

Consider a transient conduction problem with two spatial dimensions:

$$\frac{\partial T}{\partial t} = \alpha\left[\frac{\partial^2 T}{\partial x^2} + \frac{\partial^2 T}{\partial y^2}\right]. \tag{D.14}$$

In this case, the stability requirement for an explicit solution is $\alpha\Delta t[(1/(\Delta x)^2) + (1/(\Delta y)^2)] \leq 1/2$, which can be a severe constraint. However, there is an alternative. The Peaceman–Rachford or alternating direction implicit (ADI) method can be especially effective for this type of parabolic partial differential equation. Let the indices i, j, and k represent x, y, and t, respectively. The first half of the ADI algorithm is used to advance to the $k+1$ time step:

$$\frac{T_{i,j,k+1} - T_{i,j,k}}{\alpha\Delta t} = \frac{T_{i+1,j,k+1} - 2T_{i,j,k+1} + T_{i-1,j,k+1}}{(\Delta x)^2} + \frac{T_{i,j+1,k} - 2T_{i,j,k} + T_{i,j-1,k}}{(\Delta y)^2}, \tag{D.15}$$

and the second half takes us to $k+2$:

$$\frac{T_{i,j,k+2} - T_{i,j,k+1}}{\alpha\Delta t} = \frac{T_{i+1,j,k+1} - 2T_{i,j,k+1} + T_{i-1,j,k+1}}{(\Delta x)^2} + \frac{T_{i,j+1,k+2} - 2T_{i,j,k+2} + T_{i,j-1,k+2}}{(\Delta y)^2}. \tag{D.16}$$

Note that neither step can be repeated unilaterally. Let us examine a simple application. A two-dimensional slab of material is at a uniform initial temperature of 100°C. At $t=0$, one face is instantaneously heated to 400°C. Let $\Delta x = \Delta y = 1$, as well as $\alpha = 1$ and $\Delta t = 1/8$. We rewrite eq. D.15 isolating the $k+1$ terms on the right-hand side:

$$-T_{i,j+1,k} + \left(2 - \frac{(\Delta x)^2}{\alpha\Delta t}\right)T_{i,j,k} - T_{i,j-1,k} = T_{i+1,j,k+1} - \left(2 + \frac{(\Delta x)^2}{\alpha\Delta t}\right)T_{i,j,k+1} + T_{i-1,j,k+1}. \tag{D.17}$$

Now we will illustrate the process with a simple square slab: the top, left, and right sides are all maintained at 100°C.

TABLE D.3. Explicit Computation with Stable Parametric Choice(s)

t	$i=1$	$i=2$	$i=3$	$i=4$	$i=5$	$i=6$	$i=7$
0	1	0	0	0	0	0	0
Δt	1	0.4	0	0	0	0	0
$2\Delta t$	1	0.48	0.16	0	0	0	0
$3\Delta t$	1	0.56	0.224	0.064	0	0	0
$4\Delta t$	1	0.6016	0.2944	0.1024	0.0256	0	0
$5\Delta t$	1	0.6381	0.3405	0.1485	0.0461	0.0102	0
$6\Delta t$	1	0.6638	0.3872	0.1843	0.0727	0.0205	0.0041
$7\Delta t$	1	0.6859	0.4158	0.2190	0.0965	0.0348	0.0090

The bottom will be set to 400°C. The nine interior nodes are initialized at 100°C.

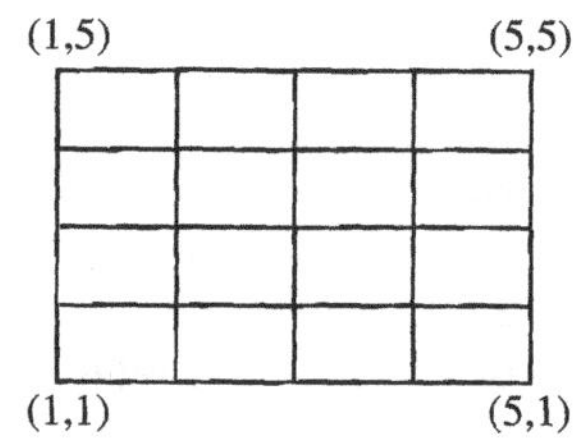

We apply (D.16) at the interior points, row by row; the first horizontal sweep results in

100	100	100
100	100	100
133.67	136.73	133.67

for the nine interior points. Now we recast (D.15) for application to the columns in order to advance to the $k+2$ time step:

$$-T_{i+1,j,k+1} + \left(2 - \frac{(\Delta x)^2}{\alpha \Delta t}\right) T_{i,j,k+1} - T_{i-1,j,k+1}$$

$$= T_{i,j+1,k+2} - \left(2 + \frac{(\Delta x)^2}{\alpha \Delta t}\right) T_{i,j,k+2} + T_{i,j-1,k+2}. \tag{D.18}$$

We solve the simultaneous equations that result from applying this equation to the columns and obtain

100.55	100.6	100.55
105.5	106	105.5
154.42	159.37	154.42

If the total number of equations is modest, then a direct elimination scheme can be used for solution. The coefficient matrix follows the tridiagonal pattern (with 1, −10, 1 for the selected parameters), so the process is easy to automate. Smith (1965) states that for rectangular regions, the ADI method requires about 25 times less work than an explicit computation. Carrying out the procedure to $t = 1.75$ yields

114.91	120.25	114.91
146.35	161.01	146.35
221.06	247.42	221.06

for the interior nodes. Chung (2002) notes that this scheme is unconditionally stable, which makes it very attractive for problems in which the time evolution is slow, that is, we can employ a very large Δt relative to the elementary explicit technique.

1. Chung, T. J. *Computational Fluid Dynamics,* Cambridge University Press (2002).
2. Peaceman, D. W. and H. H. Rachford. The Numerical Solution of Parabolic and Elliptic Differential Equations. *Journal of the Society for Industrial and Applied Mathematics*, 3:28 (1955).
3. Smith, G. D. *Numerical Solution of Partial Differential Equations*, Oxford University Press (1965).

APPENDIX E

ERROR FUNCTION

A number of significant problems in transport phenomena have the error function as part of their solution. Common examples include Stokes' first problem, transient conduction in semi-infinite slabs, and several transient absorption–diffusion processes.

The error function is defined by the integral:

$$\mathrm{erf}(\eta) = \frac{2}{\sqrt{\pi}} \int_0^{\eta} \exp(-\eta^2) d\eta. \tag{E.1}$$

The error function has the symmetry relationship, $\mathrm{erf}(-\eta) = -\mathrm{erf}(\eta)$. The complementary error function is

$$\mathrm{erfc}(\eta) = 1 - \mathrm{erf}(\eta), \tag{E.2}$$

or equivalently,

$$\mathrm{erfc}(\eta) = \frac{2}{\sqrt{\pi}} \int_{\eta}^{\infty} \exp(-\eta^2) d\eta. \tag{E.3}$$

Since $\mathrm{erf}(\eta)$ varies from 0 to 1 as η goes from 0 to ∞, it is clear that $\mathrm{erfc}(\eta)$ ranges from 1 to 0. The behavior of $\mathrm{erf}(\eta)$ is shown in Figure E.1 and a useful table of values is provided in Table E.1.

An illustrative example: Suppose we have a slab of alloy steel at a uniform temperature of 30°C. At $t = 0$, the front face is heated instantaneously to 550°C. What will the temperature be at $y = 10$ cm when $t = 200$ s?

For this problem, $\eta = y/\sqrt{4\alpha t}$, where α is the thermal diffusivity of the metal. We have $\alpha = 1.566 \times 10^{-5}$ m²/s. Therefore, $\eta = 0.8934$ and $\mathrm{erf}(\eta)$ is about 0.79. Since $\theta = (T - T_i)/(T_0 - T_i)$, we find $T \approx (520)(1 - 0.79) + 30 = 139$ °C. For $y = 5$ cm and $t = 300$ s, $\eta = 0.3647$ and $\mathrm{erf}(\eta) \approx 0.394$; consequently, $T \approx 345$°C.

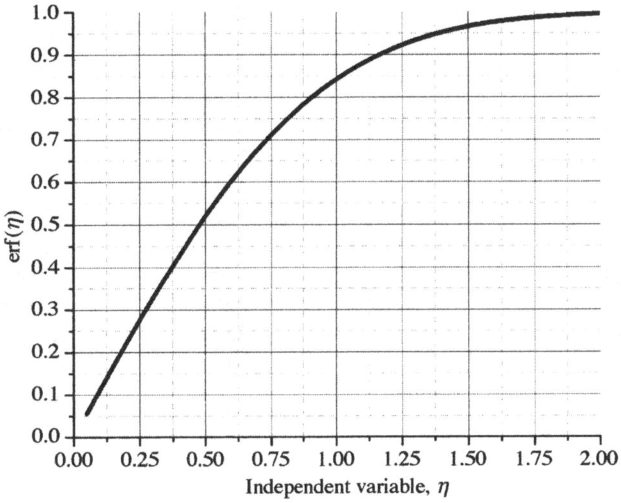

FIGURE E.1. General behavior of the error function $\mathrm{erf}(\eta)$.

E.1 ABSORPTION–REACTION IN QUIESCENT LIQUIDS

A classic application of the error function arises in the chemical engineering problem in which species "A" absorbs into a still liquid, diffuses into the liquid phase, and undergoes a first-order decomposition. The governing partial differential

Transport Phenomena: An Introduction to Advanced Topics, by Larry A. Glasgow

TABLE E.1. Error Function for Arguments from 0 to 3 by Increments of 0.05

η	$\mathrm{erf}(\eta)$	η	$\mathrm{erf}(\eta)$
0.00	0.0000	1.55	0.9718
0.05	0.0564	1.60	0.9764
0.10	0.1125	1.65	0.9804
0.15	0.1680	1.70	0.9839
0.20	0.2227	1.75	0.9868
0.25	0.2763	1.80	0.9891
0.30	0.3286	1.85	0.9911
0.35	0.3798	1.90	0.9929
0.40	0.4284	1.95	0.9942
0.45	0.4755	2.00	0.9953
0.50	0.5205	2.05	0.9963
0.55	0.5633	2.10	0.9971
0.60	0.6039	2.15	0.9977
0.65	0.6420	2.20	0.9981
0.70	0.6784	2.25	0.9985
0.75	0.7118	2.30	0.9989
0.80	0.7421	2.35	0.9991
0.85	0.7713	2.40	0.9993
0.90	0.7969	2.45	0.9995
0.95	0.8215	2.50	0.9996
1.00	0.8427	2.55	0.9997
1.05	0.8630	2.60	0.9998
1.10	0.8802	2.65	0.9998
1.15	0.8961	2.70	0.9999
1.20	0.9103	2.75	0.9999
1.25	0.9233	2.80	0.9999
1.30	0.9340	2.85	0.9999
1.35	0.9441	2.90	1.0000
1.40	0.9526	2.95	1.0000
1.45	0.9597	3.00	1.0000
1.50	0.9663		

equation is

$$\frac{\partial C_A}{\partial t} = D_{AB}\frac{\partial^2 C_A}{\partial y^2} - k_1 C_A. \tag{E.4}$$

The Laplace transform can be conveniently employed here (to eliminate the time derivative); the resulting *ordinary* differential equation is solved and the transform inverted to yield

$$\begin{aligned}\frac{C_A}{C_{A0}} = {} & \tfrac{1}{2}\exp\left(-\sqrt{k_1 y^2/D_{AB}}\right)\mathrm{erfc}\left[\frac{y}{\sqrt{4D_{AB}t}} - \sqrt{k_1 t}\right] \\ & \tfrac{1}{2}\exp\left(+\sqrt{k_1 y^2/D_{AB}}\right)\mathrm{erfc}\left[\frac{y}{\sqrt{4D_{AB}t}} + \sqrt{k_1 t}\right].\end{aligned} \tag{E.5}$$

This solution can also be adapted directly for extended surface heat transfer in which the metal (fin, rod, or pin) casts off thermal energy to the surroundings. By neglecting conduction in the transverse direction and assuming that the heat transfer coefficient h is constant, we obtain

$$\frac{\partial T}{\partial t} = \alpha\frac{\partial^2 T}{\partial y^2} - \frac{2h}{\rho C_p R}(T - T_\infty) \tag{E.6}$$

for a cylindrical rod. If we introduce the dimensionless temperature into (E.6), we can make use of the solution (E.5). However, it is to be noted that there is a potential problem with the boundary condition, as $y \to \infty$, $C_A \to 0$. In the absorption/reaction problem, the liquid may "look" as though it were infinitely deep for short duration exposures. This might not be appropriate for extended surface heat transfer, however, especially when the approach to steady state is of interest.

APPENDIX F

GAMMA FUNCTION

The gamma function arises in heat and mass transfer problems with some frequency; it is written as Γ (n) and defined by the integral:

$$\Gamma(n) = \int_0^\infty x^{n-1} e^{-x} dx. \quad \text{(F.1)}$$

The recurrence formula

$$\Gamma(n+1) = n\Gamma(n) \quad \text{(F.2)}$$

can be used to obtain needed values from abbreviated tables of $\Gamma(n)$. The functional behavior is illustrated in Figure F.1 on the interval (1,2).

A useful table for $\Gamma(n)$ follows; functional values were computed by numerical quadrature and are in agreement with those tabulated by Abramowitz and Stegun (*Handbook of Mathematical Functions,* Dover, 1965).

n	$\Gamma(n)$
1.000	1.000
1.025	0.986
1.050	0.973
1.075	0.962
1.100	0.951
1.125	0.942
1.150	0.933
1.175	0.925
1.200	0.918
1.225	0.912
1.250	0.906

n	$\Gamma(n)$
1.275	0.902
1.300	0.897
1.325	0.894
1.350	0.891
1.375	0.889
1.400	0.887
1.425	0.886
1.450	0.886
1.475	0.886
1.500	0.886
1.525	0.887
1.550	0.889
1.575	0.891
1.600	0.894
1.625	0.897
1.650	0.900
1.675	0.904
1.700	0.909
1.725	0.914
1.750	0.919
1.775	0.925
1.800	0.931
1.825	0.938
1.850	0.946
1.875	0.953
1.900	0.962
1.925	0.971
1.950	0.980
1.975	0.990
2.000	1.000

Transport Phenomena: An Introduction to Advanced Topics, By Larry A. Glasgow

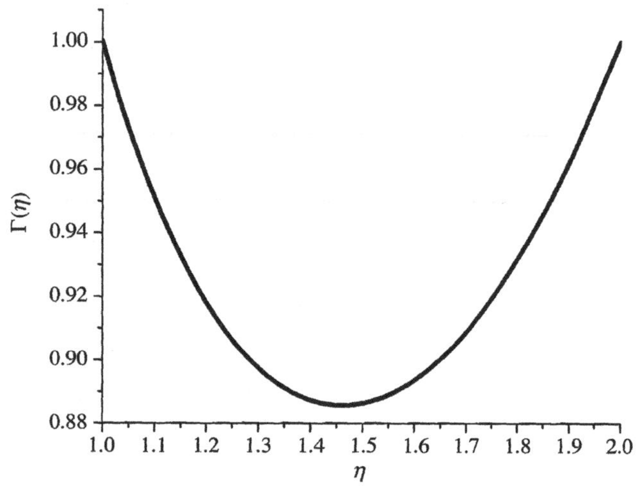

FIGURE F.1. The gamma function $\Gamma(n)$ for arguments between 1 and 2.

To illustrate how $\Gamma(n)$ comes about, we can consider the integral from the Leveque problem. Note that the limits (0–∞) correspond to the plate surface and the great distance into the moving fluid. We would expect to see these limits on η in the context of thermal or concentration boundary layers.

$$\int_0^\infty \exp(-\eta^3)d\eta. \tag{F.3}$$

Assuming $x = \eta^3$, $dx = 3\eta^2 d\eta$. Since $\eta^{-2} = x^{-2/3}$, the integral (F.3) can be written as

$$\frac{1}{3}\int_0^\infty x^{-2/3}e^{-x}dx = \frac{1}{3}\Gamma\left(\frac{1}{3}\right). \tag{F.4}$$

By the recurrence formula (F.2), this is equivalent to $\Gamma(4/3)$. And from the table above, we see that the correct numerical value is about 0.893.

APPENDIX G

REGULAR PERTURBATION

There are times when an analyst must find a functional representation for a particular transport problem, even though a numerical solution might be rapidly executed. Regular perturbation can be quite useful in such cases, particularly if the "difficult" part of the differential equation is multiplied by a parameter that has some very small value. The beauty of perturbation, as Finlayson (1980) noted, is that one can obtain the expansion of the exact solution without ever knowing what that solution is. We can best introduce the technique with an example.

Consider a slab of material that extends from $y=0$ to $y=1$. The two faces of the slab are maintained at different temperatures for all time t. The thermal conductivity of the material varies with temperature in linear fashion:

$$k = k_0 + mT. \tag{G.1}$$

The governing differential equation for this case can be written as

$$\frac{d}{dy}\left[k(T)\frac{dT}{dy}\right] = 0. \tag{G.2}$$

The problem can be cast in dimensionless form such that the boundary conditions become

$$T(y=0) = 1 \quad \text{and} \quad T(y=1) = 0. \tag{G.3}$$

Carrying out the indicated differentiation in (G.2), we find

$$m\left(\frac{dT}{dy}\right)^2 + (k_0 + mT)\frac{d^2T}{dy^2} = 0. \tag{G.4}$$

Equation (G.4) is a nonlinear differential equation for which no general analytic solution is known. We now let the temperature in the slab be represented by the series:

$$T = T_0 + mT_1 + m^2T_2 + m^3T_3 + \cdots. \tag{G.5}$$

The functions T_0, T_1, T_2, etc. are to be determined. The first and second derivatives are evaluated from (G.5):

$$\frac{dT}{dy} = \frac{dT_0}{dy} + m\frac{dT_1}{dy} + m^2\frac{dT_2}{dy} + \cdots \tag{G.6a}$$

and

$$\frac{d^2T}{dy^2} = \frac{d^2T_0}{dy^2} + m\frac{d^2T_1}{dy^2} + m^2\frac{d^2T_2}{dy^2} + \cdots. \tag{G.6b}$$

These and the series for T are inserted into (G.4):

$$\begin{aligned} &m\left(\frac{dT_0}{dy} + m\frac{dT_1}{dy} + m^2\frac{dT_2}{dy} + \cdots\right)^2 \\ &+ \left(k_0 + mT_0 + m^2T_1 + m^3T_2 + \cdots\right) \\ &\left(\frac{d^2T_0}{dy^2} + m\frac{d^2T_1}{dy^2} + m^2\frac{d^2T_2}{dy^2} + \cdots\right) \cong 0. \end{aligned} \tag{G.7}$$

Transport Phenomena: An Introduction to Advanced Topics, By Larry A. Glasgow

Now, suppose m assumes a *very small* value. We are left with merely

$$k_0 \frac{d^2 T_0}{dy^2} \approx 0. \tag{G.8}$$

Consequently,

$$T_0 = a_1 y + a_2. \tag{G.9}$$

The two constants are determined by applying the boundary conditions to (G.5), again allowing m to be small; therefore,

$$a_2 = 1 \quad \text{and} \quad a_1 = -1. \tag{G.10}$$

We determine the first and second derivatives:

$$\frac{dT_0}{dy} = a_1 \quad \text{and} \quad \frac{d^2 T_0}{dy^2} = 0. \tag{G.11}$$

These results are substituted into (G.7), and we divide by m:

$$\begin{gathered}\left(a_1 + m\frac{dT_1}{dy} + m^2\frac{dT_2}{dy} + \cdots\right)^2 \\ + \left[k_0 + m(a_1 y + a_2) + m^2 T_1 + m^3 T_2 + \cdots\right] \\ \left(\frac{d^2 T_1}{dy^2} + m\frac{d^2 T_2}{dy^2} + \cdots\right) \cong 0.\end{gathered} \tag{G.12}$$

Again, we take m to be very small, leaving

$$a_1^2 + k_0 \frac{d^2 T_1}{dy^2} \approx 0. \tag{G.13}$$

Integrating twice,

$$T_1 = -\frac{a_1^2}{2k_0} y^2 + a_3 y + a_4. \tag{G.14}$$

Returning to the boundary conditions,

$$T = a_1 y + a_2 + m T_1 + m^2 T_2 + \cdots. \tag{G.15}$$

At $y = 0$, $T = 1$; when this condition is introduced into (G.15) and we divide by m, we find

$$0 = ((T_1 + m T_2) + \cdots)|_{y=0}. \tag{G.16}$$

Accordingly, $a_4 = 0$. Of course, $T(y = 1) = 0$, so $a_3 = a_1^2/2k_0$. At this point, our approximation is

$$T \cong 1 - y + \frac{m a_1^2}{2k_0}(y - y^2)2k_0 \cdots. \tag{G.17}$$

The process illustrated here can be continued until a sufficiently accurate series is constructed or the analyst runs out of patience. Of course, we need to know whether (G.17) is going to be adequate for our purposes. Let $k_0 = 1$ and $m = 1/4$. We will find the numerical solution for comparison (Figure G.1).

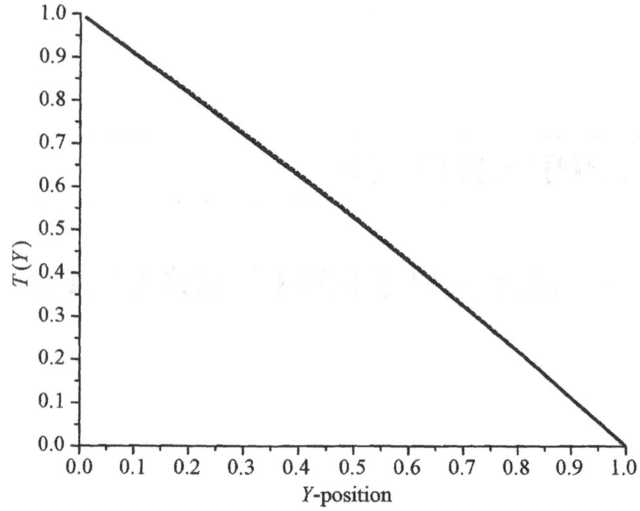

FIGURE G.1. Comparison of the exact numerical solution with the regular perturbation approximation for $m = 1/4$ and $k_0 = 1$. The results are nearly indistinguishable; admittedly, this is not a very severe test.

What has happened here needs to be noted: The perturbation expansion has resulted in a series of functions that could be determined successively by elementary methods. Thus, an intractable nonlinear problem has been solved approximately and the result is surprisingly good. However, as the parameter m becomes larger, we can expect the truncated series to represent $T(y)$ less accurately. To illustrate, let $m = 4$ (Figure G.2).

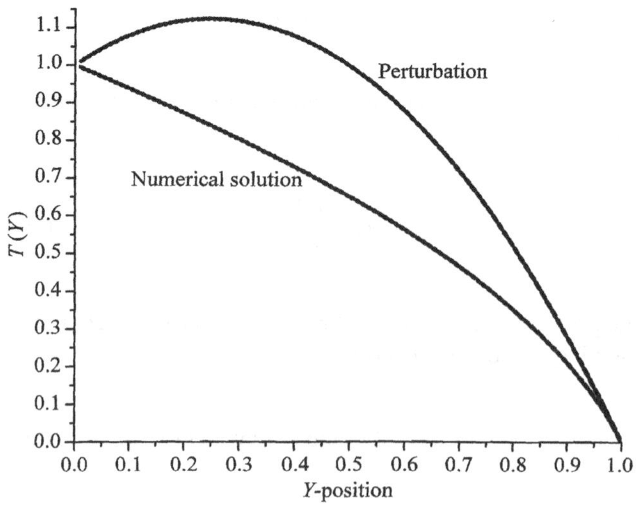

FIGURE G.2. Comparison of the exact numerical solution with the regular perturbation approximation for $m = 4$ and $k_0 = 1$. The difference between the two is now significant.

The perturbation technique described above can be applied to many other transport problems as well. By direct analogy we could imagine a diffusion problem in which the diffusivity D_{AB} was concentration dependent. Similarly, we could have a viscous flow with variable viscosity.

The conduction problem we worked through above involved a nonlinear differential equation, but it is useful to remember that perturbation methods can also be applied to both algebraic and integral equations. See Bush (1992) for additional examples. Be forewarned that there are instances in which the solution obtained as the "small" parameter $m \to 0$ is *not the same* as when $m = 0$. This situation is referred to as *singular* perturbation. Van Dyke (1964) notes that this is common in fluid mechanics, where the perturbation solution may not be "... uniformly valid throughout the flow field." This is an expected occurrence in boundary layer problems where potential flow theory does not apply near the surface. Two techniques that have been developed to deal with this difficulty are called the method of *matched asymptotic expansions* and the method of *strained coordinates*. There are many useful monographs covering perturbative techniques and a few of them are listed below:

1. Aziz, A. and T. Y. Na. *Perturbation Methods in Heat Transfer*, Hemisphere Publishing (1984).
2. Bush, A. W. *Perturbation Methods for Engineers and Scientists*, CRC Press (1992).
3. Finlayson, B. A. *Nonlinear Analysis in Chemical Engineering*, McGraw-Hill (1980).
4. Kevorkian, J. and J. D. Cole. *Perturbation Methods in Applied Mathematics*, Springer-Verlag (1981).
5. Van Dyke, M. *Perturbation Methods in Fluid Mechanics*, Academic Press (1964).

APPENDIX H

SOLUTION OF DIFFERENTIAL EQUATIONS BY COLLOCATION

The collocation technique allows the analyst to obtain an approximate solution for a differential equation; an assumed polynomial expression is required to satisfy the differential equation (in some limited sense). The technique is particularly useful for nonlinear equations for which numerical results are inconvenient or undesirable, but for which no analytic solution can be found. We illustrate the procedure in its simplest form with an example from conduction. Imagine a slab of type 347 stainless steel for which one face is maintained at 0°F and the other at 1000°F. Over this temperature range, the thermal conductivity of 347 increases (almost linearly) by more than 60%. We let $k = a + bT$ and note that in rectangular coordinates,

$$\frac{d}{dy}\left[k(T)\frac{dT}{dy}\right] = 0. \tag{H.1}$$

Therefore, the nonlinear differential equation of interest is

$$(a + bT)\frac{d^2T}{dy^2} + b\left(\frac{dT}{dy}\right)^2 = 0. \tag{H.2}$$

Our boundary conditions for this problem are

$$\text{at } y = 0,\ T = 0°\text{F, and}$$
$$\text{at } y = h,\ T = 1000°\text{F.}$$

For convenience, we set $h = 1$ ft, and we arbitrarily propose

$$T = \sum C_n y^n, \text{ such that}$$

$$T = C_0 + C_1 y + C_2 y^2 + C_3 y^3 + \cdots. \tag{H.3}$$

If we set $C_0 = 0$, the boundary condition at $y = 0$ is automatically satisfied. We form the residual by truncating (H.3) and substituting the result into (H.2):

$$[a + b(C_1 y + C_2 y^2 + C_3 y^3)](2C_2 + 6C_3 y) + b(C_1 + 2C_2 y + 3C_3 y^2)^2 = R. \tag{H.4}$$

Our task now is to choose values for C_1, C_2, and C_3 that result in the smallest possible value for R. This minimization of R can take several different forms, for example, if we select a weight function $W(y)$ and write

$$\int_0^h W(y) R dy = 0, \tag{H.5}$$

we have the method of weighted residuals (MWR). Finlayson (1980) points out that if we use the Dirac delta function for $W(y)$, then we are employing a simple collocation scheme where the residual will be zero at a few select points.

Of course, if R were identically zero everywhere on the interval, $0 < y < h$, we would have the exact solution. That seems a bit ambitious; as an alternative, we force the residual to be zero at the end points and also require (H.3) to satisfy the boundary condition at $y = h$. Thus, we have the three simultaneous algebraic equations:

$$2aC_2 + bC_1^2 = 0, \tag{H.6a}$$

Transport Phenomena: An Introduction to Advanced Topics, By Larry A. Glasgow

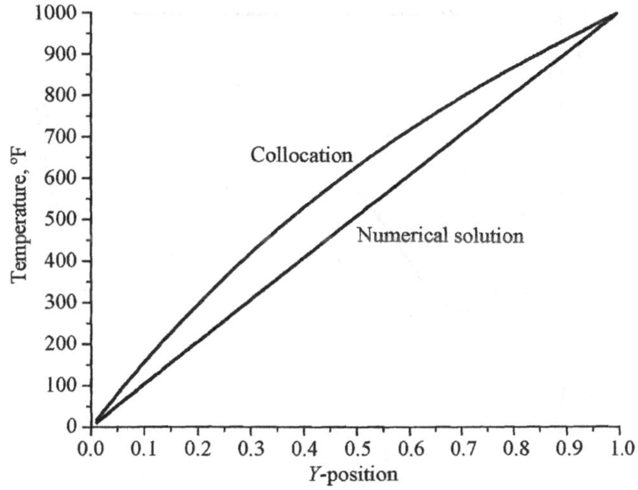

FIGURE H.1. Comparison of the exact numerical solution with the collocation result.

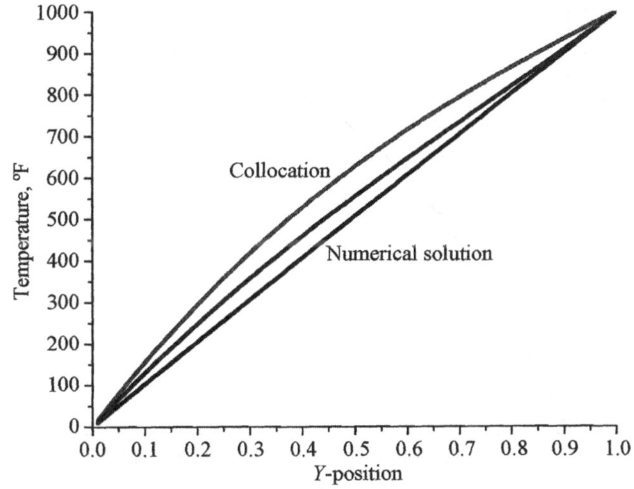

FIGURE H.2. Comparison of the exact numerical solution (bottom curve) with both the collocation results. Moving one collocation point to the center has resulted in an improved approximation, though one that is still deficient with regard to quantitative accuracy.

$$[a + b(C_1 + C_2 + C_3)](2C_2 + 6C_3) + b(C_1 + C_2 + C_3)^2 = 0, \tag{H.6b}$$

and

$$1000 - C_1 - C_2 - C_3 = 0. \tag{H.6c}$$

A solution is found by successive substitution:

$C_1 = 1641.434$, $C_2 = -920.838$, and $C_3 = 279.40$.

We will also use a fourth-order Runge–Kutta scheme to solve (H.2) numerically for comparison; see Figure H.1.

It is obvious from Figure H.1 that the collocation scheme we implemented was inadequate. Since the terminal points were chosen as the collocation points strictly for convenience, one might consider moving one (or both) of them to an interior position. Suppose, for example, we select $y = 1/2$ instead of $y = 1$. Solution of the algebraic equations now yields

$C_1 = 1351.6397$, $C_2 = -624.3936$, and $C_3 = 272.7542$.

We observe that while the additional result shown in Figure H.2 is improved, the approximate solution is really not satisfactory. A critical question concerns the placement of the collocation points—an equidistant or haphazard siting is likely to be less than optimal. Therefore, we should contemplate changes to the collocation procedure that may improve the outcome. In this connection, we draw attention to the number of *arbitrary* choices that were made in the example sketched above; these include the polynomial itself and the location of the collocation point(s). Suppose we begin by selecting a polynomial that *automatically satisfies the boundary conditions*. In addition, if we use orthogonal polynomials and place the collocation points at the roots of one or more of the terms, we will significantly decrease the burden placed on the analyst. We are now describing what Villadsen and Stewart (1967) called *interior collocation.*

Let us illustrate our first improvement with an example from fluid mechanics. Suppose we have a non-Newtonian fluid in a wide rectangular duct, subjected to a constant pressure gradient. If the fluid exhibits power law behavior, then one of the possibilities is

$$\frac{d^2 v_x}{dy^2} = -C_0\sqrt{\frac{dv_x}{dy}}. \tag{H.7}$$

The boundary conditions are

$$\text{at } y = 0,\ v_x = 0, \text{ and}$$
$$\text{at } y = 1,\ v_x = 0.$$

We can avoid any difficulties caused by the sign change on the velocity gradient by noting that at $y = 1/2$, $dv_x/dy = 0$. For this example, we choose the polynomial

$$v_x = c_1(y - y^2) + c_2(y - y^2)^2 + c_3(y - y^2)^3 + \cdots. \tag{H.8}$$

The conditions at $y = 0$ and $y = 1/2$ are automatically satisfied. We will select $C_0 = -20$ and find the exact numerical solution, so we have a basis for comparison (Figure H.3).

The reader may wish to complete this example and compare his/her result with the computed profile shown in the

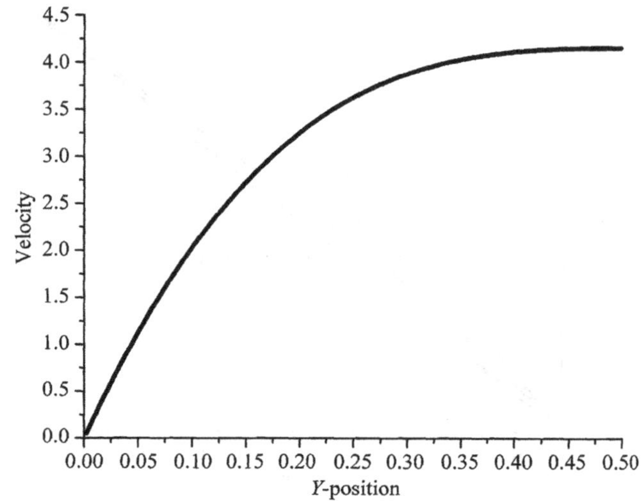

FIGURE H.3. Exact numerical solution for the non-Newtonian flow through a rectangular duct with $C_0 = -20$.

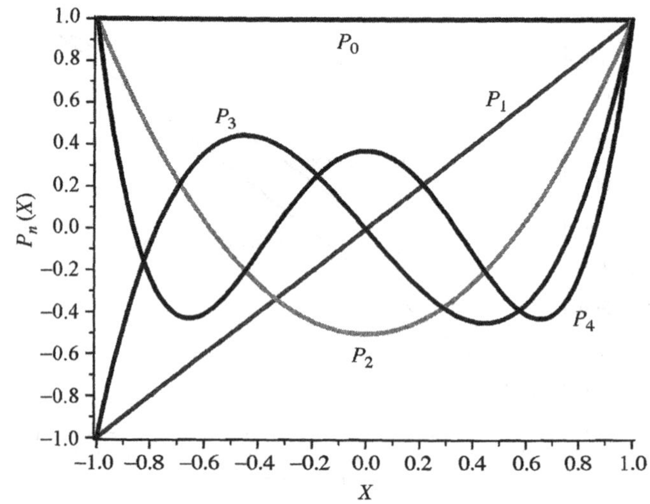

FIGURE H.4. Legendre polynomials P_0 through P_4 on the interval -1 to 1.

Figure. Note that it is necessary for $c_1 \cong 25$ (24.91347) in order for the slope at the origin to have the approximately correct value. The interested reader will find it illuminating to set $c_1 = 25$ and then attempt to identify c_2 by forcing the residual to be zero at the midpoint ($y = 1/4$). This exercise underscores one of the principal problems with the process we have employed. How many more terms must one retain in the assumed polynomial in order to get extremely accurate results? If we terminate the polynomial with the c_2-term and require the residual to be zero only at $y = 1/4$, we actually find that

$$c_1 = 46.52397 \quad \text{and} \quad c_2 = -21.68451.$$

Although the resulting shape is correct, this solution is unacceptable because the centerline velocity is roughly twice the correct value. It is clear that we should contemplate further improvements for this technique.

Polynomials are said to be orthogonal on the interval (a,b) with respect to the weight function W if

$$\int_a^b W(x)P_n(x)P_m(x)dx = 0, \quad \text{where} \quad n \neq m. \tag{H.9}$$

Let us consider the first few Legendre polynomials on the interval $(-1,1)$ for the problems that *lack symmetry*. We would like to explore how orthogonality may work to our advantage.

$$P_0 = 1, \quad P_1 = x, \quad P_2 = \frac{1}{2}(3x^2 - 1),$$

$$P_3 = \frac{1}{2}(5x^3 - 3x), \quad P_4 = \frac{1}{8}(35x^4 - 30x^2 + 3).$$

You might like to confirm, for example, that

$$\int_{-1}^{+1} P_1(x)P_2(x)dx = \frac{1}{2}\left[\frac{3}{4}x^4 - \frac{1}{2}x^2\right]_{-1}^{+1} = 0. \tag{H.10}$$

Note that if we were to locate collocation points at $x = \pm 1/\sqrt{3}$, then $P_2 = 0$. Similarly, for $x = \pm\sqrt{(3/5)}$, $P_3 = 0$. A further improvement can be obtained by making the dependent variables the functional values *at the collocation points* rather than the coefficients appearing in the polynomial representation. This modified procedure was developed by Villadsen and Stewart (1967) and it is explained very clearly by Finlayson (1980) on pages 73–74 of his book.

Let us now suppose that we have a boundary value problem *with symmetry* about the centerline where

$$\frac{d^2\phi}{dx^2} + f(x, \phi) = 0. \tag{H.11}$$

The independent variable x extends from -1 to 1 and the field variable ϕ has a set value (say, 1) at the end points. Naturally, at the centerline, $d\phi/dx = 0$. Accordingly, we propose

$$\phi = \phi(\pm 1) + (1 - x^2)\sum C_n P_n(x^2), \tag{H.12}$$

where the P_n's are Jacobi polynomials for a slab:

$n = 0\,1$	
$n = 1(1 - 5x^2)$	± 0.447214
$n = 2(1 - 14x^2 + 21x^4)$	± 0.2852315, ± 0.7650555
$n = 3(1 - 27x^2 + 99x^4 - 85.8x^6)$	± 0.209299, ± 0.5917, ± 0.87174

At this point, eq. (H.12) is substituted into (H.11) to form the residual. We can solve this set of equations for the coefficients (the C_n's) or we can develop an alternative set of equations written in terms of the functional values (ϕ_n's) at the collocation points.

H.1 PARTIAL DIFFERENTIAL EQUATIONS

Orthogonal collocation has also been used to solve elliptic partial differential equations of the form:

$$\frac{\partial^2\phi}{\partial x^2} + \frac{\partial^2\phi}{\partial y^2} = f(x, y), \tag{H.13}$$

on the unit square $x(0,1)$ and $y(0,1)$. Examples of the method's application are provided by Villadsen and Stewart (1967), Houstis (1978), and Prenter and Russell (1976). It is to be noted that an elliptic equation for any rectangular region $x(a,b)$ and $y(c,d)$, can be mapped into the unit square by employing the transformation,

$$x \to \frac{x-a}{b-a} \quad \text{and} \quad y \to \frac{y-c}{d-c}.$$

This broadens the applicability of the technique considerably. Now, let us suppose for illustration that eq. (H.13) has a solution given by

$$\phi = 3e^x e^y (x - x^2)(y - y^2), \tag{H.14}$$

which can be plotted to yield the results shown in Figure H.5.

Prenter and Russell (1976) solved this problem using bicubic Hermite polynomials, and their results indicate very favorable performance relative to the Ritz–Galerkin method.

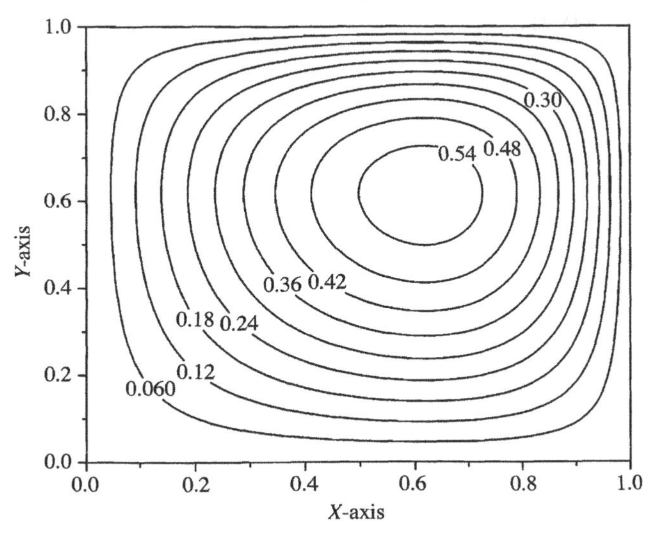

FIGURE H.5. Solution for the elliptic partial differential equation $\frac{\partial^2\phi}{\partial x^2} + \frac{\partial^2\phi}{\partial y^2} = 6xye^x e^y(xy + x + y - 3)$.

Furthermore, in some cases, the use of collocation with Hermite polynomials has outperformed the solution of elliptic equations by the finite difference method.

In an example provided by Villadsen and Stewart (1967), the Poisson equation

$$\frac{\partial^2\phi}{\partial x^2} + \frac{\partial^2\phi}{\partial y^2} = -1 \tag{H.15}$$

(for the Poiseuille flow through a duct) was solved on the square $(-1 < x < +1)$, $(-1 < y < +1)$ by taking

$$\phi = (1 - x^2)(1 - y^2) \sum \sum A_{ij} P_i(x^2) P_j(y^2). \tag{H.16}$$

If the expansion is limited to the Jacobi polynomial $P_1 = (1 - 5x^2)$ and the collocation point is placed at $(x_1, y_1) = (0.447214, 0.447214)$, then

$$\phi \cong \frac{5}{16}(1 - x^2)(1 - y^2). \tag{H.17}$$

This solution is plotted in Figure H.6 along with the correct numerical solution for easy comparison. Note that the truncated approximation is surprisingly good.

Villadsen and Stewart refined this rough solution by including $P_2 = (1 - 14x^2 + 21x^4)$ in the expansion with the three collocation points located at $(x, y) \to (0.2852315, 0.2852315)$, $(0.7650555, 0.2852315)$, and $(0.7650555, 0.7650555)$. The improved result was

$$\phi \cong (1 - x^2)(1 - y^2) \Big[0.31625 - 0.013125(1 - 5x^2 + 1 - 5y^2) + 0.00492(1 - 5x^2)(1 - 5y^2) \Big]. \tag{H.18}$$

Equation (H.18) compares very favorably with the numerical solution.

Several collocation schemes for the elliptic partial differential equations are available through a FORTRAN-based system called ELLPACK. The development of this software was initiated in 1976 and the effort was coordinated by John Rice of Purdue. Support for the project came from NSF, DOE, and ONR; collocation modules include COLLOCATION, HERMITE COLLOCATION, and INTERIOR COLLOCATION. See the ELLPACK Home Page for recent developments of this software. ELLPACK allows a user with a minimal knowledge of FORTRAN to solve the elliptic partial differential equations rapidly; even more important, the analyst can compare different solution techniques for accuracy and computational speed.

A program called HERCOL (for the solution of boundary value problems using the Hermitian collocation) was developed by John Gary of NIST; this program was tested by Welch et al. (1991) on the unsteady (start-up) laminar flow in a cylindrical tube with excellent results. The authors noted

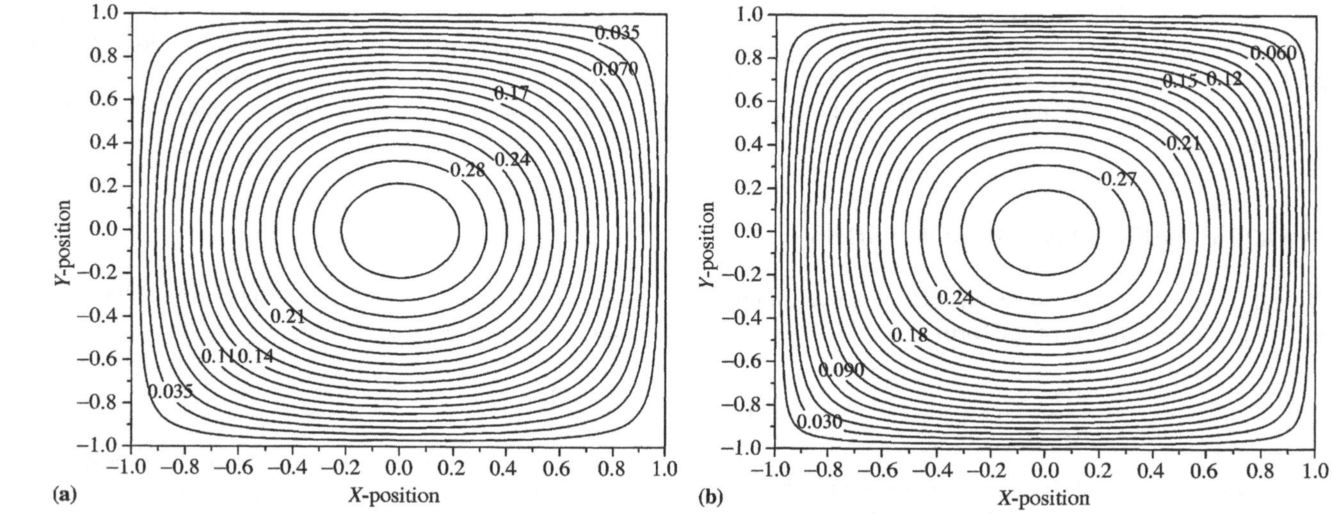

FIGURE H.6. Comparison of the approximate solution (left) with the correct numerical solution (right).

that HERCOL would be especially well suited for problems where an analytic solution was not possible, for example, for cases in which the transport properties of the fluid were not constant.

Collocation methods have been widely used in chemical engineering applications and particularly in the context of reaction engineering problems. The literature of collocation is large, but a few references useful as a starting point for further study are provided below.

1. Abramowitz, M. and I. A. Stegun. *Handbook of Mathematical Functions*, Dover Publications, New York (1965).
2. Finlayson, B. A. *Nonlinear Analysis in Chemical Engineering*, McGraw-Hill, New York (1980).
3. Houstis, E. N. Collocation Methods for Linear Elliptic Problems. *BIT Numerical Mathematics*, 16:301 (1978).
4. Prenter, P. M. and R. D. Russell. Orthogonal Collocation for Elliptic Partial Differential Equations. *SIAM Journal of Numerical Analysis*, 13:923 (1976).
5. Rice, J. R. and R. F. Boisvert. *Solving Elliptic Problems Using ELLPACK,* Springer-Verlag, New York (1985).
6. Villadsen, J. and M. L. Michelsen. *Solution of Differential Equation Models by Polynomial Approximation*, Prentice-Hall, Englewood Cliffs, NJ (1978).
7. Villadsen, J. and W. E. Stewart. Solution of Boundary-Value Problems by Orthogonal Collocation. *Chemical Engineering Science*, 22:1483 (1967).
8. Welch, J. F., Hurley, J. A., Glover, M. P., Nassimbene, R. D., and M. R. Yetzbacher. Unsteady Laminar Flow in a Circular Tube: A Test of the HERCOL Computer Code. *NISTIR 3963*, U.S. Department of Commerce (1991).

INDEX

Transport Phenomena: An Introduction to Advanced Topics, By Larry A. Glasgow

Printed in the USA/Agawam, MA
September 17, 2026

910661.038